Mississippian Oolites and Modern Analogs

Edited by
Brian D. Keith
and
Charles W. Zuppann

AAPG Studies in Geology #35

Published by
The American Association of Petroleum Geologists
Tulsa, Oklahoma, U.S.A.

Published 1993

ISBN: 0-89181-043-9

Association Editor: Susan A. Longacre
Science Director: Gary D. Howell
Publications Manager: Cathleen P. Williams
Special Projects Editor: Anne H. Thomas
Production: Custom Editorial Productions, Inc., Cincinnati, Ohio

About the Editors

The son of a geologist, **Brian D. Keith** was born in El Paso, Texas, one of many temporary residences for his parents during World War II. His father was employed at the time prospecting across the western U.S. for uranium for the Manhattan Project. Brian obtained an undergraduate degree in geology from Amherst College and a master's degree in geology from Syracuse University. He then took a much-needed hiatus from academia to work for Chevron (formerly SOTEX) in the Oklahoma City Division exploration office in 1969. Work with Silurian-Devonian Hunton reservoirs in the Anadarko basin and short courses given by Gerry Friedman and Karl Klement fueled his awakening interest in sedimentology, and in carbonate reservoirs in particular.

With the closing of the Chevron office in 1971, Brian opted to return to school at Rensselaer Polytechnic Institute to work on his Ph.D. under Gerry Friedman. There, he studied carbonate slope deposits and not critters. In 1974, he rejoined the petroleum industry at the Amoco Research Center in Tulsa. He also returned to El Paso for the first time since his birth on his way to Amoco field seminars in the Guadalupe Mountains. This and other company-sponsored seminars, and the opportunity to study carbonate and terrigenous clastic reservoirs from around the world, continued Brian's education. In 1978, he joined the Indiana Geological Survey. He is currently Basin Studies Coordinator at the Indiana Survey and a part-time faculty member in the Department of Geological Sciences at Indiana University.

Brian's 20-year fascination with carbonate reservoirs has led him from the Cretaceous of east Texas to the Ordovician (Trenton) of the eastern U.S. (*AAPG Studies in Geology #29*), and now to the Mississippian of the Illinois basin—where he expects to remain for some time. His current interest is in developing geologic models for Mississippian carbonate reservoirs that can be used by independent operators in the basin and elsewhere.

Charles W. Zuppann was born in Fort Smith, Arkansas, and was immediately introduced to the "big picture" by assuming the nomadic lifestyle of a career military officer's child. He received a bachelor's degree in business with a minor in geology from Austin Peay State University in Clarksville, Tennessee, and was inspired by Professor James X. Corgan to pursue a life of geology. There he was first introduced to Mississippian limestones. His interest in carbonate rocks was further developed at Vanderbilt University, where he earned a master's degree in geology in 1974.

Charly began his professional career with Texaco in New Orleans, where he found oil, but soon forgot about oolites. In 1976, he accepted a position with Southland Royalty Company in Houston. His geological duties concentrated on the Black Warrior basin (Alabama and Mississippi), where he discovered gas reserves in Mississippian rocks and encountered oolites in the Mississippian Bangor Limestone. In 1979, Charly returned to Tennessee with Western Reserves Oil Company. He found oil and gas in Mississippian oolite reservoirs of the Monteagle Limestone. Showing typical foresight, he moved to Evansville, Indiana, in 1984 and entered a partnership to explore for oil in the Illinois basin. He did find oil (but not nearly enough) and also began mapping oolite reservoirs in the Ste. Genevieve Limestone.

In 1986, when the price of oil fell so precipitously, Charly entered the market as a consulting geologist. During this time he looked for oil and his interest in oolitic reservoirs continued. Tiring of unemployment, he accepted his present position with the Indiana Geological Survey. Now he tries to help others find oil, and has the opportunity to study oolites to his heart's content. He is privileged to share in these studies with members of the Department of Geological Sciences at Indiana University, which is housed in the same building with the Geological Survey.

Sadly, Charly has never witnessed modern ooids growing and frolicking in their natural setting. But someday, he plans to visit their environment and experience their high energy.

Table of Contents

PREFACE

The 1989 annual meeting of the Eastern Section of AAPG was held at Bloomington, Indiana. Our involvement with planning that meeting also led to the evolution of this volume. In planning the AAPG meeting, it occurred to us that a topic in which the two of us were professionally interested, namely oolitic reservoirs of the Ste. Genevieve Limestone in the Illinois basin, would also be of interest to others attending the meeting. This led us to observe in turn that since the Mississippian in general is well known for oolitic limestones, and for petroleum reservoirs in those limestones, a separate symposium on Mississippian oolites would be a valuable addition to the meeting.

Ste. Genevieve outcrops are within a relatively short driving distance of Bloomington, so a field trip was a natural addition to the symposium schedule. At about the same time we learned that Ralph Hunter of the USGS (known for his expertise in noncarbonate eolian deposits) had begun studying Ste. Genevieve outcrops near Corydon, Indiana, because certain features in these carbonates seemed characteristically eolian to his practiced eye. The classic locality at Orleans, Indiana, studied by Carr (Indiana Geological Survey Bulletin 48, 1973), was also on the way to Corydon from Bloomington. A field trip to these two areas was announced as a complement to the symposium. The response to the announcement of the symposium and field trip was enthusiastic, resulting in submittals for 14 oral papers and seven poster presentations. The symposium was very well attended, as was the field trip.

Our hope from the beginning was to publish as many of the papers as possible in a volume following the symposium. Of the 21 symposium papers, 10 appear in this volume, three additional papers are contributions by authors who were involved in one way or another with the symposium papers and chose to make additional contributions (Dodd et al., Kelleher and Smosna, and Boardman et al.), two petroleum-related papers were solicited for inclusion (Manley et al. and Parham and Sutterlin), one paper (Hunter) was an outgrowth of the field trip, and an overview paper by the editors was requested by the AAPG. Three papers do not deal with the Mississippian per se, but are presented as models based on the Quaternary (both Holocene and Pleistocene) that can be applied to the Mississippian.

The first two papers, by Keith and Zuppann and by Ettensohn, are overviews of Mississippian carbonate deposition. Keith and Zuppann review Mississippian stratigraphy, factors controlling Mississippian ooid deposition, and the significance of ooid mineralogy and geometry of oolitic bodies on oolitic petroleum reservoirs. Ettensohn concentrates on the controls over Mississippian ooid deposition, suggesting that regional tectonic flexure related to Mississippian orogenic belts was the major factor that produced extensive shallow marine shelves in the proper paleogeographic setting for large-scale ooid production. Both papers indicate that the Mississippian in the United States was a time of coincidence of global paleoclimatic, eustatic, and geochemical conditions and regional tectonic conditions.

The next seven papers are specific to Mississippian carbonates of the Illinois basin area where most of the Meramecian consists of a continuous carbonate sequence. Four of the papers discuss the Ste. Genevieve Limestone (upper Meramecian), a major petroleum-producing formation in the Illinois basin. The paper by Hunter presents evidence of eolian carbonate deposition in grainstones interbedded with shallow marine grainstones within the Ste. Genevieve. The eolian interpretation is primarily on the basis of small-scale sedimentary structures in the carbonates that are characteristic of eolian sandstones. Documentation of pre-Quaternary eolian carbonates is relatively new in the literature, and this is the first recognition of eolian deposition in the Ste. Genevieve. Dodd et al. expand on Hunter's subject in the subsequent paper, by presenting petrologic evidence for distinguishing between marine and eolian grainstones in the same sequence of rocks studied by

Hunter. Their study provides additional lines of evidence that can be used to document other ancient eolian carbonate deposits.

Three different subsurface studies of oil fields producing from Ste. Genevieve oolitic reservoirs are presented by Bandy, Manley et al., and Zuppann. In two cases, geophysical logs are used to map oolitic facies. Bandy calibrates resistivity log response using cores, so that porous oolitic facies can be distinguished from porous dolomitic facies in the area of Lawrence field in Illinois. This study has considerable exploration and development significance in the Illinois basin, where resistivity logs are often the only source of subsurface information for many wells. Zuppann applies a slice-mapping technique using density logs to map the complex distribution of stacked oolite bodies that form petroleum reservoirs at Folsomville field in Indiana. Manley et al., in a detailed study at Willow Hill field in Illinois, used SP curves to map the distribution of permeable oolite. In addition, they were able to use porosity and permeability data and petrographic analysis from three cores to document the effect of early and burial cementation on the pore system in the reservoir zones in the field.

The last two Illinois basin papers deal with outcrop studies of Mississippian carbonates other than the Ste. Genevieve. Feldman et al. show how the composition of autochthonous benthic fossil assemblages on bedding surfaces in the Salem and Harrodsburg limestones (Meramecian in Indiana can be related to the sediment stability of the surface at the time of deposition. This stability can then be related to the day-to-day processes of particular depositional environments, including sites of oolitic deposition. The observations of this study are related by the authors to modern studies of the relationship between faunal diversity and sediment mobility, the latter being an important factor in oolite formation. Harris's study in Illinois is of the Golconda Group (Chesterian), one of the youngest Mississippian carbonate units in the Illinois basin to contain oolites. The Chesterian in the basin consists of a relatively thick terrigenous clastic succession containing several thin limestones spaced throughout. This succession immediately followed, and is in stark contrast to, the thick continuous Mississippian carbonate succession of the Meramecian in the basin.

The next three papers report on the occurrence of oolites in the sequence of Mississippian carbonates deposited in the central Appalachian basin as part of the Greenbrier Limestone (Meramecian and Chesterian). The units within the Greenbrier contain numerous shallow, prolific petroleum-producing reservoirs. Carney presents an overview of Greenbrier deposition during marine transgression over the West Virginia dome and surrounding area. The dome served to localize development of oolite shoals as other subtidal carbonates were deposited away from the uplift. Oolitic deposition was interrupted midway during Greenbrier time by a pulse of clastic deposition, then reestablished in the upper Greenbrier. Smosna and Koehler present evidence indicative of deposition of intertidal and subtidal sand waves in the Pickaway Limestone of the middle Greenbrier. Following deposition these sand waves were modified by wave activity and periodic storms. Major storms then moved the peritidal sand waves over other facies of the Pickaway. Smosna and Koehler note that this is one of the few cases of documentation of ancient tidal oolites, even though modern oolite shoals are dominated by tidal forces. The third Greenbrier paper, by Kelleher and Smosna, presents results of their study of oolitic natural gas reservoirs in the Union Limestone (upper Greenbrier) at Rhodell field, West Virginia. Geophysical logs and production data from wells were used to identify and map porous oolite reservoirs in the Union Limestone. The resulting depositional model is one of elongate tidal bars that formed in a belt along a hinge line separating a rapidly subsiding basin to the south from a stable shelf to the north.

Heydari et al. present a detailed petrographic and diagenetic study of the Pitkin Formation (Chesterian) in north central Arkansas. The ooid-skeletal grainstones of the Pitkin were deposited as a high-energy shoaling phase on a stable shallow carbonate shelf north of the subsiding Ouachita trough. Interpreted original calcite and aragonite ooid mineralogy of the Pitkin shows systematic variations related to different depositional settings on the shelf.

The paper by Parham and Sutterlin reports on productive oolite shoals in the St. Louis Limestone in southwestern Kansas. Shoaling was localized by gentle structural uplift and oolites were deposited in an unusual position leeward of island mudflats, possibly due to channeling of currents between the island mudflat and the shoreline.

The last three papers present Quaternary models for ooid deposition, which may provide additional clues for interpreting ancient oolitic limestones. These clues can in turn be useful in the exploration for petroleum reservoirs in oolitic limestones. Wanless and Tedesco describe the setting for the deposition of Holocene, wave-generated, thick and widespread oolitic sand bodies on the Caicos platform in the southeastern Bahamas. They contrast these with the tidally generated oolitic sands of the northern Bahamas, which are restricted to shoal accumulations close to platform margins. They suggest that the Caicos model may be more applicable for ancient shallow intracratonic basins and coastal areas where tidal forces may have been less significant than wave energy. Boardman et al. describe features found in both Pleistocene and Holocene ooid deposits in the northern Bahamas and evaluate their usefulness as criteria for recognition of similar deposits in the Mississippian Ste. Genevieve and Greenbrier oolitic limestones. In particular, similar grain compositions, sedimentary structures, and selected environments are commonly reported, whereas characteristics such as original topography, exposure surfaces, and eolian features are not commonly reported in Mississippian limestones. The final paper is by Caputo, who describes features of Pleistocene and Holocene eolian oolitic and skeletal grainstones from San Salvador Island, Bahamas. He finds many physical structures of these deposits are similar to those in ancient eolian sandstones. These features include grainfall, sandflow, and wind-ripple structures that might also be useful for recognizing ancient eolian limestones.

The Mississippian was, for a variety or reasons, a fascinating time of widespread shallow water carbonate deposition that produced extensive limestone deposits noted for an abundance of ooid grainstones and a paucity of organic buildups. Understanding these rocks is not only of interest for its own sake, but also has considerable significance to petroleum exploration and development. As editors, we hope that this compilation, while certainly not exhaustive, will provide valuable information on these rocks and serve as the basis and inspiration for continued study.

ACKNOWLEDGMENTS

We wish to thank the contributors to the volume for their patience and work, and the outside reviewers who generously gave of their time to review the manuscripts. The support of the AAPG Publications Department in general, and Cathleen Williams in particular, are appreciated. The support and facilities of the Indiana Geological Survey and the Indiana University Department of Geological Sciences are also gratefully acknowledged.

Brian D. Keith
Indiana Geological Survey
and Department of Geological
Sciences, Indiana University
Bloomington, Indiana

Charles W. Zuppann
Indiana Geological Survey
Bloomington, Indiana

Chapter 1

Mississippian Oolites and Petroleum Reservoirs in the United States—An Overview

Brian D. Keith
Indiana Geological Survey
and Indiana University
Bloomington, Indiana, USA

Charles W. Zuppann
Indiana Geological Survey
Bloomington, Indiana, USA

ABSTRACT

A coincidence of tectonic, eustatic, and geochemical conditions resulted in substantial deposits of oolitic limestone during later Mississippian time in the continental United States. These oolitic limestones have formed petroleum reservoirs with favorable primary and secondary recovery characteristics. Significant potential reserves in stratigraphic traps remain to be discovered and developed in these reservoirs.

INTRODUCTION

Mississippian rocks of the continental U.S. have been the subject of two previous compilation volumes by the USGS (Craig and Connor, 1979; U.S. Geological Survey, 1979), both concentrating on the Mississippian system as a whole, the former on Mississippian and Pennsylvanian stratigraphy and the latter on the paleotectonic history of the Mississippian. However, there has not been a publication specifically emphasizing Mississippian oolitic rocks. Geologists have reported oolitic limestones in Mississippian rocks in many areas of the United States for years, but Wilson (1975, p. 283) may have been the first to point out that oolites are especially common in Early Carboniferous (Mississippian) strata of the Northern Hemisphere. Wilkinson et al. (1985) first presented quantitative evidence of oolite abundance in Mississippian rocks, and Handford (1988) reported that the most widespread time of oolitic limestone deposition on the North American continent was during the Mississippian.

With the recent emphasis on global-scale geologic processes, we recognize that attention must be given to Mississippian oolitic deposition on a broader scale, seeking to understand the factors that influenced and controlled oolitic deposition at that time. A prerequisite to interpreting worldwide depositional patterns of Mississippian oolitic rocks is to study these deposits at both the regional and the continental scales. The abundant occurrence and widespread distribution of Mississippian oolitic rocks, and the wealth of available subsurface information, make the North American continent a fitting choice on which to focus this attention. In addition, Mississippian oolites are economically important for North America because they form significant petroleum reservoirs and are especially important today because they are largely stratigraphic traps, which represent the most promising exploration targets in mature basins where most of the positive structural features have been drilled.

The purpose of this chapter is to present an overview of Mississippian oolites in the continental United States based on stratigraphy, depositional setting, and their significance as petroleum reservoirs. We hope that this general treatment will lead to greater understanding of these fascinating rocks and encourage additional research.

MISSISSIPPIAN STRATIGRAPHY

Mississippian stratigraphy for most of the continental United States (Figure 1) has been generalized into the chart shown in Figure 2. This compilation was based on eight of the COSUNA (Correlation of Stratigraphic Units of North America) charts published by the American Association of Petroleum Geologists. Specific citations are noted in the figure caption. Our intent in this compilation is not to present an exhaustive analysis of Mississippian stratigraphy, but rather to review the occurrence and correlation of Mississippian carbonate rocks, especially those units reported in the literature as containing oolitic limestones. The chart in Figure 2 reflects this bias in that terrigenous clastic units are not broken out individually, but are shown only by general distribution. The widespread occurrence of the Late Devonian through Early Mississippian organic shale in most basins and the extensive Early Mississippian chert of the Appalachian basin are noted on the chart. Mississippian evaporite units, however, are named along with the carbonate units because of the often intimate association of these rock types.

Some regions where oolitic limestones are either not reported in the literature or are of minor occurrence, such as the Black Warrior basin of Mississippi and Alabama and the Great Basin of western Utah, are not represented in the chart to conserve space. The stratigraphy of the Black Warrior basin is similar to that shown for the northern Alabama portion of the southern Appalachian basin. The Great Basin

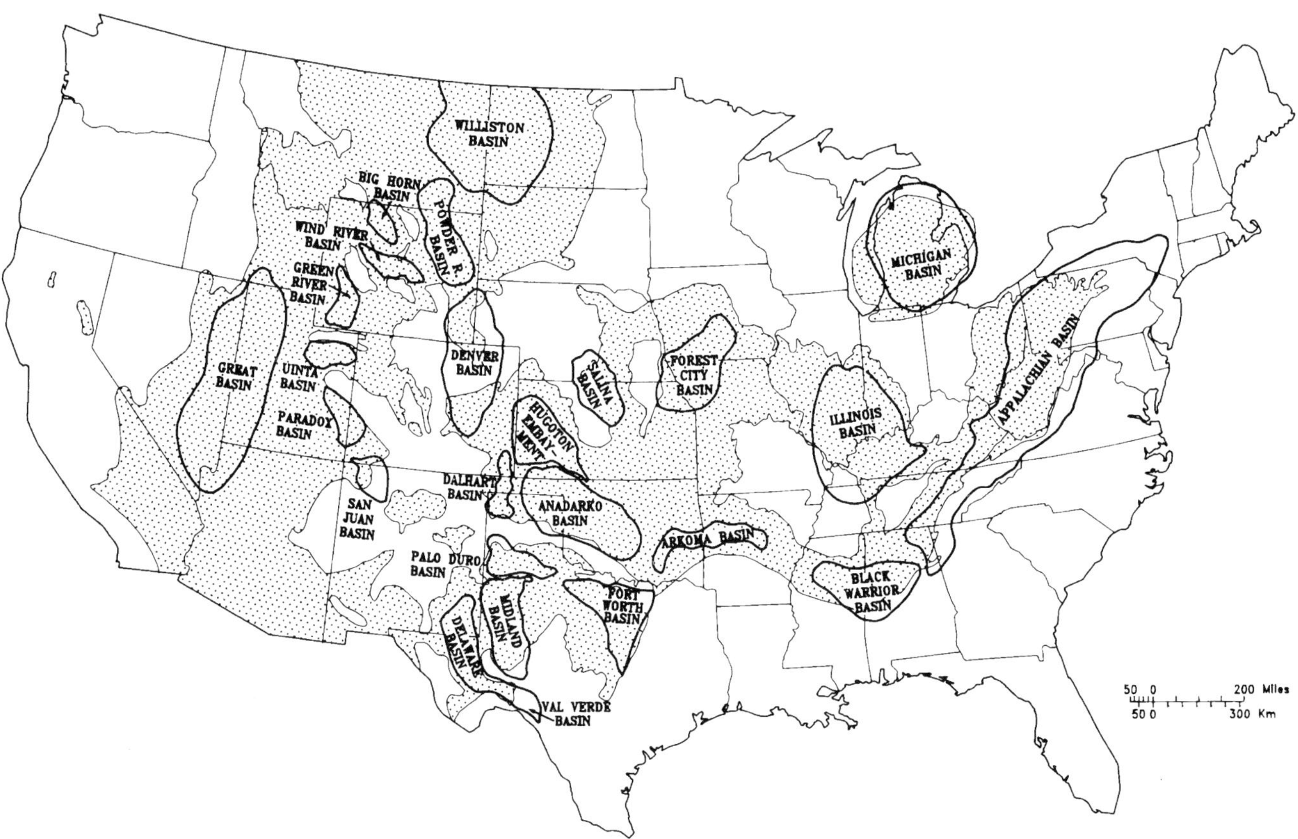

Figure 1. Map of United States showing general distribution of Mississippian rocks (stippled pattern) and location of major sedimentary basins referred to in text and on Figure 2. Basin outlines are somewhat generalized and do not necessarily correspond to Paleozoic or present configurations. Areas not labeled, but referred to on Figure 2, include: Cincinnati arch (separates Appalachian and Illinois basins), mid-continent (general area including Forest City, Arkoma, Anadarko, and Salina basins), Hardeman basin (eastern end of Palo Duro basin), Permian basin (Midland and Delaware basins and intervening area), SW platform (southern New Mexico and Arizona), and northern Rocky Mountains (area between and including Green River, Wind River, Big Horn, and Powder River basins). Compiled and modified from various sources, but primary source for outcrop distribution was Craig and Connor (1979).

stratigraphy is complex, but the western portion is primarily represented by the Joana Limestone (Osagian to Meramecian), whereas the eastern portion has terminology similar to the San Juan and Paradox basins of the Four Corners area and to the Uinta basin. Also, the Fort Worth basin of central Texas is not treated separately because it contains the Osagian Chappel Limestone that is equivalent to part of the Boone Formation and the Sycamore Limestone of Oklahoma and to the Osage Limestone and the Chappel limestones of the Palo Duro basin.

Southern New Mexico presented particular problems because of considerable stratigraphic variation, especially with regard to the large age span of the Lake Valley Formation. The column for the Sacramento Mountains was arbitrarily chosen to be representative for that area. In north-central New Mexico the Arroyo Penasco Group contains oolitic limestone in the lower Chesterian (A.K. Armstrong, personal communication, 1990) but was not included in the chart.

A limited area of oolitic rocks also occurs as a carbonate facies of the Osagian Ellsworth Shale in the southwestern part of the Michigan basin (Cohee, 1979) but is not included in the chart.

Stratigraphic names generally apply to the area under which they are shown on the chart, but column lines were intentionally omitted because names may overlap from one area to another. The placement of names on the chart should be used only as a general guide as to where a given name is used; undue significance should not be given to exact placement of names on the chart.

Following the format of the COSUNA charts, the Mississippian has been divided (from oldest to youngest) into four North American stages: Kinderhookian, Osagian, Meramecian, and Chesterian. In the Illinois basin area, Valmeyeran is used in place of the combined Osagian and Meramecian, but this usage is not shown on the chart to avoid clutter. Also, the position of the boundary between the Meramecian and Chesterian has come under review. When the COSUNA charts were compiled, the boundary was placed at the correlative position of the top of the Ste. Genevieve Limestone in the mid-continent and Illinois basin areas based on megafossils. However, Maples and Waters (1987) have recommended that the boundary be lowered to the correlative position between the Ste. Genevieve and St. Louis limestones based on microfossil zonation. Both positions are shown on the chart as dashed lines.

SETTING FOR OOLITE DEPOSITION

Any discussion of Mississippian oolites in the United States needs to consider the following characteristics of Mississippian rocks in general and Mississippian limestones in particular: (1) limestones are a volumetrically significant component of Mississippian rocks throughout most of their extent in the United States (Figure 2), and oolitic facies are an important constituent of these limestones, especially when compared to skeletal frame-building constituents that are rare in the Mississippian; (2) Mississippian oolitic limestones are more abundant in Meramecian and early Chesterian rocks (Figure 3) (Ettensohn, this volume); (3) Mississippian oolitic limestones are generally not dolomitized except locally; and (4) the texture of the ooid grains is predominantly radial, indicating that they were originally calcite (Wilkinson et al., 1985).

Even for modern ooid deposition, the exact nature of the formation of individual ooids and the role of biologic activity are uncertain. However, it is generally accepted that ooid deposits form in shallow water environments regularly agitated by waves and/or currents (see summary discussion in Tucker and Wright, 1990, p. 3–8). Extensive sequences of oolitic limestone such as those in the Mississippian required that large expanses of these high-energy environments be maintained over long periods of time. Additional environmental requisites for oolitic deposition would have been minimal siliciclastic input and generally warm temperatures (relatively low-latitude settings). Today these particular environmental conditions are favored by many organisms, especially frame-builders, but in order for extensive deposits to form during the Mississippian there must have been some controls over organic proliferation. In modern environments these controls are generally either a lack of upwelling nutrient supply or poisoning by non-normal marine waters (hypersaline or hyposaline). During Mississippian time, controls on organic proliferation may have included eustatic changes as well (see discussion below).

Modern ooids are predominantly aragonite rather than calcite. The concept that ancient ooid mineralogy might have been significantly different than the modern was raised by Sandberg (1975) and expanded upon by MacKenzie and Pigott (1981), Sandberg (1983), and Wilkinson et al. (1985), the latter being the most comprehensive. Each of these studies is concerned with the cycling through geologic time of ooid mineralogy between aragonite and calcite, and with possible controls for this cycling: climate, atmospheric, or ocean chemistry, and global sea level. Wilkinson et al. (1985) showed that there were four times of peak ooid deposition in geologic history (Figure 3)—Late Cambrian, Late Mississippian, Late Jurassic, and Holocene. When compared to a first-order sea level curve (Figure 3), it is clear that peak ooid production occurred during times when overall sea level was either rising or falling, rather than at times when sea level was at a maximum (highstands) or minimum (lowstands). Wilkinson et al. (1985) concluded that global changes in sea level, related to crustal processes and pCO_2 (vapor pressure of CO_2, which controlled carbonate concentrations in the atmosphere and the oceans), are the primary controls over both ooid mineralogy and abundance. During times of

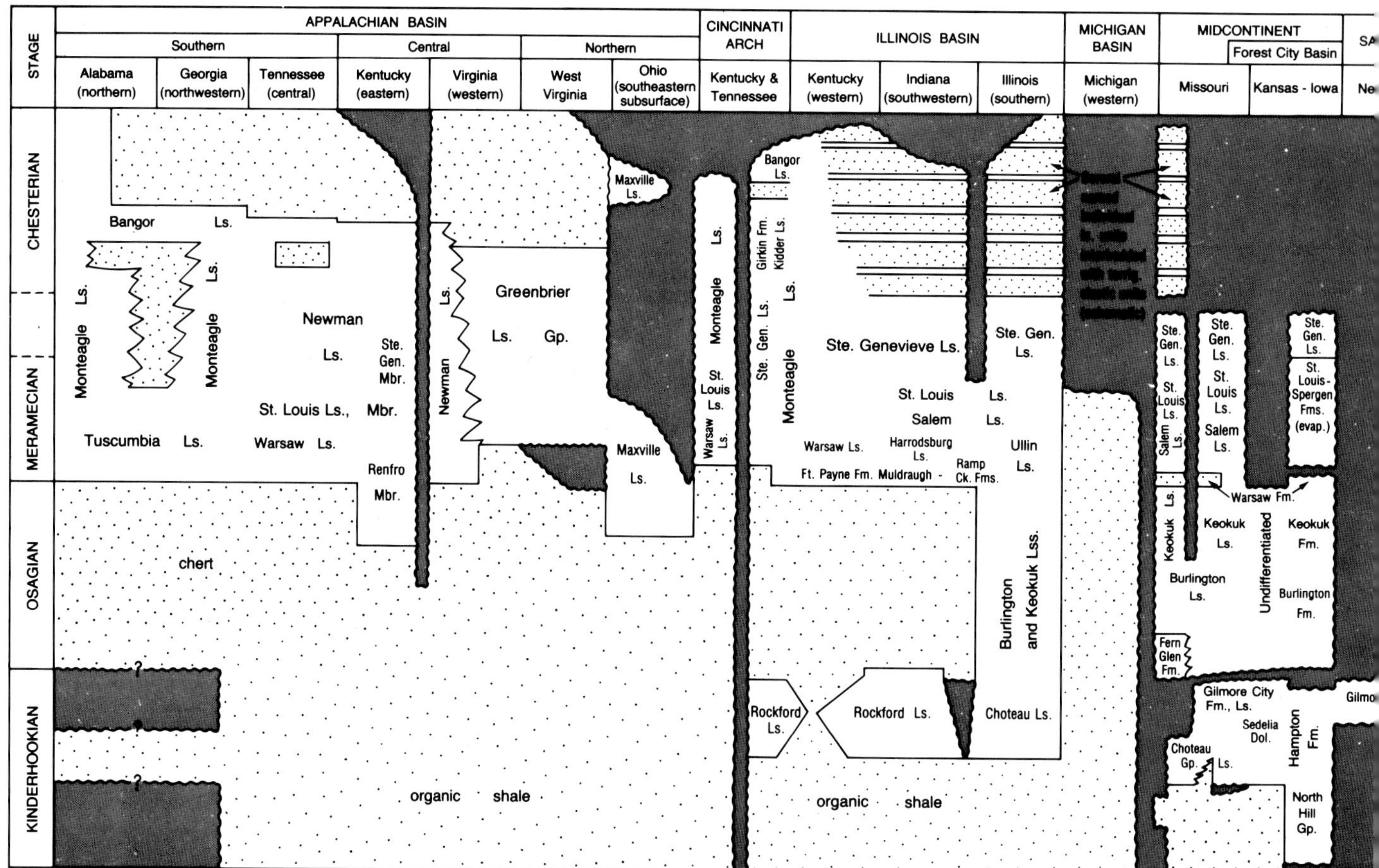

Figure 2. Stratigraphic correlation chart for Mississippian units for most of the United States. Refer to Figure 1 for location of areas noted along the top of the chart. Compiled from 219 columns contained in the following COSUNA charts: Ballard et al. (1983), Hills and Kottlowski (1983), Patchen et al. (1985a, b), Shaver (1985), Adler (1987), Mankin (1987), and Kent et al. (1988). Only names of carbonate (as well as evaporite) units are

emergence, the volume of shallow water carbonate deposited is generally smaller because the size of shallow water environments is limited. During times of submergence, organic carbonate production is high, making conditions most favorable for extensive carbonate deposition (Figure 3), and terrigenous clastic input is low; but pCO_2 levels probably limited the amount of abiotic (i.e., ooid) carbonate that could be produced (Wilkinson et al., 1985). The lack of detailed studies, however, leaves many questions about this aspect of carbonate sedimentation unanswered (see Eluik, 1987, and Wilkinson et al., 1987, for further discussion).

During Mississippian time, widespread oolitic sedimentation occurred because physical and chemical conditions were optimum for ooid production and because competition from organisms available to produce organic buildups was low (Figure 3). In addition, tectonic events during Meramecian and early Chesterian time in the United States produced broad expanses of shallow water at low paleolatitudes (Ettensohn, this volume). Thus, tectonism combined with both the lack of organic carbonate production by framework-building organisms and the apparent proper pCO_2 to control atmospheric and oceanic carbonate levels provided ideal conditions for widespread deposition of predominantly calcitic ooids.

Highstands of sea level also show general correspondence with increased dolomite abundance (Figure 3). No specific mechanism for dolomitization is implied by this relationship (Given and Wilkinson, 1987), but the relationship is consistent with the observation noted earlier that Mississippian oolites are generally not dolomitized, whereas Cambrian and Jurassic oolitic carbonate rocks commonly are. These dolomitized oolitic limestones formed during times of overall sea level rise that were followed by times of maximum emergence (with favorable conditions for widespread dolomitization), whereas Mississippian oolites formed during an overall sea level drop that was followed by maximum submergence (with unfavorable conditions for widespread dolomitization).

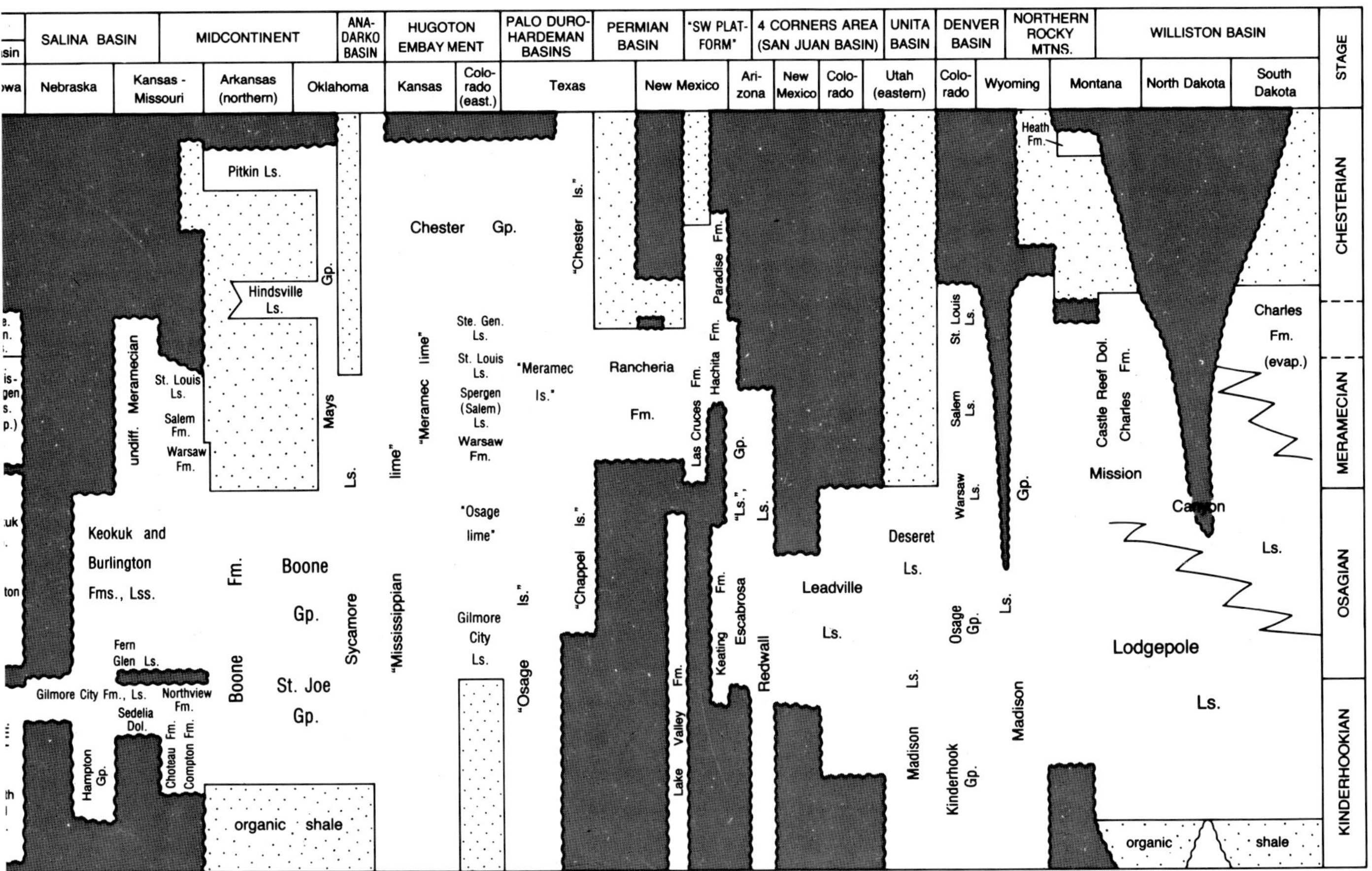

shown. Stippled pattern represents the distribution of terrigenous clastic units. The boundary between the Meramecian and Chesterian is shown dashed in two positions to reflect possible revision. See text for more detailed discussion.

PETROLEUM SIGNIFICANCE

Petroleum Occurrences in Mississippian Oolites

Significant occurrences of hydrocarbon production from Mississippian oolitic rocks are present in four areas in North America: (1) the Ste. Genevieve Limestone of the Illinois basin, (2) several Middle and Late Mississippian formations in the Anadarko basin and Hugoton embayment, (3) the Greenbrier and Monteagle limestones of the Appalachian basin, and (4) the Madison Group of the Williston basin.

A review of petroleum reservoirs in Mississippian oolites, especially contrasting their occurrence and characteristics in geographically separated basins with unique histories of petroleum industry activity, is difficult. The literature and records of petroleum production contain so many differences in terminology, available information, research motives, and interpretations, all intertwined with the historical progression of our knowledge of oolitic rocks, that this review is by necessity constrained to be general in nature and rather limited in its conclusions.

The most oil-prolific and most thoroughly studied area of Mississippian oolite reservoirs is the Illinois basin. Oolitic reservoirs are the predominant reservoir type in the Ste. Genevieve Limestone, which accounts for an estimated 18% of the basin's cumulative production, or approximately 743 million barrels of oil (Cluff and Lineback, 1981; Howard, 1991; Mast and Howard, 1991). The Ste. Genevieve oolite grainstone reservoirs are encased in impermeable limestone. The reservoirs, known informally as "McClosky sands," are widely distributed throughout the basin (see Zuppann, Figure 1, this volume). Oil in the McClosky was discovered in 1907, in Lawrence County, Illinois (Blatchley, 1913), along the regionally prominent LaSalle anticlinal belt (see Bandy, Figure 1, this volume). During the first 50 years of Ste. Genevieve oil production, it was not uncommon for wells to initially produce at rates greater than 500 BOPD. Even today some wells are still completed for more than 200 BOPD.

The Monteagle Limestone of northern Tennessee and south-central Kentucky and the roughly equivalent Greenbrier Limestone of West Virginia contain

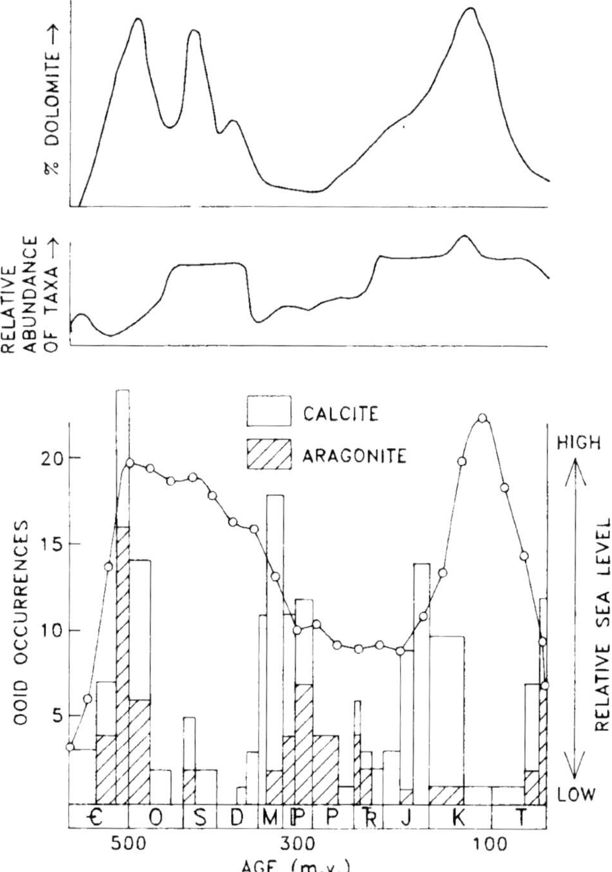

Figure 3. Plot of the distribution of documented oolite occurrences and original mineralogy through geologic time as compared to a first-order sea level curve (line connecting circles). After Wilkinson et al. (1985). Also shown (normalized to the same approximate geologic time scale) are plots of generalized relative abundance of the primary organisms (different taxa) of carbonate buildups (modified from James, 1983) and dolomite abundance plotted as calculated percent dolomite (modified from Given and Wilkinson, 1987). Of interest here are the general correlations of abundances of oolitic limestone, carbonate buildup organisms, and dolomite as they relate to the first-order sea level curve.

many oolitic hydrocarbon reservoirs that are predominantly gas-bearing. Gas wells completed in the Monteagle Limestone tend to have initial tests in the low hundreds of mcfgd per day. According to Youse (1964), some early Greenbrier gas wells had open-flow tests of 20 mmcf per day, and one well was reported with cumulative production in excess of 9 bcf of gas.

The Chester section in the Anadarko basin and the St. Louis and Ste. Genevieve limestones in the Hugoton embayment contain significant oolitic petroleum reservoirs, although they are fewer and less widely distributed than reservoirs in the Ste. Genevieve in the Illinois basin. Because hydrocarbon production in the Anadarko basin and Hugoton embayment comes from a number of different reservoir facies, and because production from oolitic reservoirs is not specifically reported as such, we cannot reasonably estimate the volume of oil production attributable to the oolitic facies. An example that illustrates the quality of these oolitic reservoirs is the Damme field in the Hugoton embayment, Kansas. The Damme field has produced more than 13 million barrels of oil, mostly from oolitic/skeletal grainstones in the St. Louis Limestone (Schmidlapp, 1959; Handford, 1988). The initial potential for the discovery well at Damme field was 1795 BOPD from an oolitic zone in the St. Louis (Schmidlapp, 1959). Asquith (1984) mapped at least four distinct oolite reservoirs within the Chester interval in Beaver County, Oklahoma, in the northwest portion of the Anadarko basin (see Figure 4). These reservoirs contained 41 wells that each have cumulative gas production greater than 1 bcf (Asquith, 1984).

Carbonate rocks of the Madison Group have been active exploration targets in the Williston basin since the early 1950s. Within this prolific oil-producing interval, oolitic rocks are common and oolitic grainstones have often been described as reservoir host rocks (early examples include Berg, 1956; Stanton, 1956; Harrison and Larson, 1958; Smith et al., 1958). However, porosity types related to dolomitization, fracturing, dissolution, and other host lithologies are also common in Madison Group reservoirs (Kent, 1987). The relative importance of specifically oolitic facies to hydrocarbon occurrence in the Williston basin has yet to be reported.

Mississippian Oolite Reservoirs

Understanding the potential shapes and distribution of target reservoirs is necessary to effectively explore for and develop any type of reservoir. Reservoir geometries of Mississippian oolites are of particular interest for several reasons: (1) patterns of reservoir geometry are readily apparent in Mississippian oolitic reservoirs; (2) differences in geometry, especially orientation of oolitic facies, have been considered a major distinguishing feature in depositional models, based on both modern and ancient oolitic deposits; and (3) oolite-body geometry has long been an important aspect of research related to Mississippian oolites, resulting in new ideas and significant contributions to the literature.

Comparing oolitic hydrocarbon reservoirs from the different basins, perhaps the most striking similarity is that those in the Illinois basin, the mid-continent area, and the Appalachian basin tend to occur in stratigraphically recurring trends of subparallel, elongate porosity bodies (Figure 4). This circumstance stems directly or indirectly from depositional controls affecting oolitic sedimentation.

A distinction must be made between a three-dimensional body of porosity (porosity body) that may form a reservoir and the three-dimensional sediment facies (sediment body) to which it may corre-

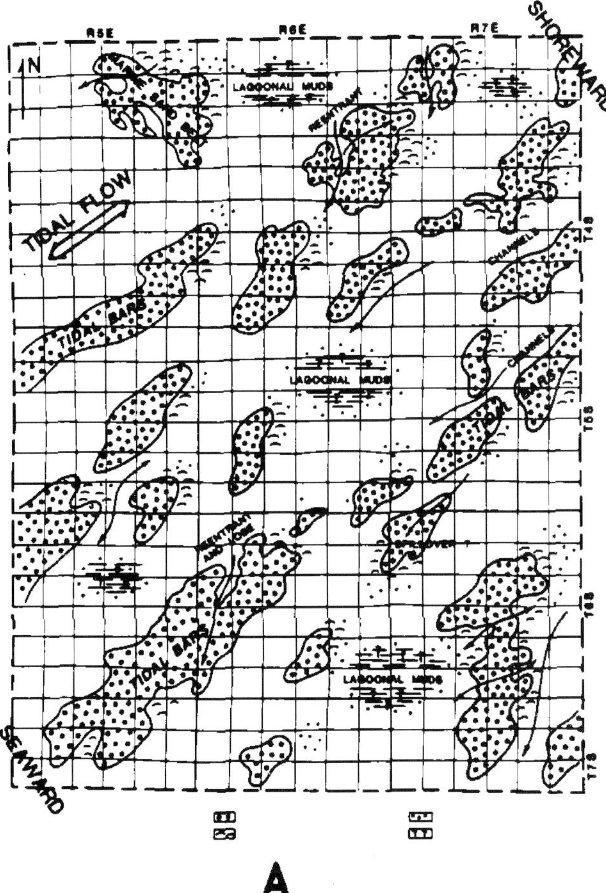

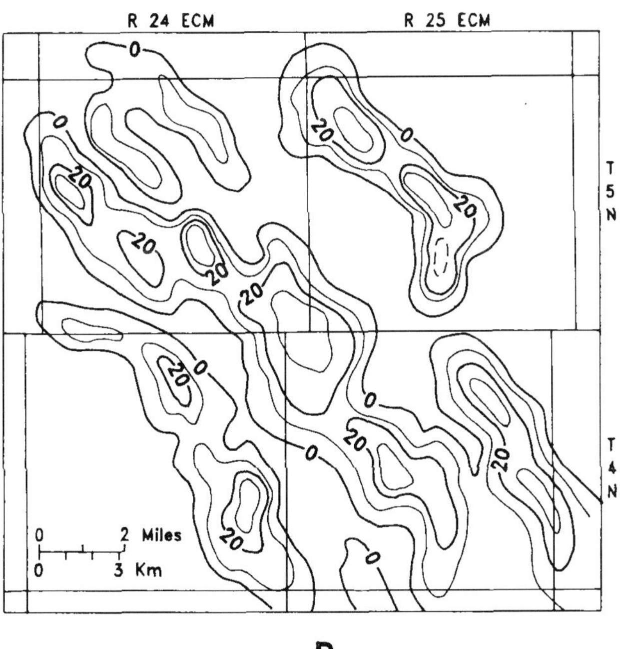

Figure 4. Examples of trends of subparallel, elongate oolite bodies typical of Mississippian oolitic sequences in the Illinois basin, the mid-continent area, and the Appalachian basin. (A) Paleogeographic map showing distribution of Ste. Genevieve oolite bodies (McClosky "B" time) in Hamilton County, Illinois (from Tharp, 1983). (B) Isoporosity (≥6%) map of upper Chester oolite bodies in north-central Oklahoma. Contour interval: 10 ft (modified from Asquith, 1984). Maps made by Zuppann in 1982 show similar trends of oolite bodies in the Monteagle Limestone of northern Tennessee in the Appalachian basin. These maps were sold, and their whereabouts are unknown.

spond. In some places the relationship between these two appears quite strong, but elsewhere it does not. In northern Tennessee, for instance, there appears to be extremely close agreement between elevated porosity values and the presence of oolitic facies in the Monteagle Limestone. This conclusion is based on examination of a core (Medeiros, 1984) and extensive examination by Zuppann of Monteagle well cuttings, many at the well site. The latter study found virtually a one-to-one correspondence between the presence of ooids in the samples and markedly increased porosity as indicated by density logs. Similarly close approximations have been documented in the St. Louis Limestone in Kansas (Handford, 1988).

In other areas, however, cementation and diagenetic alteration may occlude porosity to the point that oolitic facies may be no more porous than adjacent facies; or porosity may be developed in nonoolitic facies associated with oolitic rocks. Swann and Bell (1958), for instance, stated that most of the Ste. Genevieve oolites in the Illinois basin are "rather well cemented" and differ from the loosely cemented lenses typical of reservoir-quality oolites in that formation. Youse (1964) reported well-developed oolitic zones that were uniformly tightly cemented in the Greenbrier Limestone of West Virginia and eastern Kentucky. Tightly cemented oolites are apparently even more common in the Williston basin, where the diagenetic history is complex and significantly affects reservoir quality in Mississippian rocks (Kent, 1987). This may account for the absence in the literature of maps showing the distribution and geometries of oolite bodies in that basin.

Interpreting porosity of oolite bodies in the subsurface is also complicated because porosity may also be developed in facies adjacent to the oolitic rock. Choquette and Steinen (1980) described extensive reservoirs in dolomitized mudstones subjacent to oolitic bodies in the Ste. Genevieve at North Bridgeport field, in Illinois (see Bandy, this volume). They interpreted this non-supratidal dolomitic porosity to have resulted from influx of meteoric waters via permeability pathways present in the porous oolitic deposits. Thus, in addition to porous oolite facies, these dolomitic reservoirs are significant exploration targets.

The presence of tightly cemented oolites, as well as porosity development in facies other than the oolitic rock, presents problems in delineating oolite sediment bodies in the subsurface by mapping porosity distribution as recorded on geophysical logs.

Unfortunately, this is often the only possible method of inferring geometries of the original sediment bodies in most areas, because geophysical logs are far more commonly available than cores and reliable sample information. So this method cannot be avoided, but a certain degree of generalization must be accepted in the resulting interpretations of depositional geometries.

Oolite-Body Geometry

Controlling Factors

Ball (1967) observed that elongate oolite sand bodies are common in Holocene marine deposits in the Bahamas and proposed models that explained why they may be preferentially oriented either parallel with or perpendicular to depositional strike. He described two major types of oolite shoals: (1) marine sand belts, oriented parallel with the slope break; and (2) tidal bar belts, which, although oriented parallel with slope break, contain oolite bodies that are oriented perpendicular to the slope break because of locally amplified tidal currents directed up- and downslope.

We see no need to review the details of these and more recent models, which has been done quite effectively by Handford (1988). Ball (1967) presented these two models as end-member models of oolite shoals deposited at a shelf break, the main variable distinguishing the two being relative tidal influence. Following Ball's (1967) observations on modern oolite-body orientations, his models were soon applied to subsurface oolite bodies, and orientations of ancient bodies relative to depositional strike has been, and continues to be, discussed. However, Hine (1977) noted that in the Bahamas tidal bar belts and marine sand belts can coexist along the same bank edge in response to multiple generations of oolitic sedimentation under different environmental conditions. Therefore, the meaning of oolite-body orientations may not be all that simple to evaluate in paleoenvironmental reconstructions of ancient rocks.

Models based on modern occurrences in the Bahamas may not be entirely applicable to Mississippian rocks because the Bahamian deposits are situated on the edge of a prominent shelf break, which does not appear to have been the case for most areas of the United States during the Mississippian. Extensive Mississippian oolite sedimentation occurred on broad ramps such as the eastern shelf of the Illinois basin, which had a regional depositional slope of less than 0.5 ft per mi (0.2m per km) (Carr, 1973). However, the concept that the relative interplay between depositional topography and tidal and other currents defines the orientation and distribution of oolite shoals is entirely appropriate, and the record of Mississippian rocks suggests that these factors combined in a variety of ways to control the distribution and geometries of oolite bodies.

The main role of depositional topography in oolitic sedimentation, assuming the presence of environments otherwise suitable for ooid growth, is to bring the bottom depth up to a shallow enough level to create higher-energy environments necessary for ooids to form (typically less than 3 m in modern settings, according to Tucker and Wright, 1990). Bands of oolitic sediment should be parallel with local depositional strike, just as they are in both the marine sand belt and the tidal bar belt models of Ball (1967). If tidal currents are sufficiently strong, then the oolitic deposit might be dissected into bars with a different orientation. Also, if the crest of a depositional topographic high falls within the range of ooid production, then the oolitic sediment might be localized over the crest and generally assume the shape of the crest.

The primary factors that control the geometry of oolite bodies, i.e., depositional strike and prevailing tidal (or other) currents, tend to create elongate oolite sediment bodies with preferred orientations, explaining why trends of parallel oolite reservoirs are so common in the Mississippian. The influence of these factors would also tend to persist through time and through relative changes in sea level, explaining why parallel elongate bodies recur at different stratigraphic levels.

Of course, the relative influence of topography and tidal currents varied from basin to basin during the Mississippian and from one locality to another within each basin. Local conditions and depositional controls worked in some instances to form oolite bodies with unusual shapes or with abnormal orientations. Features such as ooid-filled channels (Bandy, 1991), spillover lobes, eolian oolitic accumulations, local structural irregularities, and shifting depositional environments account for oolite bodies with anomalous geometries and orientations.

Geometry of Reservoirs

The trail of research relating to oolite reservoir geometries has some interesting milestones in studies of Mississippian rocks. Reviewing this progress helps to gauge our present level of knowledge about the shape and distribution of Mississippian oolites and underscores the large gap between what we know and what we could know about this subject.

Prior to 1949 oolite reservoirs discussed in the literature were typically described by phrases such as "vertically and laterally variable lenses of oolitic rock." The reservoirs were generally depicted on structure maps using the top of the oolite as a mapping horizon (for example, Blatchley, 1913; Bell and Piersol, 1932; Bybee, 1948). This method does infer the oolite-body geometry to a certain extent because the structure on top of the oolite lens typically has positive relief, providing it has not been removed by subsequent structural movement. However, the true three-dimensional shape of a reservoir is not portrayed by a structure map, and details such as thickness variations, convex-downward geometries, composite reservoirs, and effects of post-depositional structural changes are not discernible on structure maps. Investigators generally acknowledged that the mapped reservoirs contained multiple oolite lenses, but they did not isolate and map the individual bodies of porosity.

Connolly (1949), in his unpublished M.S. thesis, was apparently the first to map an ancient oolite body (a continuous lens of oolite rock) in the subsurface (Carr, 1973). In his interpretation, Connolly used slice maps and fence diagrams to depict the geometry of two Ste. Genevieve oolite bodies at the Passport field in Illinois (see Zuppann, this volume). Significantly, he also compared these subsurface oolitic zones to "sub-parallel, bar-like" oolitic accumulations of the modern Bahamas as described by Rich (1948).

The first recorded use of isopach maps to show geometries of subsurface oolite bodies was apparently by Truitt (1951), who mapped parts of two oolite bodies in the Ste. Genevieve at Spencer field in southwestern Indiana. Like Connolly, his work was also part of an unpublished thesis. It was not until Nuttall (1968) and Sparks (1968) that the first portrayals of individual ancient oolite bodies were published in the literature; both reported on field studies of Ste. Genevieve reservoirs in the Illinois basin. Thereafter, isopach maps of oolite bodies and oolitic reservoirs have often been used to interpret Mississippian oolites. Among these important papers is Carr's (1973) study of Ste. Genevieve oolite bodies in the Illinois basin, a landmark work on Mississippian oolite-body geometries. A large part of an oolite body was exposed in a quarry, allowing direct observation. Carr mapped three subsurface oolite petroleum reservoirs as well and compared their geometries with the Holocene Bahamian models of Ball (1967). Another paper, by Choquette and Steinen (1980), also on the Ste. Genevieve in the Illinois basin, showed geometries of several oolite bodies occurring at two stratigraphic levels in the Bridgeport field area in Illinois. They described facies relationships between oolite grainstones and adjacent dolomite facies. Also, Handford (1988) mapped the St. Louis B-zone at Damme field in the Hugoton embayment in southwestern Kansas and discussed suitable depositional models.

Now that interest in oolite-body geometries has increased and oolite-body geometries are routinely mapped in the literature, what should be the direction of future research? Most reports that show geometries of Mississippian reservoirs are limited to local accumulations such as a single field or a small group of fields. Now research with greater geographic and stratigraphic scope is needed so that the occurrences and distribution of many oolite bodies can be related to each other and to their basin setting. For entire basins containing Mississippian oolite reservoirs, there may not be even a single oolite body mapped in the literature, the Williston basin being the most conspicuous. Probably more than 100 oolite bodies (mostly gas-bearing) occurring at eight or more stratigraphic levels have been mapped by Zuppann in the Monteagle Limestone of northern Tennessee (these proprietary maps are no longer available to the authors), but no one has described the geometry of more than a single field in any published study.

Only when the details of local oolite accumulations are applied to the larger picture will we be able to understand the controls that affect each individual oolite deposit and to understand patterns of oolite-body geometries within each basin. Until that knowledge is obtained, the geometries of Mississippian oolite reservoirs in different basins will remain difficult to compare.

Reservoir Characteristics

There has always been a fascination with oolitic reservoirs, perhaps in part because they are similar to terrigenous sandstone reservoirs. They consist of more or less round grains formed into discrete bodies by wave or current activity. Geologists may assume that oolite-body geometries and pore-system characteristics can be interpreted using terrigenous sandstone reservoirs as analogs. Although they are far more common and have been studied in more detail than their oolitic counterparts, terrigenous sandstone reservoirs may not be analogous in terms of either their geometries or their pore systems. First, ooid grains are typically deposited quite near their site of formation, whereas marine deposits of terrigenous sandstones have been transported over considerable distances. The depositional factors therefore affecting sand-body geometries of autochthonous oolite accumulations may be substantially different from those associated with allochthonous terrigenous sandstone deposits. Second, as noted in the previous section, the geometry of oolite reservoirs may be difficult or impossible to determine using available subsurface information.

An important consideration for oolitic reservoirs is the critical role played by diagenesis in preserving or modifying porosity systems. The pathway of diagenesis in a simplistic sense progresses from the reduction of primary interparticle porosity through cementation and chemical compaction to the creation of secondary moldic, vuggy, or intercrystalline porosity (or some combination) through dissolution and recrystallization. The often complex variety of porosity networks that results from this process is important not only to the formation of a reservoir but also to the quality of its performance.

Some insight into pore systems of ancient oolite reservoirs may be gained by noting simpler intergranular pore systems in Holocene ooid and peletoidal sands from the Bahama Banks that were measured for porosity and permeability (Enos and Sawatsky, 1981). Porosity values ranged from 40 to 48% and permeabilities from 25,000 to 54,000 md. These porosity values are at or near the upper limit (48%) expected for minimal packing of spherical grains (Graton and Fraser, 1935). It is interesting that capillary pressure data from these Holocene samples indicated that approximately 17% of this pore volume was related to porosity with entry diameters smaller than 1 μm. These fine pores presumably are between the aragonite needles that form the ooid and pelletoidal grains (Enos and Sawatsky, 1981). Cementation of Holocene ooid sand on Joulters Cay in the Bahamas was reported by Halley and Harris (1979). Porosities of the cemented oolite are in the same range as that of the ooid sand reported by Enos and Sawatsky (1981), between 40 and 50%. The Halley and Harris study indicated that early vadose and phreatic cementation does not substantially affect the amount of porosity, but permeabilities were substantially reduced by even relatively minor amounts of cement. Exact comparisons between the permeability measurements by Enos and Sawatsky (1981) and by Halley and Harris (1979) are not possible due to differences in measuring techniques, but the amount of permeability reduction appears to be several orders of magnitude.

Oolitic limestone reservoirs may contain fairly simple intergranular pore systems with good porosity (10 to 20%) and permeability (tens to hundreds md). However, diagenesis may also create more complex pore systems. Moldic pore systems (high porosity and low permeability) are typical of many Jurassic Smackover reservoirs which are also commonly dolomitic. Bimodal pore systems are characterized by micritized ooids with intercrystalline microporosity and coarser grained macroporosity between the grains. Oolitic bimodal porosity reservoirs have been documented in the Ste. Genevieve in Kentucky (Asquith, 1986).

General treatments of porosity in carbonate reservoirs are available (Moore, 1989; Chilingarian et al., 1992), but little information has been presented about Mississippian oolite reservoirs. The best sources of data on porosity and permeability from cores of Mississippian oolitic reservoirs are two Marathon Oil Company studies on Ste. Genevieve reservoirs in Illinois (Choquette and Steinen, 1985; Manley et al., this volume). At North Bridgeport field (Choquette and Steinen, 1985), Ste. Genevieve reservoirs occur in both microcrystalline dolomite and oolitic limestone. Porosity in the oolitic reservoir ranges from 3.8 to 28.8%, with an average of 13.7%. The permeability ranges from 0.1 to greater than 9500 md, with an average of 250 md (Choquette and Steinen, 1985). At Willow Hill field (Manley et al., this volume), only the oolitic limestone reservoir facies is present, and porosity ranges from 6.0 to 17.2%, with an average of 12.7%. Permeability ranges from 0.1 to 228 md, with an average of 113 md (Manley et al., this volume). Following the classification system of Wardlaw and Cassan (1978), the recovery efficiency for carbonate reservoirs can be estimated by knowing the amount of porosity and the nature of the pore system. For North Bridgeport and Willow Hill fields, estimated recovery efficiencies would be on the order of 40 to 45% for intergranular porosity that averages 13 to 14%. There are no published figures on recovery efficiency from Ste. Genevieve reservoirs in the Illinois basin, but this level of recovery efficiency (40 to 45%) seems to conform to the general consensus for Ste. Genevieve reservoirs. As a rule these reservoirs also seem to respond well to waterflooding, but reservoir heterogeneity is often a major problem for operators trying to establish an efficient waterflood.

SUMMARY

Mississippian oolitic limestone reservoirs represent an important petroleum resource for several reasons: (1) depositional and tectonic controls operated together to make oolitic limestones a significant component of Meramecian and lower Chesterian rocks in several petroleum-producing areas of the United States; (2) there is potentially a large undrilled petroleum resource remaining in stratigraphic traps (probably concentrated at relatively shallow depths) in Mississippian oolites; and (3) these oolitic reservoirs tend to have favorable characteristics for good recovery efficiency for both primary and secondary production. The combination of these factors makes oolitic reservoirs an especially attractive target for our domestic oil industry.

ACKNOWLEDGMENTS

Critical reviewers John B. Droste, Margaret V. Ennis, Stanley J. Keller, and John A. Rupp suggested many helpful improvements to the paper. They graciously reviewed the manuscript on unreasonably short notice. Published by permission of Norman C. Hester, Director, Indiana Geological Survey.

REFERENCES CITED

Adler, F. J., coordinator, 1987, Midcontinent correlation chart, *in* O. E. Childs, G. Steek, and A. Salvador, project directors, Correlation of stratigraphic units in North America: American Association of Petroleum Geologists, COSUNA Chart MC.

Asquith, G. B., 1984, Depositional and diagenetic history of the Upper Chester (Mississippian) oolitic reservoirs, north-central Beaver County, Oklahoma, *in* N. J. Hyne, ed., Limestones of the Mid-Continent: Tulsa Geological Society, Oklahoma, p. 87-92.

Asquith, G. B., 1986, Microporosity in the O'Hara oolite zone of the Mississippian Ste. Genevieve Limestone, Hopkins County, Kentucky, and its implications for formation evaluation: Carbonates and Evaporites, v. 1, p. 7-12.

Ball, M. M., 1967, Carbonate sand bodies of Florida and the Bahamas: Journal of Sedimentary Petrology, v. 37, p. 556-591.

Ballard, W. W., J. P. Bluemie, and L. C. Gerhard, coordinators, 1983, Northern Rockies/Williston basin correlation chart, *in* O. E. Childs, G. Steek, and A. Salvador, project directors, Correlation of stratigraphic units in North America; American Association of Petroleum Geologists, COSUNA Chart NRW.

Bandy, W. F., Jr., 1991, Recognition of oolite-filled channels, Ste. Genevieve Formation, Illinois basin (abs.): American Association of Petroleum Geologists Bulletin, v. 75, p. 537.

Bell, A. H., and R. J. Piersol, 1932, Effects of waterflooding on oil production from the McClosky sand, Dennison Township, Lawrence County, Illinois: Illinois State Geological Survey, Illinois Petroleum 22, 26 p.

Berg, C. A., 1956, Virden Rosella and North Virden fields, Manitoba, *in* Williston basin symposium: North Dakota Geological Society and Saskatchewan Geological Society, p. 84-93.

Blatchley, R. S., 1913, The oil fields of Crawford and Lawrence Counties: Illinois State Geological Survey Bulletin 22, 442 p.

Bybee, H. H., 1948, Hitesville Consolidated field, Union County, Kentucky: American Association of Petroleum Geologists Bulletin, v. 32, p. 2063-2082.

Carr, D. D., 1973, Geometry and origin of oolite bodies in the Ste. Genevieve Limestone (Mississippian) in the Illinois basin: Indiana Geological Survey Bulletin 48, 81 p.

Chilingarian, G. V., S. J. Mazzullo, and H. H. Rieke, eds., 1992, Carbonate reservoir characterization: a geologic-engineering analysis, part 1: Elsevier, Amsterdam, 639 p.

Choquette, P. W., and R. P. Steinen, 1980, Mississippian non-supratidal dolomite, Ste. Genevieve Limestone, Illinois basin: evidence for mixed-water dolomitization, *in* D. H. Zenger, J. B. Dunham, and R. L. Ethington, eds., Concepts and models of Dolomitization—A symposium: SEPM Special Publication 28, p. 163-196.

Choquette, P. W., and R. P. Steinen, 1985, Mississippian oolite and non-supratidal dolomite reservoirs in the Ste. Genevieve Formation, North Bridgeport field, Illinois basin, *in* P. O. Roehl and P. W. Choquette, eds., Carbonate petroleum reservoirs: Springer-Verlag, N.Y., p. 207-225.

Cluff, R. M., and J. A. Lineback, 1981, Middle Mississippian carbonates of the Illinois basin: Illinois Geological Society, 88 p.

Cohee, G. V., 1979, Michigan basin region, *in* L. C. Craig and C. W. Connor, coordinators, Paleotectonic investigations of the Mississippian System in the United States: U.S. Geological Survey Professional Paper 1010, p. 49-57.

Connolly, F. T., 1949, The geology of the Passport oil pool in Clay County, Illinois: M.S. thesis, University of Cincinnati, Ohio, 33 p.

Craig, L. C., and C. W. Connor, coordinators, 1979, Paleotectonic investigations of the Mississippian System in the United States: U.S. Geological Survey Professional Paper 1010, 559 p.

Eluik, L. S., 1987, Discussion—Submarine hydrothermal weathering, global eustacy, and carbonate polymorphism in Phanerozoic marine oolites: Journal of Sedimentary Petrology, v. 57, p. 184-186.

Enos, P., and L. H. Sawatsky, 1981, Pore networks in Holocene carbonate sediments: Journal of Sedimentary Petrology, v. 51, p. 961-985.

Given, R. K., and B. H. Wilkinson, 1987, Dolomite abundance and stratigraphic age: constraints on rates and mechanisms of Phanerozoic dolostone formation: Journal of Sedimentary Petrology, v. 57, p. 1068-1078.

Graton, L. C., and J. H. Fraser, 1935, Systematic packing of spheres—with particular relation to porosity and permeability: Journal of Geology, v. 43, p. 785-909.

Halley, R. B., and P. M. Harris, 1979, Fresh-water cementation of a 1,000-year-old oolite: Journal of Sedimentary Petrology, v. 49, p. 969-988.

Handford, C. R., 1988, Review of carbonate sand-belt deposition of ooid grainstones and application to Mississippian reservoir, Damme Field, southwestern Kansas: American Association of Petroleum Geologists Bulletin, v. 72, p. 1184-1199.

Harrison, R., and T. C. Larson, 1958, Oil production from the "Spearfish" and Charles in the Newburg field, Bottineau County, North Dakota, *in* Second Williston basin symposium: Saskatchewan Geological Society and North Dakota Geological Society, p. 27-32.

Hills, J. M., and F. E. Kottlowski, coordinators, 1983, Southwest/Southwest midcontinent correlation chart, *in* O. E. Childs, G. Steek, and A. Salvador, project directors, Correlation of stratigraphic units in North America; American Association of Petroleum Geologists, COSUNA Chart SSMC.

Hine, A. C., 1977, Lily Bank, Bahamas; Case history of an active oolitic sand shoal: Journal of Sedimentary Petrology, v. 47, p. 1554-1581.

Howard, R. H., 1991, Hydrocarbon reservoir distribution in the Illinois basin, *in* M. W. Leighton, D. R. Kolata, D. F. Oltz, and J. J. Eidel, eds., Interior cratonic basins: American Association of Petroleum Geologists Memoir 51, p. 299-327.

James, N. P., 1983, Reef environment, *in* P. A. Scholle, D. G. Bebout, and C. H. Moore, eds., Carbonate depositional environments: American Association of Petroleum Geologists Memoir 33, p. 354-462.

Kent, D. M., 1987, Mississippian facies, depositional history, and oil occurrences in Williston basin, Manitoba and Saskatchewan, *in* M. W. Longman, ed., Williston basin, anatomy of a cratonic oil province: Rocky Mountain Association of Geologists, Denver, Colorado, p. 157-170.

Kent, H. C., E. L. Couch, and R. A. Knepp, coordinators, 1988, Central and southern Rockies correlation chart, *in* O. E. Childs, G. Steek, and A. Salvador, project directors, Correlation of stratigraphic units in North America; American Association of Petroleum Geologists, COSUNA Chart CSR.

MacKenzie, F. T. and J. D. Pigott, 1981, Tectonic controls of Phanerozoic sedimentary rock cycling: Journal of Geological Society of London, v. 138, p. 183-196.

Mankin, C. J., coordinator, 1987, Texas-Oklahoma tectonic correlation chart, in O.E. Childs, G. Steek, and A. Salvador, project directors, Correlation of stratigraphic units in North America; American

Association of Petroleum Geologists, COSUNA Chart TOT.

Maples, C. G., and J. A. Waters, 1987, Redefinition of the Meramecian/Chesterian boundary (Mississippian): Geology, v. 15, p. 647-651.

Mast, R. F., and R. H. Howard, 1991, Oil and gas production and recovery estimates in the Illinois basin, *in* M. W. Leighton, D. R. Kolata, D. F. Oltz, and J. J. Eidel, eds., Interior cratonic basins: American Association of Petroleum Geologists Memoir 51, p. 295-298.

Medeiros, M. F., 1984, Depositional controls, geometry, and diagenetic history of a Monteagle (Mississippian) oolite reservoir: Rugby Quadrangle, northeast Tennessee: M.S. thesis, Vanderbilt University, Nashville, Tennessee, 182 p.

Moore, C. H., 1989, Carbonate diagenesis and porosity: Elsevier, New York, 338 p.

Nuttall, B. D., 1968, Hanson Pool, Hopkins County, Kentucky, *in* D. N. Miller, Jr., ed., Geology and Petroleum production of the Illinois Basin: Illinois and Indiana-Kentucky Geological Societies, p. 121-128.

Patchen, D. G., K. L. Avary, and R. B. Erwin, coordinators, 1985a, Northern Appalachian correlation chart, *in* O. E. Childs, G. Steek, and A. Salvador, project directors, Correlation of stratigraphic units in North America; American Association of Petroleum Geologists, COSUNA Chart NAP.

Patchen, D. G., K. L. Avary, and R. B. Erwin, coordinators, 1985b, Southern Appalachian correlation chart, *in* O. E. Childs, G. Steek, and A. Salvador, project directors, Correlation of stratigraphic units in North America; American Association of Petroleum Geologists, COSUNA Chart SAP.

Rich, J. L., 1948, Submarine sedimentary features on Bahama Banks and their bearing on distribution patterns of lenticular oil sands: American Association of Petroleum Geologists Bulletin, v. 33, p. 767-779.

Sandberg, P.A., 1975, New interpretations of Great Salt Lake ooids and of ancient nonskeletal carbonate mineralogy: Sedimentology, v. 22, p. 497-537.

Sandberg, P.A., 1983, An oscillating trend in Phanerozoic non-skeletal carbonate mineralogy: Nature, v. 305, p. 19-22.

Schmidlapp, R. L., 1959, Damme and Finnup Fields, *in* Kansas oil and gas fields, Volume II, Western Kansas: Kansas Geological Society, p. 8-12.

Shaver, R. H., coordinator, 1985, Midwestern basins and arches correlation chart, *in* O. E. Childs, G. Steek, and A. Salvador, project directors, Correlation of stratigraphic units in North America; American Association of Petroleum Geologists, COSUNA Chart MBA.

Smith, G. W., G. E. Summers, Jr., D. Wallington, and J. L. Lee, 1958, Mississippian oil reservoirs in the Williston basin, *in* L. G. Weeks, ed., Habitat of oil: American Association of Petroleum Geologists, Tulsa, Oklahoma, p. 149-177.

Sparks, J. B., 1968, Gard's Point Field, Wabash County, Illinois, *in* D. N. Miller, Jr., ed., Geology and Petroleum production of the Illinois Basin: Illinois and Indiana-Kentucky Geological Societies, p. 105-108.

Stanton, M. S., 1956, Stratigraphy of the Lodgepole Formation, Virden-Whitewater area, Manitoba, *in* Williston basin symposium: North Dakota Geological Society and Saskatchewan Geological Society, p. 79-83.

Swann, D. H., and A. H. Bell, 1958, Habitat of oil in the Illinois basin, *in* L. G. Weeks, ed., Habitat of Oil: American Association of Petroleum Geologists, Tulsa, Oklahoma, p. 447-472.

Tharp, T. C., 1983, Subsurface geology and paleogeography of the lower Ste. Genevieve Limestone in Hamilton County, Illinois: M.S. thesis, University of Kentucky, Lexington, 93 p.

Truitt, P. B., 1951, The geology of the Spencer oil pool, Posey County, Indiana: M.S. thesis, University of Cincinnati, Ohio, 33 p.

Tucker, M. E., and V. P. Wright, 1990, Carbonate sedimentology: Blackwell Scientific Publications, Oxford, 482 p.

U. S. Geological Survey, 1979, The Mississippian and Pennsylvanian (Carboniferous) Systems in the United States: U.S. Geological Survey Professional Paper 1110-A-DD.

Wardlaw, N. C., and J. P. Cassan, 1978, Estimation of recovery efficiency by visual observation of pore systems in reservoir rocks: Bulletin of Canadian Petroleum Geology, v. 26, p. 572-585.

Wilkinson, B. H., R. M. Owen, and A. R. Carroll, 1985, Submarine hydrothermal weathering, global eustacy, and carbonate polymorphism in Phanerozoic marine oolites: Journal of Sedimentary Petrology, v. 55, p. 171-183.

Wilkinson, B. H., R. M. Owen, and A. R. Carroll, 1987, Submarine hydrothermal weathering, global eustacy, and carbonate polymorphism in Phanerozoic marine oolites—reply: Journal of Sedimentary Petrology, v. 57, p. 186-188.

Wilson, J. L., 1975, Carbonate facies in geologic history: Springer-Verlag, New York, 471 p.

Youse, A. C., 1964, Gas producing zones of Greenbrier (Mississippian) Limestone, southern West Virginia and eastern Kentucky: American Association of Petroleum Geologists Bulletin, v. 48, p. 465-486.

Chapter 2

Possible Flexural Controls on the Origins of Extensive, Ooid-Rich, Carbonate Environments in the Mississippian of the United States

Frank R. Ettensohn
University of Kentucky
Lexington, Kentucky, USA

ABSTRACT

Widespread ooid deposition characterized much of the United States during the Mississippian and was the product of an unusual coincidence of ideal regional paleogeographic and tectonic factors, as well as global paleoclimatic, eustatic, and geochemical conditions. The contribution of tectonics, largely through flexural mechanisms, is considered to have had the greatest effect.

During the Mississippian, parts of three orogenic belts, the Antler, the Ouachita, and the Acadian, bounded southern parts of North America in the present-day United States. Various active and relaxational phases of these orogenies generated peripheral bulges that migrated within and beyond foreland basins to adjacent parts of the foreland, creating broad areas of uplift into shallow, agitated waters conducive to ooid production.

During the Early Mississippian (Kinderhookian), bulge migration and uplift were incomplete, so ooid distribution was limited. By the middle Mississippian (Valmeyeran), episodes of coeval Antler and Ouachita flexure combined with Acadian relaxation created broad belts of uplift on the inner craton, where the most prolific ooid production occurred. In the Late Mississippian (Chesterian), largely filled foreland basins and rising tectonic highlands caused siliciclastic inundation of adjacent sites of ooid production, resulting in thinner, more localized ooid deposits. By the end of the middle Chesterian, all major concentrations of Mississippian ooids had disappeared.

INTRODUCTION

Carbonate ooids not only reflect the conditions required for carbonate sedimentation in general (largely restricted tropical to subtropical seas deprived of major clastic influx) but also require unique physical conditions of agitation and shallow water to form. Although ooid deposition is not very widespread in modern seas, at times in the past, such as the middle Mississippian (Valmeyeran; late Tournaisian–early Visean), ooid sedimentation became widespread across large parts of the United States due to a unique interaction of paleogeographic, global-eustatic, atmospheric-hydrospheric, and tectonic conditions (see also discussion in Keith and Zuppann, this volume). In particular, the North American craton straddled the Mississippian paleoequator (Scotese, 1990) so that much of the United States was situated in the trade-wind belts, 5–30º north and south of the paleoequator (Figure 1). Because of the warm, dry, and evaporative conditions in these belts, carbonate becomes saturated to supersaturated in subjacent shallow seas, and calcium carbonate will precipitate directly out of sea water around nuclei to form ooids (Lees, 1975; Heckel and Witzke, 1979).

Some workers have also suggested the importance of increased salinity for ooid formation (Lees, 1975; Heckel and Witzke, 1979), but the control of salinity has not yet been proven. The control, however, may be as simple as the fact that the availability of calcium ions is proportional to the salinity (Heckel and Witzke, 1979). Whatever the relationship, it is apparent that ooid deposition can occur locally in more moist tropical or temperate zones in nearly enclosed seas and lagoons where restriction and increased salinity are present (see Milliman, 1974, p. 40; Heckel and Witzke, 1979; Tucker, 1981, p. 99). Restriction may also play a role in the confinement of ooids to high-energy environments (Wilson, 1975; Tucker and Wright, 1990).

Relative to the eustatic regime, Wilkinson et al. (1985) have demonstrated that ooid deposition was widespread during times of global transgression and regression and uncommon during times of continental emergence and submergence. Similarly, Paleozoic sea level curves (Vail et al., 1977) show that the Mississippian overall was a period of global regression during the transition from Devonian submergence to Pennsylvanian emergence, although second-order transgressions were present and quite important. Hence, the correlations by Wilkinson et al. (1985) regarding times of ooid abundance and periods of global transgression and regression agree well with published Mississippian sea level curves.

Wilkinson et al. (1985) have also suggested probable atmospheric-hydrospheric interactions as geochemical controls for times of abundant ooid deposition. They argued that the predominance of calcitic ooid cortexes during the Mississippian indicates low Mg/Ca ratios and increased pCO_2 that accompanied rapid sea floor spreading rates.

The remainder of this paper is devoted to another control—tectonism—that is equally important in explaining the abundance of Mississippian ooids. The Mississippian was a time of widespread regional uplift that generated the very shallow-water environments (normally less than 5 m) and agitated conditions necessary for abundant ooid production across large parts of the central United States. This regional uplift would appear to have been especially critical during the Valmeyeran, the time of most prolific Mississippian ooid sedimentation, because shallow-water ooid deposition immediately succeeded a period of largely deeper water, clastic sedimentation in a large black-shale basin that occupied nearly one-third of the country (Figure 1). Moreover, during a subsequent period of approximately 8–9 m.y. (see Harland et al., 1982), this basinal area was transformed into a very shallow, epeiric platform suitable for ooid deposition. The question of how this transformation occurred so quickly can be answered by looking at two effects: basin infill and regional uplift. Infilling of eastern portions of the black-shale basin with vast amounts of Kinderhookian and early Valmeyeran, post-Acadian siliciclastic sediments (Figure 1) aided this transformation. Central and southwestern parts of this basin, however, were not completely infilled, but nevertheless experienced the transformation to shallow-water, ooid-forming environments shortly thereafter—despite a general second-order rise in sea level (see Vail et al., 1977; Harland et al., 1982). Older studies commonly attributed this transformation to the second effect of broad continental warping concentrated along major regional structures like the Cincinnati arch (e.g., Kay and Colbert, 1965) or to cratonic epeirogeny (Ham and Wilson, 1967). More recent studies, however, now suggest that most of this regional uplift may be explainable in terms of lithospheric flexure (Jacobi, 1981; Beaumont, 1981; Quinlan and Beaumont, 1984) accompanying the end of the Acadian orogeny and the beginning of the Antler and Ouachita orogenies.

BASIC FLEXURAL MODELS

The advent of an orogeny along a cratonic margin drastically influences sedimentation by creating new sources of sediment and new basins to receive these sediments. The development of lithospheric flexure models (e.g., Price, 1973; Jacobi, 1981; Beaumont, 1981; Karner and Watts, 1983; Quinlan and Beaumont, 1984), however, has shown that tectonism on cratonic margins may influence the development and disposition of major unconformities, basins, areas of regional uplift, and, at least indirectly, modes of sedimentation in inland cratonic areas far removed from the active margins.

Flexural models are based on the premise that deformational loading resulting from the creation of a migrating fold-thrust belt during plate collision, along with a subordinate component of sediment loading, produces a downwarped flexural or retro-arc

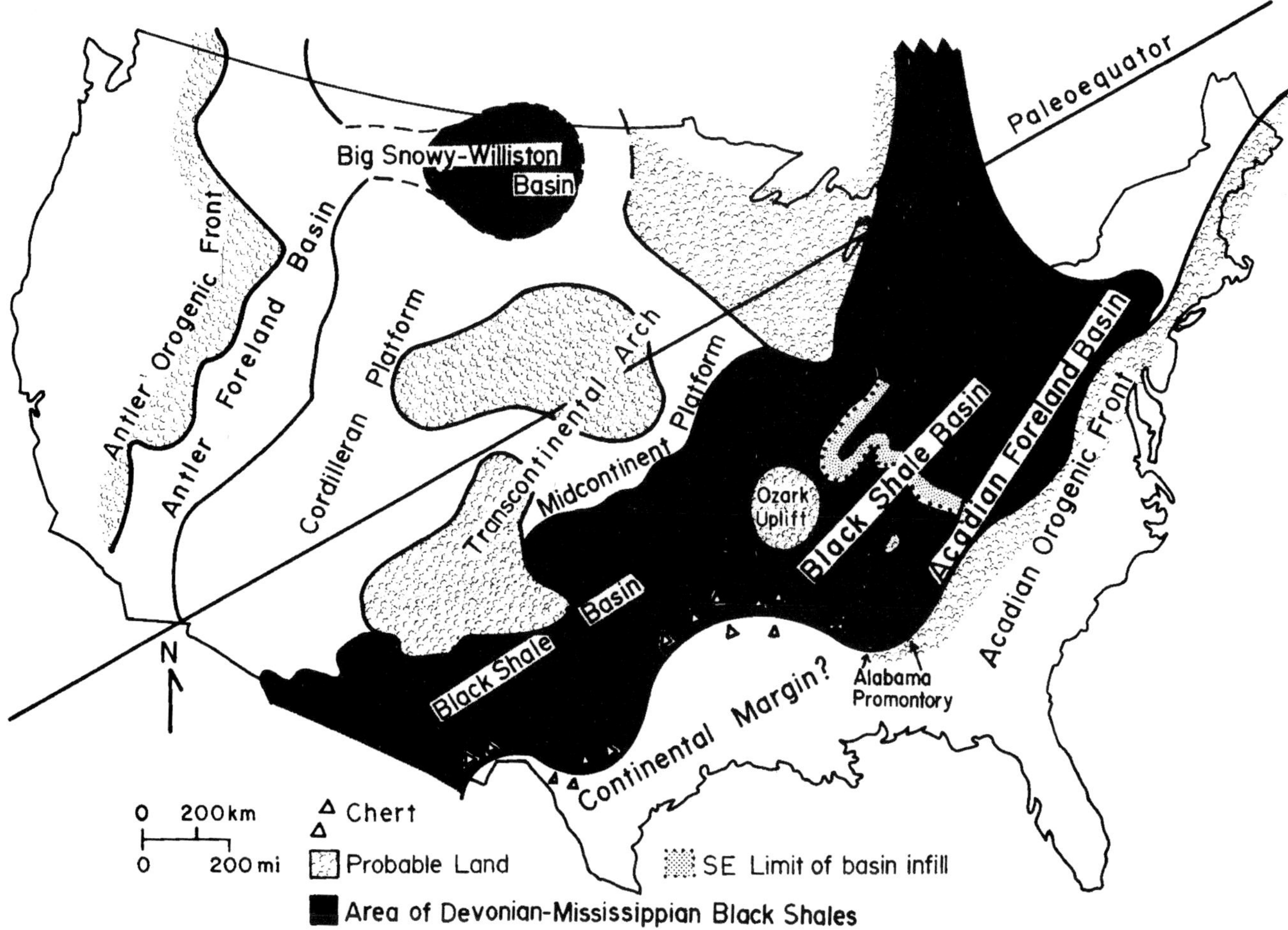

Figure 1. Location map of the United States showing the position of major paleogeographic, paleoclimatic, and tectonic features during the Late Devonian–Early Mississippian transition (adapted from Heckel and Witzke, 1979, and Ettensohn et al., 1988b). The stippled line running across the black-shale basin in the east shows the basinward limit of infilling by post-orogenic Acadian siliciclastic sediments (Borden, Grainger, Price, and Pocono formations). The occurrence of chert deposits defines the approximate position of southern continental margin.

foreland basin (sensu Dickinson, 1974). The foreland basin and its associated peripheral bulge form cratonward of the deformation zone (Figure 2) in response to isostatic compensation by the lithosphere (Beaumont, 1981; Quinlan and Beaumont, 1984). As collision continues and thrust loads shift onto the craton, the foreland basin and peripheral bulge also migrate cratonward away from the load (Figure 2). Once active deformation ceases and lithospheric relaxation ensues, the direction and nature of the bulge and basin migration change. The typical sequence of changes in flexural regime may be generalized into four parts:

1. At the inception of plate collision, cratonward bulge migration generates widespread uplift of the foreland basin and adjacent foreland, and a regional unconformity is formed (Jacobi, 1981; Quinlan and Beaumont, 1984) (Figures 2, 5A). However, Dickinson's (1974, p. 22) more general mechanism of regional uplift resulting from actual collision as subduction is initially braked or retarded may be more effective at generating such an unconformity.
2. As folding and thrusting ensue, loading and concomitant foreland basin subsidence increase to the point that an underfilled foreland basin with a stratified water column initially develops (Figure 5A, black Sunbury Shale). During periods of major ooid production, this basin may be a major source of deep, carbonate-saturated waters upwelling onto the adjacent uplifted foreland. Moreover, the adjacent bulge itself (Figure 2), which may be twice as wide as its accompanying foreland basin (Beaumont, personal communication, 1991), would be a likely place for widespread ooid deposition to

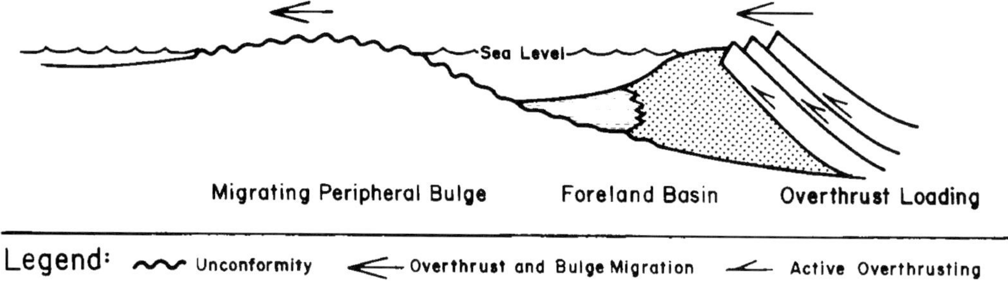

Figure 2. Schematic diagram showing development of a foreland basin and a peripheral bulge with its accompanying unconformity due to dynamic deformational loading in the orogen (adapted from Quinlan and Beaumont, 1984). Stippling represents coarse siliciclastic sediments; the lined pattern, dark shales.

occur. The bulge, when uplifted into shallow waters, would have a nearby source of carbonate-saturated water and be protected during most of its history from major clastic influx by the underfilled foreland basin, which would act as a sink for siliciclastic sediments between the bulge and source areas in the collision zone. As deformation migrates cratonward, so do the bulge and foreland basin subsidence, resulting in an overall transgressive carbonate sequence on the bulge itself. During times and places of major loading, however, flexural interactions between the foreland basin and a nearby cratonic basin may severely lower the bulge to the point of basin yoking (Figure 3). This kind of bulge attenuation during major loading episodes appears to be even more effective when a zone of structural weakness (e.g., the Big Snowy trough) connects the basins (Figure 1). Through flexural interactions, loading may influence parts of the craton up to 1000 km away from the orogen (Karner and Watts, 1983, p. 10,472).

3. Once active tectonism and deformational loading cease, the lithosphere responds to the then-static load in the orogen by relaxing stress. As a result, the foreland basin deepens, while the peripheral bulge is uplifted and migrates toward the load (Figures 4A, 5B). Erosion of the static deformational load (fold-thrust belt), however, generates so much clastic input that sedimentation exceeds subsidence and the foreland basin begins to fill with flyschlike sediments. At the same time, uplift and migration of the bulge may give rise to a shallow water platform on the distal margin of the foreland basin, where a regressive carbonate sequence generally develops (Figure 5B). Such a regressive sequence may also contain abundant ooids, suggesting that during relaxation the bulge

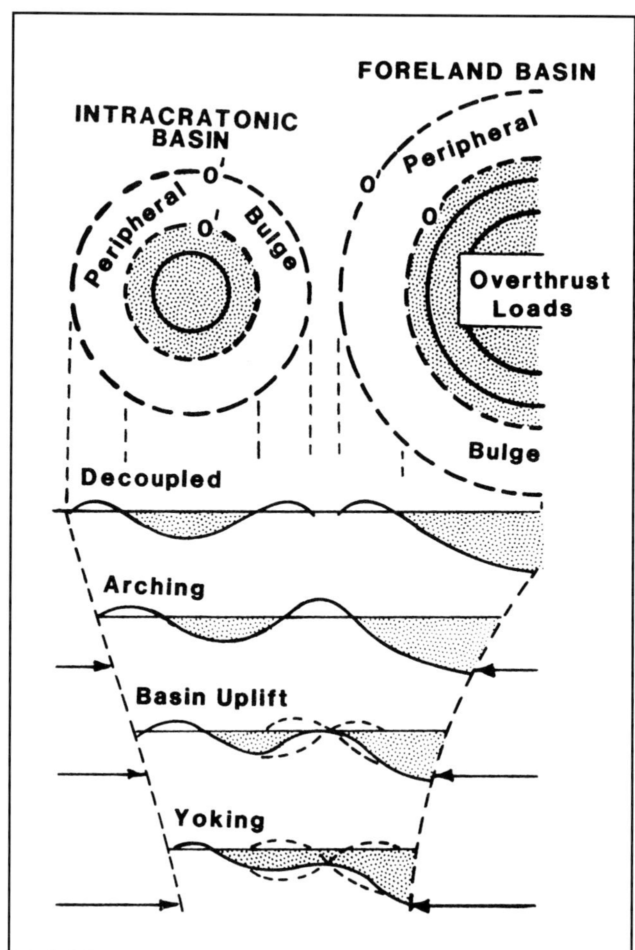

Figure 3. Flexural interaction between a foreland basin (e.g., the Appalachian or Antler foreland basin) and an intracratonic basin (e.g., the Illinois or Williston basin). The upper part of the diagram is a plan view of the basins in decoupled position corresponding to the first cross-sectional view below (0′ reflects sea level). Successive sections show the process of yoking (redrawn from Quinlan and Beaumont, 1984).

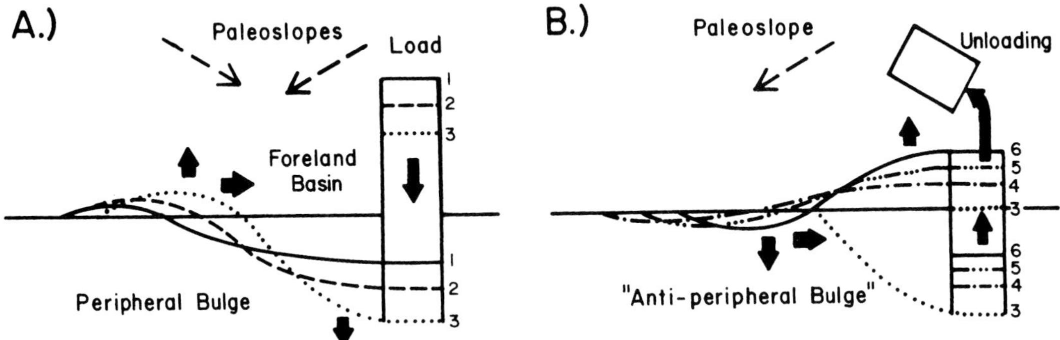

Figure 4. Two types of flexural response to lithospheric-stress relaxation. Numbers and adjacent lines refer to the position of the land, basin and bulge at various times, becoming younger with increasing value (redrawn from Beaumont et al., 1988): (A) "loading-type" relaxation: response to a static gravitational load results in a deepening foreland basin and migration of peripheral bulge toward the load; (B) "unloading-type" relaxation: response to erosional unloading results in rebound near the unloaded area and an "anti-peripheral bulge" that deepens and migrates toward the former load. Between these two stages of relaxation, a period of elevational equilibrium between eroded uplands and an infilled foreland basin occurs briefly (see Figure 5C).

once again may become a site of shallow-water, agitated conditions necessary for ooid production. As the bulge migrates toward the orogen (Figure 4A), ooid-forming environments may migrate into central parts of the foreland basin itself (Figure 5B). In contrast, if the amount of bulge uplift is large, and not overwhelmed by eustatic sea level rise, any regressive sequence may be truncated or destroyed by resulting unconformity development (Quinlan and Beaumont, 1984).

4. Once the foreland basin fills with sediments and adjacent orogenic highlands have undergone substantial lowering by erosion, a brief episode of elevational equilibrium ensues between the filled basin and eroded highlands (Figures 4B, 5C). During this period, very shallow-water conditions conducive to ooid deposition may spread across large parts of the foreland basin and adjacent parts of the craton. This period is short-lived, however, because as isostatic adjustment to unloading occurs, the former highlands begin to rebound upward and a compensating "anti-peripheral bulge" develops, deepens, and migrates toward the rebounding load (Figures 4B, 5D) (Beaumont et al., 1988). The combination of deepening associated with the anti-peripheral bulge and the cratonward progradation of molasselike marginal-marine and terrestrial sediments from rebounding areas effectively destroys the carbonate-producing environments.

Because most orogenies consist of a series of active tectophases separated by more quiescent periods, each tectophase has the potential of producing a more or less distinct four-part stratigraphic and sedimentologic response as outlined above. If tectophases succeed each other rapidly, however, final parts of the sequence may be shortened or completely eliminated. Of the three orogenies pertinent to this study, only for the Acadian orogeny are the timing and nature of individual tectophases and the resulting stratigraphic packages known well enough to present a relatively complete outline (Figure 5); only generalized accounts of orogenic history are available for the other two orogenies.

In any case, it seems likely that flexural mechanisms, especially in the form of bulge uplift and migration, could have produced the regional uplift necessary for widespread shallow-water conditions required by the present distribution of Mississippian oolite deposits. Hence, it seems reasonable that times of widespread Mississippian ooid deposition correspond to times of major bulge uplift during coeval orogenic tectophases. One test of this hypothesis is to map the distribution of unconformities and Mississippian oolites through time to see if they correlate with areas and times of probable bulge uplift. These kinds of correlations should be evident on both regional and local scales. Apparent regional controls of ooid deposition will be examined in the temporal framework of the three Mississippian epochs, and some probable cases of local tectonic control will be described as examples of what to expect.

Data concerning the occurrence and distribution of unconformities were compiled largely from the AAPG COSUNA charts and literature sources such as Craig and Connor (1979); similar data for Mississippian oolites were gleaned largely from various literature sources, especially Craig and Connor, 1979; U.S. Geological Survey, 1979; and Peterson, 1984.

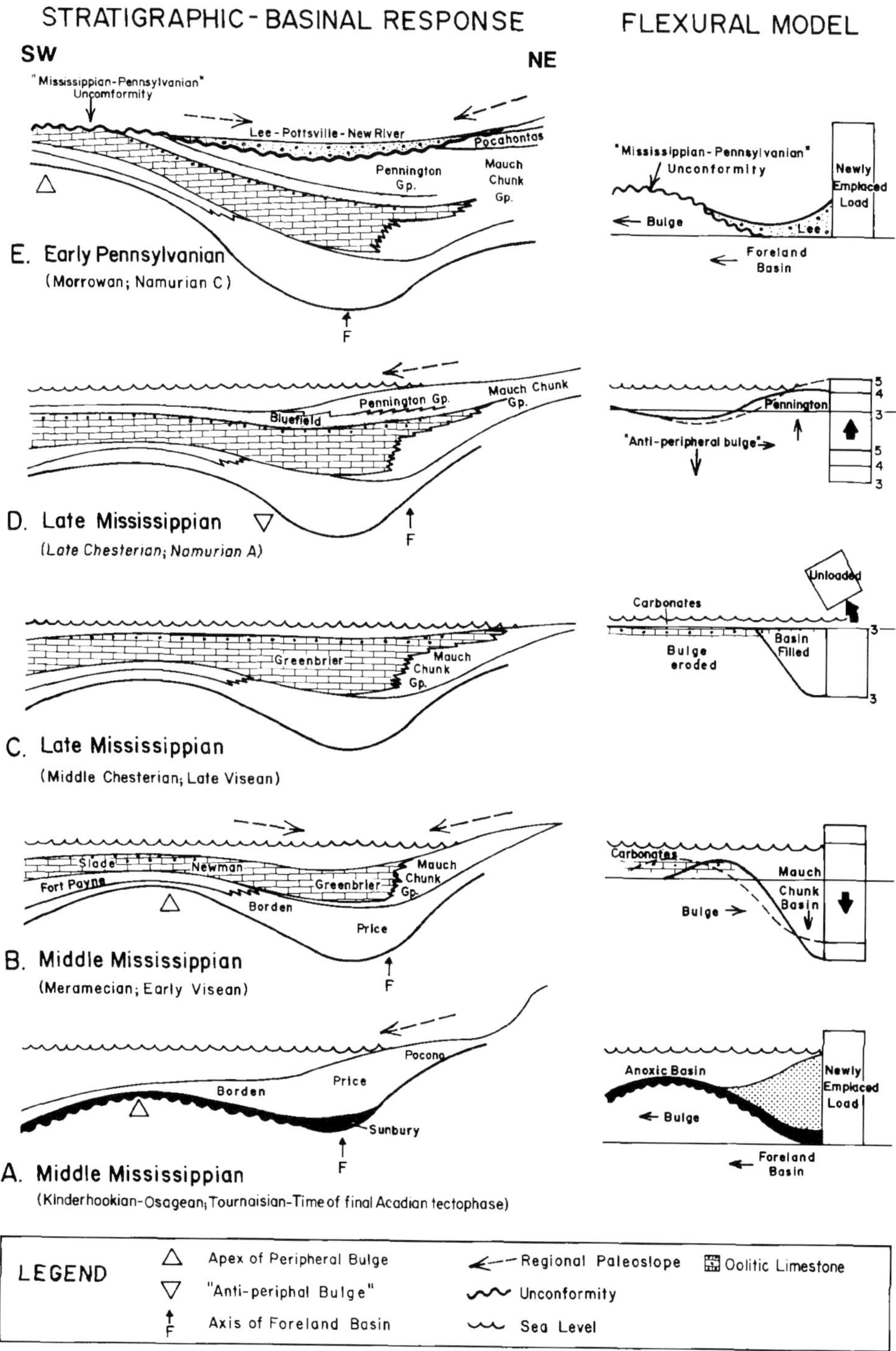

Figure 5. Probable sequence of flexural events and accompanying sedimentary responses in the Appalachian basin between the last tectophase of the Acadian orogeny (A) and the initiation of the Alleghanian orogeny (E). Line of section runs approximately northeast-southwest from southwestern Pennsylvania to west-central Kentucky. Sequence of events: (A) Bulge and basin formation and migration accompanying active loading in orogen; (B) Loading-type relaxation accompanied by basin deepening and bulge migration toward load;
(C) Relative elevational equilibrium between bulge, basin, and source area due to erosional unloading and load subsidence; brief interval of widespread shallow water environments; (D) Rebound in source area and parts of foreland basin accompanying unloading-type relaxation; cratonward prograding wedge of fine-grained, marginal-marine siliciclastic sediment; (E) Unconformity, bulge, and basin formation accompanying inception of new orogeny.

FLEXURAL CONTROLS ON REGIONAL OOLITIC DEPOSITION

Kinderhookian (Early Tournaisian)

Tectonic Framework

Although the oceanic plate-continent collision that produced the Antler orogeny may have begun as early as the late Middle Devonian (Johnson, 1971; Gutschick and Sandberg, 1983), the major tectophase apparently occurred near the Devonian–Mississippian transition (Smith and Ketner, 1968; Johnson, 1971; Nilsen, 1977) and was recorded in the emplacement of the Roberts Mountain allochthon (Poole, 1974; Dickinson, 1977; Hintze, 1985, col. 8). A widespread hiatus at or near the Devonian–Mississippian transition and extending locally into the early Kinderhookian (Figure 6) probably represents continental braking or cratonward bulge migration accompanying obduction (Poole, 1974; Dickinson, 1977) and the resultant deformational loading associated with this tectophase. Unconformity development was followed in the late Kinderhookian by a period of widespread transgression that culminated in the early Valmeyeran (Osagian).

The southern margin of the continent (Ouachita margin) was generally a quiet, shaly continental slope (Figure 6) on the margin of a possible wide-rift basin flanked on its southern side by one or more microcontinents or oceanic plateaus (Lowe, 1985; Arbenz, 1989). However, the presence of probable dark Kinderhookian shales and turbidites in the Tesnus Formation of the Marathon region (Kier et al., 1979; Ross and Ross, 1985; Ethington et al., 1989), a part of the Ouachita geosyncline in west Texas, may signal the beginning of Ouachita convergence and deformation. Paleocurrent orientations, as well as directions of stratigraphic thickening and sedimentary coarsening, suggest rising tectonic highlands to the southeast (Ross and Ross, 1985). This very localized inception of the Ouachita orogeny and accompanying bulge moveout may explain the presence of the Kinderhookian unconformity (Kier et al., 1979; Hills and Kottlowski, 1983; Ethington et al., 1989) across central Texas (Figure 6).

On the Appalachian margin, the Acadian orogeny began in the north about the time of the Early–Middle Devonian transition as a north-to-south diachronous convergence between one or more microcontinents and the North American continent (Ettensohn, 1987).

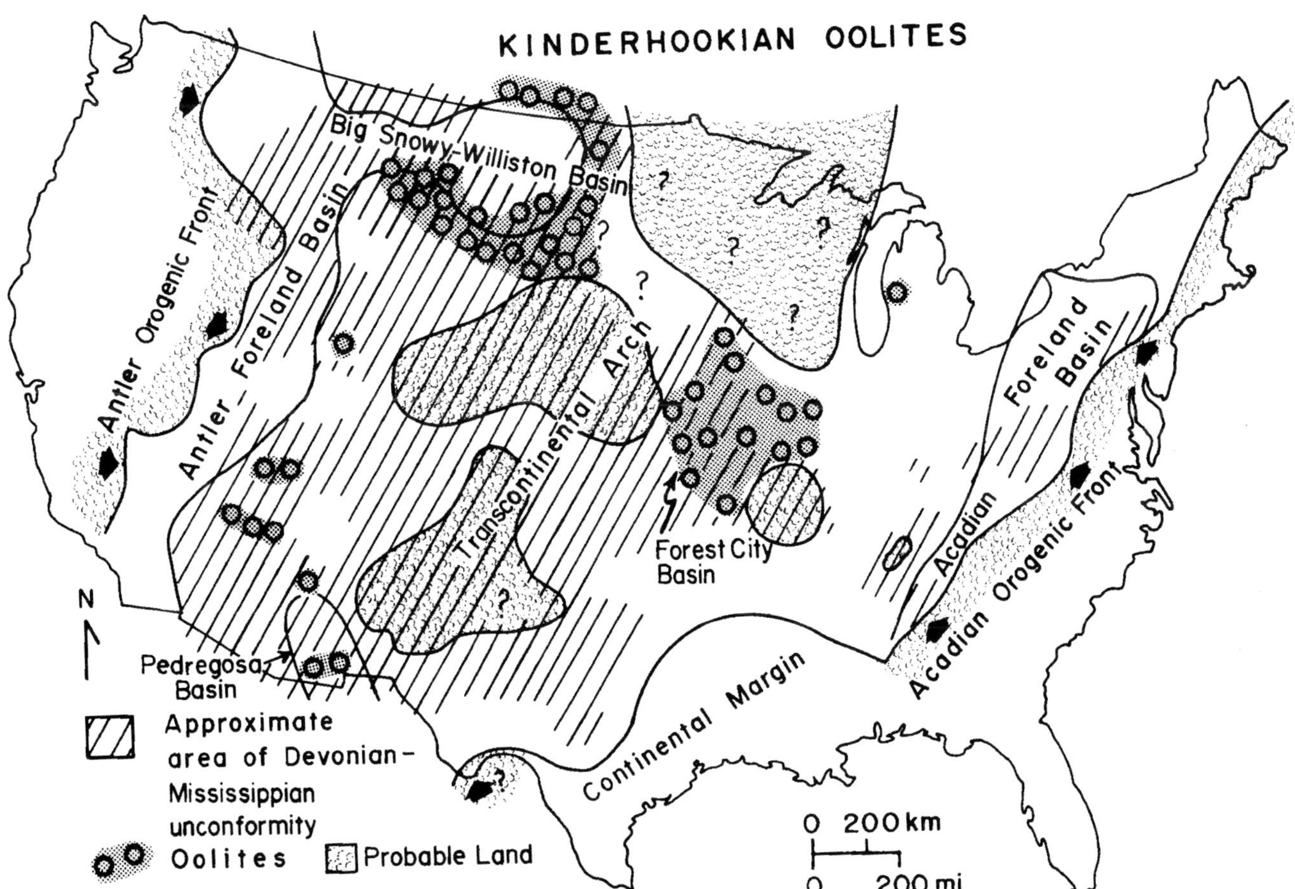

Figure 6. Map of the United States showing distribution of reported Kinderhookian (early Tournaisian) oolites and the approximate extent of the Devonian–Mississippian unconformity.

It culminated in convergence near the Alabama promontory. The final Acadian tectophase was a wholly Kinderhookian event except for the later relaxation stages (Figure 5), and bulge-generated unconformities developed only inboard of continental promontories (see Thomas, 1983) where convergence was greatest (Figure 6). Subsequent cratonic and foreland basin subsidence resulted in a large, underfilled cratonic black-shale basin (Appalachian basin) that persisted throughout much of the early Valmeyeran (Osagian) (Ettensohn et al., 1988b) despite infilling from the east (Figures 1, 5).

Ooid Occurrences

Kinderhookian ooid occurrences are concentrated in two areas: on the margins of the Williston basin and in parts of Nebraska, Kansas, Missouri, Illinois, and Iowa that correspond approximately to the Forest City basin (Figure 6).

In the Williston basin, Kinderhookian ooids are confined to limestone equivalents of the Paine Shale Member of the Lodgepole Limestone in Montana, Wyoming, and North Dakota (Sheldon and Carter, 1979; Peterson, 1984), lower parts of the Woodhurst Member of the Lodgepole in Montana and North Dakota (Smith and Gilmour, 1979), the Pahapsapa Limestone of South Dakota (Sheldon and Carter, 1979), and the Souris Valley Beds of southern Manitoba and Saskatchewan (Macauley et al., 1964). The ooids are associated with nearshore dolostones and commonly occur in lenticular masses suggesting bars and banks (Sheldon and Carter, 1979; Peterson, 1984).

The occurrences near the Williston basin probably owe their presence to both tectonic and paleogeographic factors. Immediately following the Kinderhookian uplift accompanying the inception of a new Antler tectophase, subsidence resulting from deformational loading in the orogen not only formed a foreland basin but also yoked (Figure 3) that basin to the Williston basin via the Big Snowy trough (Figure 6). As a result, Kinderhookian seas, transgressing out of the foreland basin to the west, very early migrated eastward into the Williston basin. Oolites are rare in the central part of the basin where argillaceous carbonates and dark shales predominate, but they become very common on the northeast and southern margins of the basin (Sheldon and Carter, 1979; Peterson, 1984). The thickest accumulations occur in northwestern South Dakota (Sheldon and Carter, 1979) on the shallow platform area between the Williston basin and the Transcontinental arch, probably reflecting the importance of shallow-water conditions. Nonetheless, the partially enclosed nature of the basin and its location in the evaporative tradewind belt suggest that the waters may have been slightly hypersaline (Sheldon and Carter, 1979), and this is supported at least locally by the presence of Kinderhookian evaporites in parts of the basin (Peterson, 1984). This combination of shallow transgressing seas, a nearly enclosed basin, and slightly increased salinity was almost certainly responsible for the Kinderhookian oolitic deposition in shoaling waters at the basin margins.

Farther to the south, however, along the eastern margin of the Antler foreland basin, eastward transgression onto the uplifted bulge area began somewhat later in Kinderhookian time so that oolitic deposition was limited to small areas in Wyoming, Utah, and northeastern Arizona (McKee and Gutschick, 1969; Welsh and Bissell, 1979; Lageson et al., 1979). The oolite occurrences in southeastern Arizona and New Mexico southwest of the Transcontinental arch occur within or on the margin of the Pedregosa basin (Arizona-New Mexico trough), a cratonic basin apparently yoked to the Ouachita foreland basin (see Ross and Ross, 1985, Figure 1). Minor oolites occur in the lower part of the Keating and equivalent Lake Valley formations in New Mexico (Armstrong et al., 1979; Armstrong and Mamet, 1988; Pierce, 1979). These oolitic facies apparently developed on the shoaling edge of a transgressing sea migrating into the Pedregosa basin in response to subsidence generated by the inception of tectonism in the western part of the Ouachita orogen. Except for its smaller size and apparently normal salinity as indicated by fossils (Armstrong et al., 1979; McKee, 1979), the basic causes for and development of ooid deposition in this basin appear to be very similar to those described for the Williston basin.

Another concentration of Kinderhookian oolites occurs in an embayment within the Transcontinental arch (Gutschick and Sandberg, 1983) on the eastern Mid-continent (Figure 6). The ooid-rich facies are very thin and localized in the "Glen Park" formation (informal term) of Illinois (Collinson, 1961), the Chapin Formation of central Iowa (Avcin and Koch, 1979; Carlson, 1979), the Starrs Cave and McCraney formations of Illinois, Iowa, and Missouri (Workman and Gillette, 1956; Avcin and Koch, 1979), and the Gilmore City Formation of Iowa, Kansas, and Nebraska (Avcin and Koch, 1979; Carlson, 1979; Ebanks et al., 1979). This area was uplifted and exposed at the Devonian–Mississippian transition, probably as a result of the combined effects of uplift along the Transcontinental arch and Ozark dome, marking the inception of coeval Antler and Acadian tectophases. Deeper water, shale deposition had characterized most of the embayment during the Late Devonian and continued elsewhere to the east and south well into the Kinderhookian. After Devonian–Mississippian uplift, shallow water carbonates largely replaced these shales throughout the embayment. During subsequent transgression, oolites in the lower Kinderhookian "Glen Park," Chapin, Starrs Cave, and McCraney formations apparently formed in response to shoaling conditions on the transgressing edge of the sea migrating into the embayment. On the other hand, upper Kinderhookian oolites in the Gilmore City Formation occur just below the prominent Kinderhookian–Valmeyeran unconformity and apparently reflect shoaling conditions on the regressing margins of shallow seas that receded southward accompanying regional uplift at the beginning of major Ouachita tectonism. During both the transgressive and regressive regimes, however, ooid formation was likely enhanced by the shallow-water, high-energy conditions on the margins of the Transcontinental arch and possibly by slightly increased salinities in this structurally restricted embayment within the arch (Workman and Gillette, 1956; Carlson, 1979). Based on the presence of ooids and evaporites as early as the Middle Devonian (Heckel and Witzke, 1979, Text, Figure 3), parts of this embayment had experienced episodic restriction for some time. An isolated occurrence of oolites and dolostone in the Ellsworth Shale of the western Michigan basin (Cohee, 1979) (Figure 6) may also reflect high-energy conditions on the margins of the Transcontinental arch as well as locally restrictive conditons.

Valmeyeran (Late Tournaisian–Visean)

Tectonic Framework

By Valmeyeran time, the major locus of tectonism had switched from the Antler front to the Ouachita front in the present-day Gulf Coast area. The initiation of widespread Ouachita convergence accompanied by continental braking or bulge uplift is reflected in a prominent Kinderhookian–Valmeyeran unconformity (Ham and Wilson, 1967) that approximately outlines the Ouachita foreland basin (Figure 7). On a regional scale, this is commonly an angular unconformity as far north as Kansas (Ebanks et al., 1979). The unconformity also extends northward from western Texas through western Colorado near the trend of the ancestral Rockies (Front Range uplift, Figure 7). This was an area of episodic Carboniferous uplift reflecting peculiar strain patterns generated by Ouachita convergence (Kluth and Coney, 1981). By the mid-Valmeyeran, approximately the same west Texas-Colorado area with extensions northward into Wyoming and southwestern Montana underwent uplift and unconformity development again (Figure 7). This was possibly a result of constructive interference by Antler and Ouachita bulges. As a result, the region from Arizona to Montana remained emergent until the late Chesterian or early Pennsylvanian, and this was a time of extensive karstification atop mid-Valmeyeran carbonates (McKee and Gutschick, 1969; Sando, 1988; De Voto, 1988; Meyers, 1988). The slow reemergence of the Transcontinental arch during the Valmeyeran and Chesterian (cf. Figures 7, 8) may reflect the same constructive bulge interference. Detailed knowledge of the Antler and Ouachita orogenies, however, is not sufficient to corroborate the presence of coeval tectophases, although mid-Valmeyeran unconformities are present locally near, and within, the foreland basins of both orogens (Figure 7).

Although actual continent-continent collision did not occur until the Pennsylvanian, major Ouachita convergence appears to have begun by the end of the Kinderhookian (Arbenz, 1989). Evidence includes: the prominent Kinderhookian–Valmeyeran unconformity on the cratonward margin of the foreland basin, the inception of foreland basin infilling with the dark Tesnus, Caney, Barnett, and Stanley shales and some southern-sourced volcanics. The prominence of the unconformity and the great thickness of the overlying basin-fill shales mentioned above (see Ingersoll, 1988, p. 1715), as well as the oceanic nature of preorogenic sediments and the apparent polarity of the approaching arc (Thomas, 1989; Viele and Thomas, 1989) may indicate that during the Mississippian the basin on the Ouachita margin was not a true foreland basin but rather a peripheral foreland basin. It probably formed in response to collision between a rifted continental margin and the subduction zone of an arc-trench system (Thomas, 1989). Although loading is still necessary to generate such a basin, the relatively old age of the Ouachita rifted margin initially meant a higher peripheral bulge and thicker basin fill (see Stockmal et al., 1986; Ingersoll, 1988).

By the early Valmeyeran in the Appalachian basin, all active Acadian deformational loading had ceased, and lithospheric relaxation ensued. Shallow-water, carbonate depositional environments followed the uplifted bulge eastward as it migrated toward the orogen in response to relaxation (Figure 5B).

Ooid Occurrences

During the Valmeyeran, three belts of widespread oolitic sedimentation formed, each parallel to the trend of a major orogenic front. A nearly north-south western belt was present just west of the Antler foreland basin; a west-southwest to east-northeast belt developed just north of the Ouachita foreland basin; and in the Acadian foreland basin a south-southwest to north-northeast belt was present (Figure 7).

The western belt trended approximately north to south along the cratonward margin of the Antler foreland basin. The location of this belt corresponds well with predicted bulge position after the initial Kinderhookian tectophase, but a temporal lag of about 5 m.y. separates tectophase inception from widespread ooid production. Such a time lag seems to be necessary to allow for adequate bulge migration onto the craton and for migration of shallow seas out of the foreland basin and onto the bulge area. Once these elements were in place, however, the combination of shallow, transgressing seas overlapping onto an uplifted bulge area created ideal conditions for ooid deposition, and by the early Valmeyeran (Osagian), ooid production was widespread along the cratonward margin of the Antler foreland basin (Figure 7). Parts of the foreland basin north of Colorado and Utah have been called the Madison shelf (Gutschick and Sandberg, 1983). Shallow ooid-producing seas persisted here only until the end of the early Valmeyeran (Osagian), at which time most of this area was uplifted and exposed, probably in response to a later Antler tectophase centered in the northern part of the orogen. In the Williston basin, however, which was initially farther from the bulge and would have experienced early subsidence during tectophase inception because of yoking with the foreland basin,

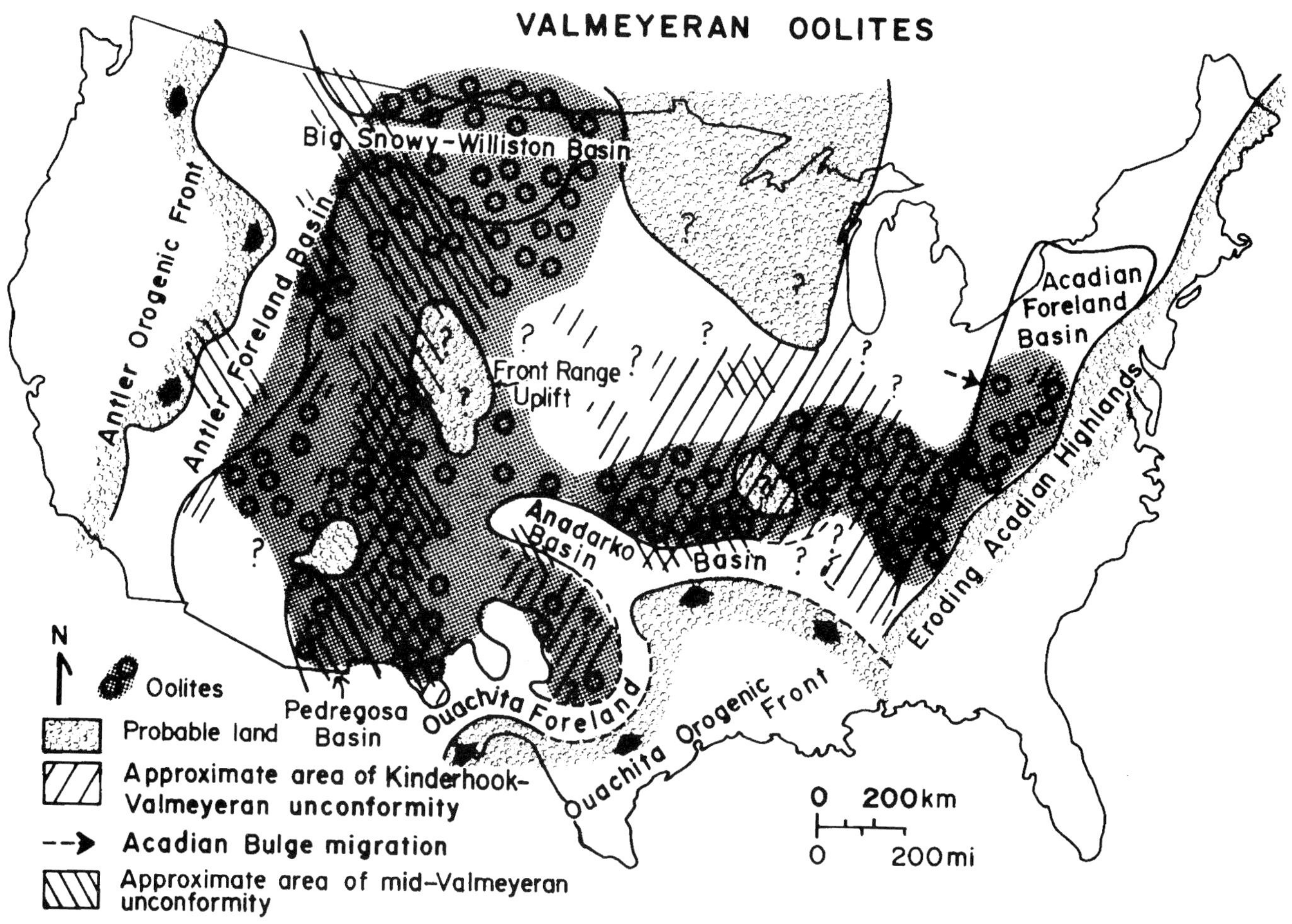

Figure 7. Map of the United States showing Valmeyeran paleogeography, the reported distribution of Valmeyeran (late Tournaisian–Visean) oolites, and the approximate areal extents of the Kinderhookian–Valmeyeran and mid-Valmeyeran unconformities.

uninterrupted deposition continued into the late Valmeyeran (Meramecian). Local oolites were deposited here in association with the Charles Salt (Peterson, 1984) until uplift and erosion halted sedimentation.

The southern part of the Antler foreland, now comprising most of the southwestern United States, was part of the Redwall-Escabrosa shelf (Gutschick and Sandberg, 1983), a peninsulalike area between two orogenic belts (Figure 9). In this area, minor ooid deposition continued throughout much of the Valmeyeran (e.g., McKee and Gutschick, 1969; Armstrong and Mamet, 1974, 1990), although a mid-Valmeyeran unconformity is present locally (McKee and Gutschick, 1969; Armstrong and Mamet, 1974, 1990). This area was apparently caught in a pincerlike movement of forces between two orogenies, so that episodes of bulge uplift and subsequent subsidence constructively interfered here. As a result, early Valmeyeran (Osagian) periods of ooid production associated with the Antler orogeny are overlapped in this area by late Valmeyeran (Meramecian) periods of ooid production associated with the Ouachita orogeny (Figure 9).

Meramecian oolites from the southwestern United States form the western end of a second belt of oolites trending west-southwest to east-northeast along the margin of the Ouachita foreland basin (Figures 7, 9). Initial convergence in this area is indicated by the Kinderhookian–Valmeyeran unconformity and the start of siliciclastic basin fill. During the early Valmeyeran (Osagian), the central and eastern parts of this future ooid belt continued as deeper water ramp and basinal areas where cherts, cherty limestones (Figure 1), and fine-grained siliciclastic sediments (St. Joe and Ft. Payne formations) were deposited. By the mid-Valmeyeran (Osagian–Meramecian transition), again after an approximately 5-m.y. lag interval, the above lithologies abruptly gave way to shallow-water, carbonate deposition. This took place even in areas where this former basin had not been infilled. The deposition of these oolites in a belt parallel to the Ouachita orogeny (Figures 7,

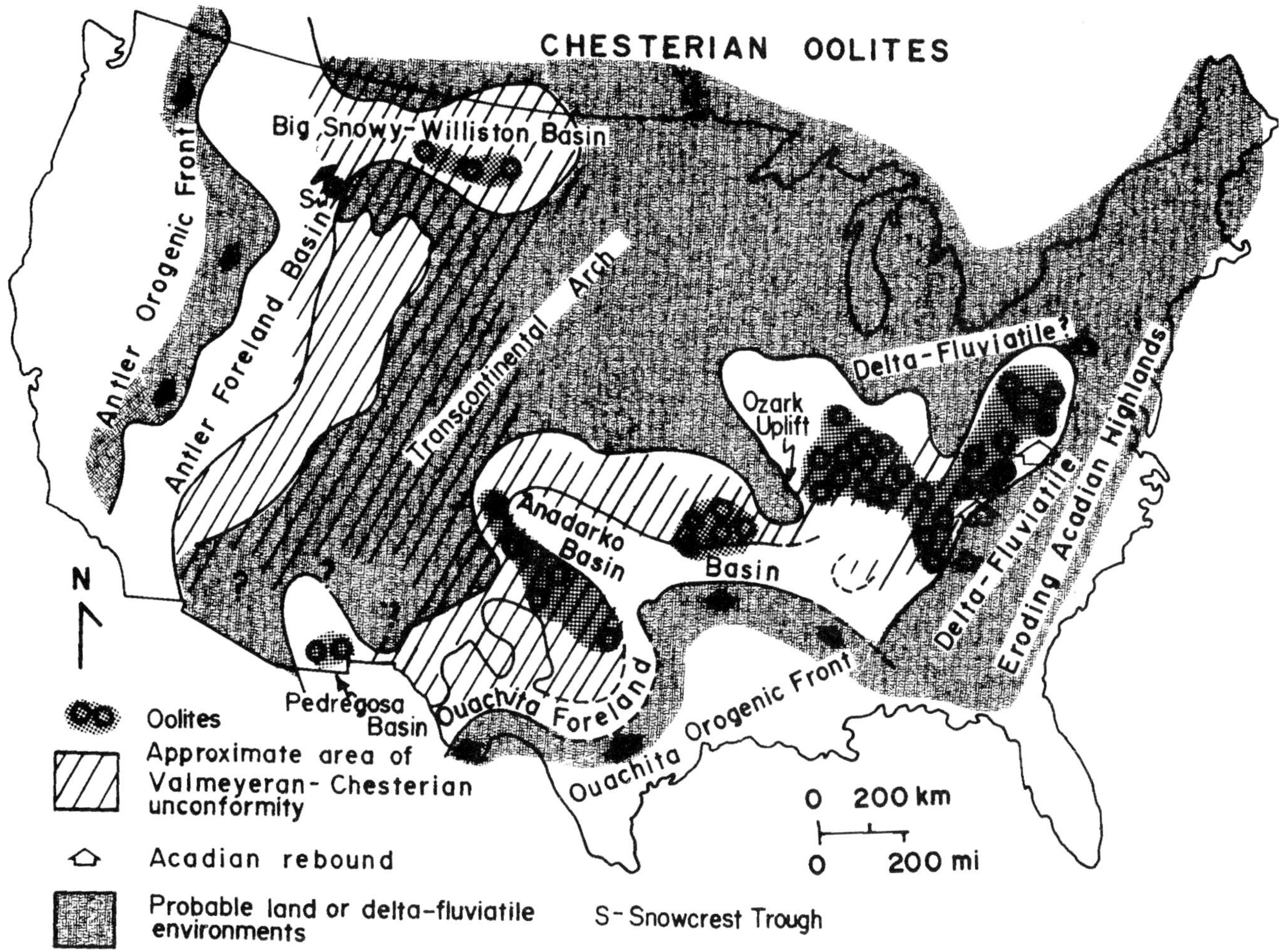

Figure 8. Map of the United States showing Chesterian paleogeography, the reported distribution of Chesterian (Serpukhovian) oolites and the approximate areal extent of the unconformity developed near the Valmeyeran–Chesterian transition.

9) suggests the likely presence and significance of a peripheral bulge in producing shallow water conditions. Moreover, it was on this bulge area (Figures 7, 9) that Mississippian ooid production attained its climax and greatest areal extent during the late Valmeyeran (Meramecian).

This prolific ooid production may also reflect the greater uplift of the peripheral bulge as suggested previously. The foreland basin probably served as a source of carbonate-saturated waters that upwelled and were transported into warm, shallow, agitated waters on the bulge area. The bulge, in combination with an Acadian bulge to the east, apparently also acted as a barrier to major siliciclastic influx (Beaumont et al., 1987). In the south-central United States, abundantly oolitic units like the Salem and Ste. Genevieve limestones and their equivalents in the Monteagle, Slade, and Newman limestones, were deposited at this time (Thomas, 1972; Milici et al., 1979; Ettensohn et al., 1984; Sable and Dever, 1990). In the southwest, Meramecian parts of the Redwall, Escabrosa, Hachita, Terrero, and upper Leadville limestones are also oolitic (Rawson and Kent, 1979; McKee and Gutschick, 1969; Welsch and Bissell, 1979; Armstrong and Mamet, 1974, 1979, 1990; Armstrong and Holcomb, 1989).

In the Appalachian basin (Acadian foreland basin), a much shorter south-southwest to north-northeast belt of oolites extends from Alabama to Pennsylvania and Ohio (Figures 7, 9). Although southern portions of this belt must have been influenced by the northward migrating Ouachita bulge, northern portions of the basin were probably more strongly influenced by an Acadian bulge (Ettensohn and Chesnut, 1989) migrating eastward in response to relaxation (Figures 5B, 7). Ste. Genevievian shallow-water, ooid-forming environments seem merely to have tracked the eastward progress of the bulge across the basin, because by the latest Valmeyeran and early Chesterian, these Appalachian environments had migrated across the basin into southwestern Pennsylvania (Brezinski, 1984). The resulting Appalachian basin oolites are

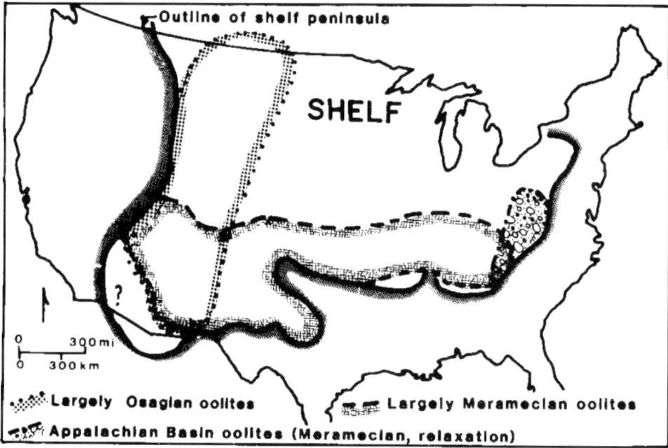

Figure 9. Map of the United States showing the peninsulalike nature of the foreland between the Antler and Ouachita orogens and the distribution of predominantly early Valmeyeran (Osagian) and late Valmeyeran (Meramecian) oolite belts relative to the orogens. Although ooids are present in extreme southern parts of the Osagian distribution and in extreme southwestern parts of the Meramecian distribution, they are rare. Clearly, major belts of oolites parallel the orogens and overlap near areas where the orogens approach each other.

common in parts of the Greenbrier, Newman, and Monteagle limestones.

Chesterian (Serpukhovian)

Tectonic Framework

By the Valmeyeran–Chesterian transition, a widespread unconformity had developed adjacent to both the Antler and Ouachita foreland basins (Figure 8), suggesting concurrent tectophases in both orogens. Constructive interference by the two bulges in the mid-continent may have also been responsible for reemergence of the Transcontinental arch (Figure 8). Bulge uplift and migration, however, probably played an important role in generating the widespread sub-Chester unconformity, although a major sea level decline at this time (Vail et al., 1977) also may have been influential. Eustasy, however, could not have been the sole cause of unconformity formation, because in parts of the Appalachian, Illinois, and Pedregosa basins, where an equal lowering of sea level would have been expected at this time, the Valmeyeran–Chesterian transition is gradational.

On the Antler foreland at this time, resulting uplift seems to have been so intense that seas did not reoccupy much of the area until the early or middle Pennsylvanian. However, in a few areas like the platform-margin basins in Utah and Wyoming and like the Big Snowy-Williston basin in Montana, where yoking with the foreland basin allowed subsidence to resume, very shallow-water, marginal-marine sedimentation recurred in the late Chesterian. Ooids were deposited locally in the Williston basin (Figure 8), but siliciclastic sediments predominate in the basin overall.

The same hiatus is also present along the cratonward margin of the Ouachita basin, but the duration of exposure was apparently much less than that experienced near the Antler orogen, based on the early Chesterian age of the overlying rocks (e.g., Ettensohn et al., 1984; Shaver, 1985; Sable and Dever, 1990).

In much of the Appalachian basin, carbonate sedimentation continued apparently uninterrupted during the Valmeyeran-to-Chesterian transition as eastward bulge migration accompanied relaxation (Figure 5B). By the middle Chesterian, however, bulge uplift and migration, basin infilling, and erosional lowering of Acadian highlands resulted in a near equilibrium of elevation between bulge, basin, and uplands that allowed shallow-water, carbonate-producing environments to develop throughout the basin (Figure 5C). These conditions were soon succeeded by rebound in the old orogen and an accompanying basinward progradation of siliciclastic sediments that inundated carbonate environments (Figure 5D).

Ooid Occurrences

In comparison with late Valmeyeran (Meramecian) oolite occurrences (Figure 7), Chesterian oolites are thinner, more localized, and more closely associated with siliciclastic deposits. Quite clearly the widespread Valmeyeran belts of ooid deposition (Figure 7) were divided into more isolated pockets of ooid deposition during the Chesterian (Figure 8). This isolation of oolites can be attributed to the uplift and subsidence of structural elements and an increased influx of siliciclastic sediments reflecting the approach of orogenic fronts migrating toward the foreland.

Along the Antler front, Chesterian oolites are found in the Lombard Limestone within the Snowcrest trough (Byrne, 1985) and in the Big Snowy-Williston basin, where they occur locally in the Kibbey and Otter formations in close association with sandstones, shales, evaporites, dolostones, and stromatolitic limestones (Rawson, 1968; Smith and Gilmour, 1979; Sheldon and Carter, 1979, Peterson, 1984). The same combination of conditions responsible for ooid deposition in the Williston basin during the Kinderhookian—a shallow, transgressing sea, restriction in a nearly enclosed basin, and slightly increased salinity—seems to have prevailed once again during the Chesterian.

Chesterian oolite occurrences along the Ouachita foreland basin occur in four isolated areas: the Pedregosa basin, the southwestern side of the Anadarko basin, and on either side of the Ozark uplift (Figure 8). Ooid deposition in these areas apparently occurred in circumstances very similar to those present during the late Valmeyeran, as shallow, transgressive seas migrated out of the foreland basin onto adjacent foreland areas that had been uplifted by bulge migration.

By the Chesterian, however, siliciclastic sediments from volcanic arcs and rising tectonic highlands to the south began to fill the Ouachita foreland basin to the point that sediments were periodically shunted across the basin and onto the cratonward margin of the foreland itself. Moreover, as older bulges progressively migrated northward, the entire north-central part of the continent was uplifted (Beaumont et al., 1987), providing a new northern siliciclastic source area (Figure 8) that drained southward via the Michigan River (Swann, 1963, 1964). As a result, most of the Chesterian oolites on the margin of the Ouachita peripheral basin are interbedded with shales and fine-grained sandstones, and at least locally, ooid deposition must have been controlled by the presence or absence of terrigenous influx. Hence, the stratigraphic and areal dispersal of Chesterian ooids, compared with the thick, uniform, contiguous ooid deposits of the late Valmeyeran, probably reflects periodic siliciclastic influx from both northern and southern sources. After the middle Chesterian, the predominance of siliciclastic-rich units like the Helms, "Caney," Fayetteville, Goddard, Barnett, Imo, and Floyd formations in former Ouachita craton-margin environments associated with ooid deposits (see Sutherland, 1988) suggests that widespread inundation by siliciclastic sediments effectively eliminated most carbonate deposition in general and ooid production in particular. In areas far removed from siliciclastic influx such as the Pedregosa basin during deposition of the Paradise Formation (Armstrong and Mamet, 1988), or in regions where siliciclastic sediments were temporarily diverted from shallow-shelf areas, as during deposition of the Pitkin Limestone in Arkansas and Oklahoma (Sutherland, 1979), ooid deposition continued intermittently into the late Chesterian.

In contrast to previous Mississippian intervals, the most prolific area of ooid production during the Chesterian was apparently the Appalachian basin, where a thin sheet of ooid sand seems to have stretched from the Cincinnati arch area to the eastern side of the Appalachian basin in Kentucky, West Virginia, Virginia, and Pennsylvania. Ooids from the Tygarts Creek and Ramey Creek members of the Slade Formation in eastern Kentucky (Ettensohn et al., 1984), part of the Maxville Limestone in Ohio (deWitt and McGrew, 1979), the "Gasper" Member of the Newman in parts of Tennessee and Virginia, the Alderson Member of the Greenbrier in parts of West Virginia and Virginia (Arkle et al., 1979; deWitt and McGrew, 1979), and parts of the Bangor Limestone in Tennessee (Milici et al., 1979) and Alabama (Thomas, 1972) are included. The widespread distribution of the shallow-water, oolitic environments in this area during the middle Chesterian may reflect a short-lived period of flexural and near-elevational equilibrium between basin and upland source areas (Figure 5C) preceding the rebound (Figure 5D) predicted in the models of Beaumont et al. (1988). Similarly, the absence of major siliciclastic influx at this time may reflect the erosionally lowered nature (unloading, Figure 5C) of formerly adjacent (Figure 5B) source areas. Slightly increased salinities in a partially enclosed embayment (Figure 8) and increased agitation on and around local structural highs (Youse, 1964; Koehler and Smosna, 1989) may have further stimulated ooid production.

At the beginning of the late Chesterian, however, a sudden influx of westwardly prograding, marginal-marine, clastic sediments, represented by upper parts of the upper Newman, Bluefield, Paragon, and Pennington formations, marked the end of major Mississippian ooid production in the Appalachian basin and the probable onset of rebound in tectonic highlands to the east (Figure 5D).

FLEXURAL CONTROLS ON LOCALIZED OOID DEPOSITION

Aside from the widespread belts of oolites that appear to have formed on elevated bulge areas paralleling the major orogens (Figures 7, 8, 9), very localized concentrations of Mississippian ooids also occur in places associated with local unconformities, immature paleosols, abrupt facies changes, and abrupt changes in thickness (Sheldon and Carter, 1979, p. 268; Ettensohn, 1975, 1980, 1981; Ettensohn et al., 1988a; Koehler and Smosna, 1989). Although these ooid deposits may have formed at the same time as some of the more widespread deposits attributed to regional bulge uplift, or even in the same general area, what is unusual about them is that they are localized along structures and generally occur in the midst of deeper water conditions where ooids do not normally form. For example, Rawson (1968) and Sheldon and Carter (1979, p. 269) described Chesterian oolite deposits from the Ray Member of the Kibbey Formation in the Williston basin localized atop the Cedar Creek anticline and surrounded by lime-mudstone facies. Similarly, Koehler and Smosna (1989) described concentrations of Valmeyeran ooids from the Pickaway Limestone of the Greenbrier Group on the crest of the West Virginia dome, while muddy skeletal sands, skeletal sands, and detrital quartz were being deposited coevally in surrounding areas. Other types of structural influence on ooid deposits have been reported from the Chesterian of eastern Kentucky near the Waverly arch and Kentucky River fault zone. Skeletal calcarenites of the Tygarts Creek Member of the Slade Formation (Ettensohn et al., 1984), which are nearly 10 m thick on the flanks of the Waverly arch, abruptly decrease to less than a meter thick and become wholly oolitic on the crest of the arch, probably due to abruptly decreasing water depths and subsequent erosion; in other places near the arch, local unconformities and paleosols developed atop these oolites (Ettensohn, 1975, 1980, 1981; Ettensohn et al., 1988a). In the overlying Ramey Creek Member, shallow open-marine shales, carbonate mudstones, and muddy calcarenites abruptly give way to thicker, oolitic skeletal calcarenites in shoals localized on the upthrown margins of fault blocks where the unit crosses the Kentucky River fault zone (Ettensohn, 1975, 1977, 1980).

In the above examples, periodic uplift on structures into shallow, agitated waters (and at times complete exposure) apparently created conditions conducive to ooid production, assuming most other prerequisites were ideal. Flexure in the form of bulge migration is a strong possible mechanism in periodic uplift of these structures because it would provide a means of reactivating the structures at times and places far removed from the area of active tectonism. Moreover, active deformation and later relaxation during each tectophase could have provided multiple phases of bulge uplift and migration in areas distant from the activating tectonism. Bradley and Kusky (1986) first suggested the possibility of such reactivations and provided evidence in the form of faults in the Taconic foreland in New York that were sequentially reactivated toward the craton in a way presumed typical of migrating bulges.

Although this kind of supporting evidence is not available for these isolated Mississippian oolite occurrences, their presence on structures, generally in the midst of unaffected deeper water environments, requires a kind of general cause for local reactivations across extensive areas that only a migrating bulge seems to provide.

SUMMARY

Mississippian oolites represent the most widespread and abundant ooid deposits in the United States, and quite clearly their occurrence was related to a unique coincidence of necessary regional paleogeographic and tectonic conditions, as well as global eustatic, paleoclimatic, and geochemical conditions. Because of this combination of regional and global factors, the mapped abundance of oolites was not uniform in time and space during the Mississippian. The mapped distribution of these oolites (Figures 6, 7, 8) suggests that the major control was the most variable of the above conditions during the Mississippian, namely tectonic regime. This is suggested by the occurrence of major concentrations of Mississippian oolites in broad belts parallel with the Antler, Ouachita, and Acadian orogens. In addition, the fullest development of these belts occurred after a time lag of approximately 5 m.y. following the advent of a tectophase. Phases of active convergence and the resulting deformational loading generated foreland basins and an uplifted peripheral bulge on the craton. Major belts of ooid deposition associated with the Antler and Ouachita orogens consistently developed on the cratonic margins of foreland basins where peripheral-bulge uplift would have predominated. Bulge uplift apparently created ideal conditions for ooid deposition in the form of shallow, agitated waters, protection from siliciclastic influx, and access to deep, carbonate-saturated waters welling up from adjacent basins. The time lag between tectophase inception and major ooid production on the peripheral bulge seems to reflect the time necessary for bulge migration and the transgression of shallow seas out of the adjacent basin and onto the bulge.

In contrast to the other oolite belts, the belt of oolites associated with the Acadian orogen occurs not on the margin of but mainly within the Appalachian basin and reflects bulge migration associated with lithospheric relaxation of the largely quiescent Acadian orogen.

During the Kinderhookian, ooid deposits were rare, apparently reflecting the fact that deeper seas covered much of the craton and that bulge uplift along basin margins was not widespread. Those Kinderhookian oolites that did develop were apparently related more to regional geographic restriction and local uplift. By the Valmeyeran, however, relaxation along the Acadian front and nearly concurrent tectophases in the Antler and Ouachita orogens generated extensive overlapping belts of bulge uplift, which became sites for the most prolific Mississippian ooid production. As orogeny progressed into the Chesterian, siliciclastic influx from rising tectonic highlands filled adjacent foreland basins and then overlapped onto bulge areas, thereby isolating and considerably reducing the size of ooid deposits. By the end of the middle Chesterian, all major ooid production had ceased, as siliciclastic sediments nearly inundated the craton from the north, east, south, and west.

Given the deeper water conditions that prevailed across much of the United States during the Kinderhookian and the rather abrupt change to shallow-water, ooid-producing conditions in the succeeding Valmeyeran, the above scenario emphasizes the likely importance of concurrent orogenies as a means of generating the widespread cratonic uplift into shallow, agitated seas necessary for major regional production of ooid deposits. The timing and distribution of these Mississippian deposits, as well as their relationship to regional unconformities, suggest that tectonism and not eustasy was the major controlling factor. Because the effects of tectonism have been offered as a possible explanation for the origin of extensive ooid-producing conditions, the reverse could also be true, namely that if the above scenario is correct, the occurrence of extensive ooid deposits and their relation to regional unconformities may have predictive value toward determining the nature and timing of major tectonic events.

ACKNOWLEDGMENTS

I wish to thank the editors of this volume, as well as A. K. Armstrong, R. C. Gutschick, and J. L. Wilson for their very helpful reviews of early phases of the manuscript. A. K. Armstrong, R. C. Gutschick, and N. Rast also provided valuable references. My special gratitude goes to B. M. Ettensohn, who did the drafting, and to M. A. Palmer, who did the word processing.

REFERENCES CITED

Arbenz, J. K., 1989, The Ouachita system, *in* A. W. Balley and A. R. Palmer, eds., The geology of North America—an overview: Boulder, G.S.A., The Geology of North America, v. A., p. 371-396.

Arkle, I., Jr., D. R. Beissel, R. E. Nuhfer, D. G. Patchen, R. A. Smosna, W. H. Gillespie, R. Lund, C. W. Norton, and H. w. Pfefferkorn, 1979, The Mississippian and Pennsylvanian (Carboniferous) Systems in the United States—West Virginia and Maryland, in the Mississippian and Pennsylvanian (Carboniferous) Systems in the United States: U.S.G.S. Professional Paper 1110, p. D1-D35.

Armstrong, A. K., F. E. Kottlowski, W. J. Stwart, B. L. Mamet, E. H. Baltz, Jr., W. T. Siemers, and S. Thompson, III, 1979, The Mississippian and Pennsylvanian (Carboniferous) Systems in the United States—New Mexico, *in* The Mississippian and Pennsylvanian (Carboniferous) Systems in the United States: U.S.G.S. Professional Paper 1110, p. W1-W27.

Armstrong, A. K., and B. L. Mamet, 1974, Biostratigraphy of the Arroyo Penasco Group, Lower Carboniferous (Mississippian), north-central New Mexico: New Mexico Geological Society Guidebook, 25th Field Conference, p. 145-158.

Armstrong, A. K., and B. L. Mamet, 1979, The Mississippian System of north-central New Mexico: New Mexico Geological Society Guidebook, 30th Field Conference, p. 201-210.

Armstrong, A. K., and B. L. Mamet, 1988, Mississippian (Lower Carboniferous) biostratigraphy, facies, and microfossils, Pedregosa basin, southeastern Arizona and southwestern New Mexico: U.S.G.S. Bulletin 1826, 40 p.

Armstrong, A. K., and B. L. Mamet, 1990, Stratigraphy, facies and paleotectonics of the Mississippian System, Sangre de Cristo Mountains, New Mexico and Colorado and adjacent areas: New Mexico Geological Society Guidebook, 41st Field Conference, p. 241-249.

Armstrong, A. K., and L. D. Holcomb, 1989, Stratigraphy, facies, and paleotectonic history of Mississippian rocks in the San Juan Basin of northwestern New Mexico and adjacent areas: U.S.G.S. Bulletin 1808, p. D1-D21.

Avcin, M. J., and D. L. Koch, 1979, The Mississippian and Pennsylvanian (Carboniferous) Systems in the United States—Iowa, *in* The Mississippian and Pennsylvanian (Carboniferous) Systems in the United States: U.S.G.S. Professional Paper 1110, p. M1-M13.

Beaumont, C., 1981, Foreland basins: Geophysical Journal of the Royal Astronomical Society, v. 65, p. 291-329.

Beaumont, C., G. M. Quinlan, and J. Hamilton, 1987, The Alleghanian orogeny and its relationship to the evolution of the eastern interior, North America, *in* C. Beaumont and A. J. Tankard, eds., Sedimentary basins and basin-forming mechanisms: Canadian Society of Petroleum Geologists Memoir 12, p. 425-445.

Beaumont, C., G. M. Quinlan, and J. Hamilton, 1988, Orogeny and stratigraphy: numerical models of the Paleozoic in the eastern interior of North America: Tectonics, v. 7, p. 389-416.

Bradley, D. C., and T. M. Kusky, 1986, Geologic evidence for rate of plate convergence during the Taconic arc-continent collision: Journal of Geology, v. 94, p. 667-681.

Brezinski, D. K., 1984, Upper Mississippian transgressive-regressive episodes, *in* R. M. Busch and D. K. Brezinski, eds., Stratigraphic analysis of Carboniferous rocks in southwestern Pennsylvania using a hierarchy of transgressive-regressive units: Guidebook to Field Trip No. 3, A.A.P.G. Northeast Sectional Meeting, p. 100-104.

Byrne, D. J., 1985, Stratigraphy and depositional history of the Upper Mississippian Big Snowy Formation in the Snowcrest Range, southwestern Montana (M.S. thesis): Corvallis, Oregon State University, 417 p.

Carlson, M. P., 1979, The Nebraska-Iowa region, *in* L. C. Craig and C. W. Connor, coordinators, Paleotectonic investigations of the Mississippian System in the United States: U.S.G.S. Professional Paper 1010, p. 107-114.

Cohee, G. V., 1979, Michigan Basin region, *in* L. C. Craig and C. W. Connor, coordinators, Paleotectonic investigations of the Mississippian System in the United States: U.S.G.S. Professional Paper 1010, p. 49-57.

Collinson, C. W., 1961, The Kinderhookian Series in the Mississippi Valley, *in* Guidebook, Kansas Geological Society, 26th Annual Field Conference, northeastern Missouri and west-central Illinois: Wichita, Kansas Geological Society, p. 100-109.

Craig, L. C., and C. W. Connor, coordinators, 1979, Paleotectonic investigations of the Mississippian System in the United States: U.S.G.S. Professional Paper 1010, 369 p.

De Voto, R. H., 1988, Late Mississippian paleokarst and related mineral deposits, Leadville Formation, central Colorado, *in* N. P. James and P. W. Choquette, eds., Paleokarst: New York, Springer-Verlag, p. 278-305.

deWitt, W., Jr., and McGrew, C. W., 1979, Appalachian basin region, *in* L. C. Craig and C. W. Connor, Paleotectonic investigations of the Mississippian System in the United States: U.S.G.S. Professional Paper 1010, p. 13-48.

Dickinson, W. R., 1974, Plate tectonics and sedimentation, *in* W. R. Dickinson, ed., Tectonics and sedimentation: S.E.P.M. Special Publication 22, p. 1-27.

Dickinson, W. R., 1977, Paleozoic plate tectonics and the evolution of the Cordilleran continental margin, *in* J. H. Stewart, C. H. Stevens, and A. E. Fritsche, eds., Paleozoic paleogeography of the western United States: Los Angeles, S.E.P.M. Pacific Section, p. 137-155.

Ebanks, W. J., L. L. Brady, P. H. Heckel, H. G. O'Conner, G. A. Sanderson, R. R. West, and F. W. Wilson, 1979, The Mississippian and Pennsylvanian (Carboniferous) Systems in the United States—Kansas, *in* The Mississippian and Pennsylvanian (Carboniferous) Systems in the United States: U.S.G.S. Professional Paper 1110 p. Q1-Q30.

Ethington, R. L., S. C. Finney, and J. E. Repetski, 1989, Biostratigraphy of the Paleozoic rocks of the Ouachita orogen, Arkansas, Oklahoma, West Texas, *in* R. D. Hatcher, Jr., W. A. Thomas and G. W. Viele, eds., The Appalachian-Ouachita orogen in the United States: Boulder, G.S.A., The geology of North America, v. F-2, p. 563-574.

Ettensohn, F. R., 1975, Stratigraphic and paleoenvironmental aspects of Upper Mississippian rocks (upper Newman Group), east-central Kentucky: (Ph.D. dissertation): Urbana, University of Illinois, 320 p.

Ettensohn, F. R., 1977, Effects of synsedimentary tectonic activity on the upper Newman Limestone and Pennington Formation, *in* G. R. Dever, Jr., et al., Stratigraphic evidence for late Paleozoic tectonism in northeastern Kentucky: A.A.P.G. Eastern Section, Field Guidebook, Kentucky Geological Survey, p. 18-29.

Ettensohn, F. R., 1980, An alternative to the barrier-shoreline model for deposition of Mississippian and Pennsylvanian rocks in northeastern Kentucky: G.S.A. Bulletin, v. 91, pt. 1, p. 130-135, pt. 2, p. 934-1056.

Ettensohn, F. R., 1981, Mississippian-Pennsylvanian boundary in northeastern Kentucky, *in* T. G. Roberts, ed., G.S.A. Cincinnati 1981 field trip guidebooks, volume 1—stratigraphy, sedimentation: Falls Church, Virginia, American Geological Institute, p. 195-257.

Ettensohn, F. R., 1987, Rates of relative plate motion during the Acadian orogeny based on the spatial distribution of black shales: Journal of Geology, v. 95, p. 572-582.

Ettensohn, F. R., and D. R. Chesnut, Jr., 1989, Nature and probable origin of the Mississippian-Pennsylvanian unconformity in eastern United States *in* J. Yugan and L. Chun, eds., Compte Rendu, Onzième Congrès International de Stratigraphie et de Géologie du Carbonifère: Nanjing, Nanjing University Press, p. 145–159.

Ettensohn, F. R., G. R. Dever, Jr., and J. S. Grow, 1988a, A paleosol interpretation for profiles exhibiting subaerial exposure "crusts" from the Mississippian of the Appalachian basin, *in* J. Reinhardt and W. R. Sigleo, eds., Paleosols and weathering through geologic time: principles and applications: G.S.A. Special Paper 216, p. 49-79.

Ettensohn, F. R., M. L. Miller, S. B. Dillman, T. D. Elam, K. L. Geller, D. R. Swager, G. Markowitz, R. D. Woock, and L. S. Barron, 1988b, Characterization and implications of the Devonian-Mississippian black-shale sequence, eastern and central Kentucky, U.S.A.: pycnoclines, transgression, regression, and tectonism, *in* A. F. Embry and D. J. Glass, eds., Devonian of the world: Proceedings of the Second International Symposium on the Devonian System, Canadian Society of Petroleum Geologists Memoir 14, v. 2, p. 323-345.

Ettensohn, F. R., C. L. Rice, G. R. Dever, Jr., and D. R. Chesnut, 1984, Slade and Paragon formations—new stratigraphic nomenclature for Mississippian rocks along the Cumberland escarpment in Kentucky: U.S.G.S. Bulletin 1605, 37 p.

Gutschick, R. C., and C. R. Sandberg, 1983, Mississippian continental margins of the coterminous United States, in D. J. Stanley and G. T. Moore, eds., The shelfbreak: critical interface on continental margins: S.E.P.M. Special Publication 33, p. 79-96.

Ham, W. E., and J. L. Wilson, 1967, Paleozoic epeirogeny and orogeny in the central United States: American Journal of Science, v. 265, p. 332-407.

Harland, W. B., A. V. Cox, P. G. Llewellyn, C. A. G. Pickton, A. G. Smith, and R. Walters, 1982, A geologic time scale: Cambridge University Press, Cambridge, 131 p.

Heckel, P. H., and B. J. Witzke, 1979, Devonian world paleogeography determined from distribution of carbonates and related lithic palaeoclimatic indicators, *in* M. R. House, C. T. Scrutton, and M. G. Bassett, eds., The Devonian System: The Palaeontological Association Special Papers in Paleontology 23, p. 99-123.

Hills, J. M., and F. E. Kottlowski, 1983, Southwest/southwest midcontinent region *in* O. E. Childs et al., eds., Correlation of stratigraphic units of North America: A.A.P.G. COSUNA chart SSMC.

Hintze, L. F., 1985, Great Basin region, *in* O. E. Childs et al., eds., Correlation of stratigraphic units of North America: A.A.P.G. COSUNA chart GB.

Ingersoll, R. V., 1988, Tectonics of sedimentary basins: G.S.A. Bulletin, v. 100, p. 1704-1719.

Jacobi, R. D., 1981, Peripheral bulge—a causal mechanism for the Lower/Middle Ordovician disconformity along the western margin of the northern Appalachians: Earth and Planetary Science Letters, v. 56, p. 245-251.

Johnson, J. G., 1971, Timing and coordination of orogenic, epeirogenic, and eustatic events: G.S.A. Bulletin, v. 82, p. 3263-3298.

Karner, G. D., and Watts, A. B., 1983, Gravity anomalies and flexure of the lithosphere at mountain ranges: Journal of Geophysical Research, v. 88, no. B12, p. 10449-10477.

Kay, M., and E. H. Colbert, 1965, Stratigraphy and life history: New York, John Wiley, 736 p.

Kier, R. S., L. F. Brown, Jr., and E. F. McBride, 1979, The Mississippian and Pennsylvanian (Carboniferous) Systems in the United States—Texas, *in* The Mississippian and Pennsylvanian (Carboniferous) Systems in the United States: U.S.G.S. Professional Paper 1110, p. S51-S245.

Kluth, C. F., and P. J. Coney, 1981, Plate tectonics of the ancestral Rocky Mountains: Geology, v. 9, p. 10-15.

Koehler, B., and R. Smosna, 1989, Mississippian oolites on the West Virginia Dome (abs.) A.A.P.G.

Bulletin, v. 73, p. 1035.

Lageson, D. R., E. K. Maughan, and W. J. Sando, 1979, The Mississippian and Pennsylvanian (Carboniferous) Systems in the United States—Wyoming, *in* The Mississippian and Pennsylvanian (Carboniferous) Systems in the United States: U.S.G.S. Professional Paper 1110, p. U1-U38.

Lees, A., 1975, Possible influence of salinity and temperature on modern shelf carbonate sedimentation: Marine Geology, v. 19, p. 159-198.

Lowe, D., 1985, Ouachita trough: part of a Cambrian failed rift system: Geology, v. 13, p. 790-793.

Macauley, G., D. G. Penner, R. M. Procter, W. H. Tisdall, 1964, Chapter 7, Carboniferous, *in* R. G. McCrossan and R. P. Glaister, eds., Geological history of western Canada: Calgary, Alberta Society of Petroleum Geologists, p. 89-102.

McKee, E. D., 1979, Arizona, *in* L. C. Craig and C. W. Connor, coordinators, Paleotectonic investigations of the Mississippian System in the United States: U.S.G.S. Professional Paper 1010, p. 1199-1207.

McKee, E. D., and R. C. Gutschick, 1969, History of the Redwall Limestone of northern Arizona: G.S.A. Memoir 114, 726 p.

Meyers, W. J., 1988, Paleokarst features on Mississippian limestones New Mexico, *in* N. P. James and P. W. Choquette, eds, Paleokarst: New York, Springer-Verlag, p. 306-328.

Milici, R. C., G. Briggs, L. M. Knox, P. D. Sitterly, and A. T. Statler, 1979, The Mississippian and Pennsylvanian (Carboniferous) Systems in the United States—Tennessee, *in* The Mississippian and Pennsylvanian (Carboniferous) Systems in the United States: U.S.G.S. Professional Paper 1110, p. G1-G38.

Nilsen, T. H., 1977, Paleogeography of Mississippian turbidites in south-central Idaho, *in* J. H. Stewart, C. H. Stevens, A. E. Fritsche, eds., Paleozoic paleogeography of the western United States: Los Angeles, S.E.P.M. Pacific Section, p. 275-299.

Peterson, J. A., 1984, Stratigraphy and sedimentary facies of the Madison Limestone and associated rocks in parts of Montana, Nebraska, North Dakota, South Dakota, and Wyoming: U.S.G.S. Professional Paper 1273, p. A1-A34.

Pierce, H. W., 1979, The Mississippian and Pennsylvanian (Carboniferous) Systems in the United States—Arizona, *in* The Mississippian and Pennsylvanian (Carboniferous) Systems in the United States: U.S.G.S. Professional Paper 1110, p. Z1-Z20.

Poole, F. G., 1974, Flysch deposits of Antler foreland basin, western United States, *in* W. R. Dickinson, ed., Tectonics and sedimentation: S.E.P.M. Special Publication 22, p. 58-82.

Price, R. A., 1973, Large-scale gravitational flow of supracrustal rocks, southern Canadian Rockies, *in* K. A. deJong, and R. Scholten, eds., Gravity and tectonics: New York, John Wiley, p. 491-502.

Quinlan, G. M., and C. Beaumont, 1984, Appalachian thrusting, lithospheric flexure, and the Paleozoic stratigraphy of the eastern interior of North America: Canadian Journal of Earth Science, v. 21, p. 973-996.

Rawson, R. R., 1968, The "Kibbey limestone" of the Williston basin and central Montana: Earth Science Bulletin, v. 1, p. 35-47.

Rawson, R. R., and W. N. Kent, 1979, Depositional dynamics of the Mississippian Redwall Limestone in northern Arizona, *in* S. S. Beus and R. R. Rawson, eds., Carboniferous stratigraphy in the Grand Canyon country, northern Arizona and southern Nevada, Field Trip 13, Ninth International Congress of Carboniferous Stratigraphy and Geology: Falls Church, Virginia, AGI Selected Guidebook Series 2, American Geological Institute, p. 81-88.

Ross, C. A., and J. R. P. Ross, 1985, Paleozoic tectonics and sedimentation in west Texas, southern New Mexico, and southern Arizona, *in* P. W. Dickerson and W. R. Muelberger, eds., Structure and tectonics of trans-Pecos Texas: West Texas Geological Society Field Conference Publication 85-81, p. 221-230.

Sable, E. G., and G. R. Dever, Jr., 1990, Mississippian rocks in Kentucky: U.S.G.S. Professional Paper 1503, 125 p.

Sando, W. J., 1988, Madison Limestone (Mississippian) paleokarst: a geologic synthesis, *in* N. P. James and P. W. Choquette, eds., Paleokarst: New York, Springer-Verlag, p. 256-277.

Scotese, C. R., 1990, Atlas of Phanerozoic plate tectonic reconstructions: International Lithosphere Program (IUGG-IUGS), Paleomap Project Technical Report no. 10-90-1, 54 p.

Shaver, R. H., 1985, Midwestern basins and arches region, *in* O. E. Childs et al., eds., Correlation of stratigraphic units of North America: A.A.P.G. COSUNA Chart MBA.

Sheldon, R. P., and M. D. Carter, 1979, Williston basin region, *in* L. C. Craig and C. W. Connor, coordinators, Paleotectonic investigations of the Mississippian System in the United States: U.S.G.S. Professional Paper 1010, p. O249-O271.

Smith, D. L., and E. H. Gilmour, 1979, The Mississippian and Pennsylvanian (Carboniferous) Systems in the United States—Montana, *in* The Mississippian and Pennsylvanian (Carboniferous) Systems in the United States: U.S.G.S. Professional Paper 1110, p. X1-X32.

Smith, J. F., Jr., and K. B. Ketner, 1968, Devonian and Mississippian rocks and the date of the Roberts Mountains thrust in the Carlin-Pinon Range, Nevada: U.S.G.S. Bulletin 1251-I, p. 1-18.

Stockmal, G. S., C. Beaumont, and R. Boutilier, 1986, Geodynamic models of convergent margin tectonics: transition from rifted margin to overthrust belt and consequences for foreland basin development: A.A.P.G. Bulletin, v. 70, p. 181-190.

Sutherland, P. K., 1979, Mississippian and Lower Pennsylvanian stratigraphy, *in* R. O. Fay et al., The Mississippian and Pennsylvanian (Carboniferous) Systems in the United States—Oklahoma: U.S.G.S. Professional Paper 1110, p. R1-R35.

Sutherland, P. K., 1988, Late Mississippian and Pennsylvanian depositional history in the Arkoma basin area, Oklahoma and Arkansas: G.S.A. Bulletin, v. 100, p. 1787-1802.

Swann, D. H., 1963, Classification of Genevievian and Chesterian (Late Mississippian) rocks of Illinois: Illinois State Geological Survey Report of Investigations 216, 91 p.

Swann, D. H., 1964, Late Mississippian rhythmic sediments of the Mississippi Valley: A.A.P.G. Bulletin, v. 48, p. 637-658.

Thomas, W. A., 1972, Mississippian stratigraphy of Alabama: Alabama Geological Survey Monograph 12, p. 44-57.

Thomas, W. A., 1983, Continental margins, orogenic belts and intracratonic sutures: Geology, v. 11, p. 270-272.

Thomas, W. A., 1989, The Appalachian-Ouachita orogen beneath the Gulf Coastal Plain between the outcrops in the Appalachian and Ouachita Mountains, *in* R. D. Hatcher, Jr., W. A. Thomas, and G. W. Viele, eds., The Appalachian-Ouachita orogen in the United States: Boulder G.S.A., The Geology of North America, v. F-2, p. 537-553.

Tucker, M. E., 1981, Sedimentary petrology, an introduction: New York, John Wiley, 252 p.

Tucker, M. E., and V. P. Wright, 1990, Carbonate sedimentology: Oxford, Blackwell Scientific Publications, 482 p.

U.S. Geological Survey, 1979, the Mississippian and Pennsylvanian (Carboniferous) Systems in the United States: U.S.G.S. Professional Paper 1010.

Vail, P. R., R. M. Mitchum, Jr., and S. Thompson, III, 1977, Stratigraphy and global changes of sea level, part 4: global cycles of relative changes of sea level, *in* C. E. Payton, ed., Seismic stratigraphy—applications to hydrocarbon exploration: A.A.P.G. Memoir 26, p. 83-97.

Viele, G. W., and W. A. Thomas, 1989, Tectonic synthesis of the Ouachita orogenic belt, *in* R. D. Hatcher, Jr., W. A. Thomas, and G. W. Viele, eds., The Appalachian-Ouachita orogen in the United States: Boulder, G.S.A., The Geology of North America, v. F2, p. 695-728.

Welsh, J. E., and H. J. Bissell, 1979, The Mississippian and Pennsylvanian (Carboniferous) Systems in the United States—Utah: *in* The Mississippian and Pennsylvanian (Carboniferous) Systems in the United States: U.S.G.S. Professional Paper 1110, p. Y1-Y35.

Wilkinson, B. H., R. M. Owen, and A. R. Carroll, 1985, Submarine hydrothermal weathering, global eustacy, and carbonate polymorphism in Phanerozoic marine oolites: Journal of Sedimentary Petrology, v. 55, p. 171-183.

Wilson, J. L., 1975, Carbonate facies in geologic history: New York, Springer-Verlag, 471 p.

Workman, L. E., and T. Gillette, 1956, Subsurface stratigraphy of the Kinderhook Series in Illinois: Illinois State Geological Survey Report of Investigations 189, 46 p.

Youse, A. C., 1964, Gas producing zones of Greenbrier (Mississippian) Limestone, southern West Virginia and eastern Kentucky: A.A.P.G. Bulletin, v. 48, p. 465-486.

Chapter 3

An Eolian Facies in the Ste. Genevieve Limestone of Southern Indiana

Ralph E. Hunter
U.S. Geological Survey
Menlo Park, California, USA

ABSTRACT

Cross-bedded oolitic grainstones in the Ste. Genevieve Limestone (Mississippian) of the Illinois basin have generally been considered to be shallow marine. However, fine- to medium-grained cross-bedded grainstones of mixed clast type in the Ste. Genevieve of Harrison County, southern Indiana, are here interpreted to be of eolian dune origin on the basis of small-scale sedimentary structures, particularly climbing-wind-ripple structures. In addition, subaerial exposure of surfaces at the tops and bases of the eolian units is indicated by pedogenic features such as in-situ breccias and rhizoliths. Associated skeletal and oolitic grainstones of marine origin are distinguished from the eolian grainstones by the presence of pebble-sized fossils. The presence of several intervals of eolian deposits in the Ste. Genevieve is probably a result of eustatic sea level fluctuations.

INTRODUCTION

Although eolian carbonate rocks of Quaternary age are well known, few pre-Quaternary examples of such rocks have been reported (Gardner, 1983; McKee and Ward, 1983). Only in the past few years have several pre-Quaternary examples been identified using reliable criteria (Loope, 1986; Loope and Haverland, 1988; Rice and Loope, 1991). Re-examination of cross-stratified carbonate grainstones in the pre-Quaternary rock record will probably reveal other eolian deposits.

In the search for eolian carbonate rocks, new knowledge of the fine structural details of eolian stratification (Hunter, 1977, 1981; Fryberger and Schenk, 1981; Kocurek and Dott, 1981) should prove to be most useful in distinguishing eolian and shallow marine carbonate rocks. Detailed examination of stratification and other sedimentary structures is the basis for the present eolian interpretation of some cross-stratified grainstones in the Ste. Genevieve Limestone (Mississippian) of Harrison County, southern Indiana.

The Ste. Genevieve Limestone of the Illinois basin is a major petroleum reservoir unit that contains a variety of carbonate rock types ranging from carbonate mudstones to grainstones (Cluff, 1984). Even the grainstones vary considerably in lithologic character. Three principal types of grainstone have been identified in Harrison County: (1) highly oolitic, medium- to coarse-grained; (2) skeletal, dominantly pelmatozoan, coarse- to very coarse-grained, and (3) mixed-clast (oolitic, skeletal, and micritic carbonate grains together with siliciclastic grains), fine- to medium-grained. All three types of grainstone may be cross-stratified, although cross-stratification is less common in the highly oolitic grainstones than in the other two types. The oolitic grainstones of the Ste. Genevieve Limestone have been regarded as classic examples of shallow marine shoal and associated channel deposits (Carr, 1973; Choquette and Steinen, 1980, 1985; Archer, 1984; Cluff, 1984), and nothing in this study contradicts that interpretation for the highly oolitic grainstones in Harrison County. The same interpretation is valid for the skeletal grainstones, which are closely associated with and intergrade with the highly oolitic grainstones. Only the fine- to medium-grained grainstones of mixed clast type are here interpreted as eolian.

GEOLOGIC SETTING

The Ste. Genevieve Limestone is a widespread Mississippian formation of the Illinois basin. The outcrops in Harrison County, Indiana, discussed in this paper are in the Mississippian outcrop belt along the east side of the basin (Figure 1). The formation is overlain by the Paoli Limestone and underlain by the St. Louis Limestone (Figure 2).

The Ste. Genevieve Limestone in Indiana is generally regarded as the uppermost formation of the Meramecian Series or locally of the Valmeyeran Series, which comprises the Meramecian and Osagian series. Some recent workers, however, believe that the formation should be considered the lowermost formation of the Chesterian Series (Maples and Waters, 1987). In the Indiana outcrop belt, the formation has been divided into (in ascending order) the Fredonia Member, the Spar Mountain Member, and the Levias Member (Carr et al., 1986) (Figure 2). However, given the degree of lateral variability and discontinuity of beds in the Ste. Genevieve in Harrison County, I regard the identification of members in this area as provisional.

Figure 1. Index map showing outcrop belt of middle Mississippian rocks (Meramecian Series) around the Illinois basin. Study area (in black) is the Meramecian outcrop belt of Harrison and eastern Crawford counties, Indiana.

LOCAL STRATIGRAPHIC SECTION

The exposures discussed here are two road cuts on Indiana State Route 135 near Corydon, Harrison County, Indiana. The northern road cut is 0.3–0.6 mi (0.5–1.0 km) south of the intersection of Indiana State Route 135 with Indiana State Route 62, and the southern one is 4.0–4.4 mi (6.7–7.3 km) south of the same intersection (Figure 3). The sections in the two road cuts have nearly the same stratigraphic extent, exposing 80–90 ft (24–27 m) of the Ste. Genevieve Limestone (Gray et al., 1983; Woodson, 1981). The sections are here divided informally into 18 lithologic units designated by letters from A at the base to R at the top (Figures 4, 5, Table 1). Some of the units are divided into subunits designated by subscripts numbered from the base up, such as E_1 and E_2. Nearly all the units are present in both sections, although the units differ somewhat in thickness and character from one section to another and within a single section. Some of the units are absent or greatly changed in character in other nearby sections.

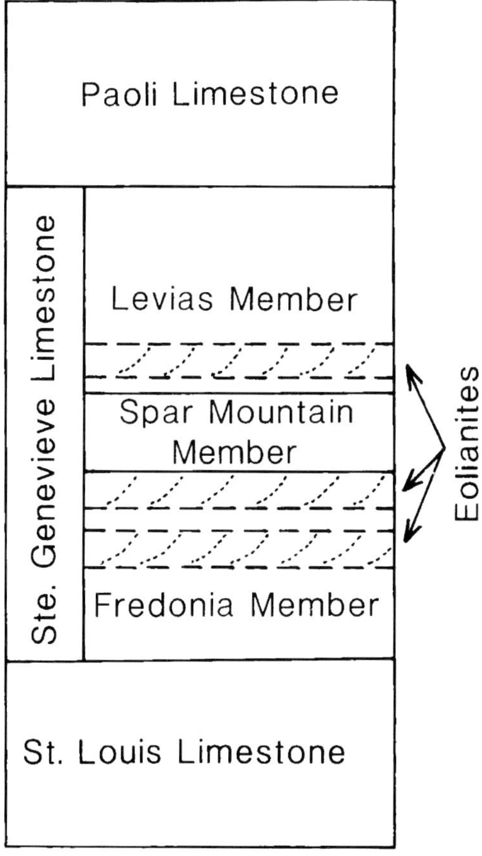

Figure 2. Schematic stratigraphic column of the Ste. Genevieve Limestone and adjacent formations, showing approximate stratigraphic positions of the eolian intervals in Harrison County, Indiana. The members are those recognized by Carr et al. (1986); see Keith and Zuppann, this volume, for broader stratigraphic context.

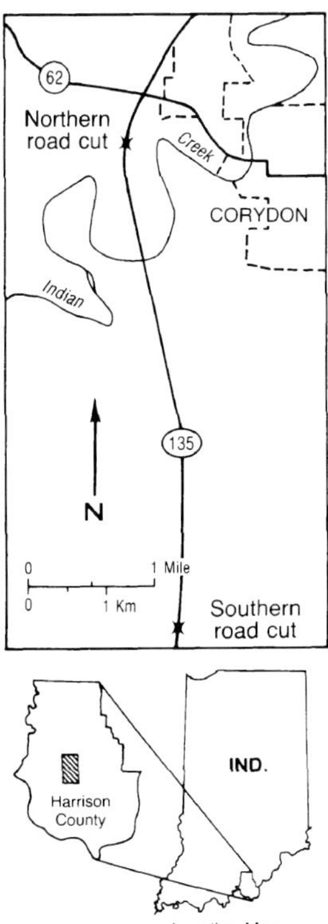

Figure 3. Index map showing outcrops of the Ste. Genevieve Limestone in road cuts along Indiana State Route 135 near Corydon, Harrison County, Indiana. Northern road cut is in E1/2 NE1/4 , Sec. 35, T3S, R3E. Southern road cut is in E1/2 SW1/4 , Sec. 13, T4S, R3E.

Correlation of the two sections near Corydon with other, more complete sections in Harrison County suggests that the contact of the Ste. Genevieve with the overlying Paoli Limestone is 10–15 ft (3–5 m) above the uppermost beds exposed in the two sections. The cherty limestone that makes up unit A has been identified by Woodson (1981) and C. B. Rexroad (personal communication, 1989) as the Lost River Chert Bed, which according to most workers is in the lower part of the Ste. Genevieve (Carr et al., 1986). However, Woodson (1981, 1982), following the redefinition of the St. Louis-Ste. Genevieve contact by Pohl (1970), has reassigned strata in the lower part of the Ste. Genevieve, including the Lost River Chert Bed and other beds up to the top of unit C, to the Horse Creek Member of the St. Louis Limestone.

The interval from the top of the Ste. Genevieve to the top of the Lost River Chert Bed is estimated to be 90 ft (27 m) thick in the two sections near Corydon. The total thickness of the Ste. Genevieve is then about 117 ft (36 m) according to the generally accepted definition or about 82 ft (25 m) according to the definition of Pohl (1970) and Woodson (1982). Along the Ohio River in southern and western Harrison County and in adjacent parts of Kentucky, the Ste. Genevieve is reported to be 140–230 ft (43–70 m) thick (McGrain, 1942; Amos, 1972; Carr et al., 1978).

The sections contain a wide variety of carbonate rock types, which are listed in Table 1 and which will be referred to throughout the text of this chapter. In addition to oolitic, skeletal, and mixed-clast grainstones, the sections contain lime mudstone, wackestone, and packstone; dolosiltstone, calcareous shale, cherty limestone, and chert; stromatolitic lime mudstone, intraclastic rudstone, and in-situ pedogenic breccia. Only the mixed-clast grainstones, which are interpreted to be of eolian origin, and associated rudstones and paleosols are described in further detail here. The mixed-clast grainstones will be referred to hereafter as eolian without the qualification that the depositional environment is interpreted. With the exception of the mixed-clast grainstones, rudstones, and paleosols, all the rocks in the two sections are interpreted as shallow marine to intertidal.

CHARACTER OF EOLIAN UNITS

Three relatively thick (up to 10.5 m) and laterally persistent grainstone units (units F, I, and N) and one thin (up to 0.9 m) unit found only in the northern road cut (unit K) are here interpreted as predominantly eolian in origin. These units have several features in common, as well as features that are unique to each unit. The units can be divided into main parts, basal rudstones, and upper parts containing noneolian exposure-surface features.

Main Parts of the Eolian Units

All the eolian limestones are grainstones, the clasts of which are petrographically of mixed type (Dodd et al., this volume). Ooids, skeletal grains, and micritic grains (peloids, intraclasts, and lithoclasts) are common carbonate grain types in units F and I, whereas the carbonate grains in unit N are predominantly micritic. All the eolian grainstones contain significant fractions (up to 26.5%) of siliciclastic grains, mainly quartz and chert fragments of silt and sand size.

Texturally the eolian grainstones are composed predominantly of fine- to medium-grained sand. A few laminae are composed of coarse-grained sand. Most of the grains are well rounded. The individual laminae are well sorted, and grains coarser than sand are absent except in the basal rudstones and along a few erosion surfaces where scattered reworked clasts of rhizoliths can be found.

The most common sedimentary structures in the eolian units are cross-stratification and subhorizontal lamination. Most of the cross-strata are inclined at low to moderate angles (5°–20°), and few of the cross-strata are inclined more steeply than 25°. The sets of cross-strata are of tabular and trough form and of medium to large scale (0.1–4.0 m thick); the mean set thickness is 0.8 m in unit I at the southern road cut. Cross-stratification is best developed in units I (Figures 6, 7) and N. Unit F is characterized by subhorizontal lamination and very gently inclined cross-lamination. The cross-stratification has a wide dispersion of dip directions. Modal directions ranging from

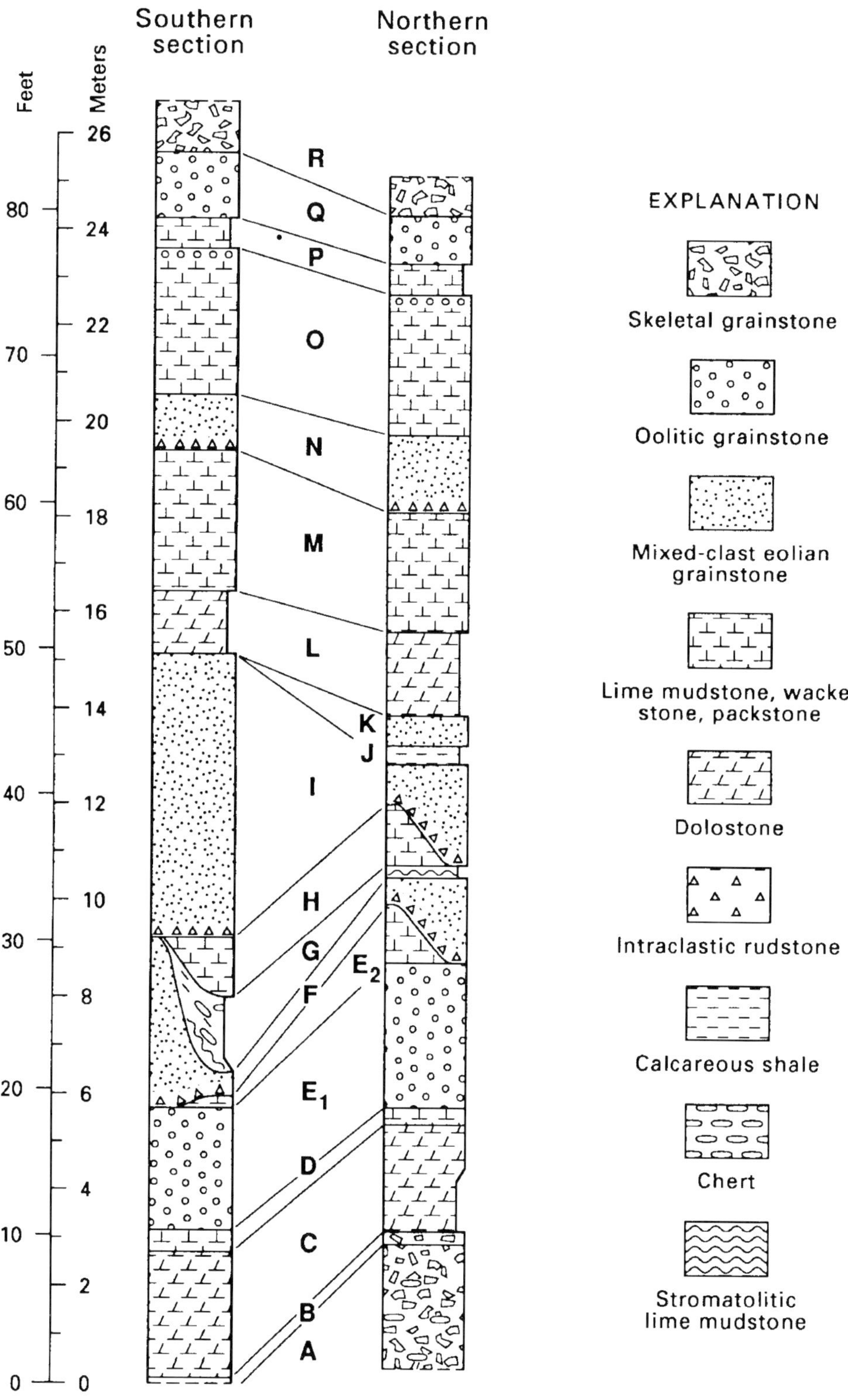

Figure 4. Stratigraphic sections of parts of the Ste. Genevieve Limestone exposed in road cuts along Indiana State Route 135 near Corydon, Indiana. The top of the Ste. Genevieve is estimated to be 12–15 ft (4–5 m) above the uppermost exposed beds of unit R. The cherty upper part of unit A is the Lost River Chert Bed, which is a short distance above the base of the Ste. Genevieve as defined by most workers.

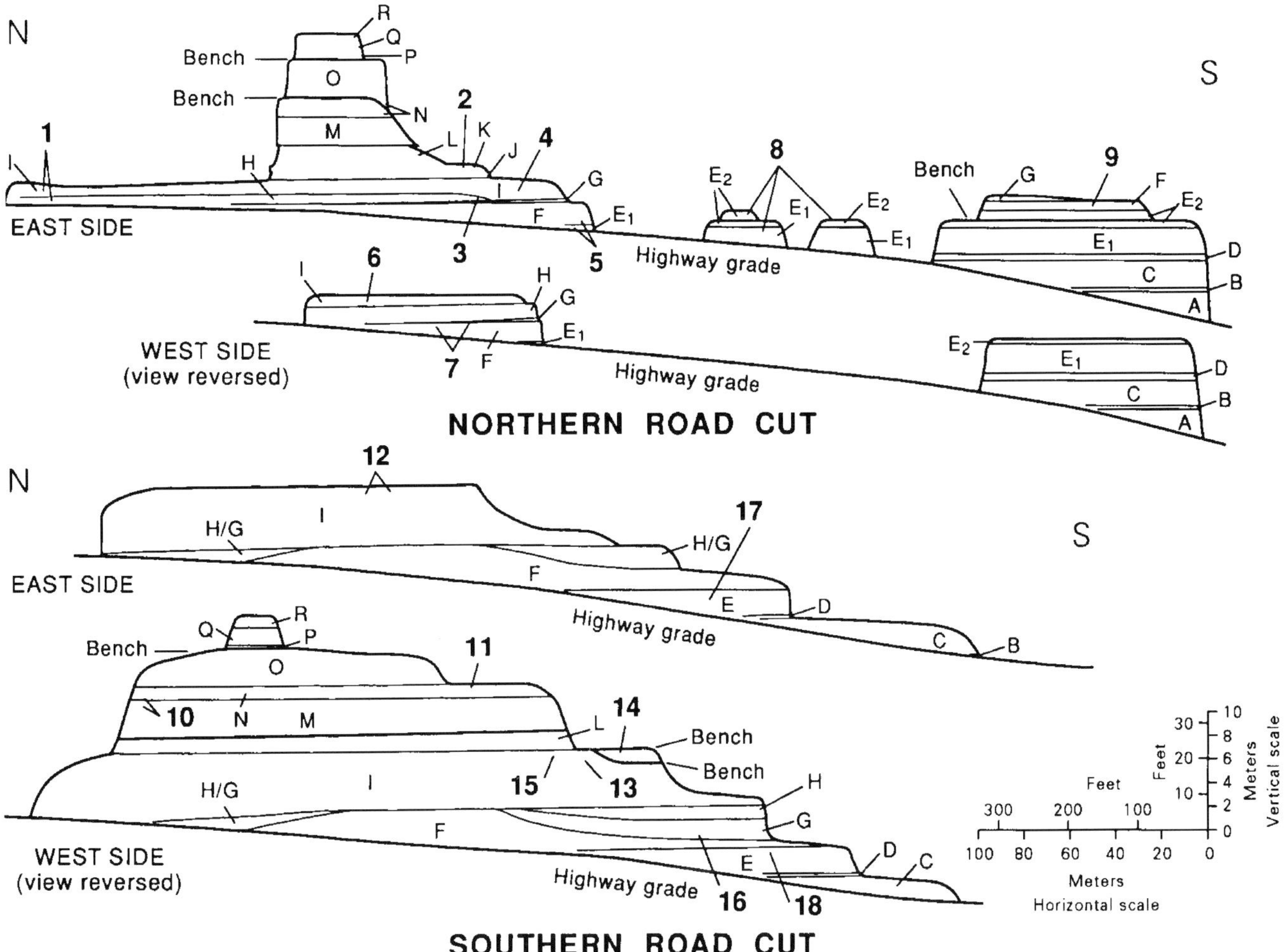

Figure 5. Cross sections of parts of the Ste. Genevieve Limestone exposed in road cuts along Indiana State Route 135 near Corydon, Indiana. Note that cross sections on both sides of highway are oriented as if viewed eastward. Letters refer to units in Figure 4. Boldface numbers refer to noteworthy features: (1) Exposure-surface features in unit I and upper part of unit H. (2) Best example of eolian cross-strata in unit K. (3) Erosion at base of unit I. (4) Best example of climbing-wind-ripple structures in unit I. (5) Basal rudstone of unit F and oolite-pebble rudstone in uppermost part of subunit E_1. (6) Basal rudstone of unit I. (7) Stromatolites in unit G. (8) Corals in growth position in subunit E_2 and oolite-pebble rudstone in uppermost part of subunit E_1. (9) Best example of eolian cross-strata in unit F. (10) Basal rudstone of unit N and in-situ breccia in upper part of unit M. (11) Exposure-surface features in upper part of unit N. (12, 13) Wind ripples on cross-stratum surfaces of unit I. (14) Exposure-surface features in upper part of unit I. (15) Trace fossils on upper contact of unit I. (16) Stromatolites in subunit G_1 and chert in subunit G_2. (17) Best example of large-scale eolian cross-strata in unit F. (18) Basal rudstone of unit F, small exposure of subunit E_2, and oolite-pebble rudstone in uppermost part of subunit E_1.

southwest through northwest to north-northeast are evident in the rose diagram (Figure 8).

Sedimentary structures other than cross-stratification and subhorizontal lamination are rare in the eolian units. Ripples are exposed on a few surfaces of gently to moderately inclined cross-strata in unit I at the southern road cut (Figure 9A). No trace fossils were found in the eolian units except near the tops of units I and N.

Basal Rudstones

Three of the eolian units (units F, I, and N) have thin (0.25 m maximum thickness) basal subunits consisting of intraclastic or lithoclastic, very fine- to fine-grained (3 cm maximum clast size) rudstone. Many but not all of the pebbles of the rudstones can be reasonably explained as having been derived from the immediately underlying beds; pebbles of oolitic grainstone derived from subunit E_1 are abundant in subunit F_1, whereas micritic pebbles are dominant in subunits I_1 and N_1. Most of the pebbles are rounded.

The basal contacts of the rudstones are commonly sharp and erosional. Erosional relief of 1 m is present at the bases of units F and I at the northern road cut. In places, especially at the base of unit N at the northern end of the southern road cut, the rudstone grades downward into breccia that is interpreted to have formed in situ.

Text continues on p. 39.

TABLE 1. Descriptions of stratigraphic sections exposed in road cuts along Indiana State Route 135 near Corydon, Indiana.

Unit/ subunit	Character	Thickness: Northern section	Thickness: Southern section
R	Grainstone, coarse- to very coarse-grained, and very fine rudstone; skeletal (predominantly pelmatozoan). Cross-stratified; cross-strata dipping gently to steeply in widely varying directions, in medium-scale sets (5–30 cm thick). Resistant. Top not exposed at either section. Base sharp. High-energy shallow marine.	2.5 ft (0.8 m)	3.3 ft (1.0 m)
Q	Grainstone-packstone, medium- to coarse-grained, oolitic. Stratification indistinct. Moderately resistant. Base poorly exposed. High-energy shallow marine.	3.3 ft (1.0 m)	4.8 ft (1.5 m)
P	Wackestone-packstone and dolosiltstone; larger grains skeletal. Stratification indistinct. Recessive, poorly exposed at both sections. Base sharp, irregular. Low-energy shallow marine.	2.2 ft (0.7 m)	1.9 ft (0.6 m)
O_2	Grainstone, medium- to coarse-grained, oolitic. Top forms bench at both sections. Base gradational. High-energy shallow marine.	0–0.8 ft (0–0.25 m)	0–0.3 ft (0–0.10 m)
O_1	Wackestone-packstone; larger grains skeletal, finer grains peloidal. Stratification horizontal overall but irregularly wavy or nodular in detail, indistinct and of medium scale in lower part, moderately distinct and thin (5–10 cm) in upper part. Some thin beds of tan-weathering dolomitic limestone. Moderately resistant. Base sharp, flat. Low-energy shallow marine.	9.0–9.8 ft (2.7–3.0 m)	9.7–10.0 ft (2.9–3.1 m)
N_2	Grainstone, fine- to medium-grained, with a few coarse-grained laminae); grains predominantly micritic (peloidal or intraclastic), some siliciclastic and skeletal grains (but none coarser than sand-sized), few ooids. Cross-laminated; cross-laminae dipping gently to moderately, in medium-scale sets, becoming indistinct, nearly horizontal, and locally disturbed by burrowing or root growth in upper 0.2–0.3 m. Rhizoliths in upper part. Resistant; outcrop surface darkened by lichen(?) growth. Base gradational into subunit N_1. Eolian.	4.8–5.4 ft (1.5–1.7 m)	3.0–3.8 ft (0.9–1.2 m)
N_1	Rudstone, very fine- to fine-grained, intraclastic or lithoclastic (clasts micritic); sand-sized matrix contains a significant fraction of siliciclastic grains. Base flat, generally sharp but locally gradational into subunit M_2. Alluvial.	0–0.2 ft (0–0.05 m)	0–0.8 ft (0–0.25 m)
M_2	Breccia (fine- to coarse-grained rudstone), intraclastic (clasts micritic). Clasts are angular, closely fitting, indicating that the breccia developed in situ. Base gradational. Probably absent in northern section. Caliche paleosol.	—	0–1.0 ft (0–0.3 m)

TABLE 1 (continued)

Unit/ subunit	Character	Thickness: Northern section	Thickness: Southern section
M_1	Wackestone-packstone; larger grains skeletal, finer grains peloidal. Stratification horizontal overall, indistinct and thick-bedded in lower part; medium- or thin-bedded (5–10 cm) and irregularly wavy-bedded in upper part. Some thin beds of tan-weathering dolomite and dolomitic limestone. Sparse quartz geodes, dolomite-lined vugs, and chert nodules in lower 0.6 m. Moderately resistant. Base flat, sharp to gradational through 1 cm. Low-energy shallow marine.	8.3–8.7 ft (2.5–2.6 m)	8.8–9.8 ft (2.7–3.0 m)
L	Dolosiltstone, calcareous in part and locally with sparse, small calcitic macrofossils, silty, highly microporous. Unstratified, with abundant faint burrows about 1 cm in diameter. Sparse dolomite-lined vugs. Recessive, tan-weathering, spalling parallel to the outcrop surface. Base sharp, flat. Low-energy shallow marine.	5.5 ft (1.7 m)	4.3 ft (1.3 m)
K	Grainstone, fine- to medium-grained, of mixed clast type. Horizontally stratified to cross-laminated; cross-laminae gently dipping, in medium-scale sets. Absent or included in upper part of unit I at southern section. Eolian in part.	2.0–3.0 ft (0.6-0.9 m)	—
J	Siliciclastic shale, calcareous, medium gray. Recessive, weathering in thin slabs and flakes. Base sharp, flat. Not present in southern section. Low-energy shallow marine.	0.7-2.3 ft (0.2-0.7 m)	—
I_2	Grainstone, fine- to medium-grained, with a few coarse-grained laminae); grains of mixed type, predominantly micritic (peloidal?) but in part skeletal (none coarser than sand-sized), oolitic, and siliciclastic. Cross-laminated; cross-laminae dipping gently to moderately or rarely steeply, in medium- to large-scale sets (as thick as 3–4 m). Stratification is locally indistinct in upper 0.1–0.3 m, and rhizoliths are present at these places. Horizontal trace fossils on top surface. Resistant. Base gradational into subunit I_1 (where present). Eolian.	3.1–5.8 ft (0.9–1.8 m)	19.2 ft (5.8 m)
I_1	Rudstone, very fine- to fine-grained, intraclastic or lithoclastic (clasts micritic); sand-sized matrix contains some siliciclastic grains. Base sharp, flat to broadly irregular. Alluvial.	0–0.1 ft (0–0.03 m)	0–0.1 ft (0–0.03 m)
H_2	Grainstone, medium- to very coarse-grained, skeletal, predominantly pelmatozoan but including horn corals 1–2 cm in diameter). Stratification subhorizontal, consisting of laminae about 1 cm thick. Present only at south end of southern section. Base gradational. High-energy shallow marine.	—	0–1.0 ft (0–0.3 m)

TABLE 1 (continued)

Unit/ subunit	Character	Thickness Northern section	Southern section
H_1	Lime mudstone-wackestone-packstone; larger grains skeletal, finer grains peloidal and in small part siliciclastic; grain content greatest in upper part. Flat-bedded, thick-bedded except for thick lamination near top. Moderately resistant. In-situ brecciation and rhizoliths indicating subaerial exposure present locally in upper 10 cm in northern section. Base gradational into unit G. Low-energy shallow marine.	0–4.0 ft (0–1.2 m)	0–4.0 ft (0–1.2 m)
G_2	Chert, medium-dark gray, nodular; shale, calcareous, medium-dark gray; lime mudstone, argillaceous. Present at south end of southern section and at south end of northern section; thickest and most cherty at south end of southern section. Recessive. Base gradational into subunit G_1. Low-energy shallow marine.	0–1.5 ft (0–0.5 m)	0–5.0 ft (0–1.5 m)
G_1	Lime mudstone, stromatolitic; locally calcareous shale at base. Stratification flat, crinkled, or with convex-up domes as much as 4 cm high and 15 cm wide, thin-bedded to finely laminated. Moderately resistant. Absent in central part of southern section and in northern part of northern section. Base sharp, flat. Intertidal(?).	0–1.0 ft (0–0.3 m)	0–2.3 ft (0–0.7 m)
F_2	Grainstone, fine- to medium-grained, with a few coarse-grained laminae); grains of mixed type, predominantly micritic (peloidal?) but in part oolitic, skeletal (none coarser than sand-sized), and siliciclastic. Subhorizontally laminated to cross-laminated, the cross-laminae gently dipping, in medium- to large-scale sets. Resistant. Thickest in central part of southern section. Base gradational into subunit F_1. Eolian.	2.0–5.8 ft (0.6–1.8 m)	1.7–10.2 ft (0.5–3.1 m)
F_1	Rudstone, very fine- to fine-grained, intraclastic or lithoclastic (clasts in part micritic, in part oolitic); sand-sized matrix contains some siliciclastic grains. Alluvial.	0–0.2 ft (0–0.05 m)	0.2–0.8 ft (0.05–0.2 m)
E_2	Wackestone-packstone; larger grains skeletal. Contains colonial corals as much as 15 cm in diameter in growth position. Stratification indistinct. Recessive to moderately resistant. Present only at southern ends of both sections. Locally contains dolosiltstone. Low-energy shallow marine.	0–4.0 ft (0–1.2 m)	0–1.0 ft (0–0.3 m)
E_1	Grainstone-packstone, medium- to coarse-grained, oolitic with common skeletal grains, some of pebble size. Stratification indistinct; flat-bedded to gently cross-stratified. Upper 15–20 cm is locally rudstone composed of rounded clasts of oolitic grainstone 2–14 cm in long dimension in oolitic-skeletal grainstone matrix. Resistant. Base sharp, broadly irregular. High-energy shallow marine.	8.3–9.2 ft (2.5–2.8 m)	8.5 ft (2.6 m)
D	Lime mudstone-wackestone; sparse skeletal grains, few larger than sand sized. Indistinctly stratified. Moderately resistant. Base sharp, flat. Low-energy shallow marine.	0.8–1.8 ft (0.3–0.6 m)	1.7 ft (0.5 m)

TABLE 1 (continued)

Unit/ subunit	Character	Thickness Northern section	Southern section
C_2	Dolowackestone; larger grains are dolomitized peloids or ooids. Indistinctly flat-bedded. Common dolomite-lined vugs. Moderately resistant. Base gradational into subunit C_1. Low-energy shallow marine.	3.8 ft (1.1 m)	3.9 ft (1.2 m)
C_1	Dolosiltstone, calcareous in part, with sparse, small, calcitic macrofossils, highly microporous; several beds of mudstone-wackestone, in part brecciated, and dolomitized oolitic grainstone. Some dolomite-lined vugs. Indistinctly flat-bedded. Recessive. Base sharp, nearly flat. Low-energy shallow marine.	3.8 ft (1.1 m)	3.9 ft (1.2 m)
B	Grainstone (fine-grained rudstone), skeletal (including conspicuous gastropod shells filled by coarse calcite spar). Base sharp, flat at northern section, not exposed at southern section. High-energy shallow marine.	0.2–1.2 ft (0.1–0.4 m)	0.3 ft (0.10 m)
A	Grainstone-packstone, medium- to very coarse-grained, skeletal (including pebble-sized macrofossils. Flat-bedded, medium- to thick-bedded, indistinctly bedded. Chert nodules common except in upper 0.3 m. Resistant. Not exposed at southern section, and base not exposed at northern section. High-energy shallow marine.	8.3 ft (2.5 m)	—

The rudstones are too coarse to be easily interpreted as eolian. They are more readily explained as thin sheets of alluvium derived from subaerially exposed, partially indurated, or calichified carbonate sediment. Similarly, the erosional relief at the bases of units F and I is less easily explained as the result of eolian deflation than as the result of marine or fluvial erosion during or following regression of the sea past the site.

Noneolian Exposure-Surface Features

The upper parts of some eolian units and of some units immediately underlying eolian units contain features that, although not suggestive of eolian action, are suggestive of subaerial exposure. These features resemble exposure-surface features reported by other workers from Mississippian limestones, including the Ste. Genevieve Limestone, of the central United States (Walls et al., 1975; Harrison and Steinen, 1978; Dever et al., 1979; Leibold, 1982; Ettensohn et al., 1988).

One feature suggestive of subaerial exposure is the brecciation found locally in the upper 0.1–0.3 m of units H and M, immediately beneath their contacts with the basal rudstones of units I and N respectively. The closely fitting angular clasts of the breccias and the downward gradation from breccia to intact limestone indicate that the breccias formed in situ (Figure 9B). A pedogenic origin by calichification is suggested by the structureless, micritic character of the breccia clasts.

Other features suggestive of subaerial exposure are cylindrical structures interpreted as rhizoliths. These cylindrical structures are 0.5–3 cm in diameter, vertically to horizontally oriented, and consist of micritic envelopes surrounding sparry calcite cores 1–3 mm in diameter. Many of the envelopes are concentrically banded or laminated, and some are replaced by chert. The structures have been found near the top of unit I, in the thin part of unit I near the north end of the northern road cut, near the top of unit H at the same locality, and near the top of unit N. They resemble rhizoliths or root structures described by Harrison and Steinen (1978), Klappa (1980), Ettensohn et al. (1988), and Loope (1988). Some pebbles in the basal rudstones are eroded fragments of these rhizoliths, and a few eroded rhizolith fragments are present along set boundaries in the main part of unit I.

Less diagnostic features suggestive of hiatus, but not necessarily of subaerial exposure, include an upward gradation from stratified to structureless limestone near the tops of the eolian units and the presence of indistinct trace fossils in the same intervals. Such features have been found in the upper 0.1–0.3 m of both unit I and unit N. The exposure and accompanying bioturbation were probably subaerial in large part, but horizontal trails and other trace fos-

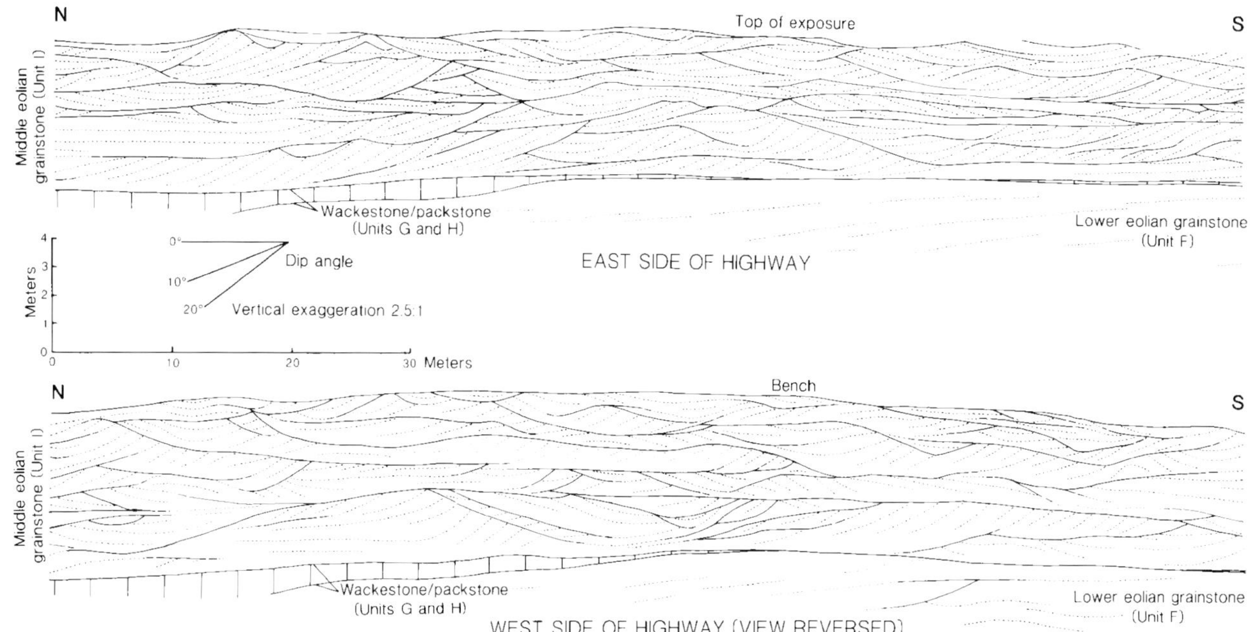

Figure 6. Cross sections of middle eolian grainstone unit (unit I) and adjacent units in the northern road cut, showing character of the cross-stratification. Prepared with the help of tracings from photographs.

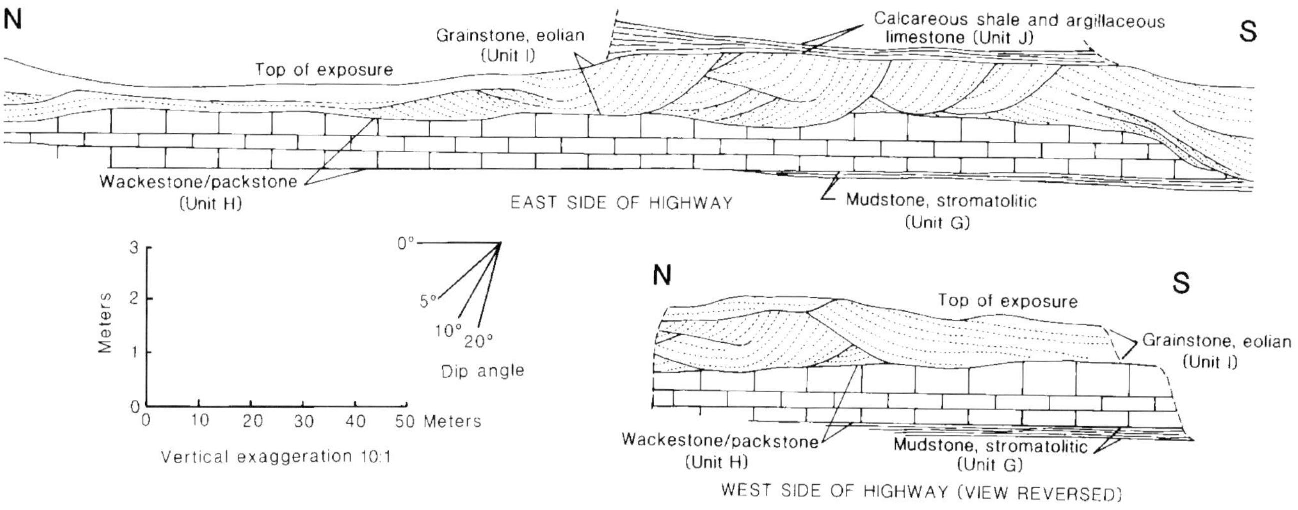

Figure 7. Cross sections of middle eolian grainstone unit (unit I) and adjacent units in the southern road cut, showing character of the cross-stratification. Prepared with the help of tracings from photographs.

sils on the upper contact of unit I probably were formed during transgression by marine waters.

It should be emphasized that stratification in the eolian grainstones commonly extends to the tops of the units without any sign of disturbance or destruction. The destruction of stratification is restricted to the upper 0.3 m of the eolian units or of the units immediately underlying the eolian units. Unless the eolian units were truncated by erosion, the evidence suggests that pedogenic processes were slow or of short duration.

IDENTIFICATION OF EOLIAN ORIGIN

Classical Types of Evidence

Certain types of evidence for identifying eolian deposits may be regarded as classical, in the sense that they have been used widely and for a long time. Some of these classical types of evidence are useful, but others are misleading and should be abandoned. Although some of the classical types of evidence can provide compelling evidence against an eolian origin,

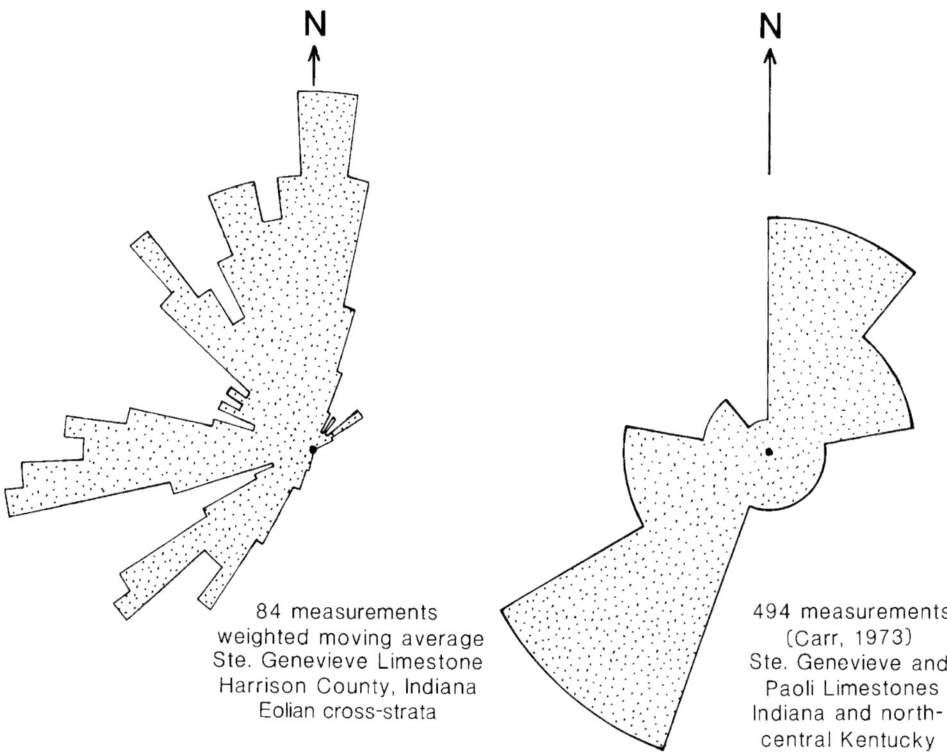

Figure 8. Rose diagrams of cross-bed dip directions. Left: Eolian cross-strata in the Ste. Genevieve Limestone of Harrison County, Indiana. Right: Cross-strata of undifferentiated origin in the Ste. Genevieve and Paoli limestones of Indiana and north-central Kentucky (from Carr, 1973).

and indeed do provide compelling evidence that the highly oolitic and skeletal grainstones of the Ste. Genevieve Limestone in Harrison County are not eolian, none can provide definitive evidence for an eolian origin.

Eolian sands are typically well or very well sorted, and eolian grainstones of the Ste. Genevieve have that character. Grains coarser than about 4 mm are extremely rare in eolian deposits except where the grains are of unusually low density or of unusually easily transportable shape (such as thin shells) or where the winds were unusually strong. The coarsest laminae in Ste. Genevieve eolian rocks are composed of grains in the coarse sand class, and no grains are coarser than 4 mm except for the few reworked fragments of rhizoliths. At the opposite extreme of grain size, interbedded shales or lime mudstones are not expected in eolian deposits except where interdune areas are occasionally flooded by muddy waters. The only interbedded micritic rock found in Ste. Genevieve eolian units is a thin lime mudstone in unit I at the north end of the northern road cut, where the unit is thin and of interdunal character.

Classical sedimentary-structural types of evidence for identifying eolian rocks include the large scale and steep dip angle of cross-stratification. Neither of these types of evidence is trustworthy. Although it is true that much eolian cross-stratification is of large scale (sets thicker than 1 m), eolian dunes can be as low as 0.3 m, and sets of eolian cross-strata can be even thinner than 0.3 m if the dunes were truncated before burial. Similarly, although it is true that much eolian cross-stratification dips at the angle of repose, much subaqueously formed cross-stratification dips at similarly steep angles and, moreover, much eolian cross-stratification dips less steeply than the angle of repose. Most of the cross-strata in Ste. Genevieve eolian rocks have low to intermediate dip angles (few dip angles greater than 25°) and occur in sets of medium scale (in the range of 0.05–1.0 m).

One useful classical type of evidence for distinguishing eolian and subaqueous deposits is based on differences between wind, current, and wave ripples. Wind ripples are relatively straight-crested and moderately asymmetric, and they have vertical-form indices (spacing-to-height ratios) typically larger than those of current and wave ripples (Tanner, 1967). In fine- to medium-grained sands, this difference is due more to the low heights of wind ripples than to their spacings, which are comparable to those of current and wave ripples. Ripples that have the characteristics of wind ripples can be seen on a few bedding surfaces of Ste. Genevieve eolian grainstones (Figure 9A). Because the ripples are so rare, however, they can hardly be accepted as definitive evidence of an eolian origin.

Paleontological evidence can be useful in ruling out an eolian origin. Although trace fossils are known

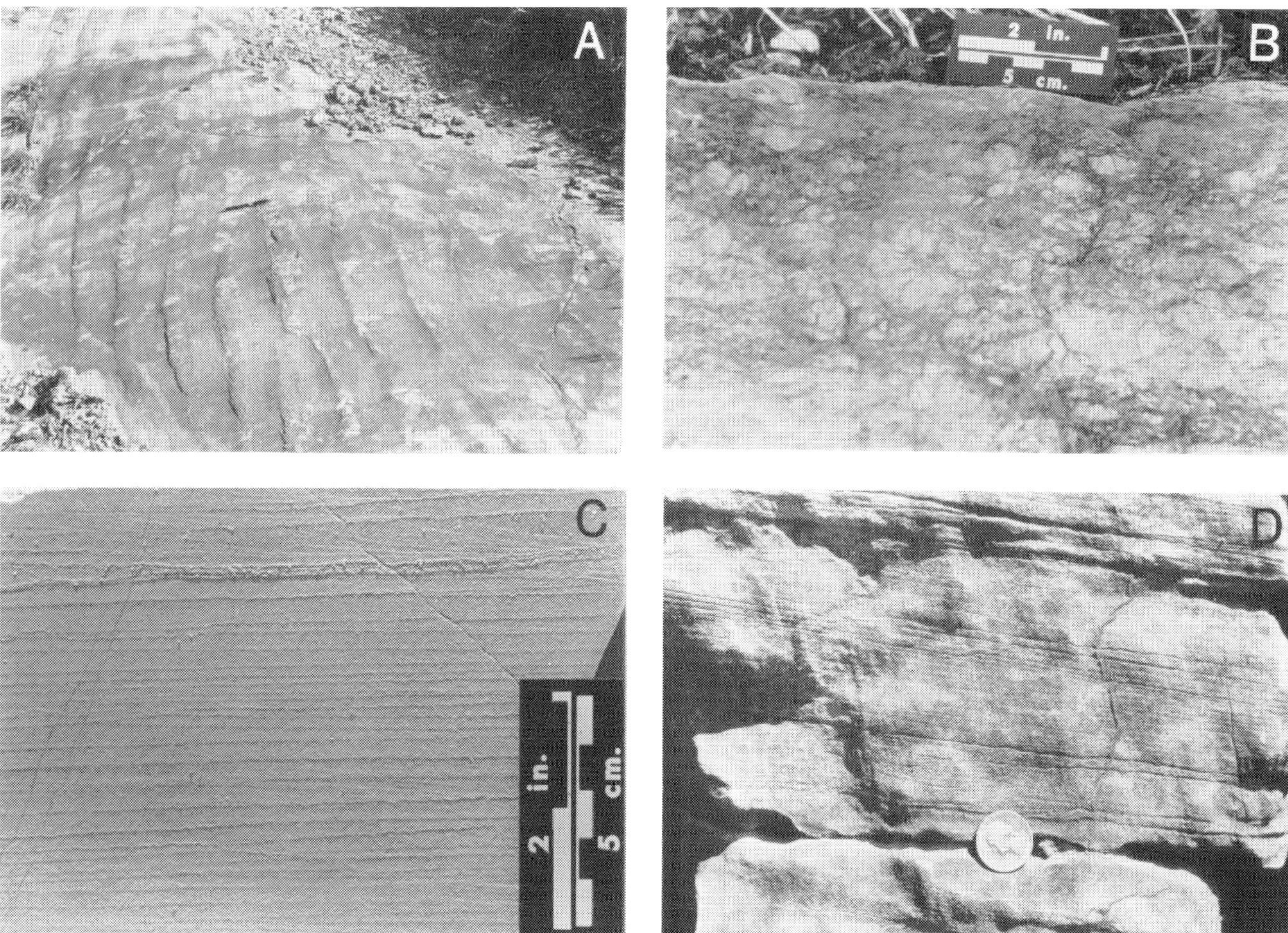

Figure 9. Sedimentary structures of the Ste. Genevieve Limestone in Harrison County, Indiana. (A) Ripples exposed on a gently inclined cross-stratum surface in unit I of the southern road cut (location 12 on Figure 5). The ripples are interpreted as eolian because of their high vertical-form index, straight crests, and moderate asymmetry. (B) Breccia in the upper part of unit M in the southern road cut (location 10 on Figure 5). Because of the closely fitting angular clasts and the downward gradation into intact limestone, the breccia is interpreted to have formed in situ as a caliche breccia. The breccia grades upward into the basal rudstone (composed of rounded and sorted pebbles) of unit N. (C) Sawed and etched sample of climbing translatent strata of the type formed by wind ripples. From the same stratigraphic interval shown in Figure 9D. The sawed surface is oblique to the stratification, and the strata therefore appear thicker than they actually are. Note the ripple-foreset cross-laminae visible in a few of the strata. Also note that relatively coarse siliciclastic grains (recognized by their protrusion above the etched surface) near the upper right corner of the photograph tend to be concentrated in the upper parts of the translatent strata. (D) Weathered exposure of climbing translatent strata of the type formed by wind ripples. Unit I in the northern road cut (location 4 on Figure 5). Note the almost varvelike regularity of the thin strata, best seen by looking at the photograph at a low oblique angle along the stratification.

in eolian deposits (Ahlbrandt et al., 1978), they are rare in unvegetated dune deposits, and an abundance of trace fossils would be strong evidence against an eolian origin. The few trace fossils found in Ste. Genevieve eolian rocks are at or penetrate downward from the tops of the eolian units and can be interpreted as the products of soil fauna and flora after dune stabilization or as the products of marine fauna after a transgression by the sea. Transported macrofaunal body fossils or fossil fragments have the same significance as nonskeletal clasts of similar size. Skeletal remains larger than 2 mm are conspicuous by their absence in Ste. Genevieve eolian grainstones. Fossils as large as 1 cm are common in the highly oolitic and skeletal grainstones, on the other hand, and rule out an eolian origin for those rocks.

Fine Structure

The fine structural details of stratification, or what has been called "fine structure" (Bagnold, 1941; Hunter, 1985), provide the best means of distinguishing eolian and subaqueous deposits. Although the first fine-structure studies were of quartz sands (Hunter, 1977, 1981), the features recognizable in quartz sands have been shown to be equally recognizable in Quaternary carbonate sands (White and Curran, 1988) and in pre-Quaternary carbonate rocks (Loope, 1986; Loope and Haverland, 1988; Rice and Loope, 1991).

Three basic types of fine structure make up the great bulk of eolian sands: (1) sandflow cross-stratification, formed by the avalanching of noncohesive sand down the slipfaces of dunes; (2) grainfall stratification, formed by the falling of previously saltating grains through the relatively calm zones of flow separation leeward of dune crests; and (3) climbing-ripple structures, formed by ripples migrating under conditions of net deposition. All three of these types of fine structure occur in subaqueous sands as well as in eolian sands, but the structural details of the climbing-ripple structures generally allow identification of the environment.

Sand-flow cross-stratification, by definition, dips at the angle of repose. Eolian and subaqueous sand-flow cross-strata have certain differences, but the differences are subtle (Hunter, 1985). Sand-flow cross-stratification is rare in Ste. Genevieve eolian grainstones and played no part in the present interpretation. Much of the more steeply dipping cross-stratification in Ste. Genevieve eolian grainstones is probably of grainfall type. However, eolian and subaqueous grainfall strata are probably not distinguishable, both being characterized by even, indistinct laminae.

The most useful fine structure for identifying eolian sands in general, and the eolian facies of the Ste. Genevieve in particular, is the translatent stratification formed by climbing wind ripples. This stratification is a type of climbing-ripple structure and forms by the same general process as any climbing-ripple structure, but climbing-ripple structures formed by wind ripples differ in several details from those formed by subaqueous current or wave ripples (Figure 10). Because wind ripples are not as steep in profile as subaqueous ripples, sand does not avalanche down the lee sides. In the absence of grain segregation due to avalanching, ripple-foreset cross-laminae are much less visible in climbing-ripple structures formed by wind ripples than in those formed by current or wave ripples (Figure 10). In the absence or rarity of ripple-foreset cross-laminae, the visually most prominent components of climbing-wind-ripple structures are climbing translatent strata, each one of which is the depositional product of a single wind ripple during its lifetime.

In addition to causing the poor development of ripple-foreset cross-laminae, the absence of avalanching on the lee slopes of wind ripples results in the concentration of the coarsest grains on the upper lee sides of wind ripples rather than on the lower lee sides as in subaqueous ripples. Because the coarsest grains collect near the crests of wind ripples, the climbing translatent strata are inversely graded (Figure 10). This inverse grading is best seen on outcrops or artificially cut surfaces that truncate the thin strata at a low angle, in effect magnifying the thickness of the strata (Figure 9C).

One of the most useful characteristics for identifying climbing-wind-ripple structures is the marked uniformity in thickness of the climbing translatent strata, both laterally along a single stratum and vertically from stratum to stratum within a set. These two types of uniformity arise because most wind ripples are highly regular in form (nearly straight-crested and with nearly identical vertical profiles at different positions along the trend of a single ripple and from one ripple to the next) and because the ripples migrate for long distances relative to their spacings before losing their identity. The lateral continuity and uniformity in thickness of the translatent strata in a set are suggestive of varves, but the translatent strata are not quite so laterally continuous and uniform as varves. The translatent strata formed by climbing current ripples are generally much more lenticular than those formed by climbing wind ripples (Figure 10).

Because wind ripples are best developed on rather gently dipping surfaces and because the ripples usually climb at low angles relative to the depositional surface, the climbing translatent stratification formed by wind ripples typically dips at a low angle. Preservation of the structure requires that the wind-rippled surface receive a net deposit, and net deposition is usually restricted to the leeward-facing sides of dunes. Gently sloping, leeward-facing wind-rippled surfaces are common on the basal aprons of dunes, especially on parts of dunes where the crest line is oblique to the wind. Most climbing-wind-ripple structures in the rock record are formed in such areas.

The climbing translatent stratification formed by wind ripples in the Ste. Genevieve Limestone is best seen on weathered outcrops, where the contacts between individual strata are defined by narrow grooves (Figure 9D). On artificially etched surfaces, the contacts stand out as sharp ridges because of concentrations of insoluble silt and fine sand along the contacts, especially where microstylolites follow the contacts (Figure 9C). The individual strata are typically only a few millimeters thick, although in some sets they are as thick as 5–7 mm. Inverse grading and ripple-foreset cross-laminae can be seen in some of the thicker strata that are exposed on surfaces that cut the stratification at a low angle (Figure 9C).

Fine structures formed by the adhesion of windblown sand to wet surfaces can be very useful in the identification of eolian deposits (Hunter, 1973; Kocurek and Felder, 1982). I have seen climbing-adhesion-ripple structures, the most easily identified eolian adhesion structure, in Quaternary eolianites at Bannerman Point, Eleuthera Island, the Bahamas, but have found no adhesion structures in eolian rocks of the Ste. Genevieve.

Petrographic Evidence

A petrographic study by Dodd et al. (this volume) provides other evidence that supports an eolian ori-

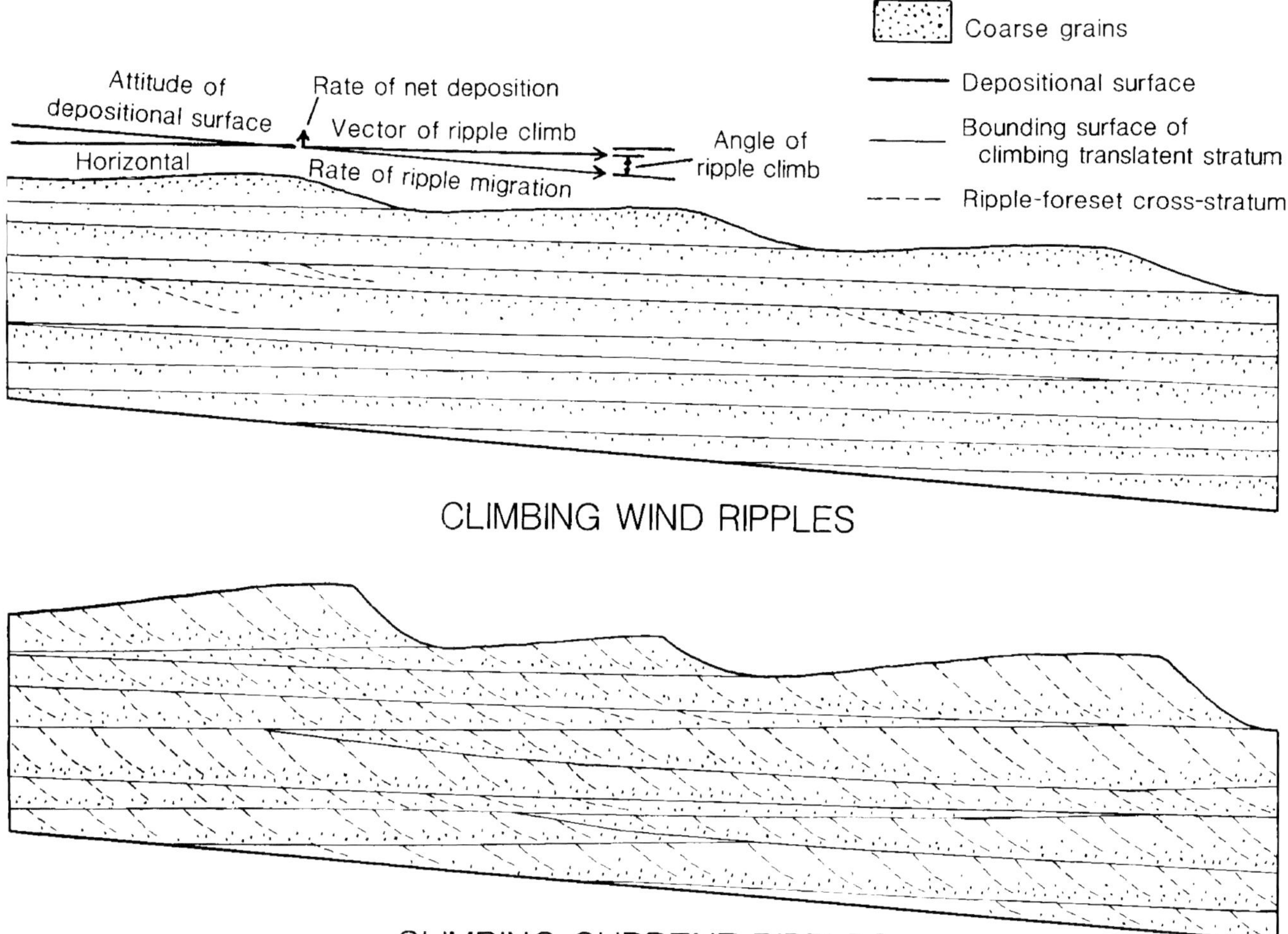

Figure 10. Schematic cross sections of climbing-wind-ripple and climbing-current-ripple structures. The angle of climb is assumed to be identical in the two cases, as is the attitude of the depositional surface. The rates of ripple migration and of net deposition are vectors measured in units of length divided by time. The rate of ripple migration is measured in a direction parallel with the plane of the generalized depositional surface, while the rate of net deposition is measured in a direction normal to that plane. The vector of ripple climb is the vector sum of the rate of ripple migration and the rate of net deposition; in the examples shown, the direction of the vector of ripple climb is slightly below the horizontal. The angle of ripple climb is the angle between the vector of ripple climb and the generalized depositional surface.

gin for the Ste. Genevieve mixed-clast grainstones in the sections near Corydon. Among the more convincing pieces of evidence are (1) the commonness of quartz silt in the eolian grainstones compared to its rarity in the marine grainstones, (2) the commonness of broken ooids in the eolian grainstones compared to their rarity in the marine grainstones, and (3) the high sphericity of skeletal fragments in the eolian grainstones compared to the low sphericity of such grains, even where they are well rounded, in the marine grainstones. Although Dodd et al. found few examples of vadose cement in the eolian grainstones, they interpreted the absence of such cement to be a result of the dryness of the sediment until it was submerged beneath the water table. The eolian grainstones underwent extensive solution packing, suggesting that no cementation occurred until burial to considerable depth, whereas the marine grainstones have little solution packing and commonly have clasts with thin isopachous rims, suggesting that some cementation occurred early.

DISCUSSION

Eolian grainstones are present in the Ste. Genevieve Limestone in Harrison County as far north as Milltown, 20 km (12 mi) northwest of the northern road cut, and as far west as eastern Crawford County, 15 km (9 mi) west of the northern road cut. The extent of the eolian facies of the Ste. Genevieve outside Harrison and Crawford counties is poorly known. I have observed no eolian grainstones in nearly complete exposures of the formation in the Orleans quarry, Orange County, Indiana, or at Cataract Falls,

Owen County, Indiana (Figure 1). On the other hand, grainstones that I interpret as eolian are present in the type area of the formation at Ste. Genevieve, Missouri (Locality 4 of Knewtson and Hubert, 1969) and at Alton, Illinois (unit M and probably unit S of Collinson et al., 1979).

Another cross-stratified stratigraphic unit that should be reexamined for evidence of eolian origin is the Aux Vases Sandstone, a partial correlative of the upper part of the Ste. Genevieve Limestone of Indiana that crops out in southeastern Missouri and southern Illinois. Although the Aux Vases has generally been interpreted to be of tidally influenced shallow marine origin (Weimer et al., 1981; Seyler et al., 1986; Cole and Fraunfelter, 1989), Cluff and Lineback (1981) inferred an eolian origin (but without evidence that I regard as strong) for some of the cross-stratified sandstones. Outcrops of the Aux Vases in the vicinity of Modoc, Randolph County, Illinois, appear to be of subaqueous origin, but other localities should be reexamined critically.

Still farther away from southern Indiana, other cross-stratified carbonate rocks roughly correlative with the Ste. Genevieve Limestone need to be reexamined for evidence of eolian origin. These include the Warix Run Member of the Slade Formation in eastern Kentucky (Dever, 1980; Ettensohn et al., 1984), the Monteagle Limestone of southern Tennessee, northeast Alabama, and northwest Georgia (Handford, 1978), and the Loyalhanna Limestone and its equivalents in southern Pennsylvania, western Maryland, and northern West Virginia (Adams, 1970; Hoque, 1975). Personal examination of the Warix Run along Interstate Route 64 between Morehead and Grayson, Kentucky, convinces me that eolian rocks are not present at that locality. Recently Handford (1990) has identified carbonate eolianites in the St. Louis-Ste. Genevieve interval in southwestern Kansas.

At least some of the Ste. Genevieve rocks whose cross-stratification was measured by Carr (1973) are highly oolitic or skeletal grainstones with macrofossils of pebble size, and these grainstones are certainly marine, not eolian. The similarity in cross-bed dip directions of the eolian grainstones in Harrison County and these other grainstones (Figure 8) calls for explanation. Although the similarity might be coincidental, the currents responsible for the shallow marine cross-stratification may not have been tidal, as has usually been thought (Knewtson and Hubert, 1969; Carr, 1973), but wind driven.

An eolian origin for parts of the Ste. Genevieve Limestone does not by itself imply an unusually arid climate, for carbonate eolianites, like siliciclastic coastal dunes, can originate in fairly wet climatic zones. However, the absence of any adhesion structures and the rather weak development of paleosols in the eolian units suggests that the climate was indeed rather dry. A dry climate is compatible with the paleolatitude of about 15°S interpreted for the Illinois basin in Mississippian time (Smith et al., 1981). Moreover, paleoclimatic evidence from Mississippian rocks over a broad geographic area is compatible with a climate that was at least fairly dry during the time of Ste. Genevieve deposition in the Illinois basin (Van der Zwan et al., 1985). Given the dry paleoclimate interpreted for the area, the basal rudstones of the eolian units can be interpreted as desert reg deposits.

The interbedding of eolian and shallow marine rocks in the Ste. Genevieve implies the following alternatives: (1) fluctuations of eustatic sea level, (2) fluctuations in the rate of subsidence, (3) fluctuations in sedimentation rate, as by the buildup of shoals to heights above sea level followed by drowning of the shoals as buildup ceased and subsidence proceeded, or (4) some combination of these factors. Broader evidence suggests that eustatic fluctuations of sea level were taking place during Ste. Genevieve time and throughout the late Paleozoic (Ross and Ross, 1985). Glacioeustatic fluctuations are a plausible cause for the interbedding of eolian and shallow marine rocks in the Ste. Genevieve, for late Paleozoic glaciations were beginning in Gondwanaland at about the time the Ste. Genevieve was being deposited (Veevers and Powell, 1987). If the fluctuations were glacioeustatic, the areas exposed subaerially may have been quite broad. Broad exposure would help to explain the transport of siliciclastic grains over wide areas by eolian saltation and would also imply that the exposure surfaces and associated eolian deposits may be useful in the correlation of beds within the Ste. Genevieve over wide areas.

CONCLUSIONS

Detailed examination of sedimentary structures in the Ste. Genevieve Limestone of Harrison County, Indiana, has revealed the presence of cross-bedded grainstones of eolian dune origin. The eolian grainstones are composed of fine to medium sand of mixed-clast type, in contrast to associated shallow marine grainstones that are medium- to very coarse-grained and of oolitic or skeletal clast type. An eolian origin of the mixed-clast grainstones is indicated by abundant translatent strata of the type formed by climbing wind ripples and by rare wind ripples on bedding surfaces. The translatent strata are identified as having been formed by wind ripples by the thinness (1–7 mm), wide lateral extent, almost varvelike uniformity, and inverse grading of the strata and by the rare presence of ripple-foreset cross-lamination. The ripples are identified as eolian by their high vertical-form indices. Subaerial exposure of surfaces at the tops and bases of eolian units is indicated by pedogenic features such as in-situ brecciation and rhizoliths. Thin rudstones at the bases of eolian units are interpreted as desert reg deposits, probably of alluvial sheet origin.

The currents responsible for the cross-bedding in shallow marine grainstones of the Ste. Genevieve have generally been thought to be tidal. The similarity of cross-bed dip directions in the eolian grainstones and associated shallow marine grainstones, however, suggests that wind-driven currents should be given consideration as a sediment-transporting agent in the shallow Paleozoic seas of the eastern interior United States.

Although the regional extent of eolianites in the Ste. Genevieve is poorly known, reconnaissance examination indicates that they are present on the west side of the Illinois basin as well as on the east side in southern Indiana. The search for carbonate eolianites needs to be extended still farther geographically and stratigraphically.

Eolianites are present in as many as four intervals in the Ste. Genevieve of Harrison County. The interbedding of eolian and shallow marine rocks could be explained in several ways but are probably a result of eustatic sea level fluctuations. A glacioeustatic origin of the interbedding is an especially attractive explanation, for the late Paleozoic glaciations in Gondwanaland were beginning at about the time that the Ste. Genevieve was being deposited.

ACKNOWLEDGMENTS

Thomas Brown of Indiana University kindly provided petrographic analyses of 15 thin sections of samples from the northern road cut, and many of his observations are incorporated in Table 1. The manuscript benefited from reviews by Don Carr, Bruce Richmond, Dave Rubin, Bob Dodd, Todd Thompson, and the editors of this volume. I am most grateful to my late stepfather and my mother, Mr. and Mrs. Ora Stepro, who provided meals and lodging during my work in Harrison County.

REFERENCES CITED

Adams, R.W., 1970, Loyalhanna Limestone—cross-bedding and provenance, *in* Fisher, G.W., F.J. Pettijohn, J.C. Reed, Jr., and K.N.Weaver, eds., Studies of Appalachian geology, central and southern: New York, Interscience, p. 83-100.

Ahlbrandt, T.S., Sarah Andrews, and D.T. Gwynne, 1978, Bioturbation in eolian deposits: Journal of Sedimentary Petrology, v. 48, p. 839-848.

Amos, D.H., 1972, Geologic map of the New Amsterdam quadrangle, Kentucky-Indiana, and part of the Mauckport quadrangle, Kentucky: U.S. Geological Survey Geologic Quadrangle Map GQ-990, 1:24,000, 1 sheet.

Archer, A.W., 1984, Preservational control of trace-fossil assemblages: Middle Mississippian carbonates of south-central Indiana: Journal of Paleontology, v. 58, p. 285-297.

Bagnold, R.A., 1941, The physics of blown sand and desert dunes: London, Methuen, 265 p. (Reprinted 1971 by Chapman and Hall, London.)

Carr, D.D., 1973, Geometry and origin of oolite bodies in the Ste. Genevieve Limestone (Mississippian) in the Illinois basin: Indiana Geological Survey Bulletin 48, 81 p.

Carr, D.D., R.K. Leininger, and M.V. Golde, 1978, Crushed stone resources of the Blue River Group (Mississippian) of Indiana: Indiana Geological Survey Bulletin 52, 225 p.

Carr, D.D., C.B. Rexroad, and H.H. Gray, 1986, Ste. Genevieve Limestone, *in* Shaver, R.H., et al., 1986, Compendium of Paleozoic rock-unit stratigraphy in Indiana—a revision: Indiana Geological Survey Bulletin 59, p. 128-130.

Choquette, P.W., and R.P. Steinen, 1980, Mississippian non-supratidal dolomite, Ste. Genevieve Limestone, Illinois basin: Evidence for mixed-water dolomitization, *in* Zenger, D.H., J.B. Dunham, and R.L. Ethington, eds., Concepts and models of dolomitization: Society of Economic Paleontologists and Mineralogists Special Publication 28, p. 163-196.

Choquette, P.W., and R.P. Steinen, 1985, Mississippian oolite and nonsupratidal dolomite reservoirs in the Ste. Genevieve Formation, North Bridgeport Field, Illinois basin, *in* Roehl, P.O., and P.W. Choquette, eds., Carbonate petroleum reservoirs: Casebooks in Earth Sciences (Springer-Verlag), p. 209-225.

Cluff, R.M., 1984, Carbonate sand shoals in the Middle Mississippian (Valmeyeran) Salem-St. Louis-Ste. Genevieve Limestones, Illinois basin, *in* Harris, P.M., ed., Carbonate sands, a core workshop: Society of Economic Paleontologists and Mineralogists Core Workshop 5, p. 94-135.

Cluff, R.M., and J.A. Lineback, 1981, Middle Mississippian carbonates of the Illinois basin: Illinois Geological Society and Illinois State Geological Survey, 88 p.

Cole, R.D., and G.H. Fraunfelter, 1989, Depositional facies analysis of the Aux Vases (Mississippian) Formation from southeastern Missouri to southeastern Illinois (abs.): Geological Society of America Abstracts with Programs, v. 21, no. 4, p. 7.

Collinson, C.W., R.D. Norby, T.L. Thompson, and J.W. Baxter, 1979, Stratigraphy of the Mississippian stratotype—Upper Mississippi Valley, U.S.A.: International Congress on Carboniferous Stratigraphy and Geology, 9th, Field Trip 8 Guidebook, p. 65-73.

Dever, G.R., Jr., 1980, Stratigraphic relationships in the lower and middle Newman Limestone (Mississippian), east-central and northeastern Kentucky: Kentucky Geological Survey, Series XI, Thesis Series 1, 49 p.

Dever, G.R., Jr., Preston McGrain, G.W. Ellsworth, Jr., and J.R. Moody, 1979, Features in the upper Ste. Genevieve Limestone of west-central Kentucky, *in* Ettensohn, F.R., and G.R. Dever, Jr., eds., Carbonferous geology from the Appalachian basin to the Illinois basin through eastern Ohio and Kentucky: International Congress on Carboniferous Stratigraphy and Geology, 9th, Field Trip 4 Guidebook, p. 261-263.

Ettensohn, F.R., G.R. Dever, Jr., and J.S. Grow, 1988, A paleosol interpretation for profiles exhibiting subaerial exposure "crusts" from the Mississippian of the Appalachian basin, *in* Reinhardt, Juergen, and

W.R. Sigleo, eds., Paleosols and weathering through geologic time: principles and applications: Geological Society of America Special Paper 216, p. 49-79.

Ettensohn, F.R., C.L. Rice, G.R. Dever, Jr., and D.R. Chesnut, 1984, Slade and Paragon Formations—new stratigraphic nomenclature for Mississippian rocks along the Cumberland Escarpment in Kentucky: U.S. Geological Survey Bulletin 1605-B, p. 1-37.

Fryberger, S.G., and Christopher Schenk, 1981, Wind sedimentation tunnel experiments on the origins of aeolian strata: Sedimentology, v. 28, p. 805-821.

Gardner, R.A.M., 1983, Aeolianite, *in* Goudie, A.S., and Kenneth Pye, eds., Chemical sediments and geomorphology: New York, Academic Press, p. 265-300.

Gray, H.H., J.L. Bassett, C.A. Munson, P.J. Munson, and G.S. Fraser, 1983, Archaeological geology of the Wyandotte Cave region, south-central Indiana, *in* Shaver, R.H., and J.A. Sunderman, eds., Field trips in midwestern geology: Geological Society of America Guidebook for Field Trips, 1983 Annual Meeting, v. 2, p. 173-213.

Handford, C.R., 1978, Monteagle Limestone (Upper Mississippian)—oolitic tidal-bar sedimentation in southern Cumberland Plateau: American Association of Petroleum Geologists Bulletin, v. 62, p. 644-656.

Handford, C.R., 1990, Mississippian carbonate eolianites in southwestern Kansas (abs.): American Association of Petroleum Geologists Bulletin, v. 75, p. 669.

Harrison, R.S., and R.P. Steinen, 1978, Subaerial crusts, caliche profiles, and breccia horizons: comparison of some Holocene and Mississippian exposure surfaces, Barbados and Kentucky: Geological Society of America Bulletin, v. 89, p. 385-396.

Hoque, M., 1975, Paleocurrent and paleoslope—a case study: Palaeogeography, Palaeoclimatology, Palaeoecology, v. 17, p. 77-85.

Hunter, R.E., 1973, Pseudo-crosslamination formed by climbing adhesion ripples: Journal of Sedimentary Petrology, v. 43, p. 1125-1127.

Hunter, R.E., 1977, Basic types of stratification in small eolian dunes: Sedimentology, v. 24, p. 361-387.

Hunter, R.E., 1981, Stratification styles in eolian sandstones: some Pennsylvanian to Jurassic examples from the western interior U.S.A., *in* Ethridge, F.G., and R.M. Flores, eds., Recent and ancient nonmarine depositional environments: models for exploration: Society of Economic Paleontologists and Mineralogists Special Publication 31, p. 315-329.

Hunter, R.E., 1985, Subaqueous sand-flow cross strata: Journal of Sedimentary Petrology, v. 55, p. 886-894.

Klappa, C.F., 1980, Rhizoliths in terrestrial carbonates: classification, recognition, genesis and significance: Sedimentology, v. 27, p. 613-629.

Knewtson, S.L., and J.F. Hubert, 1969, Dispersal patterns and diagenesis of oolitic calcarenites in the Ste. Genevieve Limestone (Mississippian), Missouri: Journal of Sedimentary Petrology, v. 39, p. 954-968.

Kocurek, Gary, and R.H. Dott, Jr., 1981, Distinctions and uses of stratification types in the interpretation of eolian sand: Journal of Sedimentary Petrology, v. 51, p. 579-595.

Kocurek, Gary, and Gordon Felder, 1982, Adhesion structures: Journal of Sedimentary Petrology, v. 52, p. 1229-1241.

Leibold, A.W., 1982, Stratigraphy, petrography, and depositional environment of the Bryantsville Breccia (Meramecian) of south-central Indiana: Indiana University unpublished A.M. thesis, 171 p.

Loope, D.B., 1986, Pennsylvanian eolian limestones, Paradox basin, U.S.A. (abs.): International Sedimentological Congress, 12th, Abstracts Volume, p. 189-190.

Loope, D.B., 1988, Rhizoliths in ancient eolianites: Sedimentary Geology, v. 56, p. 301-314.

Loope, D.B., and Z.E. Haverland, 1988, Giant desiccation fissures filled with calcareous eolian sand, Hermosa Formation (Pennsylvanian), southeastern Utah: Sedimentary Geology, v. 56, p. 403-413.

Maples, C.G., and J.A. Waters, 1987, Redefinition of the Meramecian/Chesterian boundary (Mississippian): Geology, v. 15, p. 647-651.

McGrain, P., 1942, The St. Louis and Ste. Genevieve limestones of Harrison County, Indiana: Indiana Academy of Sciences Proceedings, v. 52, p. 149-162.

McKee, E.D., and W.C. Ward, 1983, Eolian environment, *in* Scholle, P.A., D.G. Bebout, and C.H. Moore, eds., Carbonate depositional environments: American Association of Petroleum Geologists Memoir 33, p. 132-169.

Pohl, E.R., 1970, Upper Mississippian deposits of south-central Kentucky: a project report: Kentucky Academy of Sciences Transactions, v. 31, p. 1-15.

Rice, J.A., and D.B. Loope, 1991, Wind-reworked carbonates, Permo-Pennsylvanian of Arizona and Nevada: Geological Society of America Bulletin, v. 103, p. 254-267.

Ross, C.A., and J.R.P. Ross, 1985, Late Paleozoic depositional sequences are synchronous and worldwide: Geology, v. 13, p. 194-197.

Seyler, B.J., R.M. Cluff, and J. Lizak, 1986, A core workshop and field trip guide book featuring the Aux Vases and Ste. Genevieve Formations: Illinois Geological Society, Illinois State Geological Survey, and Southern Illinois University, Carbondale, 67 p.

Smith, A.G., A.M. Hurley, and J.C. Briden, 1981, Phanerozoic paleocontinental world maps: Cambridge, Cambridge University Press, 102 p.

Tanner, W.F., 1967, Ripple mark indices and their uses: Sedimentology, v. 9, p. 89-104.

Van der Zwan, C.J., M.C. Boulter, and R.N.L.B. Hubbard, 1985, Climate change during the Lower Carboniferous in Euramerica, based on multivariate statistical analyses of palynological data: Palaeogeography, Palaeoclimatology, Palaeoecology, v. 52, p. 1-20.

Veevers, J.J., and C.McA. Powell, 1987, Late Paleozoic glacial episodes in Gondwanaland reflected in transgressive-regressive depositional sequences in Euramerica: Geological Society of America Bulletin, v. 98, p. 475-487.

Walls, R.A., W.B. Harris, and W.E. Nunan, 1975, Calcareous crust (caliche) profiles and early subaerial exposure of carboniferous carbonates, northeastern Kentucky: Sedimentology, v. 22, p. 417-440.

Weimer, R.J., J.D. Howard and D.R. Lindsay, 1981, Tidal flats and associated tidal channels, *in* Scholle, P.A., and Darwin Spearing, eds., Sandstone depositional environments: American Association of Petroleum Geologists Memoir 31, p. 191-245.

White, B., and H.A. Curran, 1988, Mesoscale physical sedimentary structures and trace fossils in Holocene carbonate eolianites from San Salvador Island, Bahamas: Sedimentary Geology, v. 55, p. 163-184.

Woodson, F.J., 1981, Lithologic and structural controls on karst landforms of the Mitchell Plain, Indiana, and Pennroyal Plateau, Kentucky: unpublished M.A. thesis, Indiana State University, Terre Haute, Indiana, 132 p.

Woodson, F.J., 1982, Uppermost St. Louis Limestone (Mississippian): The Horse Creek Member in Indiana: Indiana Academy of Sciences Proceedings, v. 91, p. 419-427.

Chapter 4

Petrologic Method for Distinguishing Eolian and Marine Grainstones, Ste. Genevieve Limestone (Mississippian) of Indiana

J. Robert Dodd
Dept. of Geological Sciences
Indiana University
Bloomington, Indiana, USA

Charles W. Zuppann
Indiana Geological Survey
Bloomington, Indiana, USA

Clayton D. Harris
Karl W. Leonard
Thomas W. Brown
Dept. of Geological Sciences
Indiana University
Bloomington, Indiana, USA

ABSTRACT

Published descriptions of carbonate eolianites of pre-Pleistocene age are rare. Eolianite grainstones of the Ste. Genevieve Limestone (Mississippian) near Corydon, Indiana, contain a diverse assemblage of grains, including various skeletal grains, ooids (some broken and abraded), peloids, well-cemented intraclasts, and abundant quartz silt. Sphericity of originally tabular carbonate grains is generally high. Eolian grainstones are laminated, and in places grain size coarsens upward within laminae. Vadose cement is rare, and solution packing is extensive.

Marine grainstones, which probably formed on shallow shoals or an open platform, are also common in the Ste. Genevieve section and contain a diverse assemblage of skeletal grain types, ooids, peloids, and intraclasts; however, a single grain type (such as ooids) commonly dominates each individual unit. Detrital quartz grains are rare. Sphericity of skeletal grains not originally spherical is low. Fine laminations are not present, and no systematic grading is found on a thin-section scale. Fossils larger than 4 mm occur in the marine units. Solution packing is minor, and some marine cement is present.

Petrographic and stratigraphic data suggest that the eolianites formed due to lowering of sea level and not due to buildup of islands above sea level.

INTRODUCTION

Few descriptions of carbonate eolianites appear in the literature, although they are common in the Quaternary and can be seen forming today in many places (McKee and Ward, 1983). Hunter (1988, 1989, and this volume) has recently described eolian carbonate deposits from the Mississippian Ste. Genevieve Limestone in southern Indiana. He based his interpretation on a number of sedimentologic features including: (1) varvelike laminations produced by climbing wind ripples, (2) inverse grading of some of these laminae, (3) rarity of ripple-foreset laminae in climbing-ripple stratification, (4) low height/length ratio for rare ripples, (5) lack of grains larger than 4 mm, and (6) evidence of subaerial exposure above and below eolian units.

We studied the petrographic features of the grainstones in the Corydon sections described by Hunter in order to answer the following questions:

1. Is the petrography of eolian grainstones distinct from that of marine grainstones in this section, and if so, can these petrographic features thus be used to identify eolian vs. marine grainstones in general?
2. Do the petrographic features of these units support an eolian origin?
3. What conditions gave rise to deposition of the eolianite units?

GEOLOGICAL SETTING

The Ste. Genevieve Limestone is part of a sequence of shallow marine (and eolian) carbonate units of Valmeyeran (middle Mississippian) age in the Illinois basin (Figure 1). This unit is of particular economic interest because it contains extensive petroleum reservoirs in the subsurface. The Ste. Genevieve Limestone overlies the St. Louis Limestone (Mississippian), a shallow marine to intertidal unit, and is in turn overlain by the Paoli Limestone (Mississippian), which is also a shallow marine deposit (Droste and Carpenter, 1990). In southern Indiana, the top of the Ste. Genevieve is marked by the Bryantsville Breccia Bed, a probable paleosol horizon (Leibold, 1982).

The Ste. Genevieve Limestone contains a variety of lithologies (Carr, 1973), including: (1) oolitic, skeletal, and peloidal grainstones, all probably deposited in shallow shoals or other high-energy environments, (2) mudstones, wackestones, and packstones, probably deposited in lagoonal or intershoal settings, (3) dolomitized mudstones that were low-energy shelf deposits, and (4) grainstones with varied grain types and variable amounts of quartz silt (mixed-clast grainstone of Hunter, this volume).

METHODS

We measured and sampled the section 3.5 mi southwest of Corydon, Indiana (Figures 2, 3, southern road cut of Hunter, this volume). Units A and B were measured in a road cut 1 mi southwest of Corydon (northern cut of Hunter, this volume). A total of 37 thin sections were prepared from these samples, of which 22 were grainstones. We grouped the grainstone samples into eolian and marine (Table 1) based on the criteria used by Hunter (this volume). Samples listed as eolian were in most cases from units F, I, and N, which Hunter interpreted as eolian. Three samples considered to be eolian come from a lenticular bed within unit H. This bed has the eolian properties listed by Hunter, although the bulk of unit H is clearly of marine origin. Samples listed as marine are from units B, C, E, H, Q, R, and S, which Hunter identifies as marine.

We have also observed, but did not point count, thin sections of approximately 100 additional samples of eolian and marine grainstones from the Ste. Genevieve. All have the characteristics discussed below.

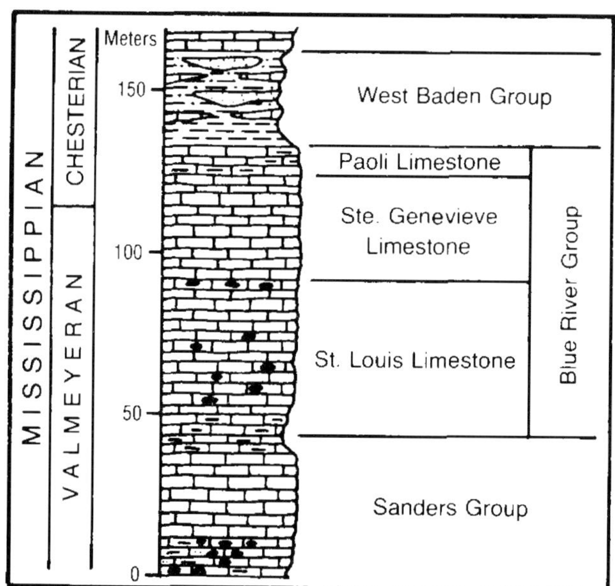

Figure 1. Middle Mississippian stratigraphic units exposed in southern Indiana. From Carbonate Seminar, 1987.

PETROGRAPHIC FEATURES OF GRAINSTONES

Eolian and marine grainstones in the sections differ in several petrographic characteristics (Table 2). In most cases these variations can be explained readily by the different modes of origin of the two rock types.

Detrital Quartz

Most eolian samples contain an appreciable amount of angular quartz silt, ranging from 0.5 to 25.5% (Figures 4, 5A). In contrast, marine grainstones rarely contain quartz silt and sand-size quartz grains (Figure 5A). One marine sample contained 0.5% quartz; all others contained none. Quartz silt does

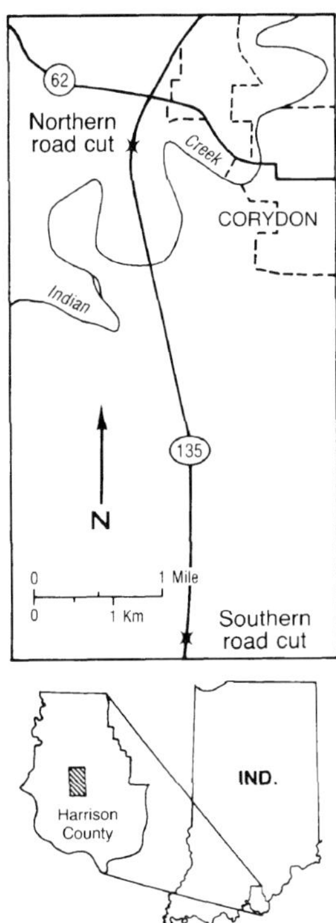

Figure 2. Locality of studied sections (marked by X). Figure modified from Hunter (this volume).

occur in some of the carbonate-mud–rich marine rocks. Although not an abundant component, scattered, well-rounded quartz sand grains occur in eolian samples.

We speculate that this difference in quartz silt and sand content results from the nature of transport of these components. Quartz silt and sand probably were transported by wind across land from a distant source. Quartz sand could be moved along the land surface by saltation, but this eolian transport of sand would stop at the shoreline. On the other hand, silt would be suspended in the air above the surface and could be transported far out to sea. Quartz silt settling into water would be winnowed from environments such as grainstone shoals with higher hydrodynamic energy. Silt would tend to collect in low-energy (muddy) environments.

Broken Ooids

Eolian samples commonly contain broken ooids, ranging from 0 to 8.5% (Figures 6, 5B). Their broken edges in many cases are well-rounded, although some still have angular edges. Broken ooids are rare in marine grainstones (Figures 7, 5B). None were encountered in point counting the sections. Where they occur, they usually have additional laminae coating the broken edges.

We are not aware of any research on the breaking of ooids, but we speculate that breakage would be more common in an eolian than a marine environment. Saltating ooids in the eolian environment

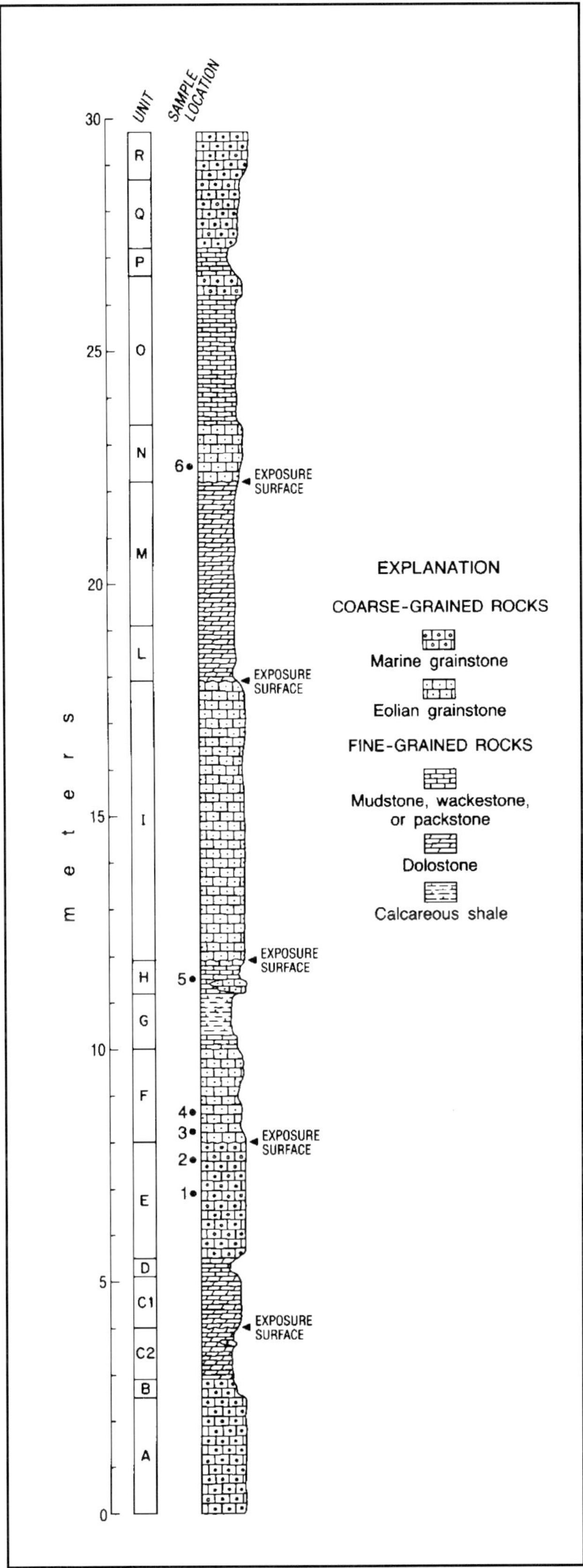

Figure 3. Measured section of Ste. Genevieve Limestone exposed at the southern road cut of Hunter (this volume). Numbered dots are sample localities for photomicrographs.

TABLE 1A. Percentage of components in eolian grainstones determined from point counting 200 grains. Samples from lithologies defined as eolian by criteria of Hunter (this volume).

SAMPLE	7-6	9-11	9-12	7-12	9-13	7-5	7-15	7-20	7-16	9-14	3-12	3-20	3-21
Echinoderm	13.0	12.0	21.0	8.0	10.0	18.5	15.5	9.5	8.0	3.0	6.0	2.0	13.0
Bryozoans	4.0	1.0	2.5	1.0		1.5	1.0	3.0	1.0	0.5	1.0	1.0	1.0
Brachiopods	0.5	0.5				0.5		0.5	0.5	0.5	1.0		
Other fossils	0.5	3.0		3.0	1.0	5.0	0.5	1.0					5.0
Total fossils	18.0	16.5	23.5	12.0	11.0	25.5	16.5	14.0	9.5	4.0	8.0	3.0	21.0
Ooids (whole)	23.0	7.0	3.0	22.0	15.0	8.0		8.0	0.5	0.5	36.0		6.0
Ooids (broken)	2.0	3.5	7.5	4.0	5.0	1.0					8.5		
Peloid	8.0	28.5	30.0	19.0	34.0	18.0	46.0	35.5	38.5	34.0	24.0	45.0	5.5
Intraclast	18.0	3.5		8.0		21.5	6.0	4.0	21.5	9.5		10.0	2.0
Quartz grains	7.0	7.5	16.0	9.0	11.0	2.0	8.0	0.5	13.5	25.5	4.0	18.0	9.5
Cement	15.0	24.0	17.0	25.0	14.0	24.0	16.0	36.0	4.5	17.5	16.5	23.0	26.0
Micrite	3.0	7.0	3.0	1.0	5.0		3.0	1.0	12.0	5.0			
Opaque					1.0		0.5			2.5		1.0	
Other grains	6.0	2.5			4.0		4.0	1.0		1.5	3.0		
TOTAL	100.0	100.0	100.0	100.0	100.0	100.0	100.0	100.0	100.0	100.0	100.0	100.0	100.0

TABLE 1B. Percentage of components in marine grainstones determined from point counting 200 grains. Samples from units defined as marine by Hunter (this volume).

SAMPLE	9-8	7-10	9-16	7-18	7-19	9-2	9-15	3-15	3-24
Echinoderm		2.5	5.5	1.5	58.5			1.0	7.5
Bryozoans			0.5		1.5			1.0	47.0
Brachiopods		0.5			1.5				1.0
Other fossils	2.0	6.0	1.0	2.0	1.0		1.5	0.5	2.0
Total fossils	2.0	9.0	7.0	3.5	63.5		1.5	2.5	57.5
Ooids (whole)	48.0		44.0	35.0	8.5		40.5	53.0	
Ooids (broken)									
Peloid	17.0	42.5	16.0	30.5	1.0		24.5	8.0	3.0
Intraclast	2.0	6.5	1.0		0.5	50.0	4.5		
Quartz grains									
Cement	30.0	32.5	31.0	31.0	26.5	33.0	29.0	36.0	38.5
Micrite	1.0	8.0	1.0			17.0			
Opaque		1.0							
Other grains		0.5						0.5	1.0
TOTAL	100.0	100.0	100.0	100.0	100.0	100.0	100.0	100.0	100.0

TABLE 2. Comparison of properties of eolian and marine grainstones in Ste. Genevieve Limestone at Corydon.

Property	Eolian	Marine
Detrital quartz	present	absent
Broken ooids	present	absent
Grain diversity	high	low
Rounding and sphericity	tabular grains have high sphericity	tabular grains rounded, not spherical
Intraclasts	well-lithified	poorly lithified
Laminations	present	absent
Sorting	poor; may be good within laminae	generally good
Inverse grading	present	absent
Large grains	absent	present
Solution packing	extensive	minor

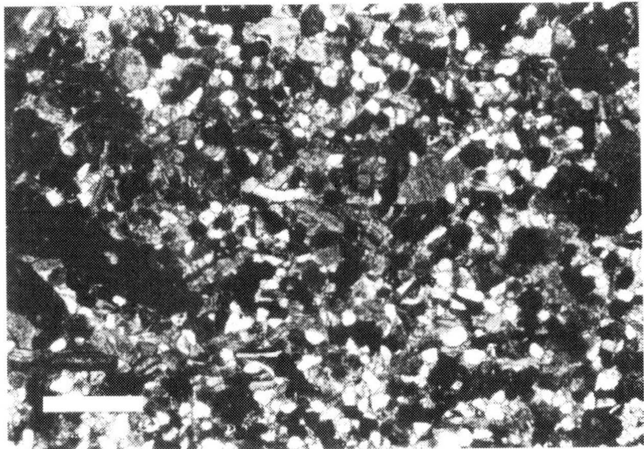

Figure 4. Abundant quartz silt (white, angular grains) in eolian grainstone (sample 6 in Figure 3). Note close packing of grains. Scale bar 0.5 mm.

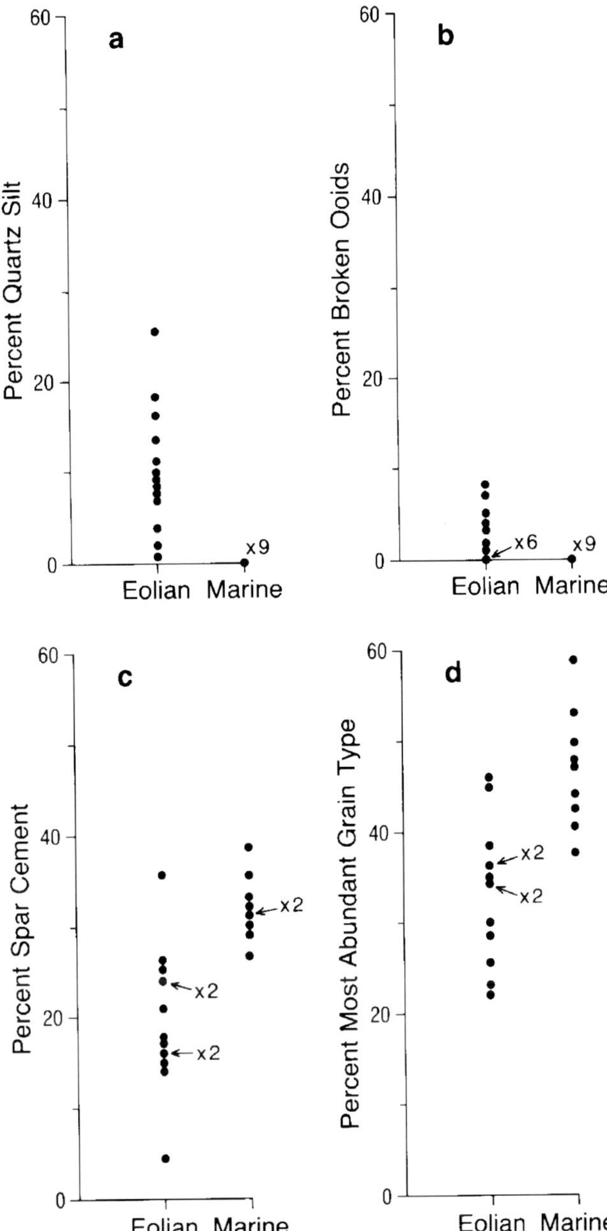

Figure 5. Comparison of abundance of (A) quartz silt, (B) broken ooids, (C) spar cement, and (D) most abundant grain type in the sample for eolian vs. marine grainstones. Percentages were determined by point counting (200 points) and are proportional to the total volume of the sample.

would impact other grains with more kinetic energy than they would in water (Pettijohn et al., 1987) and would occasionally break ooids and other grains. Also, because the viscosity of air is less than that of water, the cushioning effect on grains during eolian impact is less than in water. Wind-transported grains are therefore more likely to break on impact than are grains transported in water. Further abrasion during saltation would tend to round the broken grain edges and produce grains like those observed in eolianite units. If ooids broke in the marine environment, additional ooid coatings would generally be precipitated over broken surfaces.

Grain Diversity

The grain types in eolian samples are diverse, and usually no particular kind of grain dominates (Figure 5D). Ooids, skeletal grains, peloids, and intraclasts are all common in the samples, although their relative proportions vary. Diversity of grain types in marine samples is usually lower, and dominance by single grain types is higher in the marine than in the eolian units (Figures 7, 5D). In many samples ooids and coated grains constitute a large portion of the sample. In other cases skeletal grains are the dominant constituent. Peloids dominate in a few samples. With two exceptions, the most abundant grain types in eolian samples make up less than 40% of the total sample. On the other hand, with two exceptions, the most abundant grain types in marine samples comprise more than 40% of the sample.

Grain types are more diverse in eolian samples probably because winds blew over a variety of marine deposits exposed by lowering of sea level. Skeletal shoals, ooid shoals, tidal flats, and lagoons could act as sources for varied grains in carbonate eolianites. In contrast, grains in marine grainstones are usually derived locally. In an environment suitable for ooid formation, most grains are likely to be ooids. Skeletal grains or pellets might dominate in environments in which ooids are not forming.

Rounding and Sphericity

Although we have no numerical data on grain roundness and sphericity, our observations indicate

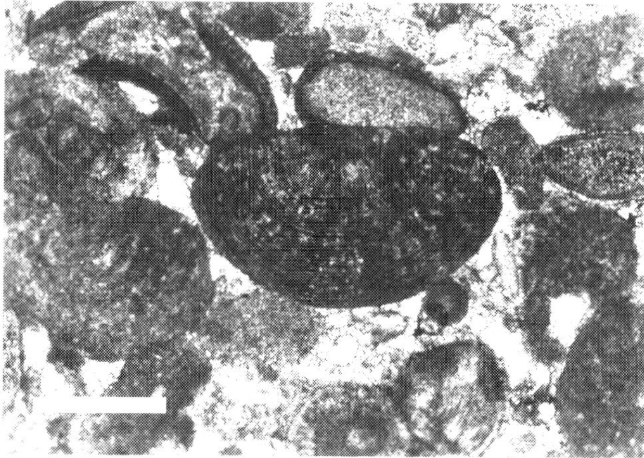

Figure 6. Broken, abraded ooid and solution at grain contacts in eolian grainstone (sample 4 in Figure 3). Scale bar 0.25 mm.

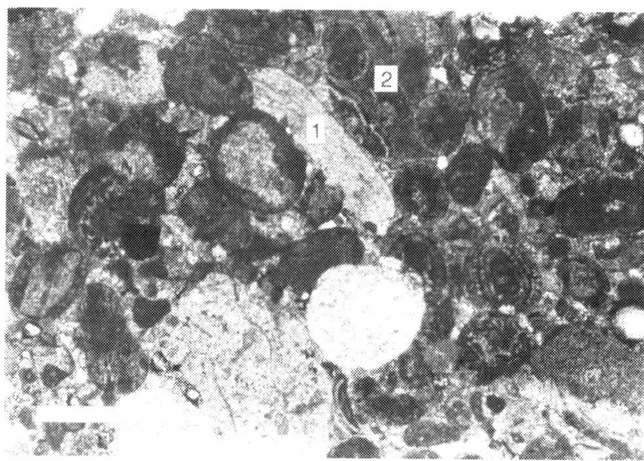

Figure 8. Very well-rounded grains with high sphericity in eolian grainstone. Note rounding of brachiopod fragment (1) and intraclast (2) and close packing of grains (sample 5 in Figure 3). Scale bar 0.5 mm.

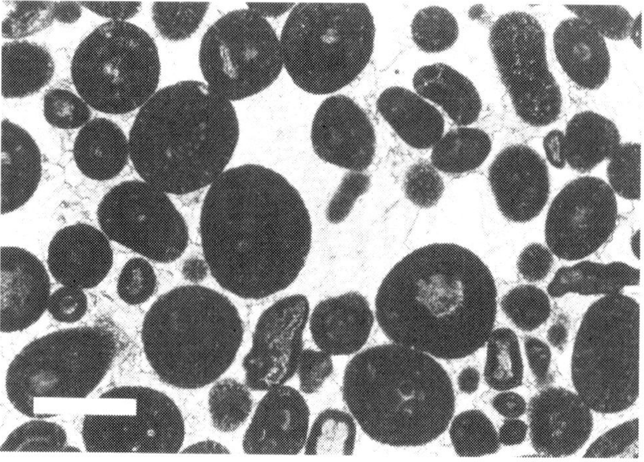

Figure 7. Sample composed almost entirely of ooids or grains with oolitic coatings in marine grainstone (sample 2 in Figure 3). Note loose packing of grains. Scale bar 0.5 mm.

that most sand-size grains in eolian grainstones are well-rounded and spherical to subspherical. This applies to grains that were originally tabular, such as fragments of brachiopod and ostracod shells (Figure 8). Most grains in the marine units also are well-rounded, but originally tabular grains retain their shape (Figure 9).

Excellent rounding has long been known as a characteristic of grains in siliciclastic eolian deposits (Kuenen, 1960; Glennie, 1970). Many carbonate grains such as ooids are naturally spherical. Although broken edges of tabular skeletal fragments such as brachiopod or mollusk shells commonly are rounded in marine deposits, the abrasion process generally does not produce spherical grains. This contrasts with carbonate eolianites in the Ste. Genevieve in which broken grains are commonly abraded to highly spherical shapes.

Intraclasts

The eolian grainstones contain several examples of intraclasts that are well-rounded and abraded equally across constituent grains and cement (Figure 10). We found one example of an intraclast that was partially silicified before being reworked into its final depositional site. Intraclasts in marine units are not so well-rounded as those in eolian units (Figure 11). We found no clear marine examples in which originally hard grains (ooids or skeletal fragments) within the intraclast had been severely abraded.

Our observations suggest that intraclasts in marine environments were not well-lithified. Thus, hard skeletal grains and ooids would not abrade as muchh as softer carbonate mud matrix. In carbonate eolianites, intraclasts may have been well-lithified so that matrix, skeletal grains, and ooids abraded to the same extent as the matrix. This suggests that in some eolianites lithification occurred soon after subaerial exposure of the eolian carbonate sediments. Some intraclasts have been partially silicified, indicating that silicification also may have occurred early near the exposure surface.

Lamination

All eolian samples are laminated at a thin-section scale (Figure 12). However, marine samples in this study are not laminated at this scale.

Lamination is produced by eolian processes (see Hunter, this volume) and is normally not destroyed by biogenic processes. Although obvious burrows are not common, we speculate that any lamination produced in marine grainstones was destroyed by burrowing organisms.

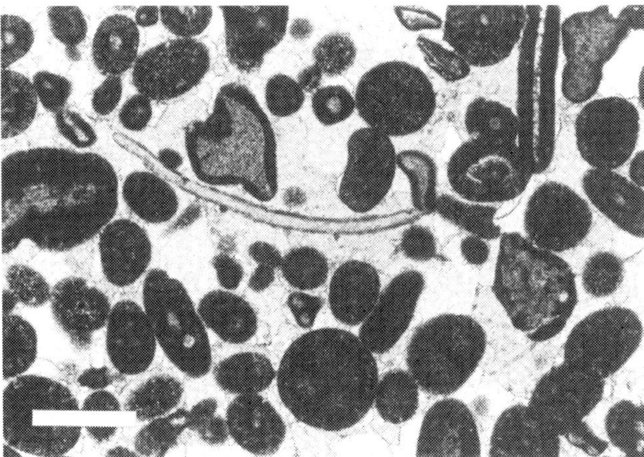

Figure 9. Tabular brachiopod shell (center) with rounded edges but low sphericity in marine grainstone (sample 2 in Figure 3). Note loose packing and preponderance of ooids and coated grains. Scale bar 0.5 mm.

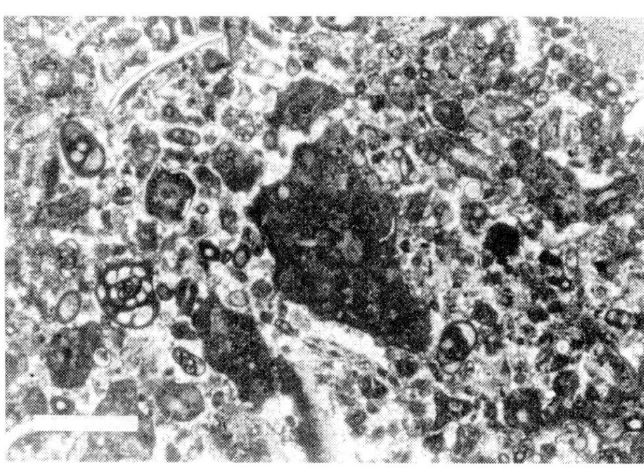

Figure 11. Poorly sorted marine grainstone with poorly rounded intraclast showing no abrasion of constituent grains (sample 1 in Figure 3). Scale bar 0.5 mm.

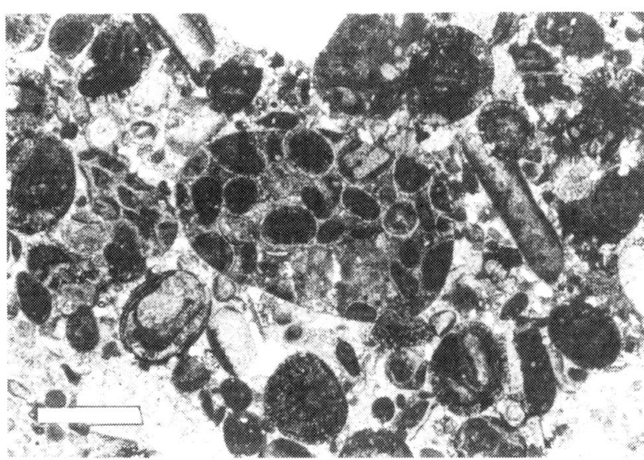

Figure 10. Spherical intraclast in eolian grainstone showing equal abrasion across constituent grains and cement (sample 5 in Figure 3). Scale bar 0.5 mm.

Sorting

Although individual laminae in eolian grainstones are well-sorted, grain-size variation in a given thin section is considerable. Sorting in marine grainstones is variable, but some samples, especially the oolitic grainstones, may be well-sorted, even better-sorted than eolian grainstones. Compare Figures 7 and 9 (typical well-sorted marine grainstones) with Figures 4, 6, 8, 10, and 12 (eolian grainstones).

Eolian processes produce good sorting within individual laminae in carbonate eolianites (Glennie, 1970), but grain size may differ considerably between laminae due to different wind velocities associated with each laminae. Marine grainstones examined in this study tend to be more uniformly sorted. A relatively narrow range of grain sizes due to derivation from a dominant source (e.g., ooids) probably contributes to good sorting in marine grainstones.

Inverse Grading

As noted by Hunter (this volume), laminae in eolian grainstones show inverse grading in places, with coarser grains at the top of the laminae (Figure 12); however, inverse grading was not observed in marine grainstones.

Inverse grading is produced by eolian processes in translatent strata formed by climbing ripples (Hunter, this volume). Under some conditions, marine beach laminations may have inverse grading (Clifton, 1969). The marine grainstones in the Corydon section are not laminated and hence cannot show inverse grading within laminations.

Large Grains

Large grains (greater than 4 mm) are absent from eolian grainstones but are found in the marine grainstones. In many cases these large grains are fossils of organisms that probably lived in situ in the sediment. Some large intraclasts also occur in the marine units.

Gravel-sized grains are too large to be carried by wind (McKee and Ward, 1983), but large grains can be transported by marine currents or waves. Also, the most common source of large grains in carbonate rocks is in-situ production, especially of large skeletons. Fossils larger than 4 mm are moderately common in a well-sorted marine grainstone matrix in marine intervals such as units E, H, O, and R.

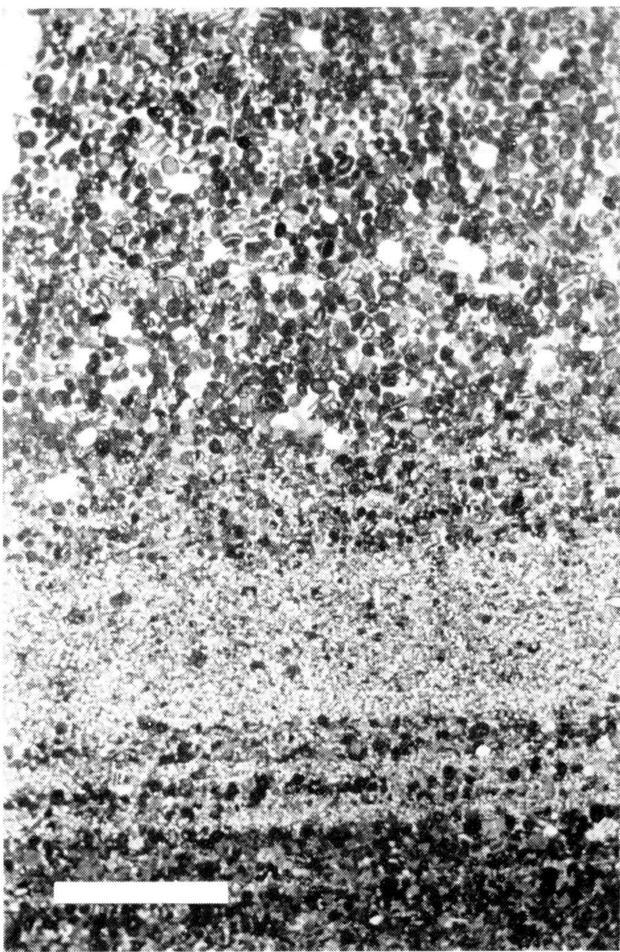

Figure 12. Laminated eolian grainstone (sample 3 in Figure 3). Note reverse grading in thick upper lamination. White grains are quartz. Scale bar 4 mm.

Solution Packing

Grain packing is tighter in the eolianite samples than in the marine units in this section. Solution and interpenetration of grains is common along grain contacts (Figures 4, 6, 8, 10). On the other hand, marine grainstones usually show minimal packing (Figures 7, 9). This is reflected in the higher proportion of spar cement in marine samples than in eolian samples (Figure 5C). Marine cement is not developed especially well in the marine grainstones, but thin isopachous rims of probable marine cement do occur.

Early vadose cementation might be expected to prevent solution packing in eolian units, and the presence of well-cemented intraclasts in the eolian units suggests that some early cementation does occur, but only to a limited extent. We have seen little evidence for vadose cements except in a few places immediately below exposure surfaces. We speculate that this is because the climate was very dry and little vadose water remained in the pores. Less solution packing in marine units suggests that they were cemented early, probably in the marine environment. Thick marine cement is not obvious in these samples, but thin isopachous coatings are present.

ORIGIN OF CARBONATE EOLIANITES

Two possible models can be proposed to explain the eolian deposits:

1. Marine waves and currents may have built shoal areas above sea level and continued to supply marine-derived grains above sea level. Once exposed, eolian processes could have caused the islands to expand both vertically and laterally (Figure 13).
2. Sea level may have fallen, exposing broad areas of carbonate sediment. Winds blowing across these exposed areas could have produced dunes and eolian deposits (Figure 14). Sea level fall may have resulted from either a worldwide eustatic fall or more local or basin-wide tectonism.

A number of tests should allow selection of the correct model (Table 3):

Correlation of Eolian Units. If eolian grainstones formed on relatively local islands (model 1) they should not correlate over large distances. If they formed by a general lowering of sea level (model 2), they should correlate over longer distances. Data collected to date by Hunter (this volume and personal communication, 1989) and Merkley (1991) suggest that eolian units can be traced over an area of a few hundred square kilometers. More research is needed to determine the extent of individual units and to adequately apply this test; however, preliminary results tend to support model 2.

Geographic Distribution. If model 1 is correct, eolian units should be found scattered throughout the basin wherever current and wave conditions could form islands. Model 2 would suggest either basin-wide occurrence of eolian units or occurrence near the basin margin. Although more research is needed, especially in the subsurface, to determine the extent of eolian deposits, present data (Hunter, this volume) suggest that the deposits may be limited to the basin margins, although islands might be concentrated around the basin margins as well.

Number of Units. Model 1 could produce a variable number of eolian units in a given area, whereas model 2 should produce the same number of eolian units, one corresponding to each lowering of sea level. If the entire basin was not exposed when sea level fell, the number of eolian units would systematically decrease toward the center of the basin. The number of units in a given section varies from four to zero, in general decreasing to the north and west. With the present state of our knowledge, this test is not definitive.

Associated Facies. If model 1 is correct, eolian units should usually overlie grainstone shoals. Once the eolian environment was established, it could expand beyond the originating shoal. In contrast, no

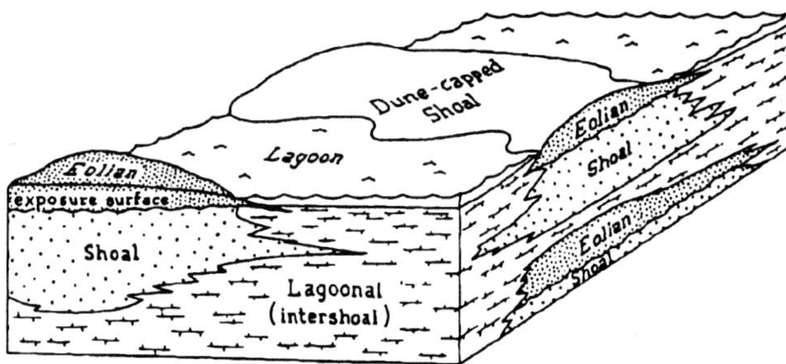

Figure 13. Schematic representation of model 1. Marine waves and currents build oolitic or skeletal shoals above sea level to form low islands. Winds blowing across these islands form dunes and eolian deposits.

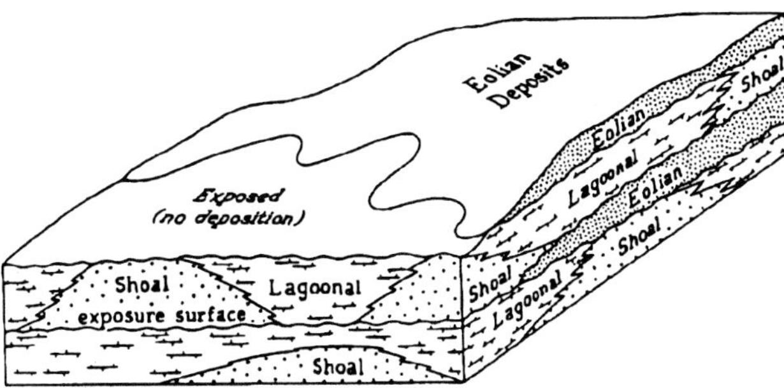

Figure 14. Schematic representation of model 2. Sea level falls, exposing broad areas of carbonate sediment. Winds blowing across these surfaces form dunes and eolian deposits.

association with a particular underlying facies is required in model 2 because the entire spectrum of originally marine environments would be exposed when sea level fell. In the Corydon exposures, eolian unit F overlies a marine grainstone shoal unit, whereas units I, K, and N overlie muddy, lagoonal units. Thus, model 2 best explains the observations.

Grain Variety. The exposed shoals in model 1 would probably have less variety in grain type than is likely in model 2. Our research in the Corydon area suggests that marine grainstone units tend to be dominated by a single grain type (Figure 5D). Eolian units derived from these marine units should also have low grain variety. As discussed above, general lowering of sea level would expose a wide spectrum of grain types, from low-energy intershoal areas as well as from shoals, for incorporation into eolian units. Grain variety in the eolian units at Corydon (Figure 5D) is best explained by model 2.

Siliciclastic Silt and Sand. Siliciclastic silt and especially sand should not be abundant in eolian units formed by the model 1 process, whereas they could be quite abundant in deposits formed according to model 2. Although silt can be transported across water by suspension in the wind, transport of sand across water to islands would be difficult. Transport by marine processes across a broad, shallow sea with quiet lagoons as well as shoals is unlikely. Transport by saltation across land is much more likely. The abundance of siliciclastic silt and sand in the eolian units supports model 2.

Many of the data needed to test these two models must come from detailed study of many basin-wide stratigraphic sections of the Ste. Genevieve Limestone and mapping of the distribution of the eolian units. Hunter (this volume) has conducted some research of this type, but much remains to be done. The last three tests are basically petrographic in nature and relate to the research we have reported here. Each of these tests would seem to support the second model. Data accumulated to date by Hunter relative to the first three tests (this volume and personal communication, 1989) also seem to be explained best by the second model. Clearly, additional work should be done to test these models more adequately.

Very few pre-Pleistocene carbonate eolianites have been described in the literature (McKee and Ward, 1983; Hunter, this volume). Examples from the Pennsylvanian and Permian limestones of the Grand Canyon, Arizona, region, recently described by Loope (1986), Loope and Haverland (1988), Rice and Loope (1987, 1991), and Blakey (1990), are especially well-documented. Loope and Haverland (1988) have also described an unusual occurrence of calcareous eolianite from the Pennsylvanian of southeastern Utah. Handford (1990) has described eolianites from the St. Louis Limestone (Mississippian) in the subsurface of southwestern Kansas. These deposits appear to be quite similar to those in the Ste. Genevieve, and are in fact of comparable age. The St. Louis and Ste. Genevieve boundary is not well defined in southwestern Kansas, so the ages of the St. Louis Limestone in southwestern Kansas and the Ste. Genevieve Limestone in southern Indiana may overlap. Other portions of the Ste. Genevieve Limestone in the Illinois basin are almost certainly eolian in origin, although the limits of eolian deposition have yet to be defined (Hunter, this volume). As Hunter has suggested, carbonate eolianites are probably much more common than the literature indicates. Stratigraphic sections containing shallow water carbonate deposits should be studied carefully for possible eolianite intervals.

Rice and Loope (1991) published a detailed description of carbonate eolianites from the Callville Limestone (Pennsylvanian), Manakacha and Wescogame formations (Pennsylvanian), and Pakoon Limestone (Permian) of northern Arizona and southern Nevada. These eolian carbonate units have many characteristics in common with those in the Ste. Genevieve Limestone, including: (1) abundant quartz silt and fine sand, (2) highly abraded intraclasts and other grain types, (3) fine, even laminations, (4) coarsening-upward of grains within laminae, (5) well-lithified intraclasts, (6) chert and silicified grains, (7) lack of grains larger than 4 mm in the eolianite proper, (8) pebbles at the base of eolianite units, and (9) exposure

TABLE 3. Test for comparison of model 1 (island model) and model 2 (general lowering of sea level model).

Test of Model	Model 1 (Island)	Model 2 (Lowered Sea Level)
Correlation of eolian units	only locally	correlatable over wide area
Geographic distribution	scattered distribution	only near basin margins
Number of eolian units	variable	constant or varies systematically
Associated facies	usually overlie marine grainstones	may overlie any marine lithology
Grain variety	usually low	usually high
Detrital quartz	uncommon	may be common

surfaces with rhizoliths, brecciation, and other exposure features at the base of eolianite units.

Features of these younger eolianites that differ from those in the Ste. Genevieve include: (1) more evidence for early lithification of eolian grainstones, (2) less evidence for early lithification of marine grainstones, (3) scarcity of ooids and thus of broken ooids, (4) grain-size decreasing upward in eolian units, (5) quartz concentration increasing upward in eolian units, (6) corrugation of upper surfaces of eolian cosets, and (7) unidirectional paleowind.

Some of the differences between the Pennsylvanian and Permian eolianites and the Ste. Genevieve eolianites may be due to climatic differences. A somewhat more humid climate with greater rainfall might explain more prevalent early cement and corrugated upper surfaces of cosets in the younger eolianites. Higher salinity and slightly greater supersaturation with $CaCO_3$ in the sea water may explain more common ooids and early marine cementation in the older Ste. Genevieve grainstones. Differences in wind patterns and source areas for carbonate and quartz grains may explain the size and quartz content trends as well as paleowind directions in the Pennsylvanian and Permian sections.

Although some of the criteria used to distinguish between eolianite and marine grainstones in the Ste. Genevieve Limestone should be generally applicable to other units, some will not be diagnostic because they are specific to the depositional setting of the Ste. Genevieve. For example, differences in solution packing will likely vary depending on climate and the degree of saturation with $CaCO_3$ of the marine waters. The diversity of grains in eolianites may be a frequently useful criterion, but if the source area is small and uniform in composition, grains in eolianites may not always be diverse. Under certain conditions marine processes may produce better laminations than those found in the Ste. Genevieve Limestone at Corydon. Inverse grading can even form under marine processes under appropriate conditions. Quartz silt and sand in the Ste. Genevieve eolianites are among the most reliable criteria for identifying eolian units, but these criteria would not apply if no quartz is available in the source area. Other criteria may be found for other units that are distinctive for a particular unit but that may not apply elsewhere. None of the petrographic criteria are absolutely diagnostic. The weight of all petrographic and sedimentologic data should be combined to increase confidence in the eolian interpretation.

If carbonate eolianites such as those in the Ste. Genevieve Limestone form as a result of general lowering of sea level (model 2), their distribution can be a useful way to document sea level history. Especially valuable would be a study of the co-occurrence of eolianites and exposure surfaces. Association of eolianites with these surfaces clearly documents their subareal origin. Exposure surfaces are likely to be more extensive than any associated eolianites. Eolian deposition can produce discontinuous deposits, but exposure conditions should leave a record wherever the surface was exposed to subareal conditions. If eolianites and exposure surfaces result from eustatic lowering of sea level, and if particular deposits and surfaces can be uniquely identified, they would be valuable tools for correlation.

Many questions remain to be answered in connection with the Ste. Genevieve eolianite deposits. How extensive are they? How far can they be traced along the outcrop belt? Can they be traced into the subsurface? How do they vary geographically? Are particular eolianite units distinctive enough to allow their regional correlation? What is the source of the quartz sand and silt in the eolianites? What are the details of the depositional setting of the eolianites? What was their diagenetic history? How general are the criteria developed for these particular carbonate eolianites?

CONCLUSIONS

Grainstones of apparently eolian origin in the Ste. Genevieve Limestone exposed near Corydon, Indiana, can be distinguished from marine grainstones in the same section based on a number of petrographic criteria, including: presence of quartz silt, greater abundance of broken ooids, diverse grain assemblage, spherical grains formed from originally non-spherical precursors, well-lithified and rounded intraclasts, fine laminations, good sorting within laminae, coarsening-upward of grains within laminae, lack of grains larger than 4 mm, and moderate-to-intense solution packing. Petrographic features of the presumed carbonate eolianites are compatible with an eolian origin and strengthen conclusions reached on the basis of sedimentary structures. The eolian units probably formed as a result of periodic eustatic sea level lowering.

ACKNOWLEDGMENTS

The authors appreciate the many helpful suggestions and comments from Ralph E. Hunter over the entire span of this investigation, as well as his critical review of the manuscript. We also thank Curtis H. Ault, Brian D. Keith, Stanley J. Keller, and David B. Loope for their critical reviews.

REFERENCES CITED

Blakey, R.C., 1990, Stratigraphy and geologic history of Pennsylvanian and Permian rocks, Mogollon Rim region, central Arizona and vicinity: GSA Bulletin, v. 102, p. 1189-1217.

Carr, D.D., 1973, Geometry and origin of oolite bodies in the Ste. Genevieve Limestone (Mississippian) in the Illinois basin: Indiana Geological Survey Bulletin 48, 81 p.

Carbonate Seminar, 1987, Ramp Creek and Harrodsburg Limestones: A shoaling-upward sequence with storm-produced features in southern Indiana, U.S.A.: Sedimentary Geology, v. 52, p. 207-226.

Clifton, H.E., 1969, Beach laminations: Nature and origin: Marine Geology, v. 7, p. 553-339.

Droste, J.B. and G.L. Carpenter, 1990, Surface stratigraphy of the Blue River Group (Mississippian) in Indiana: Indiana Geological Survey Bulletin 62, 45 p.

Glennie, K.W., 1970, Desert sedimentary environments: Developments in Sedimentology 14: Elsevier Scientific Publishers, Amsterdam, 222 p.

Handford, C.R., 1990, Mississippian carbonate eolianites in southwestern Kansas (abs.): AAPG Bulletin, v. 75, p. 669.

Hunter, R.E., 1988, Eolianites in the Ste. Genevieve Limestone (Mississippian) of southern Indiana (abs.): SEPM Midyear Meeting Abstracts, v. 5, p. 26-27.

Hunter, R.E., 1989, Eolianites in the Ste. Genevieve Limestone of southern Indiana: Field Trip Guidebook, AAPG Eastern Section Meeting, 1989 (reprinted by Indiana Geological Survey) p. 1-19.

Kuenen, Ph.H., 1960, Experimental abrasion, 4: eolian action: Journal of Geology, v. 68, p. 427-449.

Leibold, A.W., 1982, Stratigraphy, petrography, and depositional environment of the Bryantsville Breccia (Meramecian) of south-central Indiana: A.M. thesis, Indiana University, 171 p.

Loope, D.B., 1986, Pennsylvanian eolian limestones, Paradox Basin, U.S.A. (abs.): International Sedimentological Congress, 12th, Abstracts Volume, p. 189-190.

Loope, D.B. and Z.E. Haverland, 1988, Giant desiccation fissures filled with calcareous eolian sand, Hermosa Formation (Pennsylvanian), southeastern Utah: Sedimentary Geology, v. 56, p. 1-11.

Merkley, P.A., 1991, Origin and distribution of carbonate eolianites in the Ste. Genevieve Limestone (Mississippian) of southern Indiana and northwestern Kentucky: M.S. thesis, Indiana University, 147 p.

McKee, E.D. and W.C. Ward, 1983, Eolian environments: *in* P.A. Scholle, D.G. Bebout, and C.H. Moore, eds., Carbonate depositional environments: AAPG Memoir 33, p. 132-169.

Pettijohn, F.J., P.E. Potter, and R. Siever, 1987, Sand and sandstone, 2nd ed.: Springer-Verlag, New York, 553 p.

Rice, J.A. and D.B. Loope, 1987, Pennsylvanian eolian limestones, western Grand Canyon and southern Nevada (abs.): GSA Abstracts with Programs, v. 19, p. 818.

Rice, J.A. and D.B. Loope, 1991, Wind-reworked carbonates, Permo-Pennsylvanian of Arizona and Nevada: GSA Bulletin, v. 103, p. 254-267.

Chapter 5

Recognition on Wireline Logs and Mapping of Oolitic Facies in a Carbonate Sequence, Ste. Genevieve Limestone, Illinois Basin

William F. Bandy, Jr.
Department of Earth Sciences
Vincennes University
Vincennes, Indiana, USA

ABSTRACT

Oolitic facies of the Ste. Genevieve Limestone are common and prolific hydrocarbon reservoirs in the Illinois basin. However, development and exploration efforts are often hampered by the inability to recognize the reservoir on wireline logs. A comparison of cores with wireline logs in northern Lawrence field, Lawrence County, Illinois, indicates that a subjective but predictable relationship exists between carbonate lithology and log character. Mapping of the oolitic facies to determine orientation, geometry, and extent can be accomplished in areas where wireline logs are the only data available.

INTRODUCTION

Oolitic facies of the Ste. Genevieve Limestone (Mississippian) are important hydrocarbon reservoirs in the Illinois basin. Exploration and development for these reservoirs, however, are frequently hampered by a lack of data. Commonly, the only logs available for study are electrical resistivity surveys of the pre-1960s era (typically a 16″ short normal, 64″ long normal, 18′8″ lateral, and spontaneous potential measurements) or modern dual induction logs.

Many operators have failed to run any other type of log or run no logs at all. Other types of data, such as sample or core descriptions, may be hard to locate or were never performed. The purpose of this paper is to demonstrate the relationships between ooid lithology and log character, and point out some development and exploration applications. Recognition of carbonate lithology from old electric logs is crucial, for it may enable the geologist to define and correctly map reservoirs, determine favorable well locations, predict reservoir trends, and even design well completion procedures.

GEOLOGIC SETTING

The area of study is Lawrence field, Lawrence County, Illinois, located in the east-central part of the Illinois basin (Figure 1A). Since its discovery in 1906, Lawrence field has produced nearly 400 MMbbl of oil from 23 different pay zones (Huff, 1987). The majority of the field is located along the crest of the LaSalle anticlinal belt, a northwest–southeast-oriented series of structures on the northeast flank of the basin.

The Ste. Genevieve Limestone is a sequence of late Valmeyeran carbonates (Willman, et al., 1975; Keith and Zuppann, 1993) deposited in a shallow water marine environment (Droste and Carpenter, 1990). A shelf sloping gently to the southwest characterized

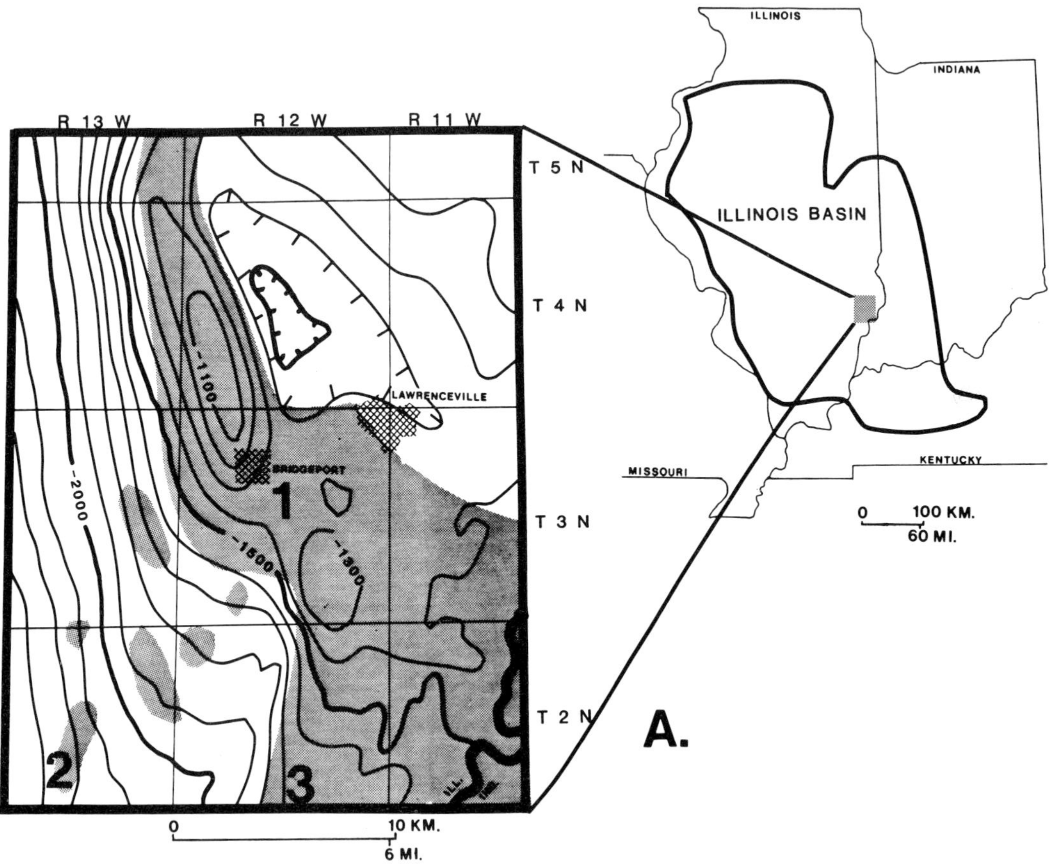

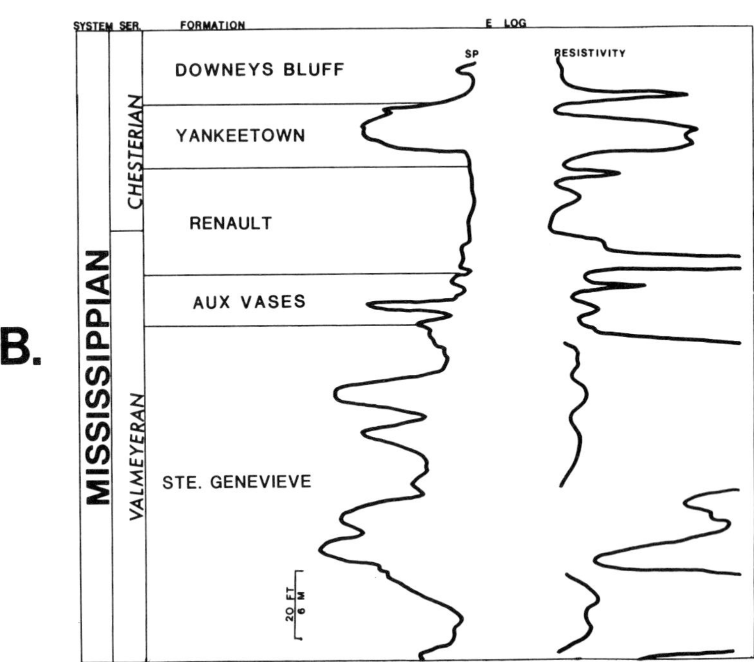

Figure 1. (A) Location of study area in the Illinois basin. Outline of basin is defined by a line marking the maximum extent of Pennsylvanian-age rocks in the basin. The structure map covers western and central Lawrence County, Illinois. Structure is on top of the Ste. Genevieve Limestone, using a contour interval of 100 ft (30 m). Gray areas are areas of oil production. Locations referred to in this report are: 1—Lawrence field, 2—northern Lancaster field, and 3—Allendale pool. (B) Schematic of electric log for upper Valmeyeran and lower Chesterian rocks in Lawrence County, Illinois.

the basin during deposition (Swann, 1963). In Lawrence field, the productive reservoirs, locally termed "McClosky" by drillers, are oolitic grainstones and microcrystalline dolomites. Choquette and Steinen (1980, 1985) have described these reservoirs in detail, including their diagenesis, cements, porosity and permeability, and log character. Early in the history of the field, initial production from oolitic grainstones was often gauged in thousands of barrels per day (Blatchley, 1913). Modern production is gauged in tens of barrels per day or less. Both oolitic and dolomitic reservoirs have been successfully waterflooded.

RECOGNITION OF THE OOLITIC FACIES ON LOGS

For a majority of wells in the Illinois basin, an electric log or an induction log is the only geophysical log available. Carbonate lithologies often prove difficult to recognize strictly from spontaneous potential (SP) and resistivity curves alone. Schematics or examples of electric logs through the Ste. Genevieve section (Figure 1B) have been provided by other authors, including Carr (1973), Caserotti (1983), Choquette and Steinen (1980, 1985), and Seyler (1986).

A comparison between cores and logs of 41 wells in northern Lawrence field (locations on Figure 2) revealed a subjective but reliable relationship between core lithology and log character. In general, both oolites and dolomites display SP deflection where permeability is present. However, no discernible differences or patterns in SP character occur that would distinguish oolites from dolomites (Figure 3). The resistivity curve, however, does respond to differences in lithology, owing primarily to differences in porosity (Figure 4). While resistivity is highest in dense limestone, it is typically lower in oolitic grainstone and lowest in dolomites. Oil-saturated dolomites display some resistivity (Figure 3A), but still maintain the lowest resistivity of any oil-saturated pay in the carbonate section. Shales likewise display low resistivity, but can be differentiated from dolomites by their lack of SP development.

The low resistivity of dolomites is attributable to several factors. Dolomite porosity in Lawrence field averages 27%, nearly twice the porosity of oolites, which average 13.7% (Figures 3A, 4). In addition, the tortuous microporosity system present in the dolomites may trap large amounts of interstitial water (Choquette and Steinen, 1985). Collectively, these features reduce the resistivity response.

Illite and chlorite constitute part of the HCl insolubles in the dolomite, which total less than 7.5 wt. % (Choquette and Steinen, 1980). Seyler and Beaty (personal communication) have shown that illite, chlorite, and mixed-layer illite-smectite, though present in amounts of only 2–5%, can effectively reduce resistivity in the Aux Vases Sandstone. Hence, the presence of illite and chlorite may contribute to the resistivity reduction in dolomites.

Using electric logs to identify oolites is not without pitfalls. A low porosity dolomite may exhibit higher resistivity and be mistaken for an oolite. In Figure 3B, a thin dolomite with a porosity of 11% lies underneath a thin oolite with a porosity of 3%. Both lithologies display high resistivity, owing to the low porosity. At first glance, an observer may conclude that only oolite is present. However, a core of this interval indicates both oolite and dolomite. A closer examination of this interval on the electric log shows a subtle difference in shallow (16″) resistivity between the two lithologies, with the dolomite displaying the lower resistivity. Conversely, oolites of unusually high porosity or water saturation may display the low resistivity of a dolomite. For example, Figure 3C shows an oolite with a porosity of over 20% and a water saturation of over 80%, producing a poor resistivity. An observer unfamiliar with Lawrence field would not know this oolite was waterflooded and would mistakenly conclude that the lithology is a thick dolomite. These examples, however, are the exception rather than the rule in Lawrence field.

Carbonate lithology of reservoirs can be distinguished more easily on gamma-ray porosity logs, but these logs are much less common for wells in the Illinois basin. The difference in porosity between oolitic grainstones and dolomites can be seen clearly on a density-neutron log, as shown in Figure 3A. The gamma ray curve, while less reliable, can serve as a double-check on lithology. Since oolitic grainstone is a carbonate sand, it generally displays a "clean" or low gamma ray (Figure 3A). In comparison, dolomites will display more radioactivity, probably because the higher porosities result in less absorption of gamma rays. The difference is usually subtle, on average 10–15 API units.

DEVELOPMENT APPLICATIONS

Offsetting a Successful Discovery

Following the discovery of oil in an oolitic grainstone, operators often drill several dry holes in a frustrating attempt to determine the orientation of the reservoir. This frustration and expense could be avoided with the use of a modern dipmeter for stratigraphic interpretation. Modern dipmeters may provide sufficient information to determine depositional environment and orientation of a reservoir. Unfortunately, few dipmeters have been run in wells in the Illinois basin. One example of a high-resolution dipmeter run by Schlumberger is from the Petroventure #1 Curlock, Belle Rive Southwest field, Jefferson County, Illinois (Figure 5A). The well penetrated oolitic grainstone at 3125- to 3136-ft (953- to 956-m) depths. The following interpretations were made from the dipmeter (Figure 5B).

1. The drape at the top of an oolite bar dips toward the pinch-out edge of the bar, because the bar is convex upward. The strike of the bar is oriented normal to this dip. Dips are

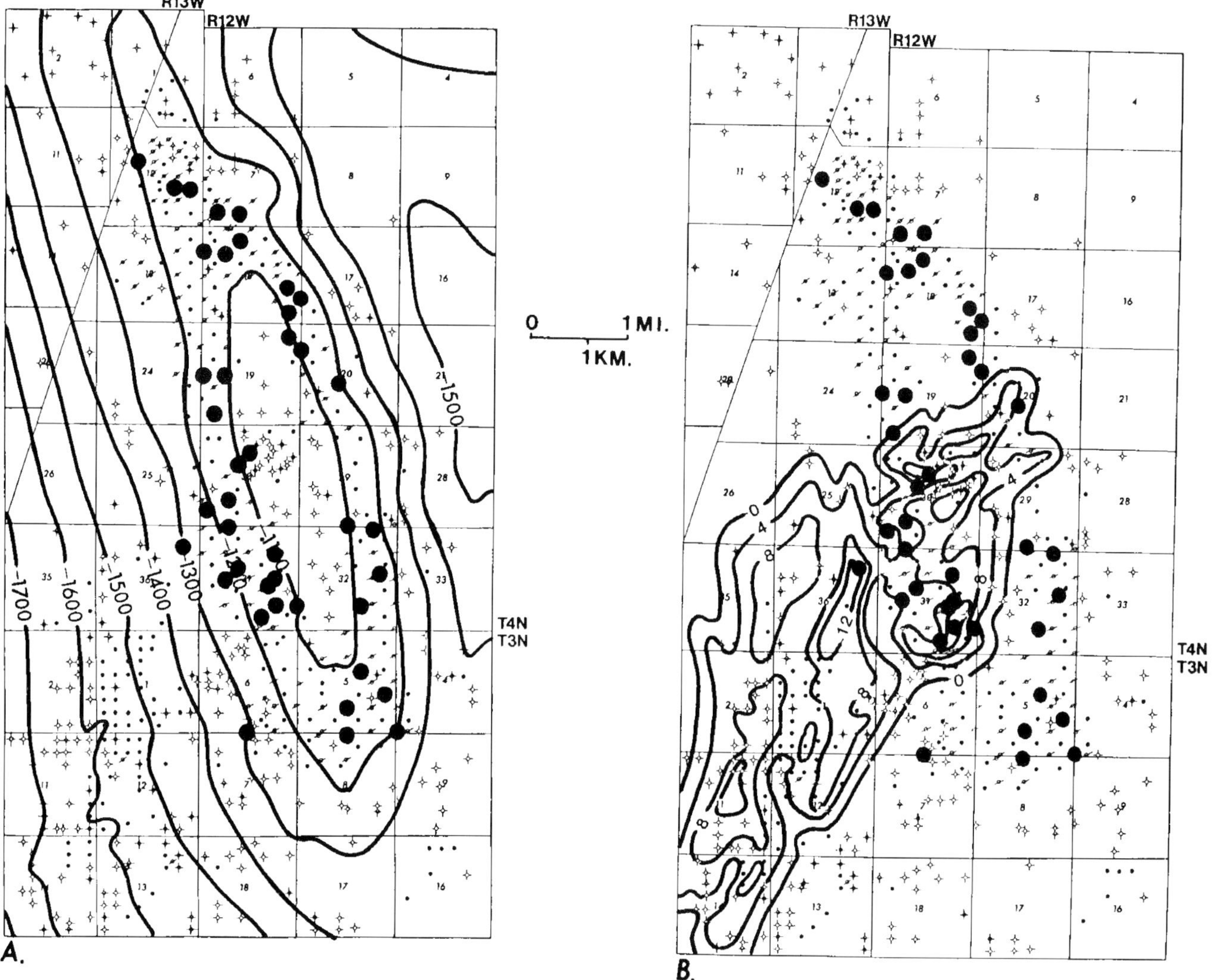

Figure 2. (A) Structure on top of the Ste. Genevieve Limestone, northern Lawrence field, Lawrence County, Illinois. Contour interval is 100 ft (30 m). The well symbols represent only Ste. Genevieve-depth wells and production. Black dots indicate the location of cores that were studied in this report. (B) Isopach map of oolite bar in northern Lawrence field, Lawrence County, Illinois. Contour interval is 4 ft (1.2 m). The well symbols represent only Ste. Genevieve-depth wells and production. Note north–south orientation of thicker reservoir, separated by thinner areas that have poor permeability. Black dots indicate the location of cores that were studied in this report.

toward the southeast (S40°E), so the bar axis is oriented northeast–southwest (N50°E).

2. Depending on which side of the bar a particular well penetrates, dips at the top of the bar could be either southeast or northwest, either indicating a northeast–southwest orientation for the bar. Since the dips are to the southeast, the well must be located on the southeast flank of the bar. The central portion of the bar should be northwest of the well.

Drilling proved this interpretation to be reasonably accurate, and successful offset wells were drilled to the northeast, northwest, and southwest. The bar axis occurs slightly northwest of the Curlock #1 (Dean Hollensbe, personal communication). More modern dipmeter tools, such as the SHDT Dual Dipmeter of Schlumberger, provide more data and greater accuracy than older tools, owing to a better ability to maintain pad contact with the borehole and the capacity to generate up to eight resistivity curves (Schlumberger, 1987).

Isopach Maps from Electric Logs

Once an oolitic grainstone is recognized, construction of an isopach map is relatively simple. Such maps have been constructed by other authors, for example Whiting (1959), Carr (1973), Choquette and Steinen (1980, 1985), Hollensbe and Tucker (1988), Norris (1988), and Handford (1988). In a carbonate section, the SP curve is a poor indicator of actual porous oolite, since it may be the collective result of

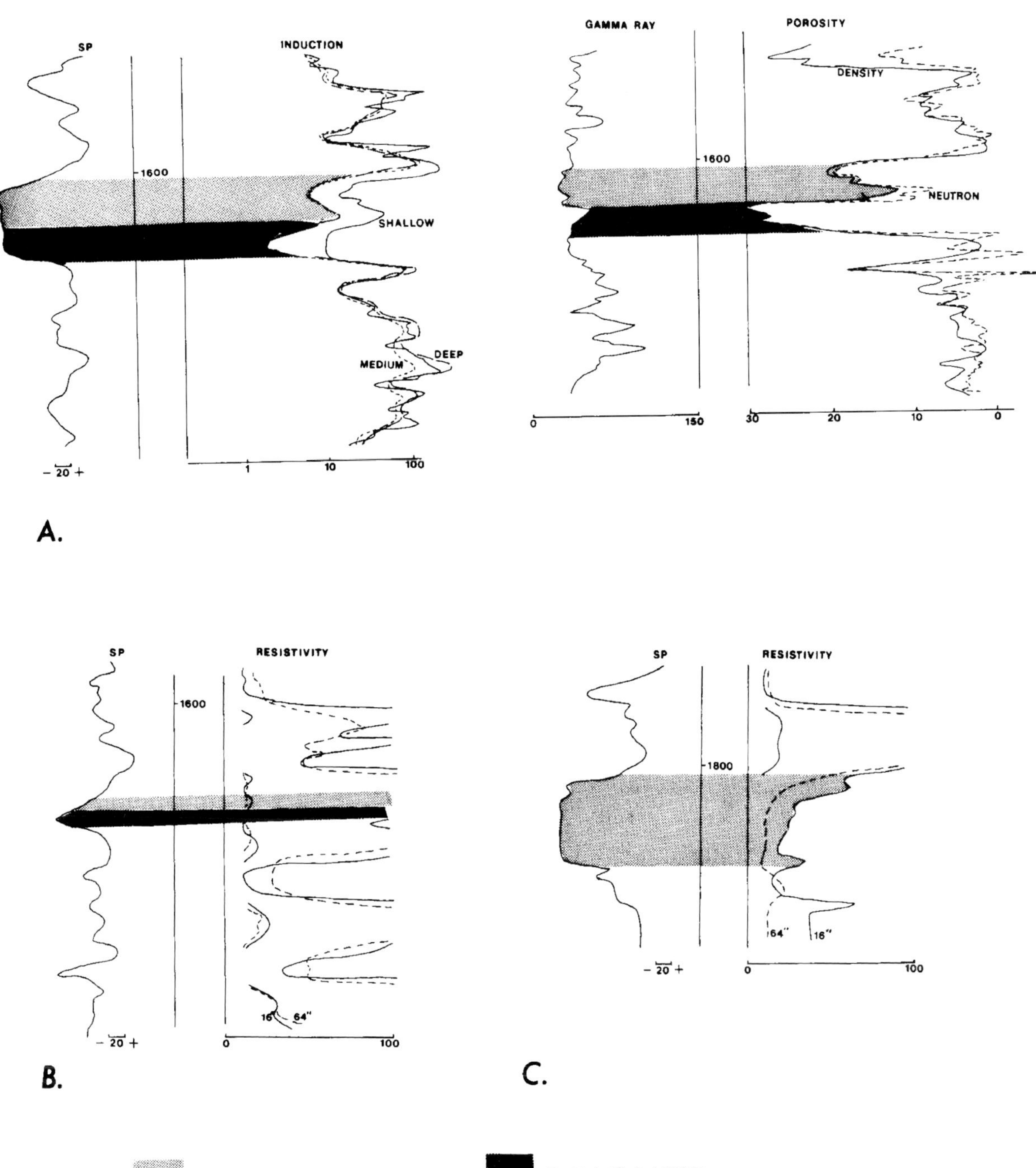

Figure 3. Log examples of the Ste. Genevieve Limestone in Lawrence field, Lawrence County, Illinois. (A) Dual induction and gamma-ray porosity logs for Marathon Oil #153 Perry King a/c 1 and 2, Sec. 18, T4N, R12W. Dual induction log illustrates the difference in resistivities among oolite, dolomite, and dense limestone located directly above the oolite. Gamma-ray porosity log illustrates the difference in porosity between oolite and much more porous dolomite. The gamma-ray curve records greater radioactivity in the dolomite compared to the oolite. Lithologies are based on core data. (B) Electric log for Marathon Oil #18 A.R. Applegate, Sec. 12, T4N, R13W. A single SP deflection, combined with high resistivity, can be mistaken for a single oolite. A core of this interval, however, contained both an oolite and dolomite. The change in lithology is revealed by a slight decrease in resistivity across the oolite-dolomite contact. The average porosity and permeability of the oolite is 5% and 9 md respectively. The average porosity and permeability of the dolomite is 16% and 17 md, respectively. (C) Log example of Illinois Oil #0-2 William Finley, Sec. 25, T3N, R12W, showing water-saturated oolite. This reservoir was once oil-productive, until it was waterflooded accidentally early in the history of the field. The reservoir could be misidentified as a dolomite, owing to the low resistivity. The oolite lithology is confirmed by nearby core and sample data.

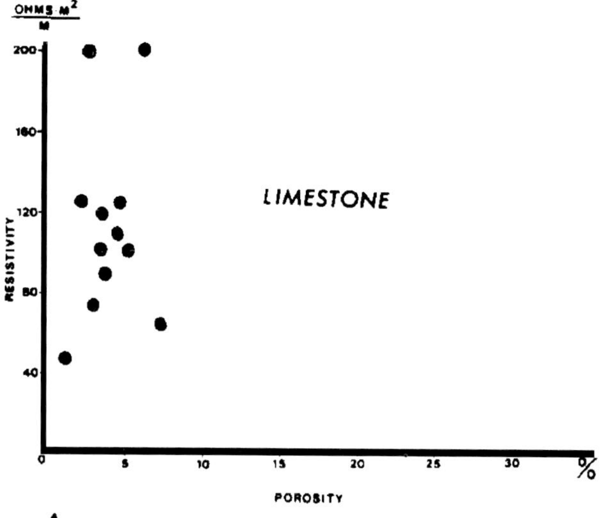

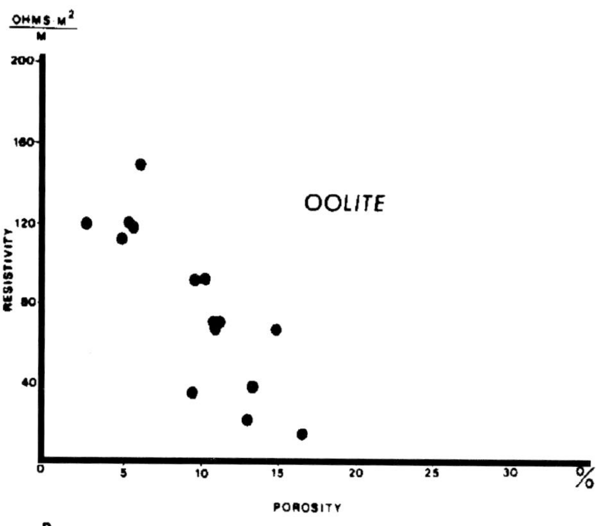

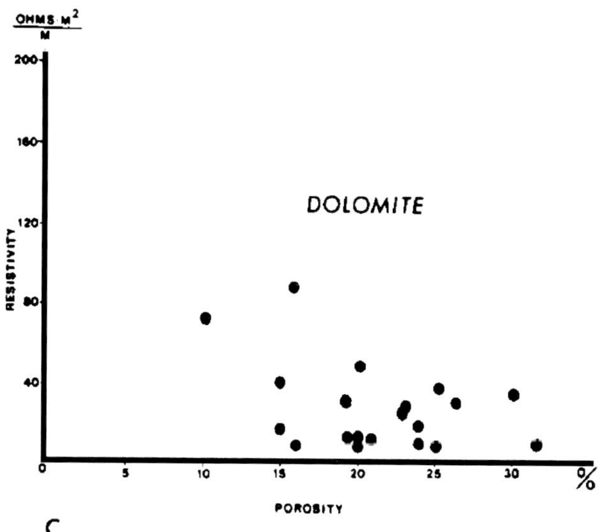

Figure 4. Crossplots of shallow (16″ normal or shallow dual induction) resistivity vs. porosity for three lithologies of the Ste. Genevieve Limestone, based on core data from 15 different wells in northern Lawrence field. Dense, non-reservoir limestone (A), exhibits the highest resistivity and, correspondingly, the lowest porosity. Oolite (B) exhibits a range of lower resistivities with increasing porosity. Dolomite (C) exhibits the lowest resistivities and highest porosities. In the absence of porosity logs, these three lithologies can be subjectively differentiated by resistivity.

several permeable lithologies. For example, in Figures 3A and 3B, the SP curve includes both dolomite and oolite. Resistivity deflections, however, provide a better indication of porous oolite thickness. Though the resistivity curve cannot provide a direct measurement of porosity, the thickness of resistivity breaks matches the thickness of porosity zones on accompanying porosity logs (Figure 3A). The electric log, however, only responds to that portion of the facies that is porous and permeable. Since oolite bars are typically cemented tightly on the edges and porous in the middle (Carr, 1973), isopach values determined from the resistivity curves tend to be net porosity values, rather than thickness of the total facies.

Isopach maps of oolitic reservoirs in three areas (Figure 1A) are shown in Figures 2B, 6, and 7. The oolite bar at Lawrence field (Figure 2B) was randomly chosen from a complex of bars in the field (Choquette and Steinen, 1980), and constructed using both resistivity and porosity logs. The bar in Lancaster field (Figure 6), however, was mapped using only resistivity logs. Nevertheless, the Lancaster map displays the same quality and detail as the Lawrence field map, despite the lack of porosity data. Both maps display an overall northeast–southwest trend, with a secondary north–south trend developed within the bar. These secondary trends reach a thickness of 8-15 ft (2.5-4.5 m) and are described as ridges (Carr, 1973), sand waves, or spillover lobes (Handford, 1988) resulting from tidal or storm influences. The thinner areas between the trends of thickness are less than 8 ft (2.5 m). Cores from these thinner areas display average oolite porosities, but poor permeabilities of less than 10 md. Therefore, injector wells placed in one trend of thickness will have little effect on production in an adjacent trend. In thinner areas of less than 4 ft (1.2 m), porosity also drops, and the majority of wells are dry holes (Figures 2B, 6B). Hence, detailed mapping of these secondary patterns is crucial to determining future well locations and waterflood patterns and can be accomplished using just electric logs.

Although oolitic reservoirs in the Ste. Genevieve Limestone are often described as bars (Carr, 1973; Cluff, 1986; Zuppann, 1988), detailed mapping may reveal a different geometry. An oolitic reservoir discovered at the south end of Lawrence County in 1986

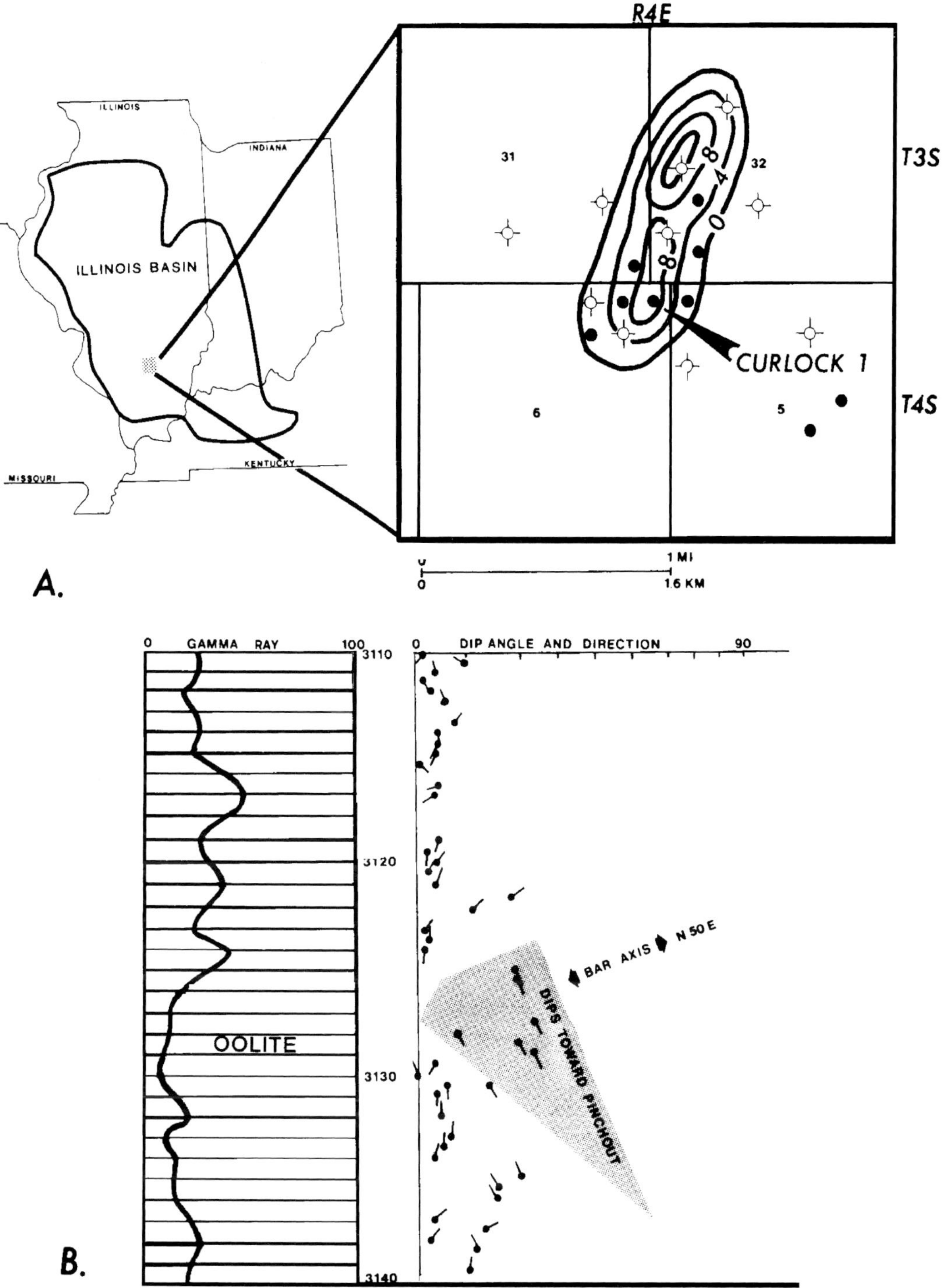

Figure 5. (A) Location of Belle Rive Southwest field, Jefferson County, Illinois. Outline of Illinois basin is defined by a line marking the maximum extent of Pennsylvanian-age rocks in the basin. The map shown is an isopach of a productive oolite bar, contour interval of 4 ft (1.2 m). The location of Petroventure #1 Curlock is shown by the arrow. (B) Schlumberger HDT dipmeter vs. gamma ray for Petroventure #1 Curlock, Sec. 6, T4S, R4E, Jefferson County, Illinois. "Tadpole" plot exhibits southeast dip direction for cross-bedding in the upper portion of the oolite bar. The bar axis would be oriented perpendicular to these dips, producing an interpreted axis orientation of N50°E, just slightly northeast of the true bar orientation.

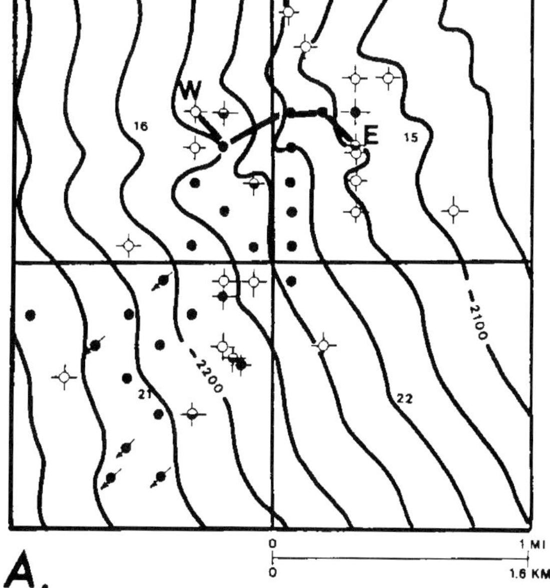

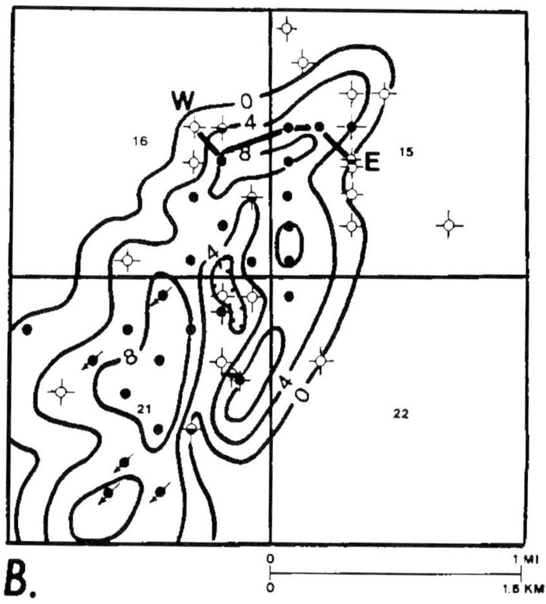

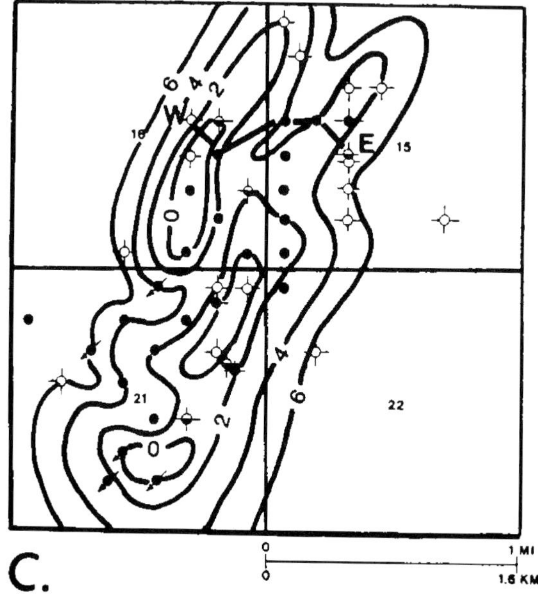

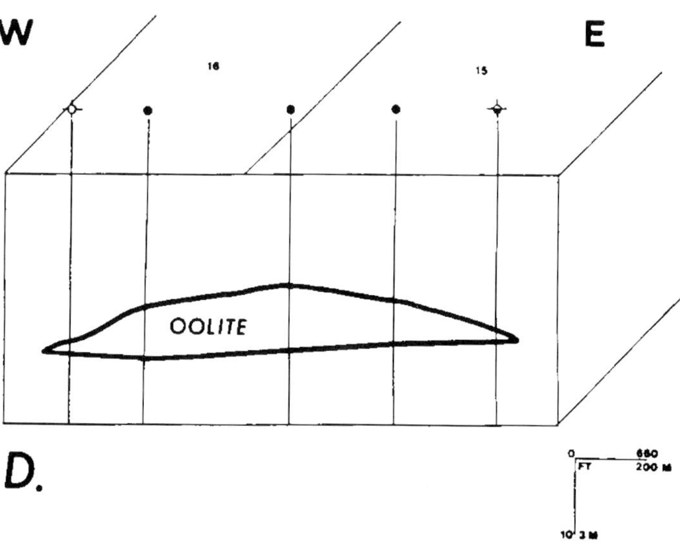

Figure 6. Maps and cross section for northern Lancaster field, T2N, R13W, Lawrence County, Illinois. (A) Structure on top of the Ste. Genevieve Limestone, contour interval 20 ft (6 m). (B) Isopach map of productive oolite bar, contour interval 4 ft (1.2 m). Note north–south orientation of thicker reservoir, separated by a thinner area across the middle of the bar. (C) Isopach map of "Rosiclare" shale, upper Ste. Genevieve Limestone, contour interval of 2 ft (0.6 m). This map, when compared to Figure 6B, exhibits thinning of shale unit above underlying oolite bar. (D) Cross section W–E, projected from the base of the Downeys Bluff Limestone (see Figure 1B). Location of cross section is shown in Figures 6A–6C. The cross section exhibits typical geometry of an oolite bar, with a flat lower surface and convex-upward upper surface.

(Figures 7A, 7B) in the area of Allendale pool displays a narrow, north–south orientation. A cross section (Figure 7C) shows that the convex upward surface normally associated with oolite bars is lacking. Instead, the reservoir displays a channel geometry. Similar channels occur elsewhere in Lawrence field (Choquette and Steinen, 1980, 1985) but are filled with nonreservoir packstones. Initial production tests

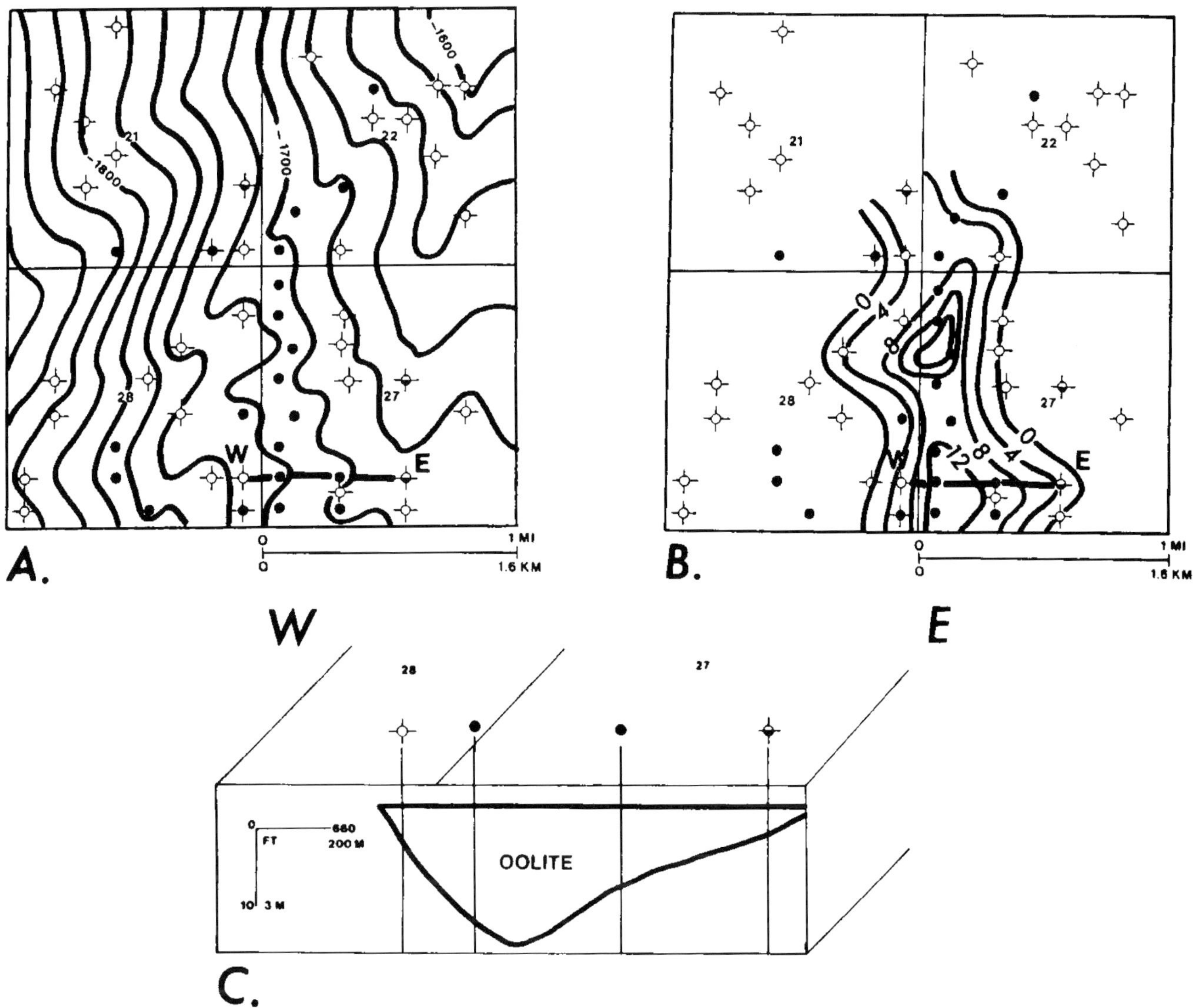

Figure 7. Maps and cross section of Allendale pool, T2N, R12W, Lawrence County, Illinois. (A) Structure on top of the Ste. Genevieve Limestone, contour interval 20 ft (6 m). (B) Isopach map of productive oolite channel, contour interval 4 ft (1.2 m). (C) Cross section W–E, projected from the base of the Downeys Bluff Limestone (see Figure 1B). Location of cross section is shown in Figures 7A–7B. The cross section exhibits a channel geometry with a flat upper surface and convex-downward lower surface.

from many wells in the channel exceeded 100 BOPD with no water. To date, the field has produced more than 700,000 bbl of oil, still with no water (Jim Blumthal, personal communication). The Allendale discovery illustrates a type of reservoir that may occur throughout the Illinois basin. Since resistivity breaks in the Ste. Genevieve section match porosity zones, a net isopach map of porosity can be drawn from electric logs. Net porosity for the Ste. Genevieve Limestone, including both oolites and dolomites, has been constructed for northern Lawrence field (Figure 8A). The map, utilizing some porosity logs as well as electric logs, reveals a northeast–southwest trend of porosity development. Such a map can reveal undrilled locations in a development setting and provide guidance for exploration in more sparsely drilled areas.

EXPLORATION APPLICATIONS

In the Illinois basin, exploration for Ste. Genevieve oolites often proceeds without considering geologic data or principles. Consequently, many dry holes are drilled, and productive reservoirs are found by accident. Since wireline logs are commonly the only data available on wells, exploration can begin by examining these logs for oolites, using the methods shown in this chapter. In addition, two other methods have been utilized in the Illinois basin successfully, using information from wireline logs.

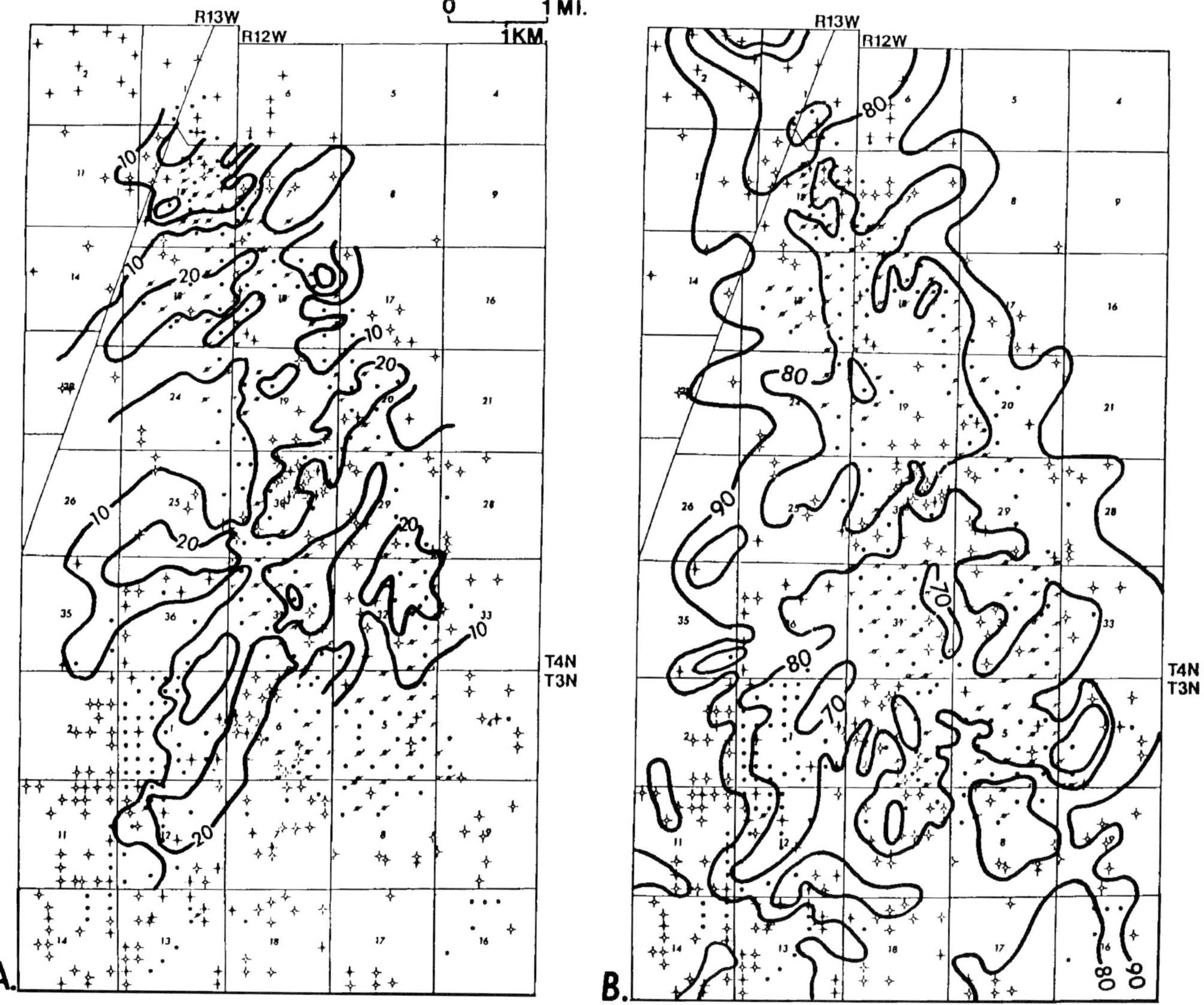

Figure 8. (A) Net porosity map for the Ste. Genevieve Limestone, Northern Lawrence field, Lawrence County, Illinois, including both oolites and dolomites. Contour interval is 10 ft (3 m). This map was constructed using both resistivity logs and available porosity logs. A northeast–southwest trend in porosity is shown. The well symbols represent only Ste. Genevieve-depth wells and production. (B) Isopach map of the interval between the top of the Ste. Genevieve Limestone and the top of the Downeys Bluff Limestone. Contour interval is 10 ft (3 m). This thinning corresponds with present-day structure (Figure 2A), suggesting the structure was present during late Valmeyeran–early Chesterian time. Such a structure could have provided a paleoshoal for oolite generation.

Many present-day structures in the Illinois basin may have been paleostructures during Mississippian time. Such structures, if they formed topographic highs, could have provided a high-energy shoal for ooid formation (Handford, 1988). The LaSalle anticlinal belt provided such a shoal in Lawrence County (Choquette and Steinen, 1980). A paleostructure map of northern Lawrence field was drawn by constructing an isopach of the interval between the top of the Ste. Genevieve limestone and the top of the Downeys Bluff Limestone (Figure 8B). A thinning of this interval occurs in the area of present-day structure. This thinning is considered to be the result of a paleoshoal, where several oolite bars developed (Choquette and Steinen, 1980). This mapping technique could be used to search for paleoshoals in other areas of the basin.

In many areas of the Illinois basin, no known structure or paleostructure exists. Oolite bars, however, such as northern Lancaster field (Figure 6A), do occur in these areas. Where a sufficient number of wireline logs exist, an isopach map of any thin shale in the upper Ste. Genevieve Limestone can be constructed. Because oolite bars are convex upward and, like siliciclastic sands, do not compact to the same degree as surrounding mudstones or wackestones, their topographic expression is revealed by the thinning of overlying shales. Because the shale interval is usually less than 10 ft (3 m) thick, thinning shows up dramatically. Such thinning has been shown to occur in Stringtown field (Cluff, 1988) and Monroe City Consolidated field (Keith, et al., 1988). A similar example is shown in Figure 6C, where a shale thins over a productive oolite bar in northern Lancaster field. No structural closure is present in the area (Figure 6A), but the geometry and orientation of the bar is shown by a thinning of the overlying "Rosiclare" shale. Although this technique may be difficult to apply in areas of sparse drilling, careful mapping of shale thickness may display general trends where the shale is thin or missing.

CONCLUSIONS

The great majority of data on oil wells in the Illinois basin is limited to electric logs. This limitation, however, should not hamper detailed evaluation of a reservoir. In the Ste. Genevieve Limestone, resistivity curves can be used to differentiate oolitic grainstones, dolomites, and dense, non-reservoir limestones. Resistivity deflections may also occur where porosity is present. Therefore, isopach maps of lithology and porosity can be constructed strictly from electric logs, even in the absence of drilling-time records, sample descriptions or cores. Oolites and dolomites are easier to differentiate on gamma-ray porosity logs, but these logs are less common. Modern dipmeters, though rarely used, can help determine orientation of an oolite bar. In addition to identifying oolites, electric logs can facilitate exploration by providing data for isopach maps of paleostructures and thin shales. Paleostructures may have provided paleoshoals for oolite generation. Where no paleostructure is discernible, oolite bars may be revealed by mapping shales in the upper Ste. Genevieve Limestone, which thin over a bar.

ACKNOWLEDGMENTS

Logs, cores, and production data were necessary for the development of this chapter, and I gratefully acknowledge the assistance of the following individuals and companies for their assistance. Marathon Oil Company, Bridgeport, Illinois supplied core and log data for Lawrence field. The dipmeter log and related maps were supplied by Dean Hollensbe of Mt. Vernon, Illinois. The isopach map of the Allendale reservoir was drawn for W. C. Padgett of St. Francisville, Illinois, who granted the use of the map in this paper. Jim Blumthal, Olney, Illinois, supplied production data for the Allendale reservoir. Beverly Seyler and Scott Beaty of the Illinois State Geological Survey supplied information on authigenic clay minerals. Donna Zerkle and Lissa Crowley, Science-Math Division, Vincennes University, assisted with figure preparation. The text was typed by Donna Zerkle. This manuscript was reviewed by Philip Choquette, Robert Cluff, Charles Zuppann, and Brian Keith.

REFERENCES CITED

Blatchley, R. S., 1913, The oil fields of Crawford and Lawrence Counties: Illinois State Geological Survey Bulletin 22, 442 p.

Carr, D., 1973, Geometry and origin of oolite bodies in the Ste. Genevieve limestone (Mississippian) in the Illinois basin: Dept. of Natural Resources, Geological Survey Bulletin 48, 81 p.

Caserotti, P.M., 1983, Geology and geophysics of the Spring Garden area, Jefferson County, Illinois, in M.K. Luther (ed.), Proceedings of the Technical Sessions, Kentucky Oil and Gas Association, 43rd Annual Meeting, 1979, p. 1-22.

Choquette, P. W. and R. P. Steinen, 1980, Mississippian nonsupratidal dolomite, Ste. Genevieve limestone, Illinois basin: evidence for mixed-water dolomitization: SEPM Special Publication No. 28, p. 163-196.

Choquette, P. W. and R. P. Steinen, 1985, Mississippian oolite and nonsupratidal dolomite reservoir in the Ste. Genevieve formation, north Bridgeport field, Illinois basin, *in* P.O. Roehl and P.W. Choquette (eds.), Carbonate Petroleum Reservoirs, Springer-Verlag, New York, p. 209-225.

Cluff, R. M., 1986, Application of modern carbonate sand models to oil and gas exploration, Mississippian Ste. Genevieve limestone, Illinois basin, in B. Seyler, a core workshop and field guidebook featuring the Aux Vases and Ste. Genevieve Formations: Illinois Geological Society, Illinois State Geological Survey and Southern Illinois University at Carbondale, p. 5-7.

Cluff, R. M, 1988, Stringtown Field, *in* C. W. Zuppann, B. D. Keith, and S. J. Keller (eds.), Geology and petroleum production of the Illinois basin, Vol. 2: Illinois Geological Society and Indiana-Kentucky Geological Society, p. 134-137.

Droste, J.B. and G.L. Carpenter, 1990, Subsurface stratigraphy of the Blue River group (Mississippian) in Indiana: Indiana Department of Natural Resources, Geological Survey Bulletin 62, 45 p.

Handford, C. R., 1988, Review of carbonate sand-belt deposition of ooid grainstones and application to Mississippian reservoir, Damme Field, southwestern Kansas: AAPG Bulletin 72, No. 10, p. 1184-1199.

Hollensbe, D. and M. Tucker, 1988, Spring Garden field, *in* C. W. Zuppann, B. D. Keith, and S. J. Keller (eds.), Geology and petroleum production of the Illinois basin, Vol. 2: Illinois Geological Society and Indiana-Kentucky Geological Society, p. 131-133.

Huff, B. G., 1987, Petroleum industry in Illinois, 1985: Illinois State Geological Survey, Illinois Petroleum 128, 140 p.

Keith, B. D., T. A. Gognat and J. P. Schnable, 1988, Monroe City Consolidated Field, *in* C. W. Zuppann, B. D. Keith, and S. J. Keller (eds.), Geology and petroleum production of the Illinois basin, Vol. 2: Illinois Geological Society and Indiana-Kentucky Geological Society, p. 138-141.

Keith, B. D., and C. W. Zuppann, 1993, Mississippian Oolites and Petroleum Reservoirs in the United States–An Overview (this volume).

Norris, R. L., 1988, Hebbardsville Field, *in* C. W. Zuppann, B. D. Keith, and S. J. Keller (eds.), Geology and petroleum production of the Illinois basin, Vol. 2: Illinois Geological Society and IndianaKentucky Geological Society, p. 145-147.

Schlumberger Educational Services, 1987, Log interpretation principles/applications: Houston, Texas, 198 p.

Seyler, B., 1986, A core workshop and field trip guidebook featuring the Aux Vases and Ste. Genevieve formations: Illinois Geological Society, Illinois State Geological Survey, and Southern Illinois University at Carbondale, 67 p.

Swann, D.H., 1963, Classification of Genevievian and Chesterian (Late Mississippian) rocks of Illinois: Illinois State Geological Survey Report of Investigation 216, 91 p.

Whiting, L.L., 1959, Spar Mountain sandstone in Cooks Mill area, Coles and Douglas Counties, Illinois: Illinois State Geological Survey Circular 267, 24 p.

Willman, H.B., E. Atherton, T.C. Buschbach, C. Collinson, J.C. Frye, M.E. Hopkins, J.A. Lineback and J.A. Simon, 1975, Handbook of Illinois stratigraphy: Illinois State Geological Survey Bulletin 95, 261 p.

Zuppann, C. W., 1988, Oolite lenses in the Ste. Genevieve limestone, in C. W. Zuppann, B. D. Keith, and S. J. Keller (eds.), Geology and petroleum production of the Illinois basin, Vol. 2: Illinois Geological Society and Indiana-Kentucky Geological Society, p. 130.

Chapter 6

Complex Oolite Reservoirs in the Ste. Genevieve Limestone (Mississippian) at Folsomville Field, Warrick County, Indiana

Charles W. Zuppann
Indiana Geological Survey
Bloomington, Indiana, USA

ABSTRACT

Correlating productive oolitic zones throughout Folsomville field is difficult because lithologies in the upper part of the Ste. Genevieve Limestone are variable both laterally and vertically. The problem is further complicated by thickness variations of this interval that result in juxtaposition of porosity zones when geophysical logs are correlated. Subsurface slice mapping, now an infrequently used method of subsurface analysis, can resolve complex geometries of oolite bodies and account for seemingly incongruous patterns of hydrocarbon production.

Interpretation of six porosity zones in the upper Ste. Genevieve Limestone and one in the overlying Paoli Limestone (Mississippian) shows that the five oldest zones contain significant oolite bodies that accumulated over an area of positive relief on the sea floor. Integrating isopach and lithofacies maps with fluid-production histories permits more accurate description of individual reservoirs and should contribute to more effective oil recovery from the field.

INTRODUCTION

The Ste. Genevieve Limestone is one of the more significant oil-producing formations in the Illinois basin, accounting for about 18% (743 million barrels) of the basin's cumulative oil production through 1986 (Cluff and Lineback, 1981; Howard, 1991; Mast and Howard, 1991). It has easily the most geographically widespread petroleum occurrence (Figure 1) of Illinois basin formations and continues to be one of the leading targets for exploration in the basin today. Most Ste. Genevieve reservoirs are found in porous oolite bodies, informally known as "McClosky sands," which are encased within fine-grained carbonate rocks. The oolite bodies have dimensions on the order of 0.125–2 mi (0.2–3 km) wide, up to 6 mi (10 m) long, and no more than 25 ft (8 m) thick (Carr, 1973; Cluff, 1986). Oolite bodies commonly occur at multiple stratigraphic positions within the formation, and many Ste. Genevieve oil fields are actually composites of separate oolite reservoirs. The Folsomville field is typical in this respect, and the irregular distribution of thin oolite lenses makes correlation of individual reservoirs extremely difficult.

Droste and Carpenter (1990) recently revised the nomenclature of the Ste. Genevieve Limestone in Indiana (Figure 2), making subsurface terminology generally consistent with named units in Illinois. The

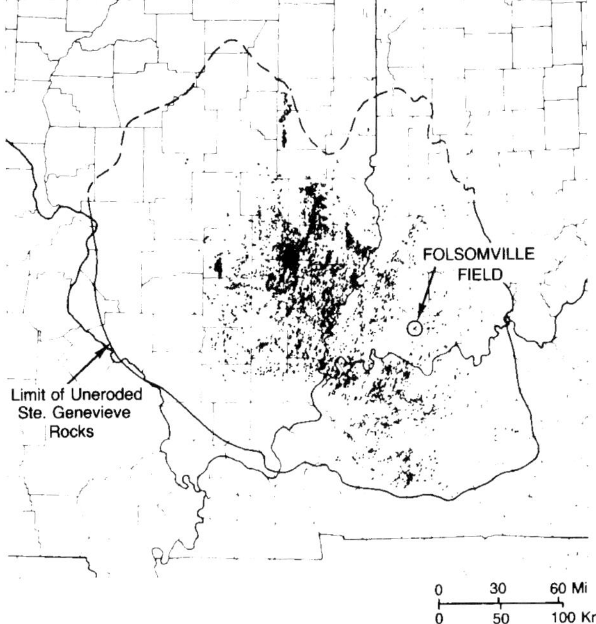

Figure 1. Map showing Ste. Genevieve oil and gas fields in the Illinois basin and location of Folsomville field. Modified from Howard (1991). Outline of uneroded Ste. Genevieve rocks from Carr (1973).

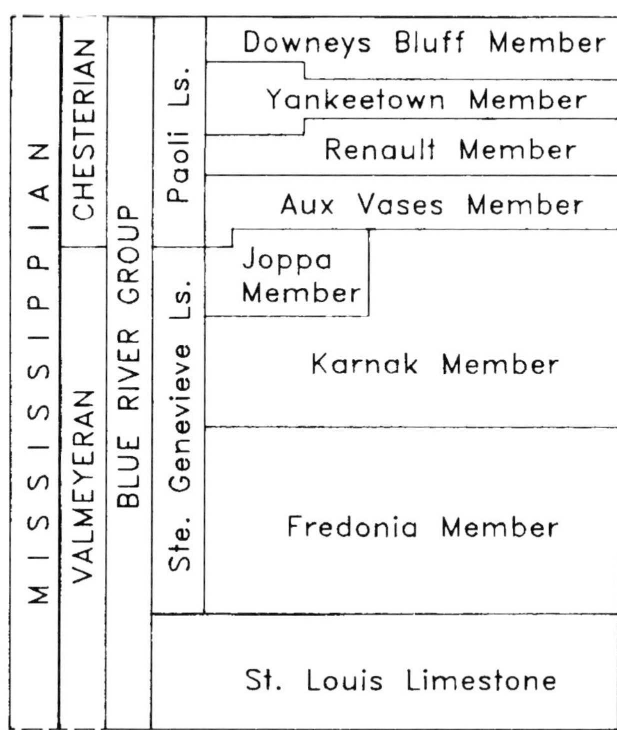

Figure 2. Blue River Group stratigraphic section. Simplified from Droste and Carpenter (1990).

Ste. Genevieve in the Indiana subsurface is the middle of three formations included in the Blue River Group (Mississippian), which is predominantly a sequence of carbonate rocks deposited in shallow marine environments. The contact with the underlying St. Louis Limestone (Mississippian) can be difficult to recognize with precision because lithologies on either side of the contact are variable and can be similar (Carr et al., 1986; Droste and Carpenter, 1990). The upper contact of the Ste. Genevieve generally is easy to recognize because in the subsurface the carbonate rocks of the Ste. Genevieve contrast markedly with the shale and sandstone of the overlying Aux Vases Member of the Paoli Limestone (Droste and Carpenter, 1990). The Ste. Genevieve is subdivided into three members, the Fredonia, Karnak, and Joppa members (Figure 2), which are all present at the Folsomville field, although only the upper two members (Karnak and Joppa) are productive there.

Folsomville field is located on the gently dipping eastern shelf of the Illinois basin. Although Blue River rocks regionally dip less than 50 ft per mile (9.5 m/km) into the basin, local structures commonly dip more steeply. Notwithstanding the typically stratigraphic nature of traps for Ste. Genevieve oolite reservoirs, most Ste. Genevieve oil fields in Indiana also are associated closely with local positive structures, suggesting a more-than-casual relationship between structural and stratigraphic controls.

This is certainly true at Folsomville field, where a prominent southwesterly plunging nose is generally aligned with Ste. Genevieve oil production (Figure 3). About 20 ft (6 m) of relief on the structure is evident on the Paoli Limestone. In the southern part of the field, the trend of producing wells is offset somewhat from the anticlinal axis, but still shows general structural alignment. Despite the stratigraphic complexity and sometimes inconsistent patterns of oil and water production, the near alignment of oil production with the plunging anticline suggests that structure is an important control on the hydrocarbon accumulation at Folsomville.

But the influence of structure may go much further than the formation of a structural trap. Bodies of oolite sand commonly accumulate in shoaling, high-energy settings that can develop in areas with persistent positive relief. If the positive structure at Folsomville showed any expression during Ste. Genevieve deposition, as was suggested for Folsomville by Johnson (1988), then the distribution of oolite bodies may have been concentrated in these areas, and the structure could actually control the position of oolite deposition.

PRODUCTION AND DEVELOPMENT HISTORY

Tamarack Petroleum Company discovered the Folsomville field in 1978 with a well completed in the Harrodsburg Limestone (Mississippian) (Figure 3, well 8-1), which is referred to locally as the Warsaw formation. Ste. Genevieve production was not established until 1981 (Figure 3, well 4-1), when active development drilling commenced; it continued

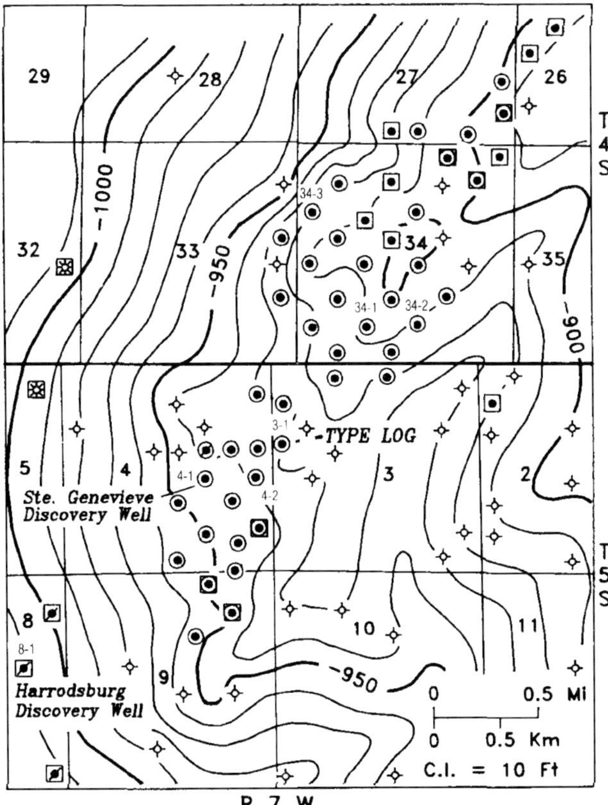

Figure 3. Structure map on top of Paoli Limestone in the Folsomville field area. The overall distribution pattern of oil wells producing from the Ste. Genevieve and Paoli (circles) and Harrodsburg (squares) aligns closely with the axis of the anticline, especially in the northern half of the field.

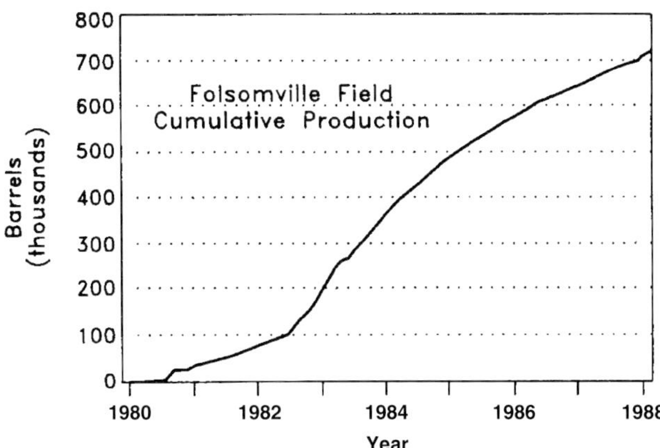

Figure 4. Graph of cumulative oil production from Folsomville field through 1988.

through 1984. Most of the field was drilled and is operated by Tamarack. For convenience, arbitrary well names have been assigned to the wells in this report. The numerical prefix indicates the section where the well is located. Only wells specifically referred to in the report are labeled on the maps.

Cumulative oil production at Folsomville through 1988 exceeded 700,000 bbl (Figure 4), and ultimate recovery is expected to approach 1 million bbl. Because a number of wells were completed in multiple horizons and production from Harrodsburg, Paoli, and Ste. Genevieve zones has been commingled in some wells, cumulative production from each formation cannot be determined accurately. Based on initial-production test data and general well histories supplied by the operator, however, the Ste. Genevieve produced an estimated 600,000 barrels of oil through the end of 1988. Water injection was begun in a few of the downdip flank wells very early in the field's history, and because injection was initiated so early, the amount of increased production attributable to waterflooding cannot be estimated reliably.

As the field was developed, a number of factors soon led to the realization that the Ste. Genevieve is a complex reservoir. For example, initial-production tests (and subsequent amounts of oil produced) from wells completed in the Ste. Genevieve varied considerably, ranging from a high of 232 BOPD to a low of less than 10 BOPD, and no systematic pattern for the variation was apparent. Water produced in association with oil in some of the wells also varied significantly, but once again no consistent pattern was evident throughout the field. Production response in wells located close to water-input wells was erratic and often unpredictable. These conditions posed a real challenge to the operators attempting to exploit the field most efficiently.

STRATIGRAPHY—A CORRELATION NIGHTMARE

At Folsomville, Ste. Genevieve oil production is from thin porosity zones that occur within the upper 50 ft (15 m) of the formation. Porosity zones are found in all three members of the Ste. Genevieve, but only the Karnak and Joppa members have produced commercial amounts of oil. Some wells did report shows of oil in the Fredonia Member, however. The distribution of porosity in the Ste. Genevieve is complex (Figure 5), and serious obstacles complicate the interpretation of the porosity geometry. Foremost is the high degree of stratigraphic variability exhibited by laterally and vertically discontinuous lenses of porosity. The number of porous lenses developed in individual wells ranges from two to more than seven. Each isolated lens of porosity can act as an individual reservoir with its own particular characteristics. To facilitate discussion of these individual lenses, the term "porosity body" will be used to designate an isolated, contiguous lens of porosity development. Not only do the porosity bodies vary in thickness across the field, but so do the thicknesses of all facies

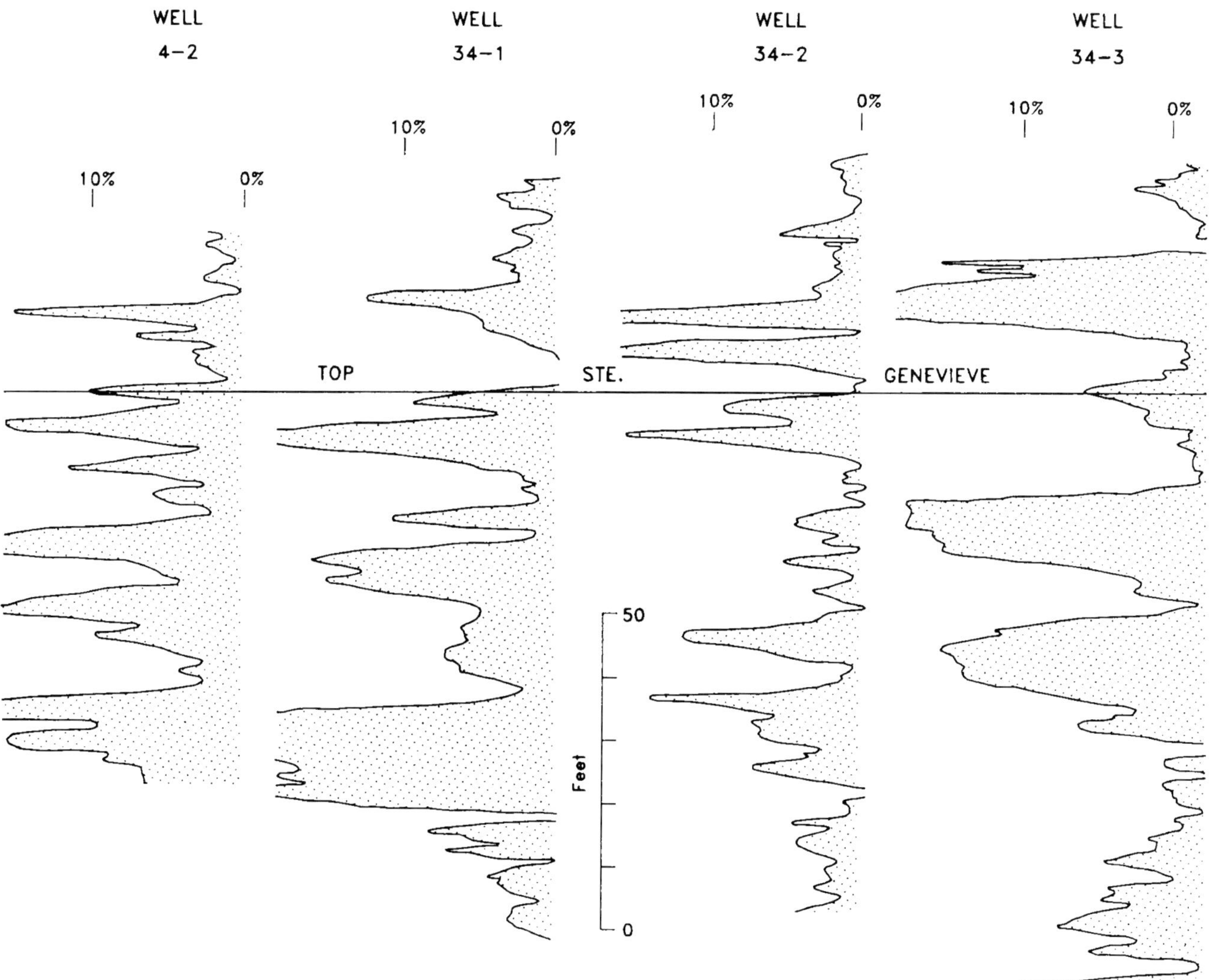

Figure 5. Density porosity logs from four representative wells illustrate the variable stratigraphic occurrence of porosity at Folsomville field. Datum is the top of Ste. Genevieve Limestone. No horizontal scale. See Figure 3 for well locations.

and members in the upper part of the Ste. Genevieve and lower part of the Paoli; thus, the stratigraphic positions of the porosity bodies appear to move up and down in the section. Porous intervals in some wells also appear to coalesce in other wells, making unique identification of porosity bodies difficult.

The main data available to identify the porosity bodies at Folsomville are density logs, which were run on all but a few wells, and the production histories of individual wells. But during the early course of this investigation, this information appeared inadequate to resolve questions about correlations of the porosity zones. Countless cross sections and logs (SP, resistivity, gamma-ray, and density) marked with detailed correlations were prepared, interpreted, and reinterpreted over the period of several months. Numerous revisions of correlations failed to yield a satisfactory interpretation that could be carried across the entire field and that would also account for patterns in oil and water production. Conventional correlation techniques involving side-by-side comparison of well logs could not be used to define confidently the geometries of porosity bodies.

If characteristic lithofacies were associated with any of the porosity zones, some of the correlation problems might be resolved. However, sample descriptions submitted with the well records suggest that lithofacies may be variable within the zones. Not enough samples are available to be of much help in this regard. Sample data are sufficient to reach some general conclusions concerning lithofacies within the porosity zones, however, and these will be discussed later.

Successful correlation is further hindered by the absence of persistent and reliable stratigraphic markers within the Ste. Genevieve. The contacts between the Ste. Genevieve Limestone and its bounding formations, the St. Louis Limestone below and the Paoli Limestone above, as well as the contacts of the three members of the Ste. Genevieve, are generally easy to define in the Folsomville area. But none of these contacts proved to be useful as stratigraphic reference horizons because of their inconsistent stratigraphic positions, which are dependent on facies variations within the units or minor discontinuities in the vertical sequence.

Useful subsurface markers do occur in formations above and below the Ste. Genevieve (Figure 6). The X marker is an informal horizon at the base of a dolomitic unit within the St. Louis Limestone and is known for its consistent electric log character throughout much of southwestern Indiana (Keller and Becker, 1980). Its stratigraphic position is approximately 150 ft (46 m) below the Ste. Genevieve. Another stratigraphic marker present at Folsomville is the top of the Paoli Limestone, which occurs about 30–40 ft (9–12 m) above the Ste. Genevieve (the Downeys Bluff Limestone, uppermost member of the Paoli, is informally known as the upper Renault limestone and is recognized as a reliable marker bed across the Illinois basin).

Adding to the correlation difficulties are the many wells that penetrate only part of the Ste. Genevieve interval, providing an incomplete stratigraphic interval to correlate.

SLICE MAPS

The slice-map method was used in this study to unravel the complex stratigraphy at the Folsomville field. The technique involves the imaginary "slicing" and "peeling away" of successive planes through the field. On each successive slice, the distribution of a mappable parameter is shown. Such a series of slice maps permits the successively shifting patterns on the

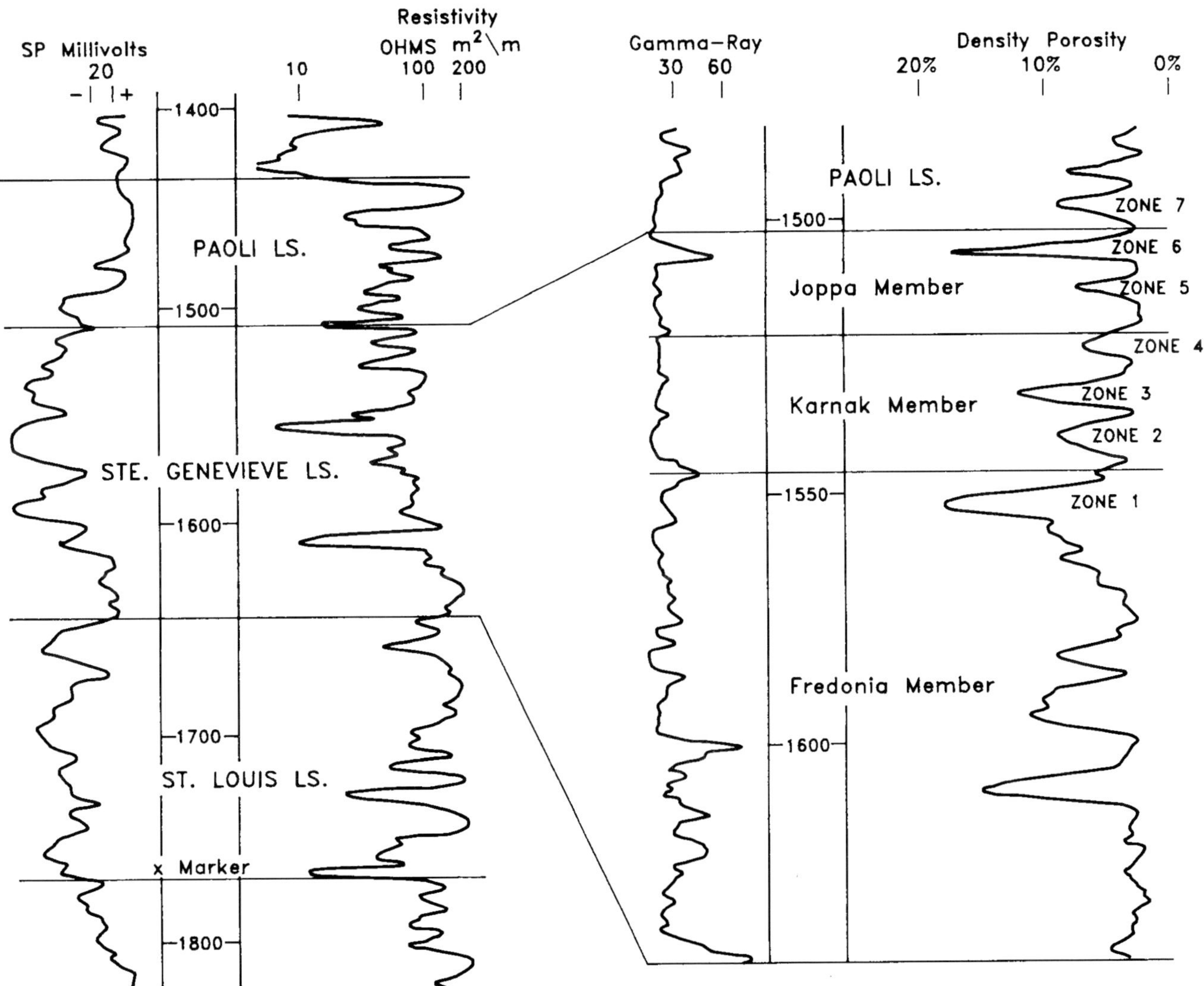

Figure 6. Type log, well 3-1 (see Figure 3 for location). Electric log (left) shows the Ste. Genevieve Limestone and adjacent units. The X marker in the St. Louis Limestone and the top of Paoli Limestone are key stratigraphic markers. Density porosity log (right) shows members of the Ste. Genevieve and seven stratigraphic levels of porosity discussed in this study. All seven zones are not present in most wells.

maps to be observed and provides a visual representation of the geometries of the mapped features. Geologists have used this method for many years as a means of portraying the three-dimensional shape of subsurface oolite bodies. Connolly (1949) showed the shifting distribution of two Ste. Genevieve oolite bodies at the Passport field, Illinois, using what he called "plans" of oolite lenses mapped at 2-ft (.6 m) intervals. Asquith et al. (1977) constructed slices at 10-ft (3 m) intervals through the Lower Permian Council Grove B-zone oolite grainstone in the Anadarko basin in the Texas panhandle. These were referred to as paleogeographic maps for different time intervals. Although the ultimate purpose of slice mapping is to depict the geometry of the units being mapped, the Folsomville study extends its use to the resolution of complex correlations by tracking the distribution of porosity bodies through the successive slices.

The first step in preparing slice maps is to select an appropriate interval between the slices. Although the most suitable interval depends on the degree of stratigraphic complexity, the interval must be on a scale that permits identification of the beds involved. Thinner beds require a smaller slice interval. For the very thin units investigated at Folsomville, a slice interval of 2 ft (.6 m) was used. This is the smallest practical interval given the resolution of the density logs used in the interpretation.

The next step is selection of a suitable reference datum from which to measure the map layers or slices. This datum must be persistent and uniform. Although sea level might seem an ideal datum, it may not serve as well as a stratigraphic datum because structural tilting skews beds to intersect a larger number of map slices. As a result, several porosity bodies can intersect a given map slice simply because they occur at the same structural depth in different parts of the field. Natural groupings of posted porosity values on any map are therefore difficult to identify. If a stratigraphic marker datum is chosen as an alternative, it also must be considered carefully. The marker must be present in all the wells under study and should exhibit uniform character and stratigraphic position, and additionally should be as close as possible to the beds of interest. If the position of the reference marker varies, the positions of the mapped beds will be shifted to different slices, potentially obscuring map patterns. The only marker meeting these criteria at Folsomville is the top of the Paoli Limestone, which appears exceptionally consistent in the study area.

When used in the traditional manner, slice maps have generally been prepared as paleogeographic or lithofacies maps. However, any mappable parameter can be evaluated in a series of slices. The best information available to evaluate the porosity zones at Folsomville is provided by density logs, and calculated porosities derived from these logs can be shown effectively using the slice-map method.

For each well in the study area, measured porosity from the density log was posted on the slice map for the corresponding depth below the top of the Paoli reference marker. As contours were interpreted for each map, patterns or trends became evident and so helped to delineate porosity bodies intersecting that slice. During the interpretation process, the porosity distribution was compared with adjacent slices, and an attempt was made to maintain continuity of the porosity bodies between the slices, in effect correlating the vertical patterns between slices. This process of continual feedback from adjacent slices adds another dimension (vertical) to the correlation process. By tracing porosity bodies through successive slices, it was possible to determine whether different areas of porosity intersecting a particular map slice represent separate porosity bodies or are parts of the same body that converge on another map slice.

A total of 35 slice maps were prepared for Folsomville field, covering a 68-ft (21-m) stratigraphic interval. For these maps and subsequent maps presented in this report, the convention will be followed of showing only those wells that were used in the interpretation. Wells not shown include those with no available data and those not reaching the mapped horizon. For the locations of all wells drilled in the study area, see Figure 3.

A representative series of six slice maps (Figure 7) illustrates the slice-map method. This partial set of maps is presented to conserve space, and because the slice maps are merely a tool used to resolve correlations and have no genetic meaning. Slice maps used to interpret the geometry of oolites have been considered to be time planes (Connolly, 1949; Asquith et al., 1977), and the outline of an oolite body on a slice has been understood to indicate the shape of the oolite body at that particular time. Although this interpretation may approach reality in certain cases in which strata have uniform thicknesses and the zones of interest are relatively thick, generally the patterns observed on the slices do not describe the shape of a body at any specific time. Instead, they are an artifact of the mapping process and represent only a way of visualizing the geometry of facies relative to a specific reference datum. This is exemplified on the Folsomville slice maps where vertically separate porosity bodies intersect the same map slice in different parts of the field. Thus the patterns observed on slice maps do yield a picture of the geometry of the porosity bodies, but they give a relative picture only, useful in depicting the continuity of the porosity bodies in three dimensions. Additional geological techniques such as isopach maps and cross sections must be used to describe their geometry further.

The validity of the slice-map method can be checked against the reasonableness of the resulting interpretation of the porosity bodies. If the interpretation is correct, certain questions must have acceptable answers. Do the correlations result in viable cross sections? Is oil/water production consistent with structural position? Are the geometries of porosity bodies and their lithofacies distribution patterns logical? These questions were answered affirmatively in the Folsomville study and will be addressed in subsequent sections of this chapter.

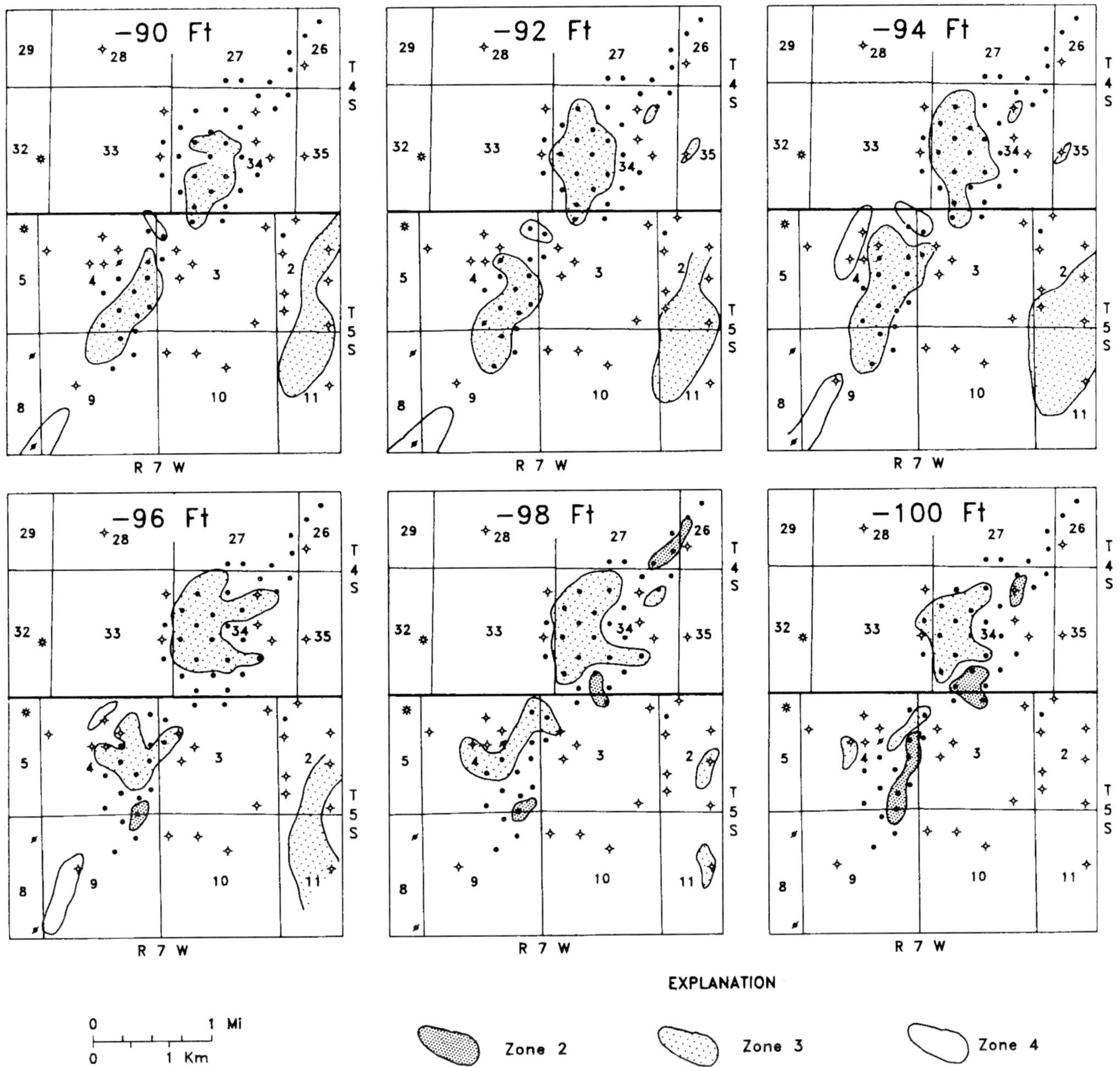

Figure 7. Slice maps for 2-ft (0.6-m) intervals from 90 ft (27 m) below the top of Paoli Limestone to 100 ft (330 m) below. Patterned areas represent porosity equal to or greater than 6%.

POROSITY ZONES

The slice-map interpretation revealed that laterally discontinuous porosity bodies tend to occur within specific stratigraphic levels, which in this discussion are referred to as "porosity zones." Six stratigraphic levels of porosity zones were identified in the upper part of the Ste. Genevieve and one zone in the Paoli. They are labeled zones 1 through 7 (Figure 6). Porosity zones below zone 1 and above zone 7 were not investigated.

Using the correlations established by the slice-map method, the geometries of porosity bodies within in each porosity zone were further described by isopach maps (Figures 8–14). When used in conjunction with structure maps, the isopach maps permit the reservoir performance of each porosity body to be evaluated. If the correlations determined by the slice-map method are correct, then the distribution and geometry of the porosity bodies should make sense and should be consistent with patterns of oil/water production. In addition, lithofacies patterns noted for each porosity body should be reasonable.

To facilitate comparison of reservoir geometry with performance, a minimum porosity cutoff of 8% was used to construct the net-porosity isopach maps.

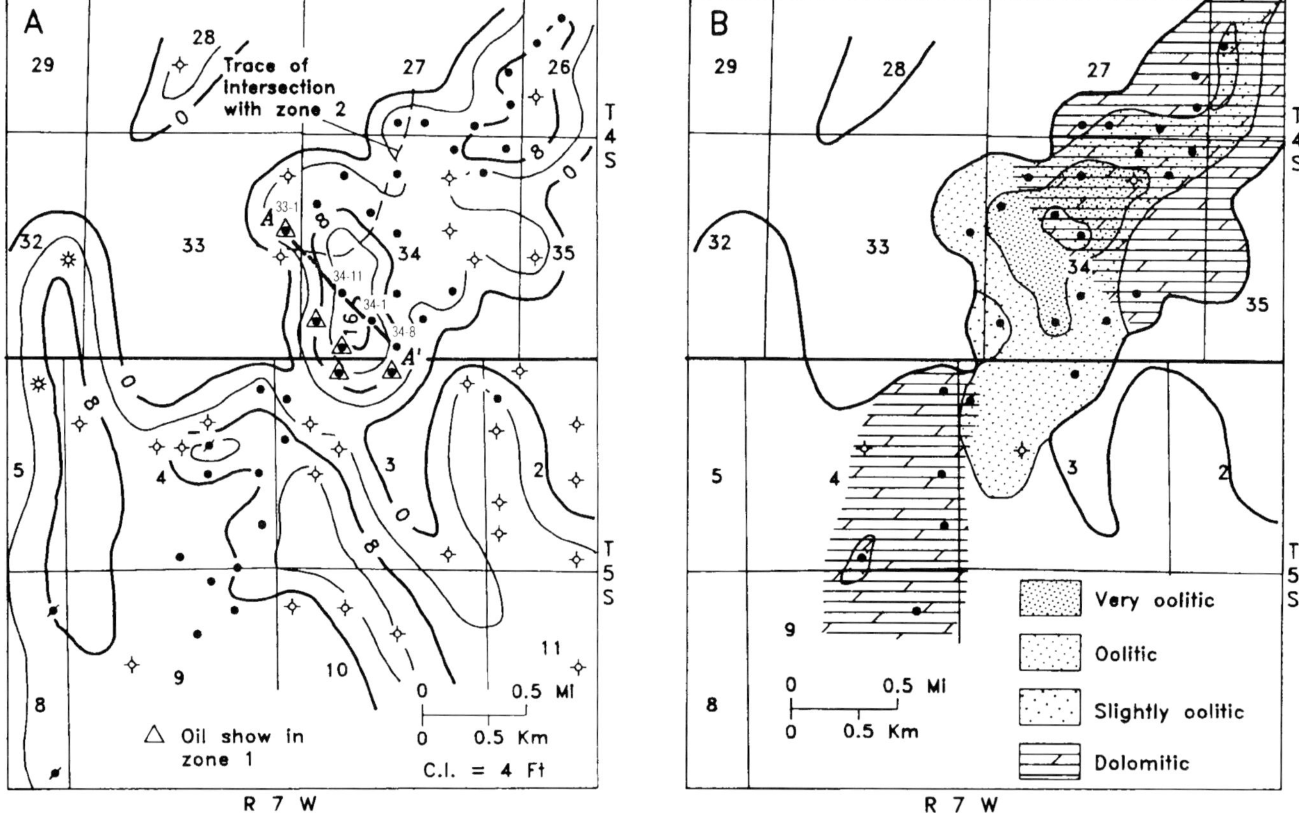

Figure 8. Zone 1. (A) Isopach map of net porosity greater than 8%. Cross section A–A′ is shown on Figure 15. (B) Gross lithofacies map.

This value approximately corresponds to the effective porosity value below which Ste. Genevieve zones do not produce at Folsomville.

Limitations imposed by available production information and by limited rock-sample data affected interpretations of the porosity zones. Production in some wells was commingled from several formations, and in some wells multiple completions were made in separate porosity zones within the Ste. Genevieve as well, often making it impossible to determine cumulative production for a specific zone. In addition, oil sales from the field were combined into two production units in 1984 and 1987. Production from these units is reported in lump sums, and production totals for individual unit wells have not been reported since the units were formed.

Evaluation of lithofacies associated with the porosity zones at Folsomville is complicated by several factors. The samples available for study at Folsomville are inadequate. Although the Ste. Genevieve was cored in at least three wells at Folsomville, none of the cores were preserved for study. Existing descriptions of these cores are brief and general. Drill cuttings are available for only seven wells that produce from the Ste. Genevieve, and for a similarly small number of the nonproductive wells, and even these few sets of sample cuttings are not reliable. Samples were washed commercially prior to delivery to the Indiana Geological Survey, and comparison of the stored samples with original well-site descriptions suggests that fine-grained ooids (typically 0.2–0.5 mm) were commonly washed out of samples from zones having appreciable porosity and containing more loosely cemented grains. In other words, the ooids are missing from many of the processed drill cuttings obtained from the better porosity zones.

The most useful source of information regarding lithologic makeup of the porosity bodies comes from the relatively few available sample descriptions that were prepared by well-site geologists when the wells were drilled and from the extremely brief sample descriptions given on the state well-completion forms. Admittedly, use of these data, which can be highly generalized and subjective, introduces a significant margin of error to the lithofacies interpretations, because the descriptions were prepared by several different geologists, at different times, and under different circumstances. Some adjectives taken directly from the available sample descriptions and well-completion forms ("very oolitic," "oolitic," "slightly oolitic," and "dolomitic") were deemed suitable for constructing lithofacies maps, but no specific quantitative basis for the lithofacies designations is implied. The resultant maps using these subjective lithofacies designations are appropriately titled "gross" lithofacies maps (Figures 8B–12B). Also, use of the term "dolomitic" is uncertain because samples probably were not stained for dolomite content at the well site.

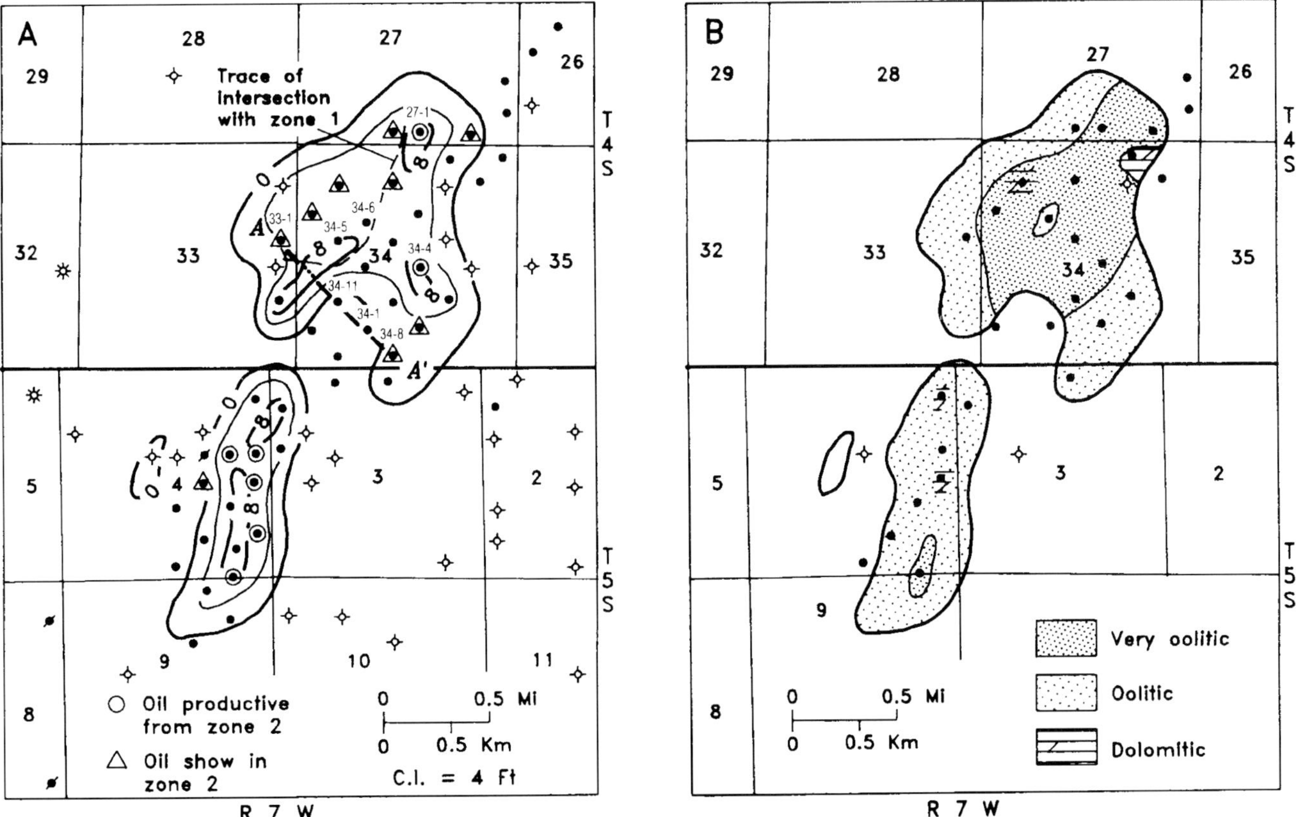

Figure 9. Zone 2. (A) Isopach map of net porosity greater than 8%. Cross section A–A′ is shown on Figure 15. (B) Gross lithofacies map.

"Dolomitic" most likely refers to sucrosic texture observed in the samples and in most cases probably does indicate truly dolomitic rock. The brief sample reports describe many samples as being both dolomitic and oolitic. This is inconsistent with detailed petrographic descriptions of the Ste. Genevieve elsewhere in the Illinois basin, where dolomite is commonly associated with muddy carbonate rocks but rarely with oolitic grainstones (for example, Carr, 1973; Choquette and Steinen, 1980). Whether mixing of the drill cuttings, as might happen if thin dolomitic mudstone beds lay adjacent to oolite beds, is responsible for the samples being described as both dolomitic and oolitic or whether some of the rocks at Folsomville field are actually dolomitic oolites is not known.

But even taking into account all the above problems and recognizing a high margin of error, certain patterns are evident in the oil production and lithofacies distribution. These patterns can be related to the individual porosity bodies that were determined by the slice-map method. In the following sections, these factors are considered for each porosity zone.

Porosity Zone 1

Zone 1 is the uppermost interval of porosity in the Fredonia Member of the Ste. Genevieve (Figure 6). An elongate area of porosity development, oriented southwest–northeast, covers most of Sec. 34 (for convenience, Township and Range designations will not be given in the discussion, as they are evident on the maps) and extends to the southwest and northeast (Figure 8A). This linear trend appears to merge with a broad area of porosity development that covers most of the south half of the study area. Net porosity thins markedly in the northwest quarter of Sec. 3, and possibly these two areas of porosity development pinch out and do not connect. Zone 1 is the thickest porosity zone at Folsomville, being at least 16 ft (5 m) thick at its maximum in the southwest quarter of Sec. 34. This thickness is consistent with oolite bodies in the Fredonia Member in Indiana, which are generally thicker and more extensive than those in the Karnak Member (Droste and Carpenter, 1990). In the northern part of the field (northwest quarter of Sec. 34), zone 1 appears to coalesce with zone 2 (see porosity cross section A–A′, Figure 15), and here the apportionment of net porosity to zones 1 and 2 was entirely subjective.

No commercial oil has been produced from zone 1, but oil shows were reported in five wells clustered around the southwest part of Sec. 34. These wells are structurally lower than other wells that are wet and had no oil shows. This apparent discrepancy in the oil/water column may be due to permeability variations within the zone, but a definitive cause could not be determined. In the broad area of porosity covering

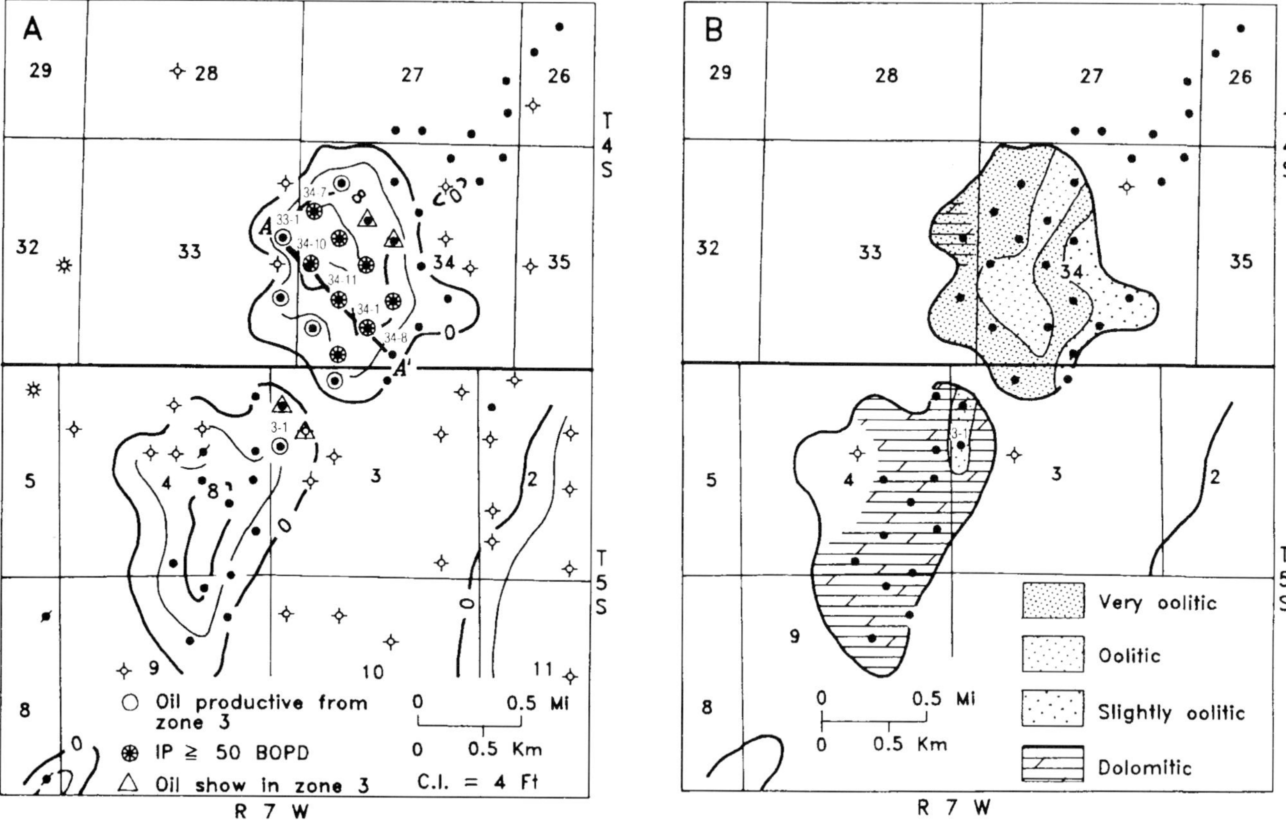

Figure 10. Zone 3. (A) Isopach map of net porosity greater than 8%. Shaded wells reported initial-production tests greater than 50 BOPD. Cross section A–A′ is shown on Figure 15. (B) Gross lithofacies map.

the south half of the study area, no oil shows were reported and log calculations indicate the interval is wet. No wells were tested in this area.

Gross lithofacies are distinctly different in the elongate northern area of porosity development and the broad area of porosity to the south (Figure 8B). The northern area is clearly associated with an extensive elongate oolite body that encompasses much of the porous area. The southern area of porosity, on the other hand, is dominated by dolomitic porosity (although well control is limited) and is appreciably oolitic only in the northwestern part of Sec. 3.

Within the oolite associated with the elongate northern area of porosity, relative ooid concentration is greatest along its central part and diminishes toward the flanks (Figure 8B). This logical arrangement tends to validate the geometry of this porosity as derived by the slice-map method. Although significant portions of the zone 1 porosity are dolomitic, the thickest part of the elongate porosity trend (southwest quarter of Sec. 34) is not.

Porosity Zone 2

Two main porosity bodies, each about 9 ft (3 m) thick, are present in zone 2 (Figure 9A). The southern body is slightly more than a mile (1.6 km) long, and less than 0.5 mi (0.8 km) wide. The northern body is not elongate and measures slightly more than a mile (1.6 km) across. Porosity is absent from wells in the southwest quarter of Sec. 34, supporting a two-body interpretation, but the two bodies could easily be mapped as one by connecting them across the southeast quarter of Sec. 33, in the area of no well control. Such an alternate interpretation would define a single elongate porosity body extending along the entire field.

Only two wells in the northern porosity body are oil-productive from the zone. Well 34-4 is an excellent producer of water-free oil and is also structurally the highest well in the porosity body. Well 27-1 produces oil from a similar structural level that drill-stem tested large amounts of water in wells 34-5 and 34-6. This apparent discrepancy can be accounted for readily by heterogeneous permeability in zone 2. Evidence for heterogeneity is provided by many wells that recovered mostly small amounts of drilling mud on drill-stem tests, indicating that they are in areas of lower permeability. These include wells that separate the oil-productive from the water-productive wells.

Even though heterogeneous permeability renders oil/water patterns difficult to interpret, the occurrence of prolific oil production from well 34-4 updip from the two wells that tested copious water is a logical pattern that adds some support to the correctness of the slice-map correlations.

Fluid-recovery patterns are difficult to interpret in the southern porosity body because several wells

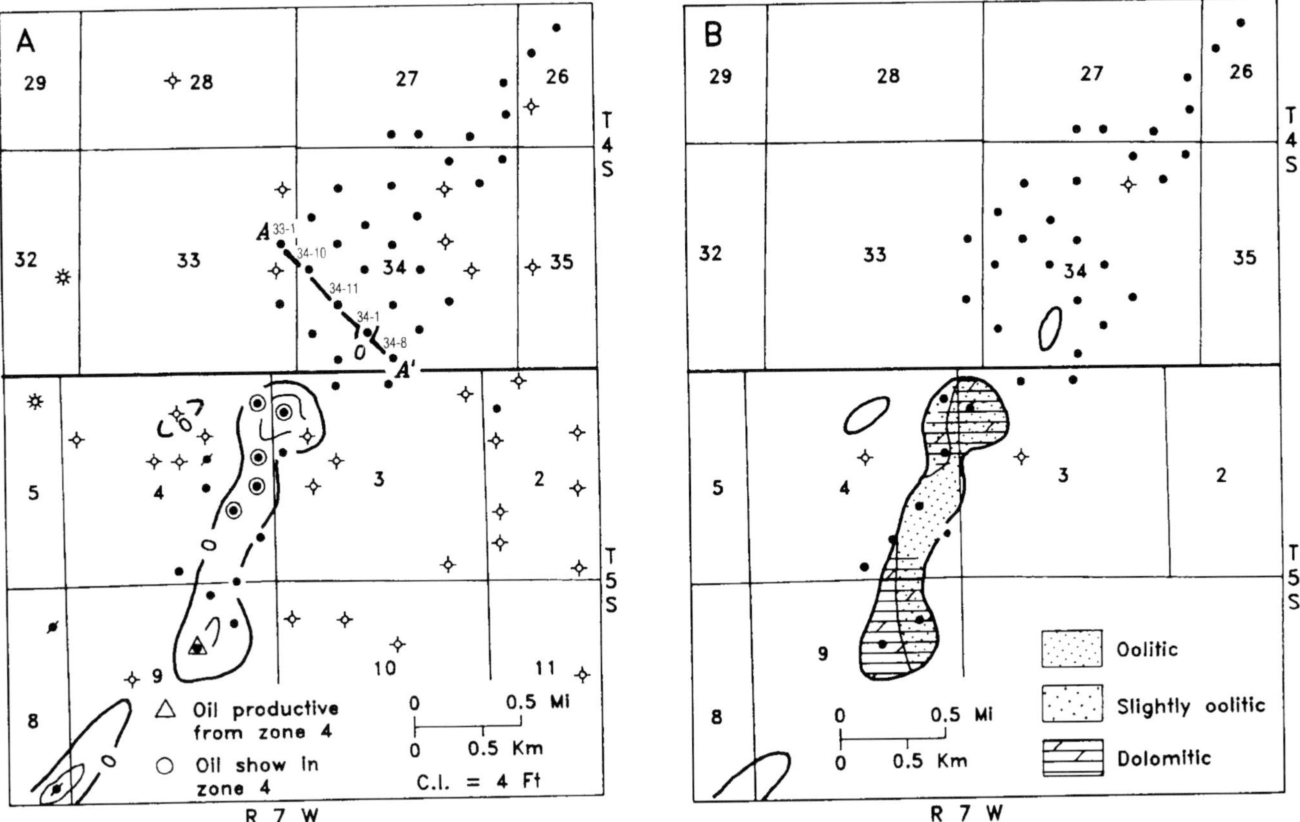

Figure 11. Zone 4. (A) Isopach map of net porosity greater than 8%. Cross section A–A′ is shown on Figure 15. (B) Gross lithofacies map.

were completed dually with zone 5 and because the test data are particularly sketchy for some of the wells. Some minor discrepancies in the structural positions of water-free oil wells vs. wells that produce both oil and water could not be reconciled. Given the available data, these discrepancies suggest that revisions in the interpretation of the southern porosity body likely would be required if additional information were available. Revisions might include permeability barriers or separation of the southern porosity body into at least two smaller bodies, but data presently are not sufficient to make such a determination.

Zone 2 porosity bodies are predominantly oolitic (Figure 9B). The southern porosity body has a southwest–northeast orientation and, although the northern porosity body shows no elongation, the "very oolitic" facies forms an elongate southwest-northeast trend flanked by "oolitic" facies on either side. Only minor dolomitic porosity was reported in the porosity bodies in zone 2, unlike zone 1, which has the extensive areas of dolomitic porosity body described earlier.

Porosity Zone 3

Two prominent porosity bodies are present in zone 3 (Figure 10A). Their positions roughly correspond to those of bodies in underlying zone 2, and the northern body corresponds to the position of the elongate porosity trend noted in zone 1. These relationships suggest that some continuing or recurring control may be responsible for the stacked porosity distribution. The northern porosity body of zone 3 is at least 14 ft (4 m) thick, thickest of all the oil-productive reservoirs and nearly as thick as zone 1, which is nonproductive.

The northern porosity body of zone 3 is the most productive petroleum reservoir in the field, containing 13 oil wells. Cumulative oil production from this reservoir through 1988 is estimated at 202,000 barrels. Considering the reasonableness of fluid-production histories for wells in the northern body as a measure of support for the slice-map interpretation, a few minor inconsistencies are present in which water was produced from some wells that are structurally higher than other water-free oil wells. A well-by-well description of production, structural position, porosity, and permeability is beyond the scope of this paper, but even taking the few oil/water discrepancies into account, the porosity-body geometry interpreted from the slice maps is supported.

Unlike the other Ste. Genevieve porosity zones at Folsomville, the northern body of zone 3 is thick, persistent, and relatively easy to correlate among wells. These correlations conform with the slice-map interpretation. Also, the general pattern of oil production is consistent with the slice-map interpretation. The

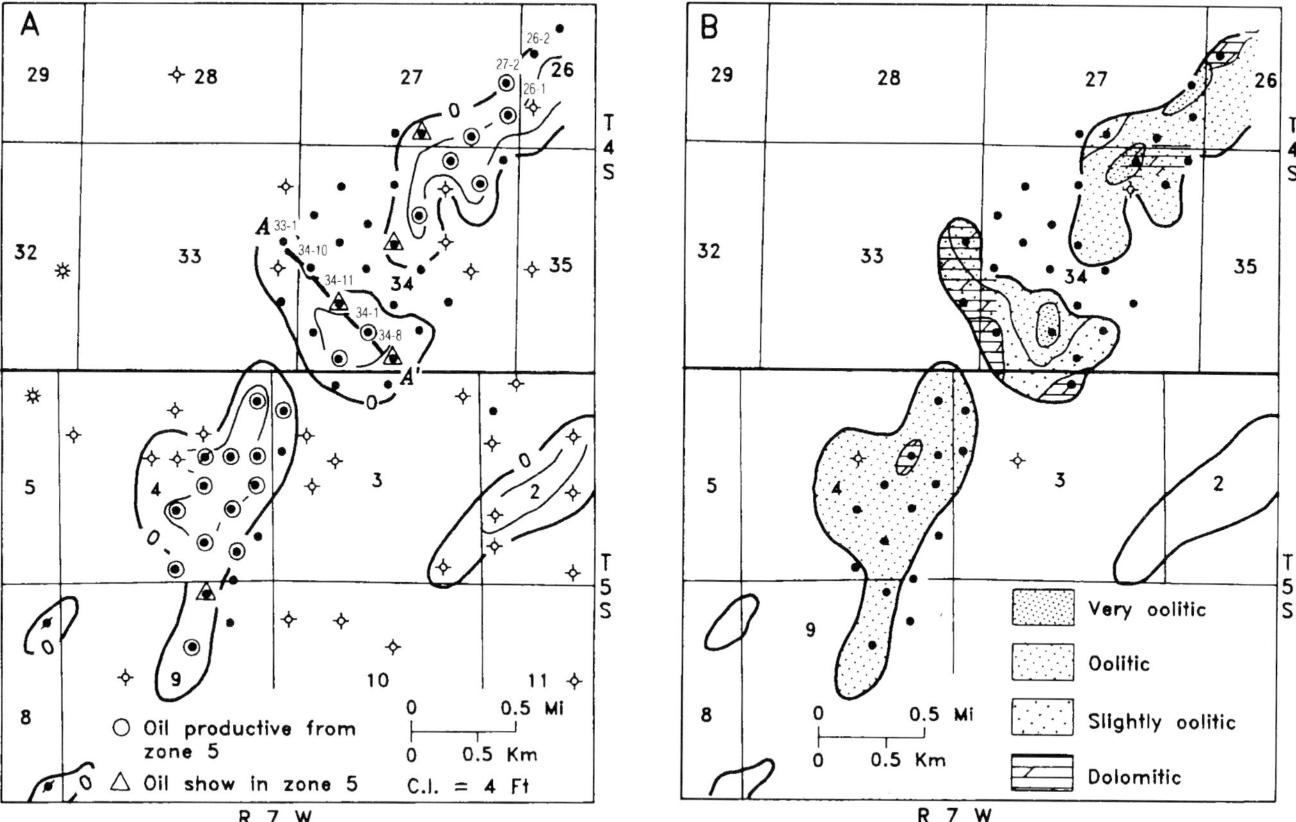

Figure 12. Zone 5. (A) Isopach map of net porosity greater than 8%. Cross section A–A′ is shown on Figure 15. (B) Gross lithofacies map.

most productive oil wells (here arbitrarily identified as wells with initial-production tests greater than 50 BOPD) are concentrated in the thickest part of the porosity body (Figure 10A). These wells are also the structurally highest producing wells in the reservoir, with the exception of well 34-7. All of the wells that initially tested more than 50 BOPD were originally water-free producers, but some of the structurally higher wells did begin producing various amounts of water during their histories. Well 34-7 is the main oddity, being at least 10 ft (3 m) downdip from the other wells that initially tested more than 50 BOPD, yet it continued water-free oil production for more than five years, long after some of the higher wells began producing significant amounts of water. No particular reason for this discrepancy is evident, but possibly water has channeled into some of the structurally higher wells. The few other oil/water inconsistencies all are located around the fringe of the porosity body, where poor drill-stem test recoveries and low porosity values on density logs indicate that heterogeneous reservoir conditions are present, possibly providing a logical explanation for the apparent inconsistencies.

The southern porosity body of zone 3 produces oil in only one well (Figure 10A), which had cumulative production of only 3000 bbl through 1988. Being much thinner than its northern counterpart, the southern body is more difficult to correlate among wells, and its delineation depends more on the slice maps. The interpretation is supported by the fact that most of the wells appear by log analysis to be in the water column, whereas many of these same wells are within the oil column in zone 2 below and zone 4 above. The only productive well in zone 3 is among the highest structurally, and the other two structurally high wells have low permeability.

The northern and southern porosity bodies of zone 3 have distinctly different lithofacies (Figure 10B). The northern body is predominantly oolitic, with a southwest–northeast orientation to the facies pattern, similar to that observed in the northern body of zone 2 (Figure 9B). In contrast, the southern body is almost entirely dolomitic. Only two wells in the southern body were described as oolitic. One of these is well 3-1, the lone oil-productive well from the southern body, and one of only two wells in the southern body in which dolomite was not reported. Based on the lithofacies patterns noted in zone 3, whether oil accumulation somehow preferentially occurs in porosity zones that are both oolitic and undolomitized is a matter of speculation. Although there is tendency for this relationship to be the case in the other porosity zones at Folsomville field, enough exceptions exist to cast doubt on whether such a simple relationship exists between lithofacies and production.

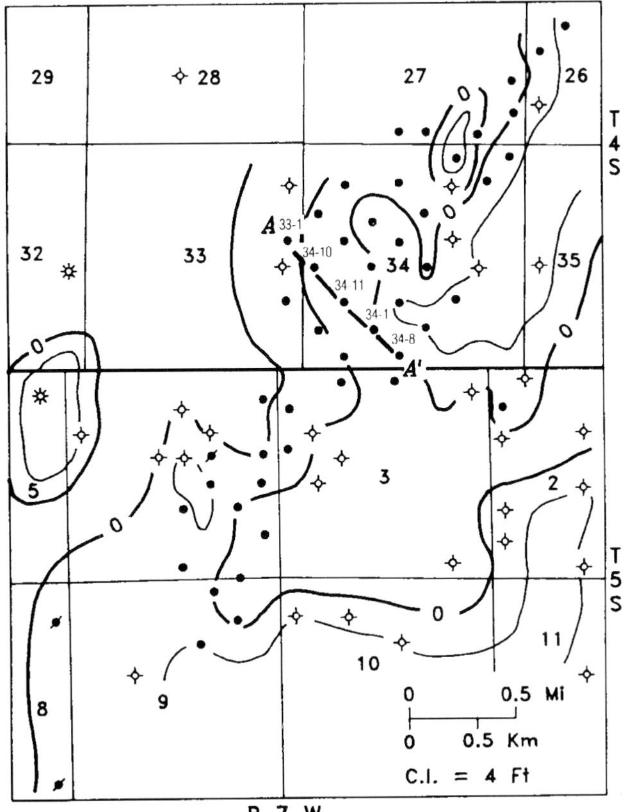

Figure 13. Zone 6. Isopach map of net porosity greater than 8%. Cross section A–A´ is shown on Figure 15.

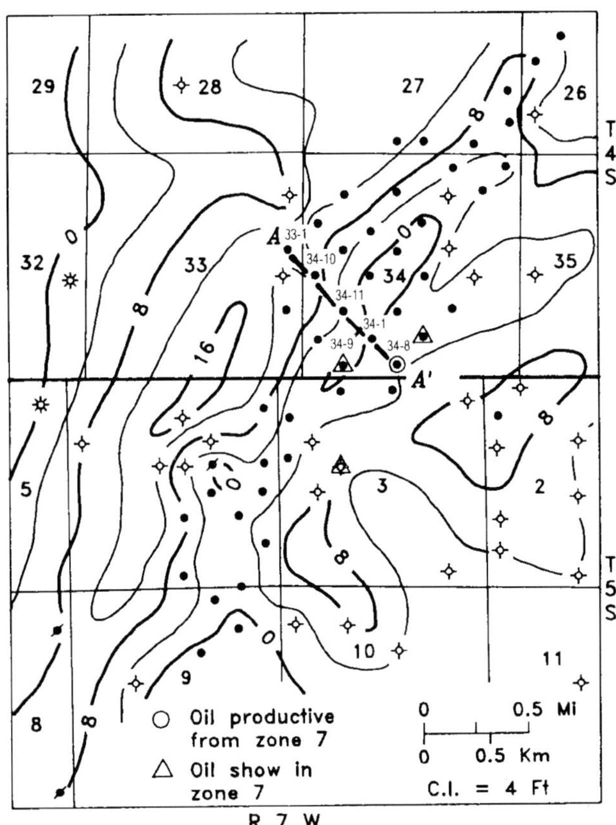

Figure 14. Zone 7. Isopach map of net porosity greater than 8%. Cross section A–A´ is shown on Figure 15.

Porosity Zone 4

Porosity development is very limited in zone 4. Unlike the other porosity zones, only a single significant porosity body is present (Figure 11A). Located in the southern part of the field, this body is thinner and more narrow than those in the deeper porosity zones. One well does have an 8-ft-thick (2-m-thick) porous interval, but the porosity body is otherwise no more than 4 ft (1 m) thick. The body is generally less than 0.25 mi (0.4 km) wide and slightly more than a mile (1.6 km) long. The direction of elongation is southwest–northeast, and though of smaller dimensions, the body maintains the pattern of stacked porosity development observed in the southern porosity bodies of zones 2 and 3 (Figures 9A, 10A). Why a corresponding northern porosity body did not develop as in the deeper zones is not known. Perhaps the greater thickness of the northern body in underlying zone 3 (Figure 10A) in some way inhibited subsequent oolitic deposition that might otherwise have occurred.

Five wells were completed as oil producers from the zone 4 porosity body (Figure 11A). Their logical grouping into a single area in the structurally highest part of the reservoir supports the slice-map interpretation.

Lithofacies in the zone 4 porosity body are variable (Figure 11B). The body appears to be largely oolitic, but the few available sample descriptions also indicate that much of the rock has been dolomitized.

Porosity Zone 5

Porosity zone 5, the uppermost zone of oolitic porosity in the Ste. Genevieve at Folsomville, is nowhere greater than 7 ft (2 m) thick (Figure 12A). Zone 5 differs from each of the underlying zones in that it contains three distinct porosity bodies in the field (additional porosity bodies away from the field proper are also present). The porosity body in Sec. 4 is similar in both location and southwest–northeast orientation to porosity bodies in each of the three immediately underlying zones that were described earlier (zones 2, 3, 4; see Figures 9A, 10A, 11A). But in the northern part of the field, where single porosity bodies are developed in zones 1, 2, and 3, and no significant porosity is developed in zone 4, two separate porosity bodies are present in zone 5 (compare Figures 8A, 9A, 10A, 11A, 12A). The northernmost of these two porosity bodies in zone 5 (centered in the southeast corner of Sec. 27) also has a strong southwest–northeast orientation, but the remaining porosity body in the southwestern part of Sec. 34 is unique

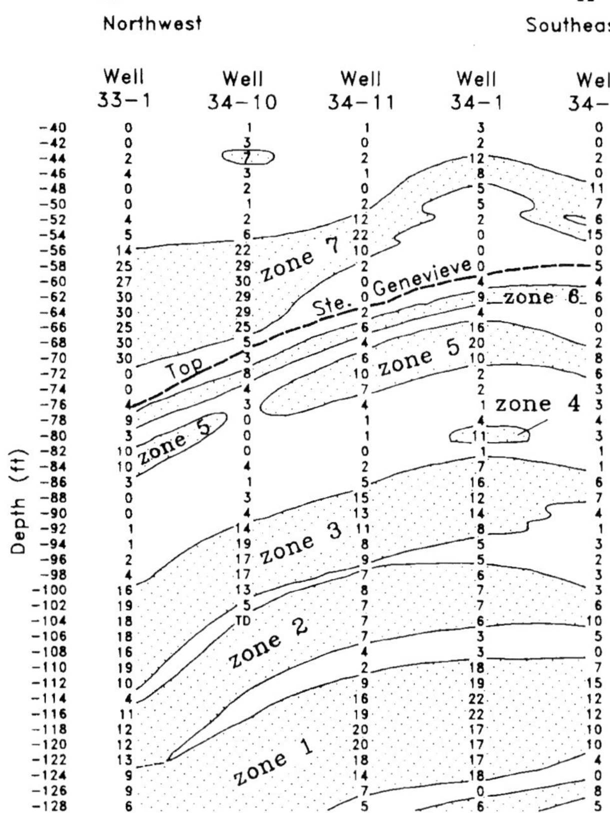

Figure 15. Porosity cross section A–A´. Depths below the top of Paoli Limestone datum are indicated. Porosity values corrected from density porosity logs are listed for each well. Wells are spaced about 900–1000 ft (274–304 m) apart but are not shown to horizontal scale. Vertical exaggeration is approximately 55×. Location of cross section is shown on Figures 8–14.

in having an elongate northwest–southeast orientation. Evidently the conditions that favored stacking of the porosity bodies in the deeper zones had changed for the northern part of the field.

All three porosity bodies in zone 5 in Folsomville field are oil productive, but nearly all the productive wells were completed as producers commingled with other zones in the Ste. Genevieve or Harrodsburg, so reasonable estimation of cumulative oil production for this reservoir is not possible.

Oil/water-production patterns relative to structural position within each porosity body tend to corroborate the slice-map interpretation. One small area is inconsistent with oil/water patterns. All of the wells in the northernmost porosity body (centered in Sec. 27) were completed as water-free oil producers except well 27-2, the structurally highest producing well, which tested 82% water. This apparent discrepancy could easily be accounted for because production from this well was commingled with Harrodsburg production. However, wells 26-1 and 26-2, which are even higher structurally, appear to be water-productive by log calculations although they were never tested in zone 5. The reason for this apparent oil/water discrepancy is not known. One possibility is that the slice-map interpretation of this zone is wrong, even though the zone appears to correlate quite well. Reservoir heterogeneity in this area, or some other type of permeability barrier such as a small fault, might also be the culprit. Regardless of the explanation, the slice-map interpretation of zone 5 throughout the rest of the field is consistent with production patterns.

For the most part, all three porosity bodies in zone 5 in Folsomville field are oolitic, but dolomite was reported in some samples as well. The lithofacies distribution pattern is oriented southwest–northeast (Figure 12B), with greater ooid concentration along the axis, but this interpretation is not conclusive.

Porosity Zones 6 and 7

Zones 6 and 7 were deposited after the main oolitic units and represent different depositional environments. A brief discussion of these units is included to acquaint the reader with all of the zones that were interpreted by the slice-map study of Folsomville field.

Porosity zone 6 is not oil productive and in fact does not contain reservoir-quality rock. Both gamma-ray logs and sample descriptions indicate that the zone is shale or shaly carbonate wherever it occurs. The shale content alone may account for the higher porosity recorded on density logs of this zone. Although zone 6 may contain no carbonate porosity, it can be difficult to distinguish from other porosity zones in the field, even using all of the available geophysical logs and sample descriptions. It was therefore necessary to treat this porosity interval in the same manner as the others in constructing the slice maps. The unit is never more than 6 ft (2 m) thick (Figure 13).

No gross lithofacies map was prepared for zone 6 because the zone is everywhere shale or shaly carbonate and because the oolitic and dolomitic facies mapped in the other porosity zones are not present in zone 6.

Porosity zone 7 (Figure 14) is the lowermost interval of porosity in the Paoli Limestone and has a distribution distinctly different than the underlying zones in the Ste. Genevieve. It is thin to absent along the axis of the field (the structural crest). On the adjacent flanks its thickness increases to as much as 16 ft (5 m).

A possible trap is formed where the zone terminates against the crest of the field in the south half of Sec. 34. Three wells in Sec. 34 (Figure 14) recovered free oil from zone 7, but only well 34-8 was completed in the zone. Initial production from this well was reported at 111 BOPD, but production was apparently commingled with production from zone 5. Information on cumulative production attributable solely to this zone is not available.

Another possible explanation for the hydrocarbon reservoir and trap is change in lithofacies. Except for one well, zone 7 is described as dolomitic throughout the field. (Because the dolomitic lithofacies occurs nearly everywhere, a gross lithofacies map of zone 7 is not presented.) The exception is well 34-9 (Figure 14), which was described as oolitic. The well is one of the three previously mentioned wells in Sec. 34 that recovered free oil from zone 7. Of these three wells, only well 34-9 has a sample description available, so each of the oil-bearing wells may be oolitic, and oil may be preferentially trapped in this lithofacies. Well control is insufficient to identify the true trapping mechanism or reservoir geometry.

Summary of Porosity Zones

Cross section A–A´ (Figure 15) depicts the complete vertical sequence of porosity zones identified by the slice-map method. On the cross section, the porosity zones appear relatively uncomplicated, and their correlations from well to well seem reasonable.

The slice-map interpretation accounts for variations in porosity, fluid production, and lithofacies distribution. Only a few minor inconsistencies in oil/water-production patterns are evident, whereas the overall pattern is internally consistent within each zone. Thus, the slice-map interpretation presented here should be regarded with a fairly high level of confidence, at least in the immediate vicinity of the field where well control is most dense. But the minor inconsistencies observed and the inherently complex nature of the porosity zones require recognition of possible error.

A few general observations can be drawn from the series of porosity isopach maps and lithofacies maps (Figures 8–14). Thicknesses of the bodies range from 2 ft (0.6 m) to greater than 16 ft (5 m). The overall trend of porosity development for all the oolitic zones is southwest–northeast, essentially parallel to the anticlinal axis (Figure 3), and to varying degrees this orientation is also characteristic of the individual porosity bodies. None of the oolitic porosity zones in the Ste. Genevieve appear to be continuous across the entire field; instead, isolated bodies of porosity occur within each zone (except possibly in zone 1, where porosity development is more extensive and dolomitic porosity is more common). A stratigraphically recurring pattern of porosity development is apparent from zone 1 through zone 5, giving rise to a series of stacked porosity bodies. For instance, the northern porosity trend in zone 1 (Figure 8A) is similar to those occurring in zones 2 and 3 (Figures 9A, 10A). Both zones 2 and 3 have similar porosity-body distributions in the northern and southern parts of the field. Moving upward through the column, this pattern is no longer found in zones 4 and 5 (Figures 11A, 12A), where only the southern area of porosity development noted in the underlying zones recurs. Zones 6 and 7 (Figures 13, 14) had different depositional controls that developed after the main series of oolitic porosity zones.

Oil occurrence and lithofacies distribution (Figures 8–12) are variable from zone to zone and from porosity body to porosity body within each zone. The main petroleum reservoirs are predominantly oolitic. The porosity zones have been variably enhanced or extended by dolomitization.

An obstacle to understanding the distribution and depositional history of oolitic bodies at Folsomville field is the relatively small amount of available sample information. This obstacle could be circumvented to some extent if the oolite-body geometries could be related directly to geometries of the porosity bodies, for which considerably more data are available. Close approximation of oolitic facies and porosity development in Ste. Genevieve rocks in Indiana was noted by Carr (1973). But at the Folsomville field the relationship between oolitic facies and porosity is not so clear. We know from the lithofacies maps that most of the porosity zones contain both oolitic and dolomitic facies, and porosity that is solely oolitic cannot be distinguished. Another problem is that tightly cemented oolitic rocks are present, though not abundant, in sample cuttings from wells within the field and more rarely from wells outside the field. These samples might be derived from the oolite-body edges that are more highly cemented than the bodies' interiors (Carr, 1973), or they could be from nonporous oolite lenses.

Given these limitations, precise delineation of oolite-body geometries at Folsomville field is not possible. However, porosity can generally be said to be preserved preferentially within the oolite bodies, and the oolite bodies are most certainly concentrated in a southwest–northeast trend within the immediate field area.

DEPOSITIONAL MODEL

The general alignment of present-day structure (Figure 3) with the distribution of oolite bodies at Folsomville suggests the two may be related. To determine whether oolite deposition was localized by shoaling, depositional relief rather than present-day structure must be considered. The paleorelief at Folsomville and northeastern Warrick County is inferred in Figure 16, an isopach map of the interval from the X marker to the top of the Paoli Limestone (see type log, Figure 6). This interval thins by as much as 20 ft (6 m) over the Folsomville field, in subparallel alignment with the present-day structure on top of the Paoli (Figure 3). Thinning of the interval between these two persistent stratigraphic markers suggests that this area did not subside as much as did surrounding areas during deposition of that interval. Johnson (1988) reached similar conclusions by mapping the interval from the top of the Ste. Genevieve to the top of the Paoli.

The most extensive area of oolite development was located somewhat off the highest area of depositional relief, as delineated by the isopach map (Figure 16), but still along the crest of the southwesterly trending ridge. Possibly water depth and energy conditions in this position favored oolite accumulation. The oolite

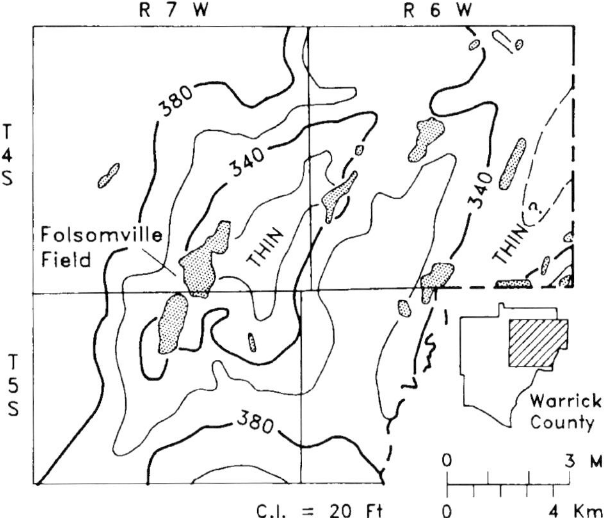

Figure 16. Isopach map of the interval from the X marker to the top of the Paoli Limestone. Isopach thins reflect positive relief that was present at sometime during deposition of the interval. Solid areas are oil fields that produce from the Ste. Genevieve Limestone.

bodies apparently developed immediately basinward from the highest area.

The locus of oolitic deposition was in the north part of the field (Sec. 34), where zones 1–3 are thicker and the lithofacies are described as more highly oolitic. These porosity bodies are stacked vertically. Each of these porosity bodies has an equivalent in the south part of the field, but only zone 2 is appreciably oolitic there. Why the southern porosity trends in zones 1 and 3 are predominantly dolomitic rather than oolitic, as are zones 2, 4, and 5, is not readily apparent. Oolitic deposition diminished in the younger strata represented by zones 4 and 5, which contain thinner and smaller porosity bodies. The "very oolitic" lithofacies is also less common in these younger zones. Overall, oolitic deposition decreased from zone 1 to zone 5.

Oolitic deposition stopped before zone 6 was deposited, indicating lower energy conditions resulting in shaly deposition. The shaly sediment was probably deposited in shallow, protected environments near sea level.

Except for the small pocket of oolitic facies, porosity zone 7 is dolomitic and thickens away from the crest of the field. The dolomitic facies may be low-energy carbonate mud deposits that accumulated in shallow lagoons. The absence of zone 7 along the crest could indicate subaerial exposure.

CONCLUSIONS

The oolite bodies at Folsomville field are shoal deposits that accumulated near the axis of a regionally prominent, 10-mi-long (16-km-long), southwest–northeast-oriented ridge in eastern Warrick County. A stacked sequence of relatively thin, stratigraphically complex oolitic and dolomitic porosity zones developed. These zones serve as petroleum reservoirs. Porosity was additionally enhanced by dolomitization that differentially affected the porosity bodies at various places in the field, also further complicating the porosity distribution. The complicated pattern of reservoirs that resulted proved difficult to interpret.

The slice-map method employed in this study resolved well-correlation problems and accounted for most of the apparent discrepancies in fluid-production patterns. Application of this method should prove useful not only for interpreting oolitic reservoirs, but for any thin, stratigraphically complex units, whatever their depositional setting or diagenetic history. The main limitation of using slice maps to unravel complicated correlations is the need for closely spaced well control, making the method useful only as a development geology tool. In this capacity, however, it could play a critical role in more accurately defining reservoir geometries and could ultimately lead to more effective plans for reservoir management and better designs for enhanced recovery programs.

From the standpoint of petroleum exploration, whether in the Ste. Genevieve Limestone in the Illinois basin or in oolitic rocks anywhere, recognition of depositional highs analogous to the feature described at Folsomville might point the way to localized concentrations of potentially hydrocarbon-bearing oolite bodies.

ACKNOWLEDGMENTS

Bob Liem and the Tamarack Petroleum Company were extremely helpful in allowing access to geological and engineering records of Folsomville field. Bob spent many hours relating the field's production history to me and sharing his ideas about the nature of the oolitic reservoirs.

Many improvements to the manuscript were suggested by critical readers J. Robert Dodd, John B. Droste, Brian D. Keith, Stanley J. Keller, Carl B. Rexroad, and John A. Rupp.

REFERENCES CITED

Asquith, G. B., S. R. Russell, and J. E. Drake, 1977, Depositional history of the Lower Permian Council Grove B-zone oolite shoal, Ochiltree County, Texas: The Compass of Sigma Gamma Epsilon, v. 54, no. 2, p. 20-28.

Carr, D. D., 1973, Geometry and origin of oolite bodies in the Ste. Genevieve Limestone (Mississippian) in the Illinois basin: Indiana Geological Survey Bulletin 48, 81 p.

Carr, D. D., C. B. Rexroad, and H. H. Gray, 1986, Ste. Genevieve Limestone, *in* R. H. Shaver, C. H. Ault, A. M. Burger, D. D. Carr, J. B. Droste, D. L. Eggert,

H. H. Gray, D. Harper, N. R. Hasenmueller, W. A. Hasenmueller, A. S. Horowitz, H. C. Hutchison, B. D. Keith, S. J. Keller, J. B. Patton, C. B. Rexroad, and C. E. Wier, eds., Compendium of Paleozoic rock-unit stratigraphy in Indiana—a revision: Indiana Geological Survey Bulletin 59, 23 p.

Choquette, P. W., and R. P. Steinen, 1980, Mississippian non-supratidal dolomite, Ste. Genevieve Limestone, Illinois basin: Evidence for mixed-water dolomitization: SEPM Special Publication 28, p. 163-196.

Cluff, R. M., 1986, Application of modern carbonate sand models to oil and gas exploration, Mississippian Ste. Genevieve Limestone, Illinois basin, *in* Aux Vases and Ste. Genevieve Formations—a core workshop and field trip guidebook: Illinois Geological Society and Illinois State Geological Survey (available from Illinois Oil and Gas Association, Mt. Vernon, Illinois), p. 5-7.

Cluff, R. M., and J. A. Lineback, 1981, Middle Mississippian carbonates of the Illinois basin: a seminar and core workshop: Illinois Geological Society, 88 p.

Connolly, F. T., 1949, The geology of the Passport oil pool in Clay County, Illinois: [M.S. thesis]: Ohio, University of Cincinnati, 33 p.

Droste, J. B., and G. L. Carpenter, 1990, Subsurface stratigraphy of the Blue River Group (Mississippian) in Indiana: Indiana Geological Survey Bulletin 62, 45 p.

Howard, R. H., 1991, Hydrocarbon reservoir distribution in the Illinois basin, *in* M. W. Leighton, D. R. Kolata, D. F. Oltz, and J. J. Eidel, eds., Interior cratonic basins: American Association of Petroleum Geologists Memoir 51, p. 299-327.

Johnson, J. J., 1988, A subsurface study of the Ste. Genevieve Limestone (Mississippian) in the northeastern Warrick Co., Indiana [M.S. thesis]: Bloomington, Indiana University, 275 p.

Keller, S. J., and L. E. Becker, 1980, Subsurface stratigraphy and oil fields in the Salem Limestone and associated rocks in Indiana: Indiana Geological Survey Occasional Paper 30, 63 p.

Mast, R. F., and R. H. Howard, 1991, Oil and gas production and recovery estimates in the Illinois basin, *in* M. W. Leighton, D.R. Kolata, D. F. Oltz, and J. J. Eidel, eds., Interior cratonic basins: American Association of Petroleum Geologists Memoir 51, p. 295-298.

Chapter 7

Paleogeography and Cementation in a Mississippian Oolite Shoal Complex: Ste. Genevieve Formation, Willow Hill Field, Southern Illinois Basin

Ronald D. Manley
Marathon Oil Company
Petroleum Technology Center
Littleton, Colorado, USA

P. W. Choquette
Dept. of Geological Sciences
University of Colorado at Boulder
Boulder, Colorado, USA

M. B. Rosa
Marathon Oil Company
Petroleum Technology Center
Littleton, Colorado, USA

ABSTRACT

Lenticular bodies of porous, permeable ooid grainstone in the upper Ste. Genevieve Formation (Valmeyeran, middle Mississippian) form small oil reservoirs in Willow Hill field, Jasper County, Illinois. The ooid grainstones occur in a swarm of elongate, convex-upward bodies 0.2–2.5 mi (0.32–4.0 km) long by 500–3000 ft (150–1000 m) across and up to 15 ft (5 m) thick. They are interpreted as largely marine shoal deposits that developed along a bioclastic carbonate-sand shoal of low relief and NW–SE trend. Some show evidence of exposure and residence of freshwater lenses. Lime mudstones form capping and updip seals.

Early cements, mainly meteoric in origin, prevented physical compaction and "propped" original pores. Fringing isopachous cements, interpreted as phreatic, resulted in systems of well-connected pores. Ooid grainstones with this combination of cements have porosities (ϕ) of 6.0–17.2% (average 12.7%) and permeabilities (k) of <0.1–228 md (average 113 md). In other ooid grainstones, meniscuslike cements (vadose) also propped original pores, but were overgrown by later blocky burial cement that resulted in less well-connected pores and poor to non-reservoir facies with ϕs of 5.2–17.9% (average 12.8%) but ks of <0.1–1.0 md, with an average of 0.13 md. In associated bioclastic grainstones, ϕ and k values are somewhat less than those of the reservoir-quality ooid grainstones due to syntaxial cement around uncoated crinoid grains.

INTRODUCTION

Willow Hill field is located just south of the town of Willow Hill in Jasper County, Illinois. It is situated on the gently dipping (approximately 100 ft/mi, 19 m/km) southwest flank of the LaSalle anticlinal belt and produces from the Ste. Genevieve Limestone of middle Mississippian (Valmeyeran) age (Figure 1). Oolitic carbonate reservoirs within the field occur in the upper 100 ft (30 m) of the Ste. Genevieve. Here, as elsewhere in the Illinois basin, the oolites are referred to locally as "McClosky zones." The Ste. Genevieve Limestone at Willow Hill is a relatively thin (135 ft, or 40 m) sequence of shallow marine carbonates. Elsewhere in the Illinois basin the Ste. Genevieve ranges up to 400 ft (120 m) thick at the southernmost extent of the basin in Kentucky. For a more complete discussion of regional Ste. Genevieve deposition, see Choquette and Steinen (1980, 1985).

Willow Hill and other nearby Ste. Genevieve reservoirs are stratigraphic traps formed by the capping of localized oolite bodies by "dense," impermeable lime mudstones. Historically these reservoirs have made good waterflood candidates because they are solution-gas drive reservoirs that generally respond well to renewed reservoir pressure.

The field was initially developed in the mid- to late 1940s by Pure Oil Company and other independent operators. Initial development took place in Sec. 6, T6N, R11E (3rd Prime Meridian). Subsequent drilling began in the mid-1970s and continued through 1989, with the result that the initial producing area in Sec. 6 was expanded and a new area for development was discovered immediately to the southeast in Sec. 7, T6N, R14W (2nd Prime Meridian) (Figure 2).

Figure 2 displays the available electric logs, porosity logs, cores, and sample-cuttings data that were available for this study. E-logs only were available for 65 wells in the field, and porosity logs for an additional seven wells. Six of these latter were wells drilled as a part of the waterflood development project, so the field was essentially developed without the use of porosity logs. Cores from three waterflood injector wells were available for study and have been described in detail. One or two additional cores were cut during the primary development of the field, but neither the cores nor core-analysis data are available. Sample cuttings from 29 wells were available, and approximately 450 thin sections were made and described from both the sample cuttings and cores. In addition, drilling-time records, primary production data, and waterflood data were available on selected wells in the field.

The nature of this database significantly limited our ability to define accurately the complexities of the reservoir. The greatest limitation was the paucity and areal restriction of core data in the field, since the three available cores were from wells in Sec. 7 of T6N, R14W (Figure 2). Because core control was sparse, facies were interpreted without core data in 69 of the 72 logged wells in the field and from electric logs only in 40 of the 69 wells. Although criteria were established that allowed facies interpretations to be made from SP/resistivity log responses, this approach can never be as accurate as direct core observations.

Although analysis of thin sections made from cuttings proved helpful, there were limitations to the use

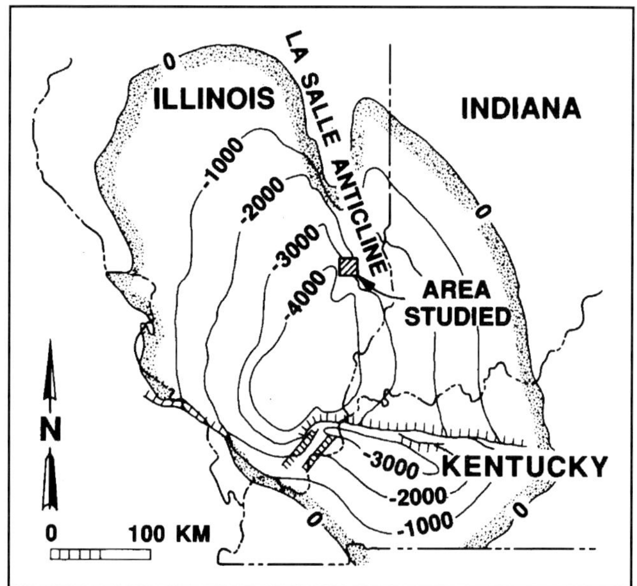

A

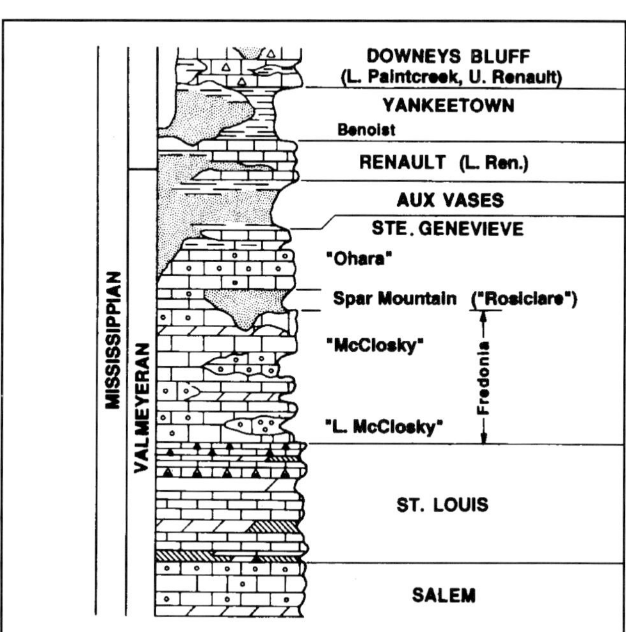

B

Figure 1. (A) Regional New Albany Shale structure map showing location of study area, from an unpublished map by J. H. Buehner, Marathon Oil Company. CI = 1000 ft (304.8 m). (B) Generalized stratigraphic column showing Ste. Genevieve Limestone and other units (modified from Huff, 1987).

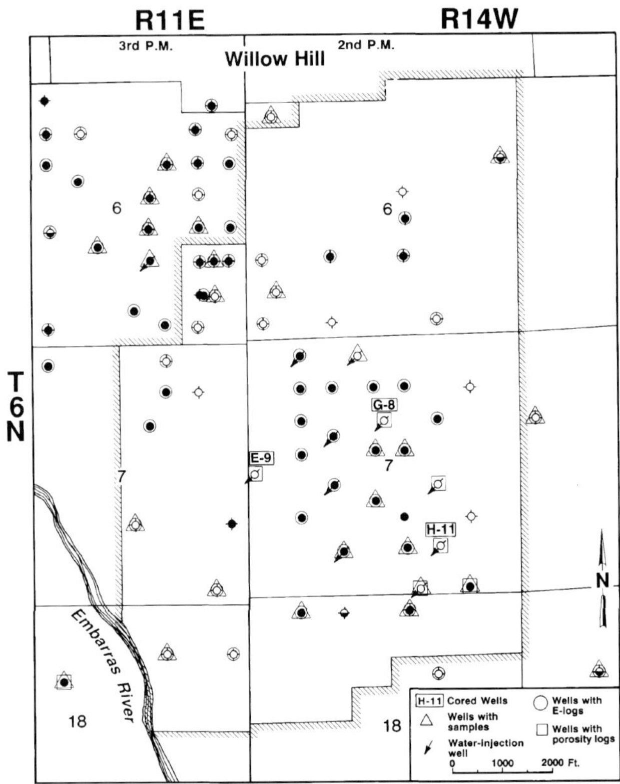

Figure 2. Base map of Willow Hill field showing well locations and data types.

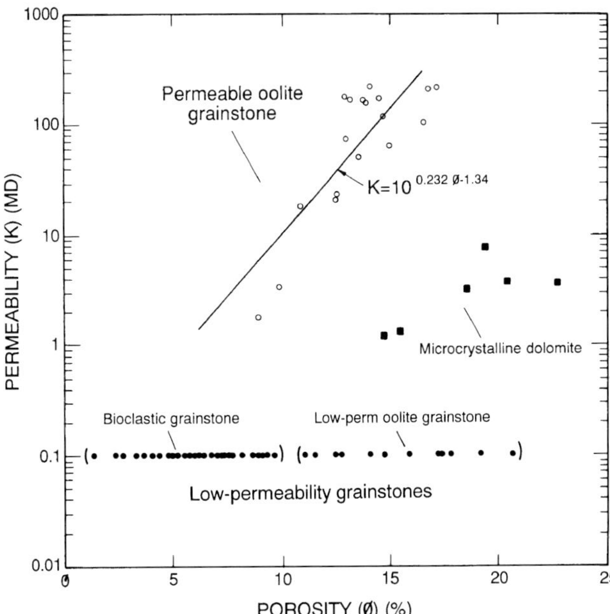

Figure 3. Porosity-permeability core data crossplot. (Note three groupings of data points corresponding to permeable oolite grainstone, low-permeability grainstones, and microcrystalline dolomite. 0.1 md is the minimum permeability that was measured by core analysis, so the 0.1 md values shown are the maximum permeabilities possible for those samples.)

of such data as well. Many of the sample fragments were small, making accurate descriptions difficult. Also, porosity was difficult to determine because porous grainstones were often reduced to individual grains. In fact, one of the criteria used to determine loosely cemented grainstones was abundance of loose ooids. In more tightly cemented grainstones, individual grains or crystals of intergranular cements were often plucked during the thin-sectioning process, making it difficult to determine porosity accurately. Another problem encountered with the cuttings was that of sample mixing. This caused difficulty in using the sample descriptions to assign facies to the logs. Often it was necessary to use the samples to confirm the presence of a particular facies while using the SP and resistivity responses to assign facies intervals.

METHODS OF STUDY

Log Analysis

Because SP/resistivity logs were the most common type available in the field, an effort was made to use them as much as possible. Core-analysis data were limited to 86 analyses from three cored wells, the E. Urfer a/c 3 #H-11, J. A. Eubank a/c 1 #G-8, and F. Ochs #E-9. (Locations of cored wells are shown in Figure 2.) Core porosities and permeabilities were crossplotted to identify groupings and trends of data. From this crossplot, shown in Figure 3, three general rock groupings emerged. These rock groupings are related to, but separate from, the seven lithofacies based on core description and log analysis that are presented in Table 1:

1. *permeable oolite grainstone*—This main reservoir lithofacies is represented by the points with highest permeabilities on the crossplot.
2. *low-permeability grainstones*—The grainstones in this group include both oolite and bioclastic grainstone facies (Table 1) and, as discussed later in this paper, their intergranular cements differ in distribution and type.
3. *microcrystalline dolomite*—This group is not a major producing reservoir at Willow Hill field and will not be discussed in detail in this article.

From these results, porosity alone can be seen to be a poor indicator of permeable grainstone at Willow Hill field. In addition to the limited porosity data available, it is impossible to distinguish between groups 1 and 2 using porosity alone. Without the group 2 nonpermeable oolite, it would be possible to estimate permeability based on log-derived porosity and a known oolite grainstone lithology.

Without porosity to estimate permeable grainstone, we were limited to the use of SP/resistivity logging devices. Resistivity was first considered a

TABLE 1. Lithofacies of Ste. Genevieve Limestone distinguished on SP-resistivity logs using information from cores and well samples. McClosky Zones include carbonates of facies 1 and 2.

ROCK FACIES NAME	CHARACTERISTICS AND INFERRED DEPOSITIONAL ENVIRONMENTS
Oolite Grainstone [Facies 1]	Main reservoir facies; minimum (most negative) SP deflection and variable but relatively low resistivity; lenticular or tabular-lenticular bodies, generally with flat bases and convex-upward tops; mostly very shallow-marine bars and shoals but some oolite-sand bodies exposed and probably modified by eolian processes.
Bioclastic Grainstone and Bioclastic-Oolite Grainstone [Facies 2]	Secondary reservoir facies; slightly less negative SP deflection and higher resistivity compared with facies 1; tabular-lenticular carbonate sands accumulated as shallow-marine shoals; commonly formed substrate for oolite bars and shoals.
Mixed Siliciclastic-Carbonate Sand (Grainstone) [Facies 3)	Generally non-reservoir and cemented tight; SP quite positive (despite very low clay content) and characteristically hump-shaped in association with facies 4 owing to very low porosity and permeability; low-angle swash-type cross-stratification in F. Ochs #E-9 core; tabular-lenticular in geometry; interpreted as beach-foreshore sands, periodically exposed and possibly wind-reworked.
Mixed Siliciclastic-Carbonate Silt and Sand [Facies 4]	Non-reservoir facies; SP as above in association with facies 3; non-cross stratified, tabular; interpreted as shoreface carbonate-terrigenous sand, muddy sand, and silt.
Limestone-Wackestone [Facies 5]	Non-reservoir facies, SP rather positive owing to near-zero porosity and permeability; moderately fossiliferous to barren, and very fine-grained; tabular in geometry; may separate, overlie and enclose facies 1 or 2; interpreted as subtidal-lagoonal carbonate muds.
Mudstone-Shale [Facies 6]	Non-reservoir facies; most positive SP and low resistivity; tabular in geometry and may drape facies 1 and 2 bars; organic-argillaceous lagoonal muds.
Microcrystalline Dolomite [Facies 7]	Potential reservoir facies but generally has very low resistivity due to high porosity and high water saturation; occurs in at least 4 units 2–14′ thick; highest unit (1st Dolomite) is datum for stratigraphic cross sections; interpreted as shallow subtidal-lagoonal deposits; some have tidalite caps. SP deflection moderately negative.

permeability indicator but was judged unreliable due to extreme variations caused by a wide range of tool types run during the field-development period, 1946–1989. However, modern logging devices with shallow, medium, and deep resistivity curves distinguish the low- and high-permeability oolites fairly well. This distinction might have been useful except for the fact that the great majority of resistivity logging devices used in the field had only shallow and deep normal resistivity curves, which do not distinguish between the two ooid grainstone types. In addition, depletion effects could seriously alter resistivity curves of the later drilled wells while not identifying a well's potential for waterflooding. Therefore, the only consistent log data that could be used to estimate permeable grainstone throughout the field was the SP curve.

Core Description

Cores were available for this study from three waterflood injection wells in Sec. 7, T6N, R14W. The cored intervals and thicknesses are as follows: J. A. Eubank a/c 1 #G-8 (2644–2718 ft, 74 ft; 806–828 m, 23 m), F. Ochs #E-9 (2669–2740 ft, 71 ft; 814–835 m, 22 m), and E. Urfer a/c 3 #H-11 (2657–2687 ft, 30 ft; 810–819 m, 9 m). In these cores 12 carbonate lithofacies were distinguished in the upper Ste. Genevieve.

Of these lithofacies, seven could be identified with some confidence on the SP/resistivity logs. Graphic core summaries for the three wells are included as Figures 4–6.

Thin sections prepared from the cores proved indispensable in identifying and interpreting the intergranular cements. As is standard practice in carbonate-reservoir studies of this scope, thin sections made from the cores were impregnated with blue-dyed epoxy to accentuate connected pores and were stained with Dickson's solution of alizarin-red and potassium ferricyanide (Dickson, 1966) to distinguish calcite from dolomite cements and to identify qualitative differences in Fe content between generations of calcite cements. The same techniques were used in making thin sections of cuttings.

Cuttings and Log Analysis of Lithofacies

Thin sections of well samples representing 5- or 10-ft intervals from 30 uncored wells were analyzed with a low-power petrographic microscope (6–50x). Information including percentages of the seven log-identified lithofacies, porosity data, and information on cements was tabulated. These data were used extensively when assigning lithofacies to log intervals and in evaluating cement types. Both the newer Dual-Induction Laterologs (DILs) from recently drilled wells and the older SP and normal-resistivity logs from a majority of the wells were interpreted qualitatively in assigning facies. It was feasible to identify and determine the distributions of the seven major lithofacies with some confidence using this combination of approaches (Table 1).

Mapping Techniques

Mapping techniques were used extensively in attempting to define permeability distribution in the field. In addition to standard structure and thickness maps, many of which are included in this report, slice maps were constructed in an effort to define the areal geometry of oolite sand bodies and reconstruct the sequential development of the Ste. Genevieve shelf in this area. The maps were made following the general approach described by Asquith (1979) from facies interpretations recorded for each well. The slices were composed of successive 5-ft intervals measured up from the datum horizon. The datum horizon was chosen as the top of a particular interval of microcrystalline dolomite, informally called the First Dolomite in this study, which in cores from the J. A. Eubank #G-8 well (2690–2696 ft; 820–822 m) and F. Ochs #E-9 well (2716–2722.6 ft; 828–828 m) is capped by a thin mud-cracked and/or pellet-rich fenestral dolomite interpreted to be of tidal-flat origin. Log characteristics also suggested that a lithologically similar interval was encountered in some of the uncored wells, leading us to interpret the intertidal carbonates as part of a coastal or island tidal-flat system. The thickness distribution of the First Dolomite (Figure 7) suggests that the top may have had as much as 10–12 ft (3–3.7 m) of relief without correcting for compaction.

The main conclusions from constructing the slice maps were that the general trend of the oolite bodies is northwest–southeast and that the mapped oolites, with some exceptions, consist of a series of overlapping and coalescing lenses of oolite. The greatest limitation to the use of slice maps at Willow Hill, however, is that the intervals as defined are not strictly time slices across the entire area.

DEPOSITIONAL SEQUENCE

In the Willow Hill area the upper 30–40 ft (10–12 m) of the Ste. Genevieve is sampled only in cuttings and on E-logs. The cores available from Willow Hill wells sampled the McClosky oolite zones and associated and underlying beds of the Fredonia Limestone Member, as well as overlying sandy limestone beds that are probably equivalent to the Rosiclare Sandstone (Figures 1, 4). The probable contact between Rosiclare and Fredonia in the Eubank #G-8 well is suggested in Figure 4 (R. M. Cluff, personal communication, 1992).

The cored interval in the Eubank #G-8 well consists essentially of three sedimentary packages (Figure 4): a *lower* interval (2690.2–2718.5 ft, 820–829 m) composed mainly of interbedded lime mudstone/wackestone and microcrystalline dolomite (facies 5 and 7, Table 1); a *middle* interval (2656.1–2690.2 ft, 810–820 m) made up almost entirely of oolitic grainstone and bioclastic grainstone (facies 1 and 2); and an *upper* interval (2644.0–2656.1 ft, 806–810 m) of mostly subtidal lime mudstone and dark gray calcareous shale (facies 5 and 6).

The lower interval contains rare to common elongate nodules of anhydrite and two thin units, at and near its top, of fenestral pellet-rich dolomite boundstone interpreted as carbonate tidal-flat deposits. These two tidalites are probably correlative with tidal-flat carbonates encountered in the Ochs #E-9 (Figure 5) core at the tops of units 9 and 11. Both wells also encountered a single bed of dolomitic lime mudstone containing scattered oncolites (unit 18 in the Eubank #G-8 core and unit 13 in the Ochs #E-9). The lower interval is distinctly evaporitic in the Eubank #G-8, where it is interpreted as mostly of lagoonal, shallow-subtidal origin but punctuated by at least two episodes of shoaling to emergence. The interval is less evaporitic in the Ochs #E-9 well.

Lime mudstone and calcareous shale beds of the upper interval are interpreted to comprise in both wells the main lithologies above the cored sequence, in the upper Ste. Genevieve. These lime mudstones closely resemble those in the lower interval but are more argillaceous than any of the lower interval mudstones.

The McClosky oolites are represented in all three cored wells. Their larger-scale characteristics are summarized in Table 1, and details of their petrography are outlined below.

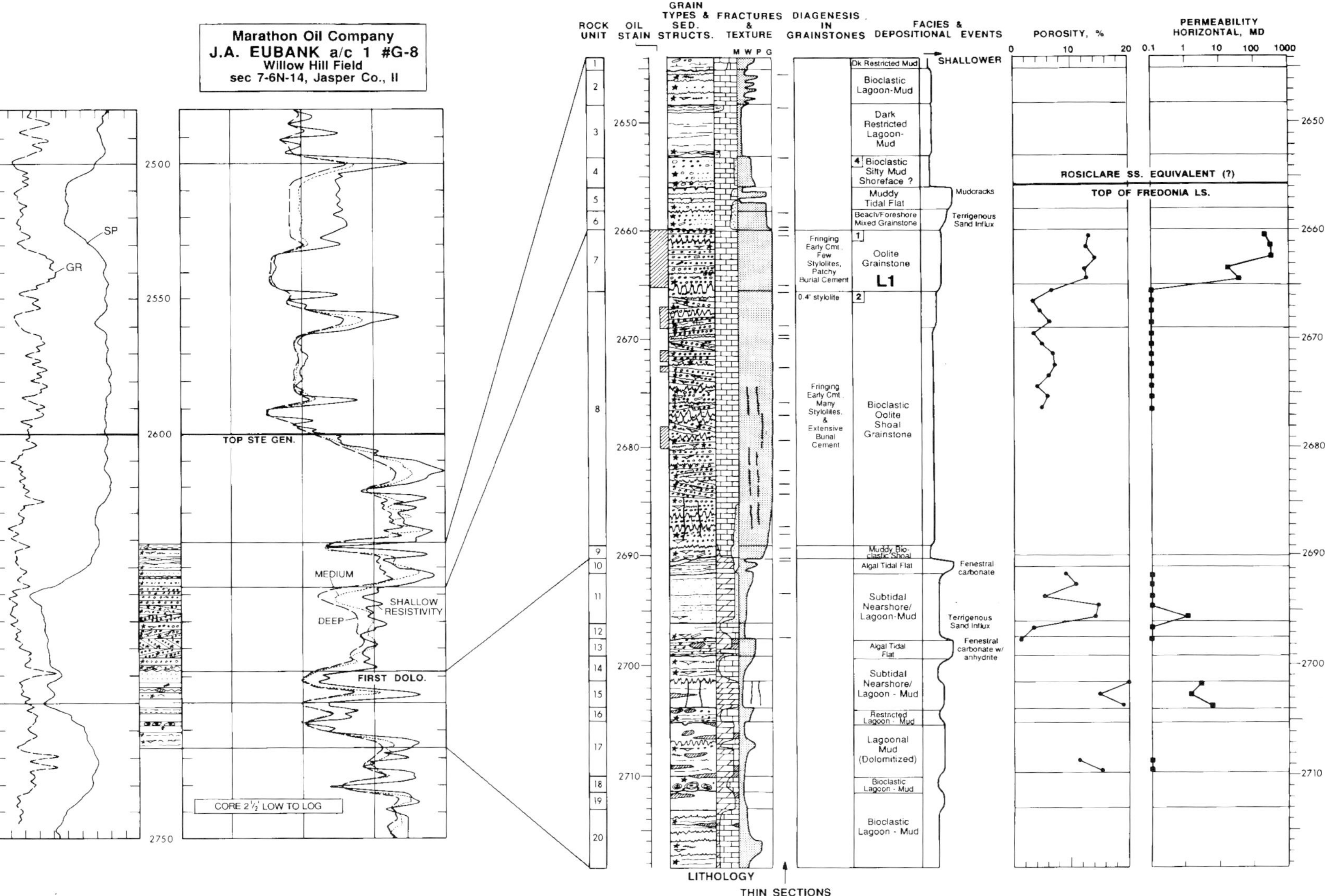

Figure 4. Graphic core summary for the Marathon J. A. Eubank a/c 1 #G-8. The oolite grainstone interval cored in this well, unit 7, is part of the L1 oolite. Along with units 8 and 9, it makes up the Lower Grainstone Shoal sequence, while units 10 and 11 comprise the First Dolomite.

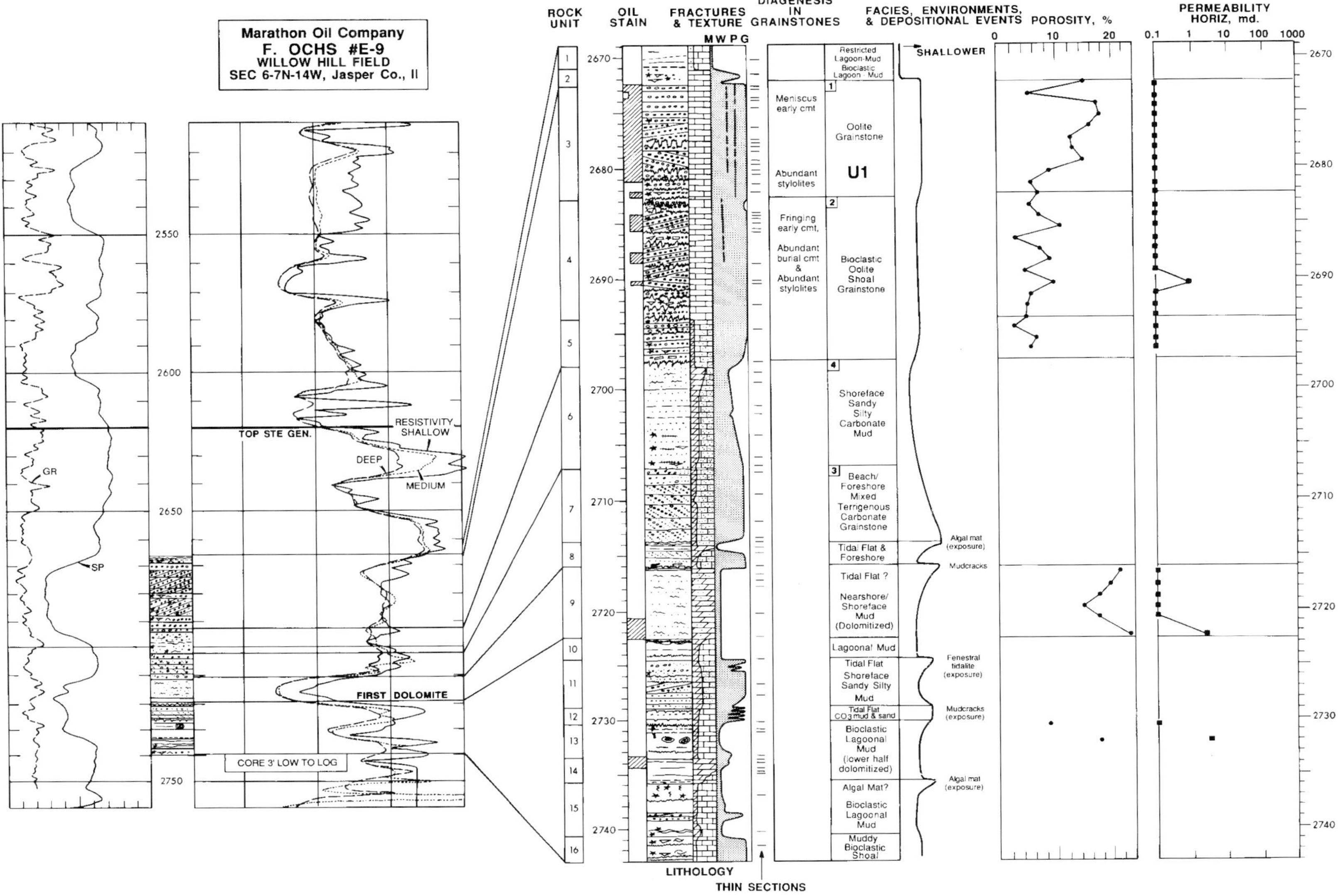

Figure 5. Graphic core summary for the Marathon F. Ochs #E-9. The oolite grainstone interval cored in this well, unit 3, is part of the U1 oolite. Along with units 4 and 5 it composes the Upper Grainstone Shoal sequence while units 8 and 9 comprise the First Dolomite.

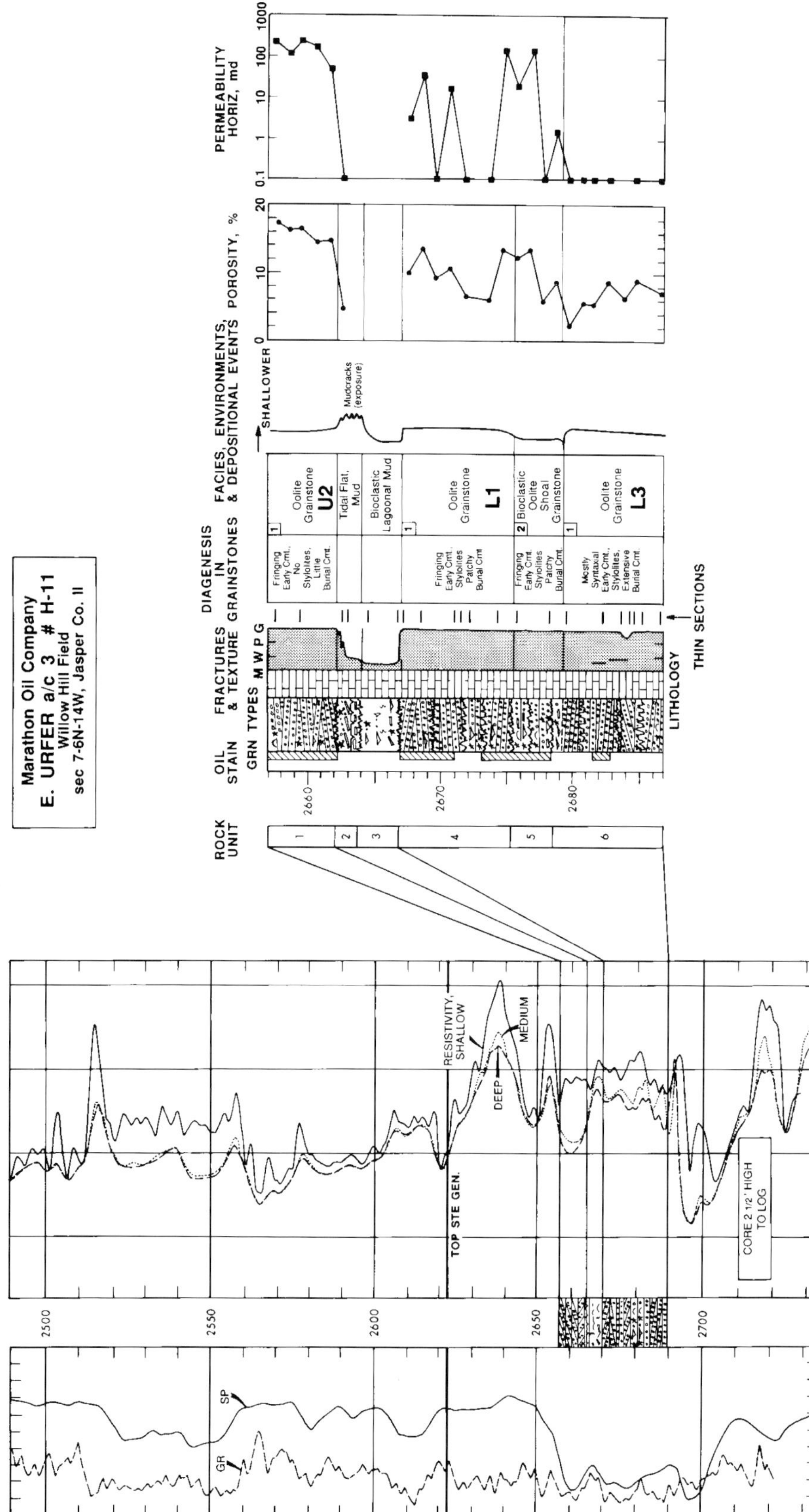

Figure 6. Graphic core summary for the Marathon E. Urfer a/c 3 #H-11. The three intervals of oolite grainstone cored in this well, units 1, 4, and 6 are interpreted to sample respectively the U2, L1, and L3 oolites. The First Dolomite was not cored.

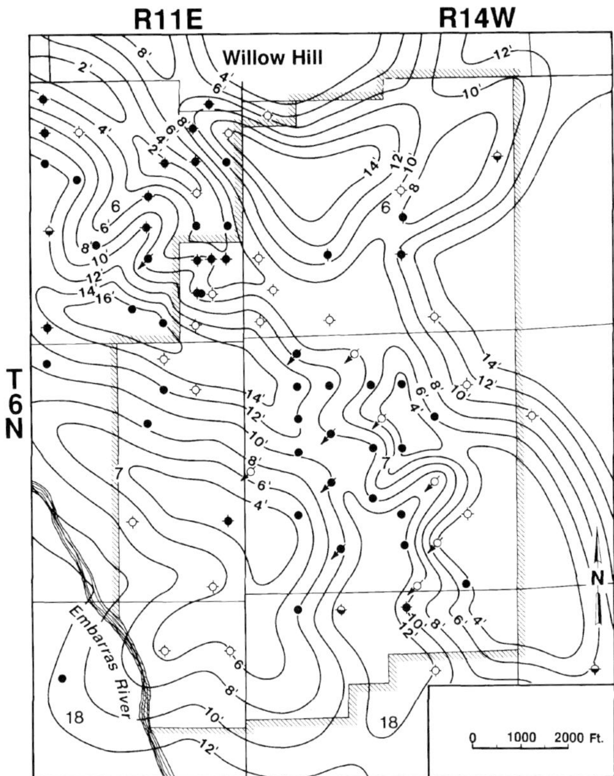

Figure 7. Thickness of First Dolomite. Contour interval: 2 ft (0.61 m).

Oolite Grainstone (Facies 1)

Facies 1, the primary productive facies at Willow Hill, closely resembles the oolite grainstones described by other workers on the Ste. Genevieve and McClosky (e.g., Carr, 1973; Choquette and Steinen, 1980, 1985; Cluff and Lineback, 1981; Cluff, 1984). The grainstones are made up dominantly (30–100%) of ooids 0.1–0.4 mm in diameter with a variety of nuclei, including pellets, quartz sand grains, or fragments of brachiopods, bryozoans, or crinoids, with a few thin to multiple thick oolitic coatings. Varying but small percentages (<10%) of larger, nonoolitically coated grains, such as brachiopod shells and large crinoid fragments, may also occur. Crinoid grains, if uncoated, commonly have overgrowths of syntaxial calcite cement.

In the cores, most oolite grainstone with porosity greater than about 6–7% is partly to uniformly oil stained (Figures 4, 5, 6). Intervals of oolite grainstone are characterized by relatively thick sets of cross-beds (up to 3–4 ft, 1–1.2 m) composed of cross-stratified grainstone. Cross-stratification dips as great as 22° were recorded. Dip directions vary through oolite intervals, although the strongly bidirectional cross-bedding reported in a Ste. Genevieve oolite bar on outcrop in Indiana (Carr, 1973) is not seen in these cores.

Bioclastic Grainstone and Bioclastic Oolite Grainstone (Facies 2)

Sand-sized and some larger fragments of brachiopods, crinoid ossicles, and fenestrate bryozoans are the dominant grains in this carbonate-sand facies. Ooids are rare to common, comprising up to 60% of the grains. As a result, distinctions between facies 1 and 2 are often hard to draw in cuttings. Generally the bioclastic grainstones contain more crinoid fragments and fewer oolitically coated crinoids compared with facies 1. Consequently, these grainstones are more extensively cemented and less oil-productive than the main producing reservoir, facies 1. A more detailed discussion of cementation is found later in the report. A tendency for oolite to overlie bioclastic grainstone observed in the two longest cores (Eubank #G-8 and Ochs #E-9) is inferred from SP/resistivity logs in numerous other wells.

Geometry of Grainstone Bodies

In the cores, contacts between the two grainstone facies appear gradational (over a few centimeters at most) and are commonly obscured by contact stylolites. From the vertical relationships in cores (Figure 6) we infer that the two facies may in places interfinger.

More generally, however, oolite grainstone sand bodies appear to have more or less flat bases, convex-upward tops, and distal ends that taper rather than interfinger with beds occupying similar intervals in nearby wells. Similar geometry has been interpreted by Carr (1973) in a partly exposed oolite sequence near Orleans, Indiana, and can also be seen in continuous exposures at Alton Bluffs, Illinois (Collinson et al., 1979).

Well-to-well correlations of bioclastic grainstone (facies 2), together with interpretation of the slice maps, suggest that this facies occurs in lenticular units, oriented NW–SE, which are as much as 25 ft (7.5 m) thick, a mile (1.6 km) or more across, and 2 mi (3.2 km) or more in length.

Mixed Siliciclastic and Carbonate Facies (Facies 3 and 4)

Rocks in which detrital quartz (also minor chert and feldspar) sand is a major constituent (up to 65% of total rock) are important to an understanding of the geology of the Ste. Genevieve at Willow Hill. No core analyses were available for these facies at the time of our study, but thin sections of cores and cuttings of these rocks reveal small amounts (<3–5%) of poorly connected or isolated porosity, suggesting low permeabilities. This observation and the lack of oil staining in rocks of these facies in cores of the Ochs #E-9 well (units 6 and 7) indicate that they are not reservoir rocks in this area.

These rocks of mixed mineralogy are clearly divisible in the Ochs #E-9 core into two facies that intergrade vertically (Figure 5, units 6 and 7): a lower interval (facies 3) of moderately to very sandy, cross-bedded, medium-grained ooid-pellet grainstone alternating in cross-laminations with fine- to very fine-grained pelletal sandstone (also grainstone texturally); and an upper interval (facies 4) of fine- to very fine-grained, sandy pelletal packstone which becomes finer-grained, less sandy, and more dolomitic wackestone-packstone upward toward the top.

In the Ochs #E-9 core, facies 3 consists of eight sets, 0.3–1.9 ft thick (0.1–0.6 m) of gently dipping (5–7°) cross-laminated sands about 1 mm–1.5 cm thick. Core recovery in units 6 and 7 was complete in this well, and when the cores were fitted back together, it was apparent that the cross-laminations in all eight sets totaling 6.5 ft (~2 m) dip in essentially the same direction within about 15°azimuth. Photos of slabbed cores and thin sections (Figures 8, 9, 10) illustrate the thinness and gentle dips of the cross-lamination.

Petrographically, the mixed siliciclastic and carbonate grainstones of facies 3 resemble in some respects a succession of Ste. Genevieve rocks exposed in road cuts near Corydon, Indiana, that Hunter (1989 and this volume) and others (Dodd et al., this volume) have interpreted as eolianites. The facies 3 carbonates at Willow Hill are dominantly fine- to medium-grained, with coarse sand in some laminae as illustrated in Figure 8, and in some laminae reverse grading with the coarsest and best-rounded grains at tops of laminae. Such reverse grading may occur in translatent stratification of the sort described by Hunter (1989 and this volume) and suggests the possibility that translatent stratification exists in facies 3 as sampled in the F. Ochs #E-9 core (unit 7). Hunter (1989 and this volume) has attributed such reverse grading to accretion of saltating grains on wet surfaces of eolian bed forms. Possible root casts occur at one horizon in the core (Figure 9; 2708.6 ft, 825.6 m) and a thin veneer of possible soil at another horizon (Figure 9; 2709.5 ft, 825.9 m). These features suggest that some bed forms were at least briefly exposed.

Other features of facies 3 as represented in the F. Ochs #E-9 core (Figure 10) seem more compatible with a subaqueous origin. These features include apparent burrows in the upper part of unit 7, consistently low-angle, planar(?) cross-stratification, and vertical intergradation with locally bioturbated, finer-grained wackestone and mudstone of unit 8. Marine bottom or infaunal invertebrates, which might help resolve the question of eolian vs. marine origin, have not been observed in either unit. The abundant debris of marine invertebrates in facies 3, chiefly of fenestrate bryozoans, crinoids, and brachiopods, is nondiagnostic of an eolian or subaqueous marine origin, as is the occurrence of facies 3 and 4 beneath a marine shoal grainstone sequence.

The cross-stratification in unit 7 (facies 3) is provisionally interpreted to be of wave-swash origin as interpreted in other settings by McCubbin (1982, pp. 249–258) and others (Harms et al., 1982). Wave swash is best known in beach/foreshore environments, where it runs up the beach slope, occasionally overtopping the berm, then washes back downslope, depositing sand in thin seaward-dipping cross-laminae. In such beach settings, tides tend to shift the swash and surf zones along shore as well as vertically. Beach profiles vary in width, depending in part on gradient and sediment supply, from the order of 100

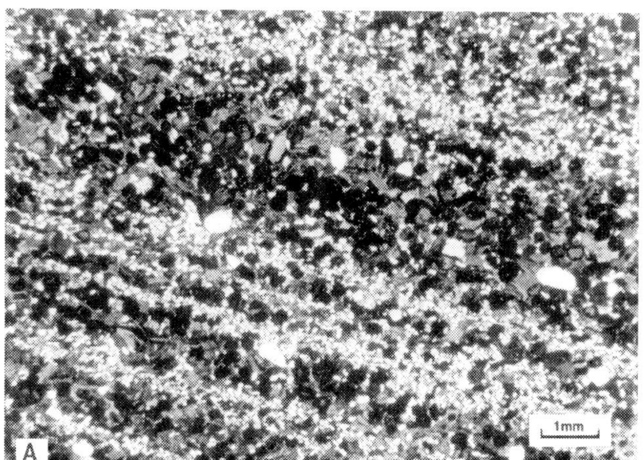

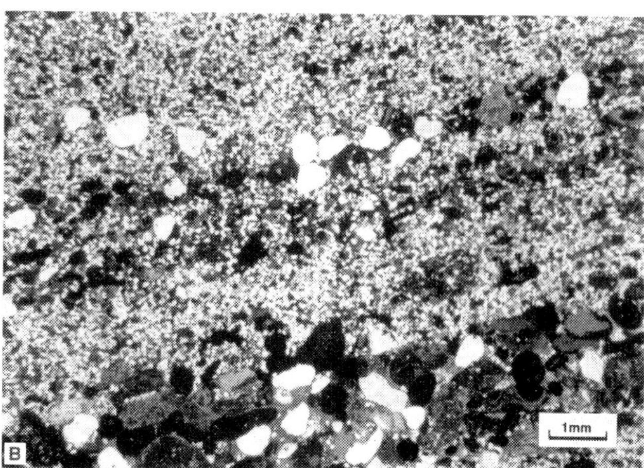

Figure 8. Photomicrographs of cross-bedded siliciclastic/carbonate grainstone of facies 3. Plane-transmitted light calcite constituents look gray because of staining with alizarin-red. (A) Cross-laminae of calcareous fine-grained quartz sandstone alternate with laminae of bioclastic pellet grainstone containing scattered coarse sand grains (F. Ochs #E-9, 2711.8-ft [826.56-m] core depth). (B) Cross-laminae of fine- to very fine-grained quartz sandstone alternate with grainstone, that has scattered medium-coarse quartz grains. Some of these may mantle cross laminae; others occur within cross-beds and laminae (F. Ochs #E-9, 2709.5-ft [825.86-m] core depth).

Figure 9. Core photo of mixed siliciclastic/carbonate facies 3, in F. Ochs #E-9 well. The darkened surfaces of cross-stratification sets at ~2708 ft (825.4 m) and ~2709 ft (825.7 m) are interpreted as thin soil zones with root traces. Photos are closer views of cores shown in Figure 10.

ft (30.48 m) to broad and gently dipping beaches a mile or more in width, and can form strand plains 10 mi (16.09 m) or more in width. Beach-accretion wedges along very low-relief carbonate shorelines can be as wide as the upper dimensions just cited (see, for example, Figure 5A of Inden and Moore, 1983). The nature of the cross-stratification in facies 3 is compatible with a beach/foreshore origin (Table 1).

The rather widespread distribution of facies 3 and 4 in the Willow Hill area (Figure 11) and known occurrences of wave-swash cross-stratification on the seaward faces of some offshore bars (D. G. McCubbin, personal communication, 1990) suggest this as an alternative to a beach/foreshore origin. In the Ochs #E-9 well where facies 3 was cored, it rests on a tidal-flat deposit that caps the First Dolomite (Figure 5, units 8 and 9) and grades upward into finer-grained, commonly muddy (now dolomitic) carbonate and terrigenous sandy wackestones and packstones of facies 4 (unit 6). The rocks of facies 4 are inferred to be of somewhat deeper shoreface or possibly more offshore origin. Either a beach/foreshore or an offshore-bar origin for facies 3 would be consistent with this sequence of deposits. The apparent soil and root zones, however, and the reverse grading and signs of adhesion surfaces indicate that these cross-stratified sediments of mixed provenance were at times exposed and reworked by wind. Indeed, the very existence of beach/foreshore sediments carries with it the strong likelihood that dunes also were developed nearby as on typical barrier islands. The fact that we did not see compelling evidence of wind-transported sediments by no means excludes the possibility that they existed in the immediate area at that time. Certainly it is conceivable that eight thin (0.3–1.9 ft, 0.09–0.58 m) packages of very low-angle, unidirectionally dipping cross-laminations accumulated in an eolianite complex and then were onlapped gradationally by sandy, silty carbonate muds of subaqueous origin. Close juxtaposition of shoreline and shallow offshore facies with wind-transported sediments is to be expected in sequences like this, particularly where sediment packages shift and relative sea level varies.

STRATIGRAPHY OF McCLOSKY ZONES

Detailed stratigraphic correlations of facies assigned to all wells based on log character, cores, and samples resulted in the interpretations shown on four cross sections (Figure 12). Two series of oolite grainstone bodies have been distinguished, a lower series designated the L1, L2, and L3 oolites from youngest to oldest, and an upper series composed of the U1, U2, and U3 oolites. The lower series of oolite bodies occurs in close association with bioclastic grainstones, with which they are referred to collectively as the lower grainstone shoal. Their gross-thickness distributions (Figures 13A,B) indicate that the youngest oolite (L1) is the most extensive. The lower grainstone shoal trends NW–SE and is approximately .25 mi (0.4 km) wide and 2.5 mi (4 km) long. The southwest margin and northwest end of this shoal complex are defined by well control, whereas the northeast margin and southeast end are approximate, except in the SE quarter of Sec. 6, T6N, R14W, where they are also controlled by well penetrations.

The upper series of oolites seems to have accumulated largely around the margins of the lower grainstone shoal, except for the uppermost oolite, U1, which extends partway onto the shoal. The cored interval in the Ochs #E-9 well is representative of this upper sequence and illustrates that it begins with mixed siliciclastic and carbonate sands of facies 3 and 4. Gross thickness isopach maps of the U1, U2, and U3 oolites are shown in Figures 13C and D.

The interpretation of the Willow Hill McClosky zone stratigraphy just outlined was developed from

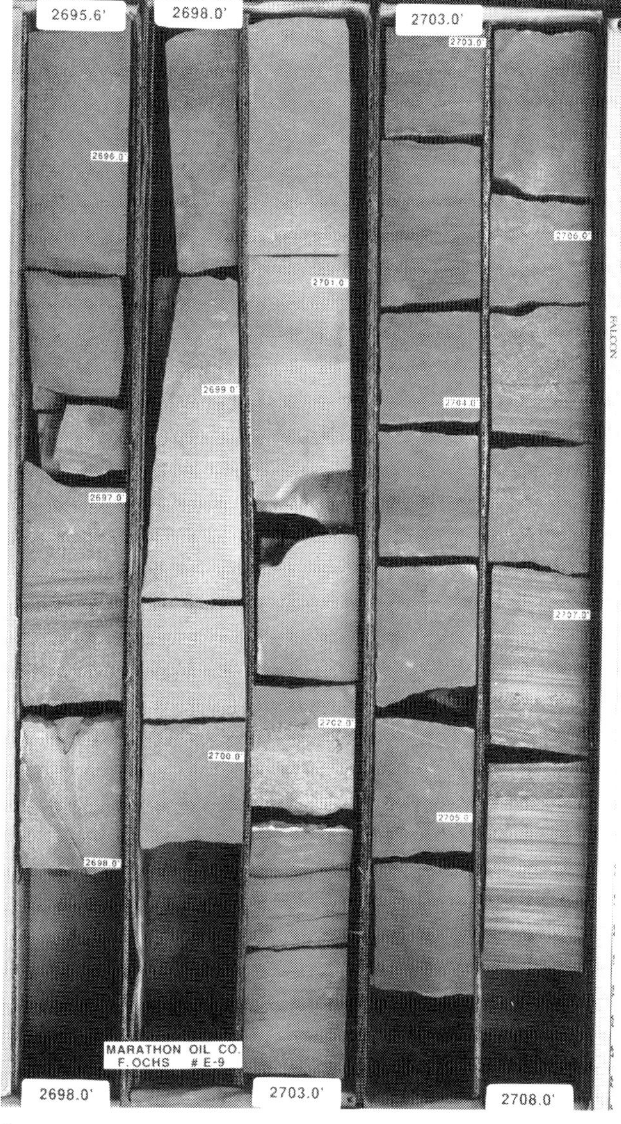

A

B

Figure 10. Core photos of part of Ste. Genevieve Limestone (2695.6–2718.5 ft, 821.59–828.6 m) in the F. Ochs #E-9 well. Cross-stratification dips are essentially unidirectional; changes in dip direction, as in the interval 2711–2712 ft (826.8–827.2 m), are apparent only, due to non-coincident orientation of the saw cuts.

the following considerations. First, the two oolite grainstone sequences are quite different and do not appear to interfinger. For example, bioclastic and oolite grainstones of the lower oolite series rarely contain more than a few percent of terrigenous sand and do not appear to represent facies equivalents of the terrigenous-sand-rich facies 3 and 4. Second, lines of section normal to the depositional trends (Figure 12) indicate that there are at least two oolite bodies, U1 and U2, which partly or entirely postdate the uppermost L1 oolite. One of these, the U2 oolite, has been recognized in almost every well beyond the limits of the lower grainstone-shoal trend, but is found in only two edge wells on the trend itself. Even the U1 oolite has been recognized in only two wells on the lower shoal trend (Figure 13C). Third, the upper and lower series of oolites, as represented in the Ochs #E-9 and Eubank #G-8 cores respectively, contain different series of early pore-lining cements. Early cements in the #E-9 well (U1 oolite) are primarily of meniscus type, whereas those in the #G-8 well (L1 oolite) are chiefly fringing isopachous cements. Fourth, the

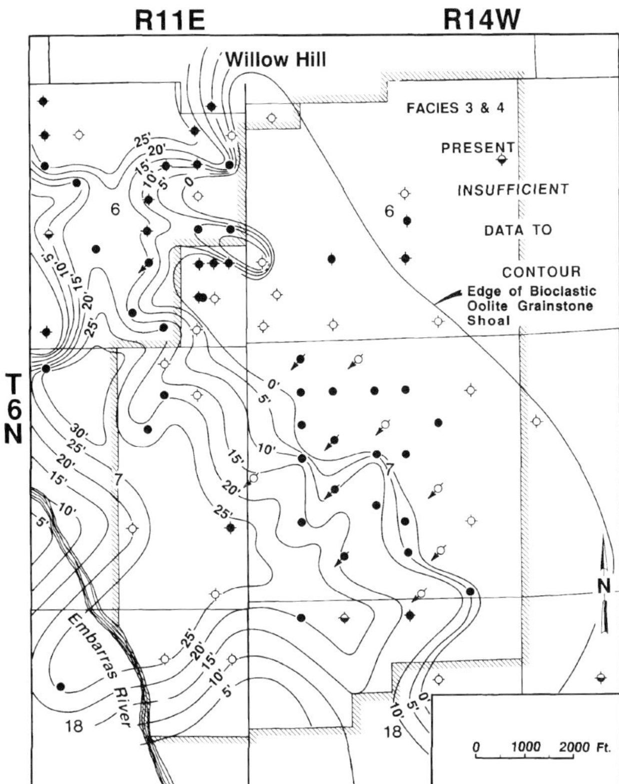

Figure 11. Thickness map of terrigenous/carbonate nearshore deposits (facies 3 and 4).

youngest oolites in the two series, L1 and U1, were drowned by onlap of different lime mudstones, as illustrated in the St. Pierre J. A. Eubank #3 (#F-8) on cross section B–B′ of Figure 12B.

CONTROLS ON POROSITY AND PERMEABILITY OF GRAINSTONES

Porosity in the grainstones consists of intergranular pores that have been preserved to varying degrees. Solution porosity is rare, and moldic, small-vug, and solution-enlarged intergranular pores together make up less than 5% of the pore volume.

Although core-analysis data were limited and restricted to small portions of the field, data for 36 ooid grainstone samples in the three cored wells indicate an average porosity of 12.7% (5.2–17.9% range), and for 29 bioclastic grainstone samples, an average porosity of 5.6% (2.3–11.0% range). Permeabilities vary widely within the ooid grainstone depending on the presence of meniscus-style cements (Dunham, 1971). The average permeability of ooid grainstones that lack meniscus cement is 113 md (0.1–228 md range), whereas in ooid grainstones that have meniscus cement it is only 0.13 md (<0.1–1.0 range). The average permeability for the bioclastic grainstone is 0.1 md. Detailed study of thin sections from cores and cuttings shows that porosity occlusion is a result of cementation and in some intervals of grain-on-grain physical compaction. The original porosity of the carbonate sands at deposition is likely to have been 40–55%, judging from the best available data on Holocene carbonates (Enos and Sawatsky, 1981).

DEPOSITIONAL MODEL AND EVOLUTION OF STE. GENEVIEVE OOLITE RESERVOIRS, WILLOW HILL AREA

Based on the stratigraphic relationships just outlined and isopach maps of numerous carbonate sand bodies and other intervals, the evolution of the Ste. Genevieve carbonate shelf in the Willow Hill area has been reconstructed in a succession of stages, portrayed by the block diagrams of Figure 14.

Stage 1. Deposition of First Dolomite

Prior to the deposition of the First Dolomite, the area was dominated by shelf and lagoonal lime-mud deposition with local low-lying islands fringed by evaporative tidal flats. Mud-cracked beds of stromatolitic algal mats and associated carbonate sediments accumulated on the tidal flats, along with minor nodules and thin beds of anhydrite (Ochs #E-9 core). The First Dolomite, presently about 2–16 ft (1.2–4.8 m) in thickness uncorrected for compaction, represents the final episode of sedimentation in this early Ste. Genevieve regime. It accumulated as a thin sequence of pelletal carbonate muds mixed with very fine-grained siliciclastic sand and silt. These sediments were deposited under very shallow offshore to low intertidal conditions followed by shoaling to a low-lying coastal or island system veneered in places by tidal flats. The tidal flats and associated "highs" are interpreted to have occupied areas where the First Dolomite is thick, whereas lagoons and tidal estuaries occupied areas where it is thin (Figure 7).

Stage 2. Establishment of Lower Grainstone Shoal

A sizable marine shoal composed mostly of bioclastic carbonate sands developed along the NW–SE trend roughly defined by the low islands across Secs. 6 and 7, T6N, R14W, illustrated as First Dolomite thicks in Figure 7. These grainstones consist mainly of crinoid, brachiopod, and bryozoan debris, with varying amounts of ooids and small benthic foraminifers. Bioclastic debris may have been supplied from the shoal itself, or by storm transport from off-shoal areas where lime-mud sedimentation occurred and tracts of crinoids with associated fauna existed. At the southeast end of the grainstone shoal, the L3 oolite accumulated as a shallow-marine bar, shown in cross section D–D′ (Figure 12D) and Figure 13B. At the northwest end of the shoal the L2 oolite accumulated first as a system of separate bars, which later coalesced into a nearly continuous blanket of amalgamated bars.

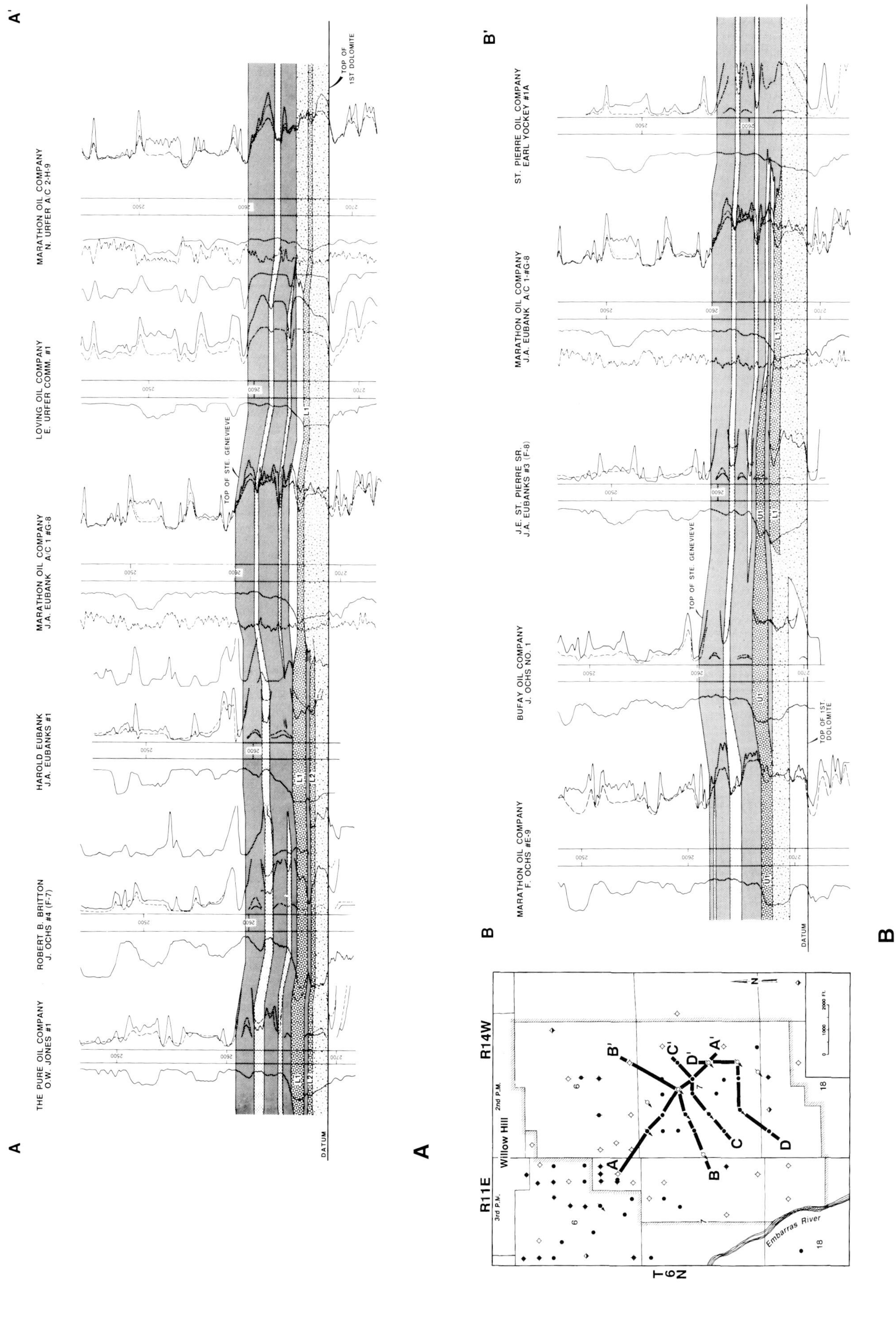

A
A'
THE PURE OIL COMPANY
O.W. JONES #1
ROBERT B. BRITTON
J. OCHS #4 (F-7)
HAROLD EUBANK
J.A. EUBANKS #1
MARATHON OIL COMPANY
J.A. EUBANK A/C 1 #G-8
LOVING OIL COMPANY
E. URFER COMM. #1
MARATHON OIL COMPANY
N. URFER A/C 2-H-9
TOP OF STE. GENEVIEVE
TOP OF
1ST DOLOMITE
DATUM
B
B'
MARATHON OIL COMPANY
F. OCHS #E-9
BUFAY OIL COMPANY
J. OCHS NO. 1
J.E. ST. PIERRE SR.
J.A. EUBANKS #3 (F-8)
MARATHON OIL COMPANY
J.A. EUBANK A/C 1-#G-8
ST. PIERRE OIL COMPANY
EARL YOCKEY #1A
TOP OF STE. GENEVIEVE
TOP OF 1ST.
DOLOMITE
DATUM
R11E
R14W
3rd P.M.
2nd P.M.
Willow Hill
T6N
Embarras River

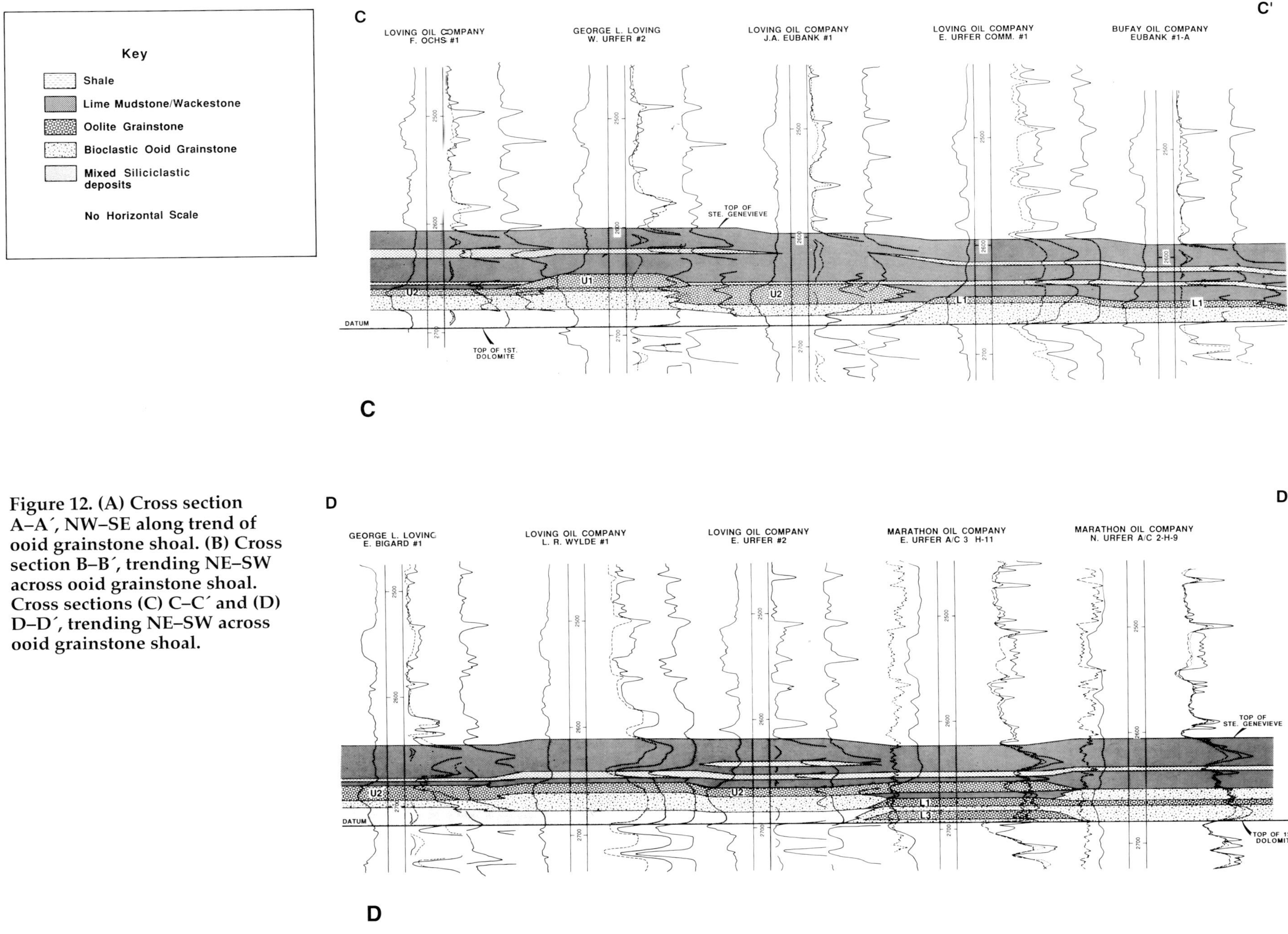

Figure 12. (A) Cross section A–A′, NW–SE along trend of ooid grainstone shoal. (B) Cross section B–B′, trending NE–SW across ooid grainstone shoal. Cross sections (C) C–C′ and (D) D–D′, trending NE–SW across ooid grainstone shoal.

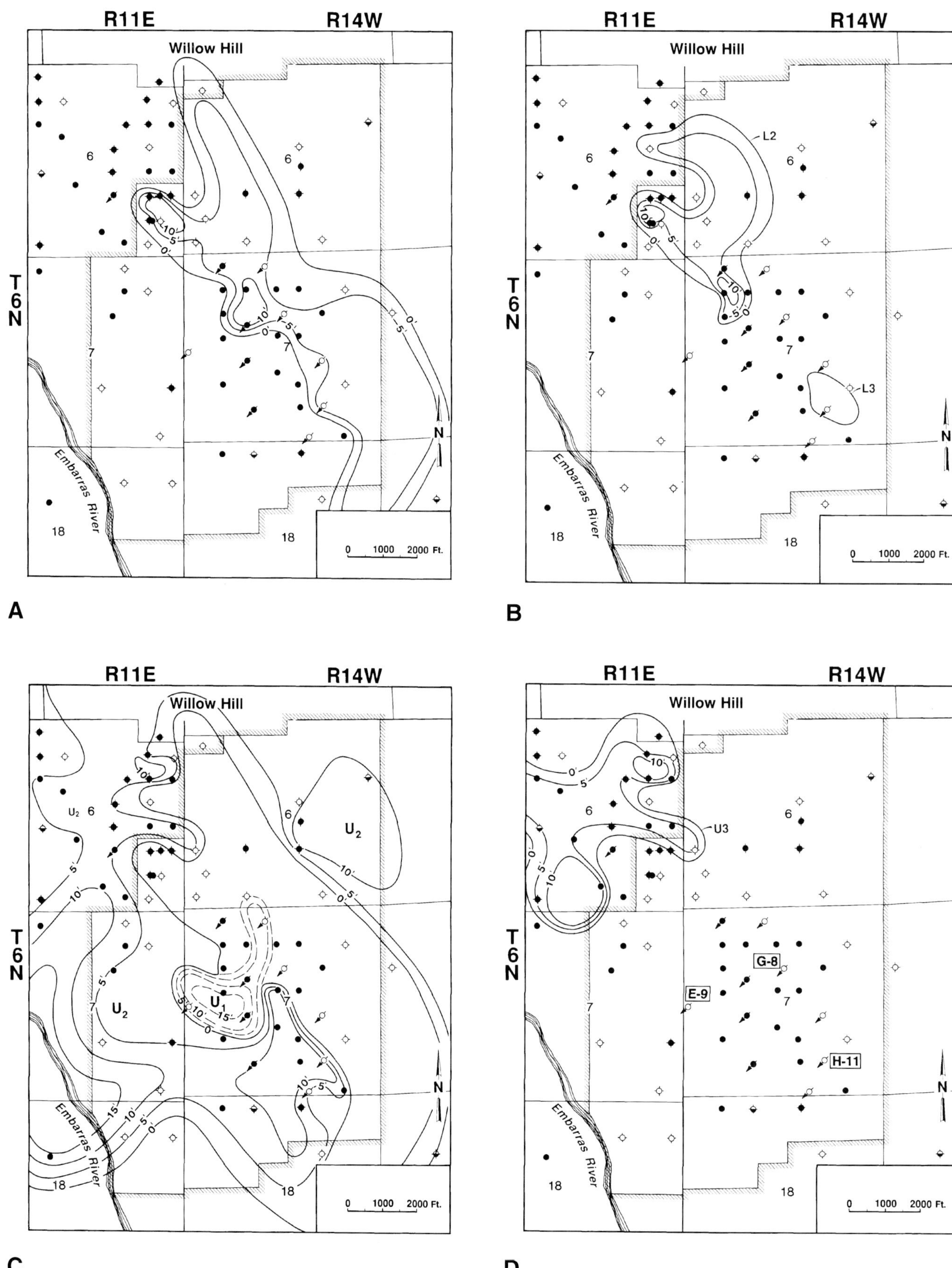

Figure 13. (A, B) Thickness maps of lower series of ooid grainstones (A: L1 oolite; B: L2 and L3 oolites). (C, D) Thickness maps of upper series of ooid grainstones (C: U1 and U2 oolites; D: U3 oolite).

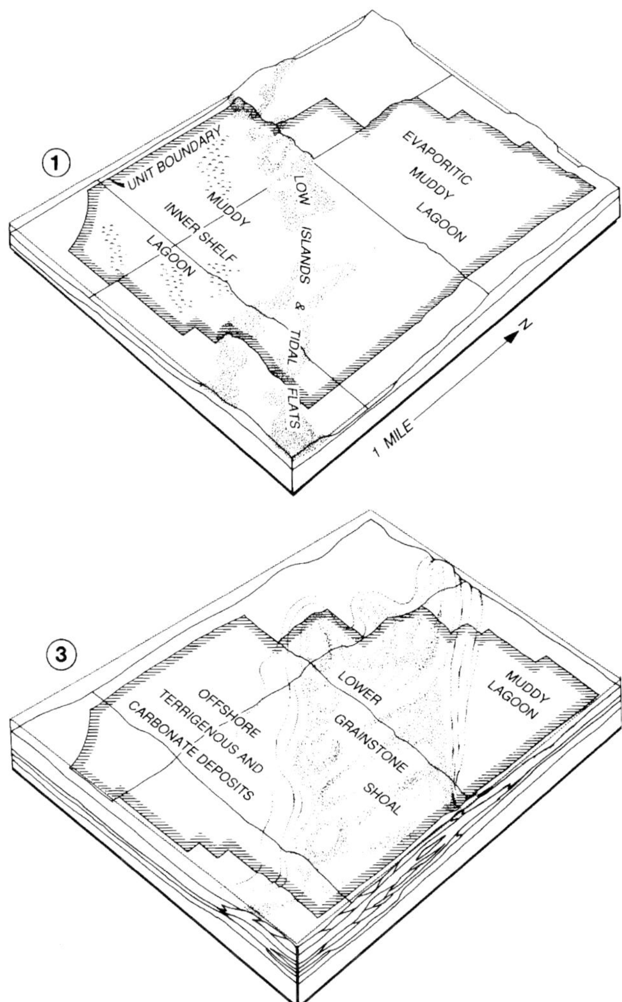

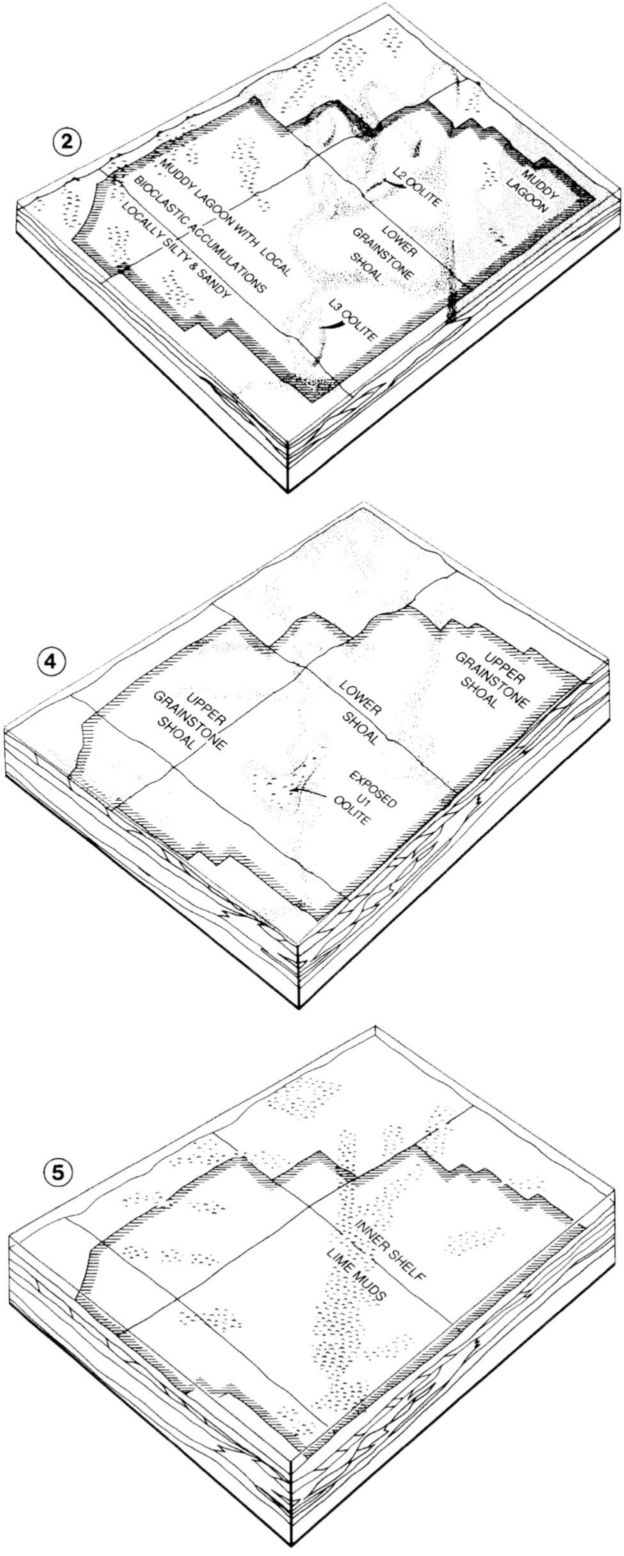

Figure 14. Block diagrams (stages 1–5) showing the Willow Hill area as it may have looked at various stages of Ste. Genevieve deposition. *Stage 1:* Area at the end of First Dolomite deposition (datum surface). Northwest-trending islands with low relief were partly covered by moderately evaporitic algal-mat encrusted tidal flats, and adjoined by shallow water in which marine lime muds accumulated. The island sediments and nearby muds all were pervasively dolomitized. *Stage 2:* Area during an early stage in accumulation of the lower grainstone shoal. Entire area is submerged to shallow depths of perhaps 1–2 ft (0.3–0.6 m) to 20–30 ft (6.1–9.1 m). Northwest-trending oolite bars of L2 and L3 oolite have developed on the shoal. Offshoal, thin fossiliferous muds and mudbanks may have supplied bioclastic debris to the shoal. *Stage 3:* Area at maximum extent of the L1 oolite on the lower grainstone shoal. Most of area is submerged to shallow depths as in stage 2. Offshoal, mixed siliciclastic/carbonate sediments either have begun to accumulate or soon will. Oolite is exposed as small islands on which local freshwater lenses occur. *Stage 4:* Area at maximum extent of upper grainstone shoal (U1 and U2 lower shoal is exposed and has developed an active oolites). Most of area is submerged to shallow part-way across the depths (1–10 ft?), but the U1 oolite bar extending freshwater lens. *Stage 5:* Willow Hill area after grainstone shoals have been drowned by shallow transgression and area-wide accumulations of shallow-marine lime muds. These muds, and offshoal lime muds deposited earlier, eventually form updip and capping barriers for the oolite reservoirs.

Stage 3. Culmination of the Lower Grainstone Shoal

At this stage the oolite bars and shoals of the lower oolite series had their maximum extent, covering most of the shoal area (L1 oolite, Figure 13A). Parts of the shoal where oolite bars were thickest may have been subaerially exposed at the end of L1 oolite deposition. There is some indication of this in cross section B–B´ (Figure 12B), where the L1 oolite in the J. A. Eubank a/c 1 #G-8 is capped by facies interpreted as beach and tidal-flat deposits. There is, however, no direct evidence in the upper part of the L1 oolite of exposure such as caliche, root casts, soils, or vadose meniscus cements. Features like these would be difficult to detect in cuttings and might be developed in local areas not sampled by cores. There is a strong possibility that the L1 oolite was exposed in some areas where it was thickest, such as the area represented by the middle well (Eubank #F-8) on B–B´.

The lower grainstone shoal may have had relief of anywhere from 10 to 35 ft (3–10 m) relative to nearby shelf areas. As the shoal built vertically, mixed siliciclastic and carbonate sands began to accumulate along its margins. The abundant terrigenous material in deposits like these may have been supplied via the Michigan River system and its seaward extensions from a land mass that existed to the northeast of this general area (Swann, 1963), then perhaps was spread by longshore currents. This sediment was mixed with carbonate produced on either the lower grainstone shoal or the shelf nearby.

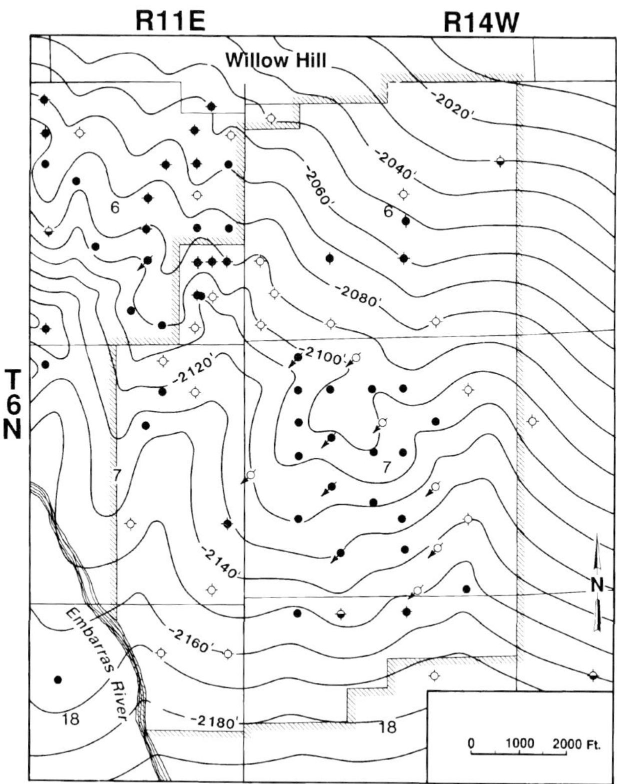

Figure 15. Structure on top of Ste. Genevieve Limestone. CI = 10 ft (3.05 m)

Stage 4. Development of the Upper Grainstone Shoal

Once the offshore mixed terrigenous and carbonate sands accumulated in an upward-deepening sequence (units 8, 7, and 6 in the F. Ochs #E-9 core), a second grainstone shoal developed outboard (SW) of the lower shoal. Initially the shoal sediments were dominated by bioclastic grainstones in the south and by the U3 oolite in the northwest (Figure 13D). With progressive shoaling, however, most of the area surrounding the lower shoal developed into an extensive ooid sand shoal referred to as the U2 oolite. The final depositional episode of this stage was the development of the U1 oolite, which may be nothing more than a diagenetic lens representing the vadose zone of the U2 oolite and which formed at the highest part of the shoal complex, cross section C–C´ (Figure 12C) (R. M. Cluff, personal communication, 1992). As can be seen in Figure 13C, this oolite grainstone body partially overlaps the lower grainstone shoal.

A structure map on the top of the Ste. Genevieve (Figure 15) shows a prominent nose developed in Sec. 7, T6N, R14W and suggests that "thicks" (depositional highs) on the U1 oolite may have affected the deposition of overlying sediments.

The episode of exposure that ended the development of the upper shoal, and affected the U1 oolite, may or may not have affected other parts of the upper shoal complex. The main effect of exposure was the development of meniscus cements of vadose-zone origin that essentially closed off permeability. These cements are most evident in the F. Ochs #E-9 well, where oolite grainstone permeabilities were reduced to 0.1 md or less.

Stage 5. Drowning of the Grainstone Shoals

Following the deposition and exposure of the U1 oolite, the areas occupied by both the lower and upper shoals deepened and became covered by shallow seas that deposited mud-rich sediments. These sediments are now represented by lime mudstone and wackestone, which are bioclastic in places and are interbedded with thin dark gray carbonate-rich shales of the upper Ste. Genevieve. The muds blanketed the lower shoal (e. g., units 3, 2, and 1 in the Eubank #G-8 core) and the upper shoal (e. g., units 2 and 1 in the Ochs #E-9 core). Lime mudstone and wackestone formed the capping and updip seals for the Ste. Genevieve oolite bodies in this area.

CEMENTS AND CEMENTATION

All grainstones at Willow Hill have varying amounts of cement. An estimated 98% of this cement is calcite; the remainder includes coarse dolomite spar, quartz-overgrowth cement on detrital quartz grains, and a few occurrences of poikilitic fluorite.

The rest of this discussion deals only with the calcite cements. These can be divided into two main groups according to the diagenetic stages in which they are believed to have formed. The two groups differ in their effects on porosity and permeability in Willow Hill reservoirs.

Early Cements

Cements of early, near-surface origin bonded the original carbonate sands into partly indurated grainstones and prevented or inhibited later compaction while preserving most original pore volume. The early cements (Figures 16, 17, 18) include: (1) grain-contact cements of meniscus type, (2) grain-fringing, isopachous cements, and (3) syntaxial overgrowth cements on single-crystal carbonate grains, most of which are crinoid segments. Under cathodoluminescence (CL), the early grain-fringing and meniscus cements are dominantly nonluminescent and dark, with thin zones of bright yellow luminescence.

The early cements show evidence of having been precipitated before physical compaction took place. Where fringing, meniscus, or syntaxial cement exists, little or no sign of compaction can be seen and grainstones have open textures (Figures 17A, 18B). Conversely, where grainstones are extensively compacted, little or no early cement can be seen. Also, there are instances in which fringe-cemented brachiopod grains have been broken or cortical layers of ooids together with their cement fringes have been spalled apart by compaction.

In morphology, the fringing and meniscus cements closely resemble varieties that have been observed in Holocene carbonate sands (e.g., Halley and Harris, 1979; Budd, 1988) and Pleistocene carbonates (review by James and Choquette, 1984) that were exposed long enough to develop freshwater lenses. The early syntaxial cements seem to have formed at least as early as the other cements. Where syntaxial cement on a crinoid grain extends out to nearby grains, it interrupts and excludes any fringing and early meniscus cements from coating those grains. This relationship, as well as the observation that uncoated crinoid grains have only syntaxial cement, suggests that the syntaxial cement was at least as early as the other cement types and may have grown faster.

Burial Cement

Cements inferred to be of later, burial origin were responsible, together with early syntaxial cement and physical compaction, for the most significant porosity/permeability reduction (Figure 16). Exact amounts of pore-space occlusion by burial cements might be determined, if necessary, by back-stripping based on point-count analyses. Burial cements include overgrowths on early meniscus cements, overgrowths on early syntaxial cements, and mosaics of calcite-spar cement composed of crystals 50–600 μ in size. The

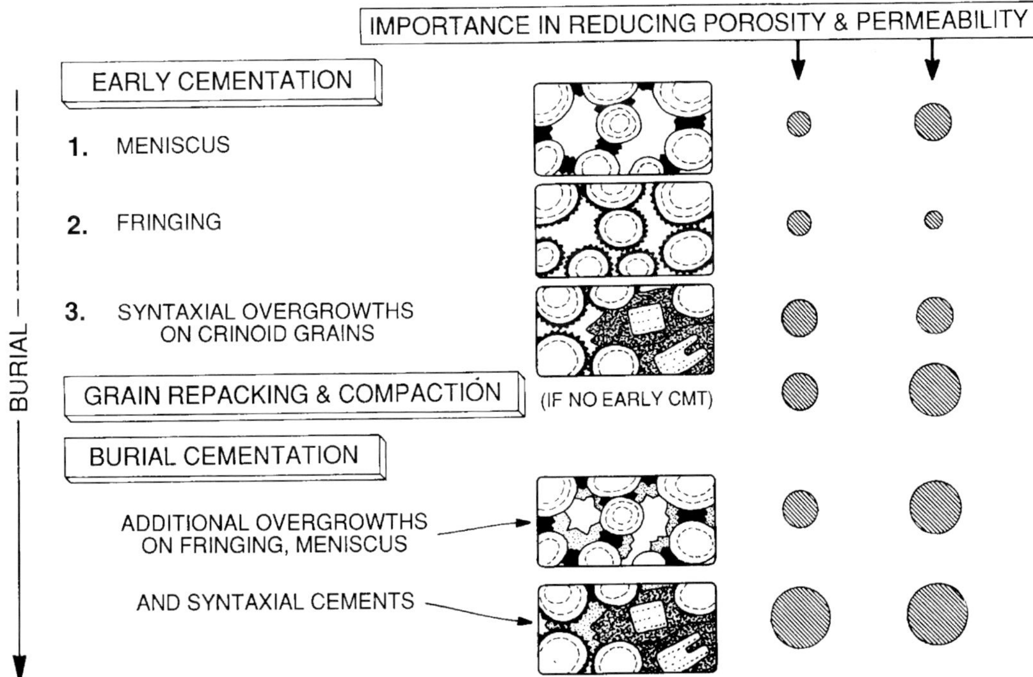

Figure 16. Commonly occurring types of early and burial cement in Ste. Genevieve grainstone reservoir rocks at Willow Hill. Hachured circles at right indicate the relative importance of these cements, and compaction, in reducing porosity and permeability (largest circles indicate greatest effect).

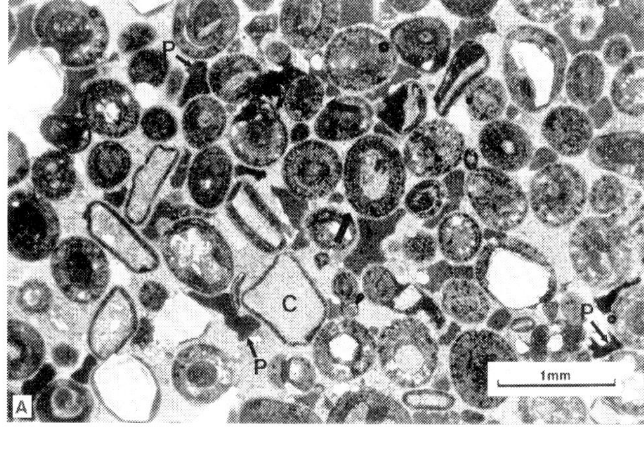

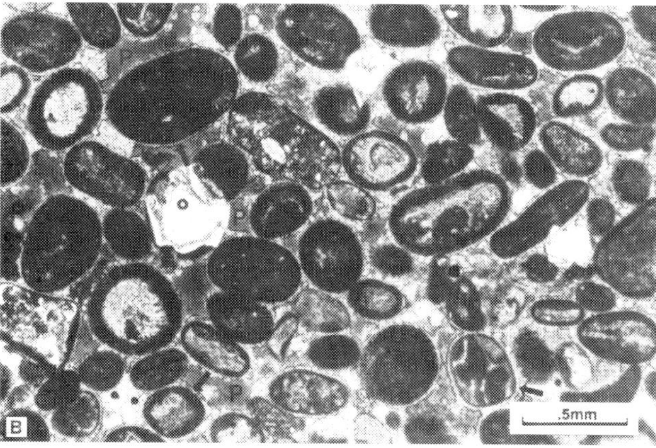

Figure 17. (A) Meniscus-style cement in ooid-rich grainstone. (U1 oolite, F. Ochs E-9, 2674.6-ft [815.22-m] core depth.) Meniscus cement precipitated (probably as calcite) at grain contacts (arrows) and was overgrown by coarse syntaxial calcite cement precipitated around crinoid fragments (C, lower center). 17% porosity remains, while permeability has been reduced to 0.1 md. White objects are quartz grains, with overgrowths. (B) Poorly developed fringing cement (arrows), interpreted as phreatic, with later patches of coarser calcite cement. (U2 oolite, E. Urfer #H-11 well, 2657.5-ft [810-m] core depth.) 17% porosity and 218 md permeability remain after some compaction and moderate cementation. Pores are medium gray (P).

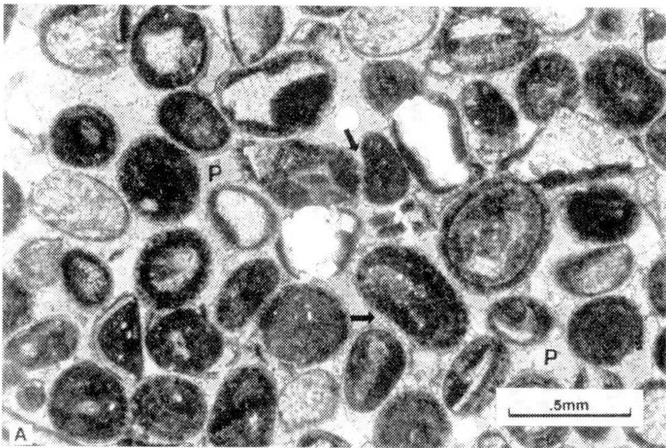

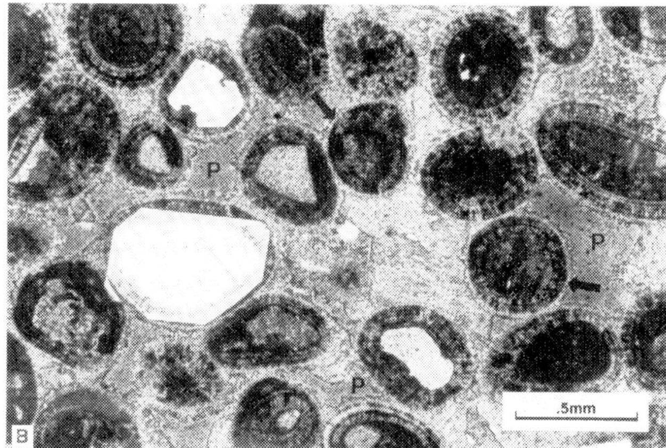

Figure 18. (A) Poorly developed fringing cement in slightly compacted grainstone. (U2 oolite, E. Urfer #H-11, 2659.4-ft [810.59-m] core depth.) Not enough early cement precipitated to prevent some grain-on-grain compaction. A small amount of meniscus cement can be seen at some grain contacts (arrows), suggesting the existence of a vadose zone. Pores (P) are light gray. 17% porosity and 213 md permeability remain after cementation. (B) Well developed early-fringing cement (arrows), as well as abundant, coarse ferroan calcite cement believed to be syntaxial on crinoid fragments. Pores (P) are the darkest gray components between the ooids (L1 oolite, J. A. Eubank a/c 1 #G-8 well, 2660.4-ft [810.89-m] core depth). Porosity 13%, permeability 172 md. Quartz overgrowth on quartz nucleus of ooid at left.

rare dolomite-spar cement in these grainstones is generally associated with burial cements.

Figure 16 illustrates burial calcite cements in diagrammatic form, and the photomicrographs in Figures 17 and 18 illustrate variations in both burial-cement types and extents of burial cementation. These cements can be distinguished from the earlier cements by their CL, which is generally characterized by vague zonation in hues of yellowish gray to grayish yellow. They can also be differentiated more easily but not as clearly by staining with Dickson's (1966) solution. Upon staining, the early cements acquire a light pink to red color depending on duration of staining, and the later cements a pale to intense mauve or purplish red color reflecting higher contents of Fe^{+2}.

Distribution of Early Cements

Because early cements rarely occur alone in the grainstones, it is necessary to "look past" the associated burial cements to see patterns of early-cement distribution. Stratigraphic distributions were determined in the cored sections and are shown in Figure 19. In the Ochs #E-9 well, where the U1 oolite was cored, meniscus cement is prominent in the upper 6 ft (1.83 m) (log depth 2670–2676 ft, 813.82–815.65 m) and disappears downward at 2681 ft (817.7 m). Fringing cement becomes abundant at 2675 ft (815.34 m), and disappears in turn at about 2692 ft (820.52 m). Early cement other than syntaxial does not occur below 2691 ft (820.22 m), and compaction effects become intense from that depth downward to the base of the bioclastic grainstone below the U1 oolite. In the Eubank #G-8 well, no meniscus cement occurs in the L1 oolite. Instead, fringing cement is the dominant form from the top of the L1 oolite, at about 2658 ft (810.16 m) log depth down to 2675 ft (815.34 m), diminishing there and disappearing below 2680 ft (816.87 m). In the Urfer #H-11 core, a similar vertical distribution exists in the L1 and L3 oolites together with the intervening bioclastic grainstone.

These vertical distributions are similar to cement distributions seen in vertical profiles of Holocene

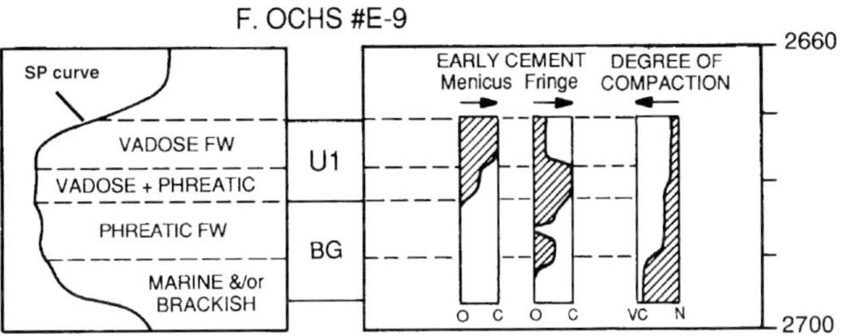

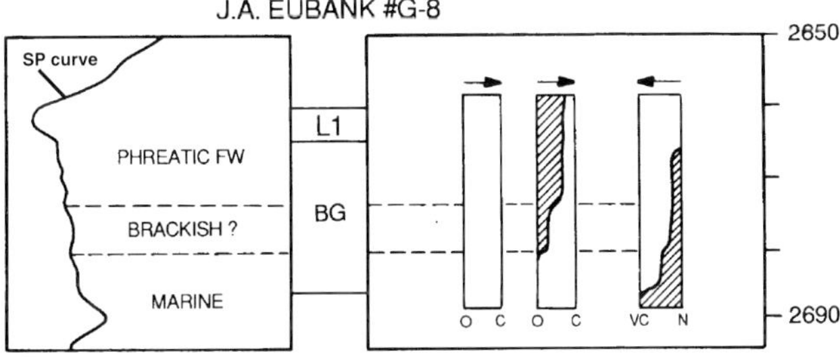

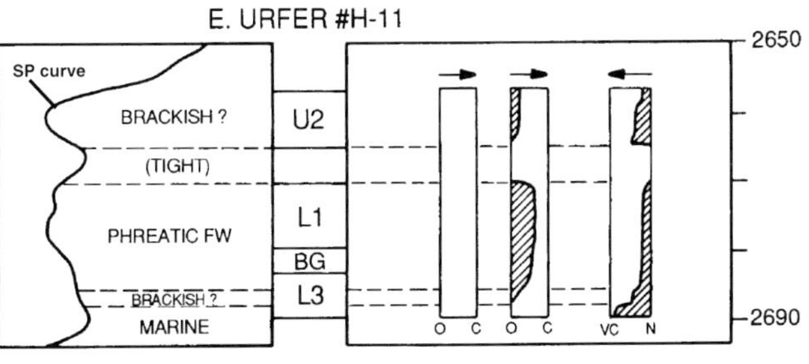

Figure 19. Profiles of cored grainstone sequences showing relative abundances of early cement and relative intensity of compaction vs. depth, and interpreted early-diagenetic zones. Based on observations in thin sections. Letter symbols: 0 = none, C = common, N = essentially uncompacted, VC = intensely compacted, BG = bioclastic grainstone, FW = fresh water.

ooid-grainstone islands in the Bahamas within which freshwater lenses exist (Halley and Harris, 1979; Budd, 1988). The idealized cement-distribution model pieced together from these studies is as shown in Figure 20. Although beachrock cement and even marine synsedimentary cement might be anticipated and may exist in other areas of Ste. Genevieve grainstone, neither was identified in the Willow Hill area.

Two points implied by this idealized model may be of use in predicting Ste. Genevieve oolite reservoir quality elsewhere in the basin. One is that meniscus cement, because it forms at grain contacts, will tend to have relatively adverse effects on permeability, whereas fringing cements of phreatic origin may have a positive effect in preserving porosity. The second point is that, below the freshwater lens, early cementation that would have prevented compaction is so limited that all but a small fraction of the porosity will generally be destroyed by compaction.

The vertical profiles identified in the three cored sequences (Figure 19) can be extrapolated to nearby wells. The assumptions made when doing this are as follows: (1) the datum horizon at the top of the First Dolomite approximated a horizontal plane when freshwater lenses were established in the grainstone sequences; (2) the water table and base of the freshwater lens in a particular oolite body and its underlying bioclastic grainstone were essentially horizontal; and (3) zones of early vadose and phreatic cement also had similar extent and continuity. Although the approach just discussed is valid in principle as a predictive method, its validity could not be fully evaluated in the Willow Hill area due to the limited core data.

CONTROLS AND PREDICTION OF BURIAL AND/OR SYNTAXIAL CEMENTATION

Two sedimentologic features of Ste. Genevieve grainstones in general, crinoid grains and stylolites, emerge as important factors in determining the abundance and distribution of the phreatic and burial cements.

Thin-section study clearly shows that wherever nonoolitically coated crinoid grains occur or where coated grains have been breached by erosion or breakage, these grains were nucleation sites for syntaxial cement. Successive zones of syntaxial cement, judging from their changes in CL from dark/bright to zoned dull and changes from low to higher Fe^{+2} content inferred from staining colors, began precipitating in near-surface freshwater lenses and continued precipitating with increasing burial. The more abundant such crinoid grains are when compared with ooids, the greater the volume percent of syntaxial cement and the smaller the percent of surviving porosity. Because of these relationships, the best porosity and permeability within the Ste. Genevieve tend to occur in the most ooid-rich and crinoid-poor grainstones.

The second important sedimentologic feature, stylolites, is related to the intensity of burial cementation

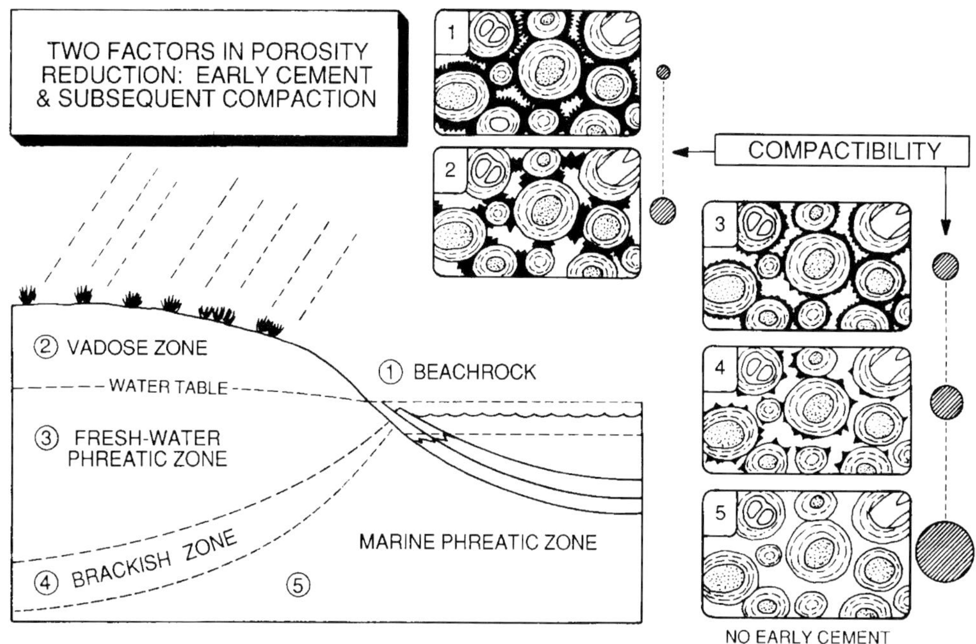

Figure 20. Idealized distribution of early near-surface cements in and near a freshwater lens. Cements are black and pore space white. Hachured circles to right of panels indicate relative degrees of compaction likely upon later burial (largest circles indicate greatest compaction). Little or no cementation will occur outside the freshwater lens (cases 4 and 5), in which cases later compaction may substantially reduce original porosity. Compaction will be slight to nil in cases 1–3, where grains are "propped" by early cementation.

as inferred from two lines of evidence. One is that some oolite grainstone intervals, made up of nearly 100% well-coated ooids, have been completely cemented next to stylolites. Their main pore-occluding cement is of mosaic type. The other line of evidence is a strong inverse relationship between the amount of porosity and both the frequency of stylolites and the amount of section lost at stylolites (Figure 21). These data were compiled by counting and measuring in the cores all of the stylolites in a given grainstone interval, then summing their apparent maximum amplitudes. The summed amplitudes provide a minimum estimate of the amounts of section dissolved by pressure-solution along the stylolites. Although the apparent relationship between stylolites, burial cementation, and (inversely) percent porosity appears to be real, as has been observed in other studies (e.g., Harms and Choquette, 1965; Wong and Oldershaw, 1981), its usefulness for prediction of reservoir quality is limited at Willow Hill due to the lack of core data.

CONCLUSIONS

Two series of elongate ooid grainstone lenses form reservoirs in the Ste. Genevieve at Willow Hill: a lower series related to a bioclastic grainstone shoal that developed along an earlier trend of low-lying islands, and an upper series of smaller lenses that accumulated around the perimeter of the lower shoal complex.

The ooid grainstones were affected by both early cementation (whereas freshwater lenses were developed in exposed oolite bodies) and burial cementation as a consequence of pressure solution. Early cementation, probably in upper parts of freshwater lenses, occluded only small amounts of porosity but inhibited later physical compaction by "propping" ooid grains. Where the early cement was meniscus in type, overgrowths of burial spar cement blocked pore throats, resulting in grainstones of fairly high porosity but low permeability and poor reservoir quality. Where the early cement was phreatic, pore throats were not selectively blocked and oolite of good reservoir quality resulted after burial.

ACKNOWLEDGMENTS

We thank Marathon Oil Company and Reserve Oil Corporation for granting permission to publish this paper. We also acknowledge editorial reviews by D. M. Cooper and K. P. Williams of Marathon Oil

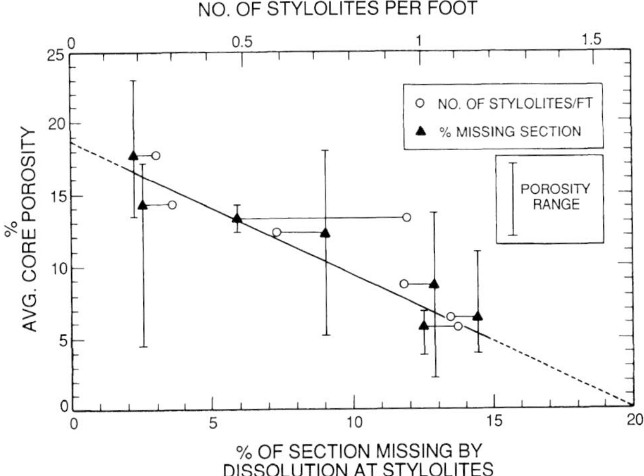

Figure 21. Graph of averaged core porosity (by core analysis) vs. both stylolite frequency and thickness (%) of section loss by dissolution at stylolites. Based on data from the three cored wells in Willow Hill field and from one cored well in the Applegate 5-M waterflood in North Lawrence field, Lawrence County, Illinois, the P. King a/c 1&2 #153 (data points farthest left).

Company's Petroleum Technology Center and critical comments offered by R. M. Cluff and J. R. Dodd.

REFERENCES CITED

Asquith, G. B., 1979, Subsurface carbonate depositional models: a concise review: Penn Well Publishing Company, Tulsa, OK, 121 p.

Budd, D. A., 1988, Aragonite-to-calcite transformation during fresh-water diagenesis of carbonates: insights from pore-water chemistry: Geological Society of America Bulletin, v. 100, p. 1260-1270.

Carr, D. D., 1973, Geometry and origin of oolite bodies in the Ste. Genevieve Limestone (Mississippian) in the Illinois basin: Indiana Geological Survey Bulletin 48, 81p.

Choquette, P. W. and R. P. Steinen, 1980, Mississippian non-supratidal dolomite, Ste. Genevieve Limestone, Illinois basin: evidence for mixed-water dolomitization, *in* D. H. Zenger, J. B. Ahham, and R. L. Ethington, eds., Concepts and models of dolomitization: SEPM Special Publication 28, p. 163-196.

Choquette, P. W. and R. P. Steinen, 1985, Mississippian oolite and non-supratidal dolomite reservoirs in the Ste. Genevieve Formation, North Bridgeport field, Illinois basin, *in* P. O. Roehl and P. W. Choquette, eds., Carbonate Petroleum Reservoirs: Springer-Verlag, New York, p. 209-225.

Cluff, R. M. and J. A. Lineback, 1981, Middle Mississippian carbonates of the Illinois basin: a seminar and core workshop: Illinois Geological Society, 88 p.

Cluff, R. M., 1984, Carbonate sand shoals in the Middle Mississippian (Valmeyeran) Salem-St. Louis-Ste. Genevieve limestones, Illinois basin: *in* P. M. Harris, ed., Carbonate sands—a core workshop: SEPM Core Workshop No. 5, p. 94-135.

Collinson, C., R. D. Norby, and J. W. Baxter, 1979, Stratigraphy of the Mississippian stratotype, Upper Mississippi Valley, USA: 9th International Congress of Carboniferous Stratigraphy and Geology, Field Trip 8 Guidebook, 109 p.

Dickson, J. A. D., 1966, Carbonate identification and genesis as revealed by staining: Journal of Sedimentary Petrology, v. 36, p. 491-505.

Dunham, R. J., 1971, Meniscus cement, *in* O. P. Bricker, ed., Carbonate cements: Johns Hopkins University Studies in Geology No. 19, p. 297-300.

Enos, P., and L. H. Sawatsky, 1981, Pore networks in Holocene carbonate sediments: Journal of Sedimentary Petrology, v. 51, p. 961-985.

Halley, R. B., and P. M. Harris, 1979, Fresh-water cementation of a 1,000-year-old oolite: Journal of Sedimentary Petrology, v. 49, p. 969-988.

Harms, J. C. and P. W. Choquette, 1965, Geologic evaluation of a gamma-ray porosity device: Society of Professional Well Log Analysts, 6th Annual Symposium, Transactions, v. 2, p. C1-C37.

Harms, J. C., J. B. Southard, and R. G. Walker, 1982, Structures and sequences in clastic rocks: SEPM Short Course #9, Lecture Notes.

Huff, B. G., 1987, Petroleum industry in Illinois, 1985: Illinois State Geological Survey, Illinois Petroleum 128, 140 p.

Hunter, R. E., 1989, Eolianites in the Ste. Genevieve Limestone of southern Indiana: Field Trip Guidebook, American Association of Petroleum Geologists Eastern Section Meeting, 1989 (reprinted by Indiana Geological Survey), p. 1-19.

Inden, R. D., and C. H. Moore, 1983, Beach, *in* P. A. Scholle, D. G. Bebout, and C. H. Moore, eds., Carbonate depositional environments: American Association of Petroleum Geologists Memoir 33, p. 211-265.

James, N. P. and P. W. Choquette, 1984, Diagenesis 9.–Limestones–The meteoric diagenetic environment: Geoscience Canada, v. 11, no. 4, p. 161-194.

McCubbin, D. G., 1982, Barrier-island and strandplain facies, *in* P. A. Scholle, and D. R. Spearing, eds., Sandstone depositional environments: American Association of Petroleum Geologists Memoir 31, p. 247-281.

Swann, D. H., 1963, Classification of Genevievian and Chesterian (Late Mississippian) rocks of Illinois: Illinois State Geological Survey Report of Investigations 216, 91 p.

Wong, P. K. and A. Oldershaw, 1981, Burial cementation in the Kaybob reef complex, Alberta, Canada: Journal of Sedimentary Petrology, v. 51, p. 507-520.

Chapter 8

Benthic Assemblages as Indicators of Sediment Stability: Evidence from Grainstones of the Harrodsburg and Salem Limestones (Mississippian, Indiana)

Howard R. Feldman
Department of Geology and Kansas Geological Survey
University of Kansas
Lawrence, Kansas, USA

Mark A. Brown
B. P. Exploration Inc.
Houston, Texas, USA

Allen W. Archer
Department of Geology
Kansas State University
Manhattan, Kansas, USA

ABSTRACT

Within the Mississippian (Valmeyeran) strata of Indiana, the uppermost Harrodsburg and lowermost Salem limestones contain a variety of grainstone facies that were deposited within a tide- and wave-dominated, shallowing-upward sequence. Compositions of autochthonous fossil assemblages in each facies are interpreted to have been constrained primarily by sediment stability. For example, echinoderm-bryozoan-brachiopod assemblages apparently thrived on stable substrates. Conversely, gastropod-dominated assemblages lived on frequently reworked, tidally influenced substrates. Low-diversity assemblages lived on substrates that were almost continually reworked. Faunal diversity decreased with increasing sediment mobility because of the increased (biological) energy required to maintain a life position at the sediment-water interface. Within modern analogs, a decrease in faunal diversity commonly correlates with increased sediment mobility.

INTRODUCTION

Traditionally, interpretations of sedimentary facies and depositional environments of shallow marine systems primarily have been based on sediment characteristics such as sedimentary structures, sediment texture, and composition. However, these characteristics may be more representative of rare, high-energy events than of normal day-to-day conditions. Many shallow marine stratigraphic sequences now are recognized to have been dominated by storm-induced sedimentation (Ager, 1981; Dott, 1983; Brett et al., 1986; Feldman and Maples, 1989). Within these sequences there may be little sedimentologic record of normal day-to-day processes. In contrast, benthic organisms living in these environments are influenced profoundly by daily processes occurring at the sediment-water interface and, therefore, potentially can be used to determine a variety of environmental parameters. Although shelly fossils commonly have been used to distinguish between freshwater and marine environments, fossil assemblages in normal-marine, non-reef limestones largely have been overlooked by sedimentologists as a means for recognizing and interpreting depositional processes and environments.

Ecological structures of modern, benthic marine communities respond to environmental stability, magnitude of environmental perturbations, and frequency of perturbations. In-situ fossilized benthic assemblages, therefore, may provide information not only about the normal environment, but additionally may yield clues to the frequency and magnitude of storms or environmental perturbations. Careful study of fossil assemblages then could be used to reconstruct aspects of depositional environments not easily discerned by traditional facies analysis.

This chapter describes the relationship between preserved fossil assemblages and sedimentary facies in grainstones, including oolitic lithologies, of the uppermost Harrodsburg and lowermost Salem limestones in Indiana (Figures 1, 2). These relationships reflect original sediment stability and can be used to interpret depositional conditions. Relationships between facies and fossil assemblages noted within the Harrodsburg and Salem limestones also occur in other Mississippian grainstones throughout the Midwest, suggesting that the observed relationships in the Harrodsburg and Salem limestones may be representative of Mississippian shallow marine communities on carbonate substrates in general. If similar relationships can be demonstrated for grainstones of other ages as well, then analysis of fossil assemblages may become a significant addition to facies analysis.

DATA COLLECTION

The Salem Limestone and adjacent strata were studied throughout the outcrop belt from Bloomington, Indiana, south to the Ohio River (Figure 3). Most exposures occur in quarries within the building-stone districts and road cuts exposing

Sanders Group	St. Louis Limestone
	Salem Limestone
	Harrodsburg Limestone
	Ramp Creek Formation

Figure 1. Stratigraphic nomenclature for units considered in this study (see Keith and Zuppann, this volume, for regional stratigraphic context).

lateral facies near Bloomington and Bedford, Indiana. Petrographic analysis is based on thin sections taken from quarry exposures. Macrofossil collecting was limited to weathered exposures (localities 2, 3, 4, 5, and 11). (All localities referred to this chapteer are listed in Table 1.) Analysis of macrofossil distributions also is based on qualitative observations of fresh and weathered exposures and on data collected by Cumings et al. (1906). Trace fossils primarily were observed on the weathered faces of dimension stone in buildings (mostly on the Indiana University campus) and in quarry and mill spoil piles.

THE PROBLEM OF TRANSPORTED FOSSILS

Cross-bedded grainstones analyzed in this study primarily are composed of broken or abraded, sand- or granule-sized fossil debris. These fossils obviously have been transported. Paleoecologic analysis requires estimation of the contribution to the total fossil assemblage of shells indigenous to the environment and those transported from other environments. The specific taphonomic evidence used to discern allochthonous from autochthonous assemblages will be provided with the analysis of each facies, but the general principles applied in such analyses will be reviewed here.

Autochthonous, as used herein, designates skeletal material derived from the local environmental facies. Conversely, *allochthonous* refers to skeletal material derived from outside the local facies. The best possible evidence indicating autochthonous assemblages is preservation of fossils in original growth position. Care must be taken in interpreting fossils as in situ because some organisms, such as disk-shaped coral colonies, can be transported and deposited in apparent life orientations. Other fossils, such as crinoid root systems or large productid brachiopods with long delicate spines, probably could not be transported and deposited intact. Other important evidence of autochthonous fossils includes the general preservational state of the fossils. Many studies have demonstrated that preservational quality is directly linked to the time between death and final burial; long exposure time on the sea floor leads to decay, disarticulation, and scattering of skeletal elements, whereas

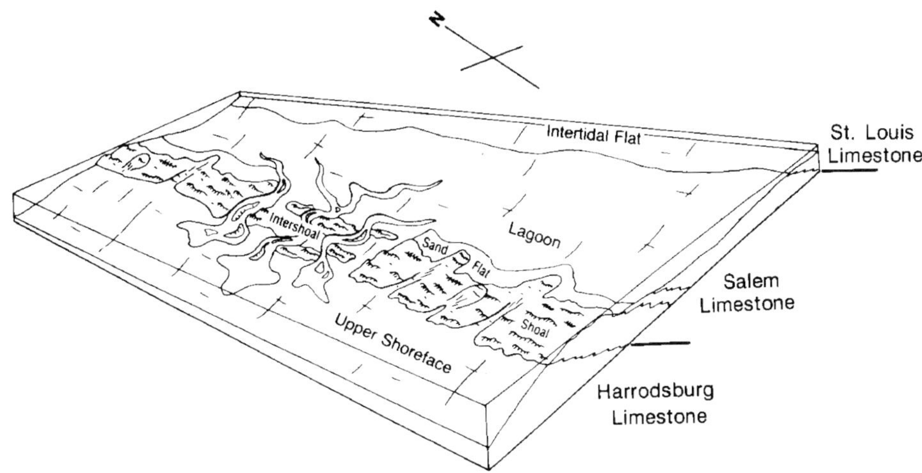

Figure 2. Facies model for the uppermost Harrodsburg, Salem, and lowermost St. Louis limestones (modified from Brown, 1990).

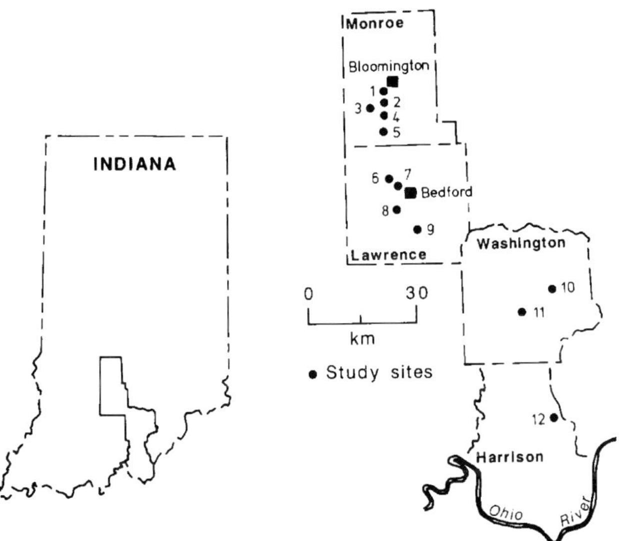

Figure 3. Salem Limestone exposures examined in this study. See locality list (Table 1) for specific locality information.

rapid burial commonly assures good preservation (Brett and Baird, 1986). Preservational quality also may be linked to transport distance; increased transport distances generally increase disarticulation, abrasion, and fragmentation of skeletons. This evidence must be used with caution because some skeletons may be transported long distances with insignificant detrimental effects.

Sedimentological parameters also can be used to assess transport direction and distance. Generally, well-sorted fossil assemblages are suspect because sorting implies sedimentological rather than biological processes. Cross-bedding can be used to indicate current direction and, thereby, suggest the direction to possible sources from which transported material was derived. Lastly, studies of modern environments can be instrumental. Several studies of the distribution of living and dead organisms in modern shelf settings suggest that aerial distributions of fossils, but not their relative abundances, are preserved (e.g., Stanton, 1976).

One additional line of evidence was used to indicate the relative intact nature of fossil distributions in the study area. The relationships between facies and fossil assemblages first seen in the study area have been observed in Mississippian grainstones throughout the Midwest (Indiana, Illinois, Iowa, and Missouri). We can never be certain that the fossil distributions were not effected by taphonomic processes and have no relation to the original patterns, but that these affects would be repeated in a wide range of local circumstances in stratigraphically and geographically different locations seems unlikely.

RELATIONSHIP BETWEEN FACIES AND FAUNAS

The sequence from uppermost Harrodsburg through lowermost St. Louis limestones generally fines upwards. The lower half of the sequence (uppermost Harrodsburg and lowermost Salem limestones) is composed of cross-bedded grainstones. The upper half of the sequence (uppermost Salem and lowermost St. Louis limestones) consists of micritic, shallow subtidal to supratidal deposits (Brown, 1987, 1990). The succession of stacked facies suggests a shallowing-upward sequence that is assumed to have prograded to the southwest into the Illinois basin (Brown, 1987, 1990; Dodd, 1987) (Figure 2).

Overall compositions of the grainstone facies are very similar, and all are dominated by various combinations of echinoderm, bryozoan, and foraminifer skeletal debris. Each grainstone facies is distinguished based on differences in grain size, sorting, development of oolitic coatings, and fossil content. The various grainstone facies are interpreted as having been deposited in upper shoreface, intershoal, shoal, and sand-flat environments (Brown, 1990).

TABLE 1. Locality list.

1. Maple Hill Quarry. NW1/4, SE1/4, Sec. 18, T8N, R1W, Bloomington Quadrangle.
2. Road cuts on new and old State Highway 37. Sec. 29, 32, T8N, R1W, Clear Creek Quadrangle.
3. Shawnee Quarry. NW1/4, SW1/4, Sec. 36, T8N, R2W, Clear Creek Quadrangle.
4. Road cut on State Highway 37. SE1/4, SE1/4, Sec. 17, T7N, R1W, Clear Creek Quadrangle.
5. Road cut on State Highway 37 at bluffs of Clear Creek. SE1/4, SE1/4, Sec. 20, T7N, R1W, Clear Creek Quadrangle.
6. Dark Hollow Quarry. S1/2, Sec. 5, and N1/2, Sec. 8, T5N, R1W, Oolitic Quadrangle.
7. Road cut on State Highway 37 at bluffs of Salt Creek. SE1/4, SE1/4, Sec. 9, Oolitic Quadrangle.
8. Road cut on State Highway 37 at bluffs of White River. SE1/4, SW1/4, Sec. 27, and N1/2, Sec. 34, T5N, R1W, Bedford West Quadrangle.
9. Lehigh Portland Quarry. SW1/4, SE1/4, Sec. 19, T4N, R1E, Bedford East Quadrangle.
10. Spergen Hill railroad cut. NW1/4, SE1/4, Sec. 24, T2N, R4E, Salem Quadrangle.
11. Road cut near Lake Salinda. NW1/4, NE1/4, Sec. 32, T2N, R4E, Salem Quadrangle.
12. Georgetown railroad cut. SE1/4, SE1/4, Sec. 31, T2S, R5E, Georgetown Quadrangle.

These environments will be discussed along an offshore-to-onshore transect. Throughout the rest of this paper, these interpretive environmental categories are used for facies names for ease of discussion. The transect from upper shoreface to sand-flat environments generally defines a gradient of decreasing sediment stability. The principal evidence for sediment stability is the relative abundance of autochthonous sessile benthic fossils, because such organisms are intolerant of shifting sediment. This evidence is consistent with other evidence of sediment reworking, such as degree of sorting and abrasion and development of oolitic coatings.

UPPER SHOREFACE FACIES

Facies Description

The upper shoreface facies directly underlies the Salem Limestone (Figure 2) and consists mostly of coarse skeletal grainstone and minor amounts of wackestone and packstone in the uppermost Harrodsburg Limestone. The grainstone is moderately to poorly sorted and dominated by coarse sand- to granule-sized clasts, although individual fossils, such as bryozoan fronds, may be much larger (several tens of centimeters). Most grains are not significantly abraded. Allochems are composed of primarily echinoderm and fenestrate bryozoan debris with lesser amounts of foraminifers, brachiopods, peloids, and corals (Figures 4A, B). Foraminifers as well as coated and abraded grains increase in abundance as grain size decreases towards the top of the facies near the gradational Harrodsburg-Salem contact.

Cross-bedding is prominent on weathered grainstone exposures (Figure 5). Trough cross-beds predominate, and tabular planar cross-beds occur locally. Most cross-bed sets are less than 30 cm thick and terminate with low-angle truncations. Horizontal bedding and rippled beds are present but rare. Some thick (up to about 1 m) tabular planar cross-bed sets have basal lags of coarse skeletal debris. Grain size and degree of sorting vary within individual cross-bed sets.

Composition and Interpretation of Fossil Assemblage

The preserved macrofossil assemblage is composed of typical normal-marine Mississippian benthic organisms (Table 2). Preservational qualities of fossils range from abraded shell fragments to whole and articulated skeletons. Generally the most easily disarticulated or fragmented components of fossils, such as crinoid arms and fenestrate bryozoans, are rarely found intact, whereas the more sturdy fossils, such as crinoid and blastoid calyces and brachiopods, are well-preserved. The sand-sized fragmentary fossil debris is dominated volumetrically by echinoderm and fenestrate bryozoan pieces, but the most abundant whole and nearly whole macrofossils in descending order of abundance include articulated brachiopods, rugose corals, articulated crinoid and blastoid calyces, fenestrate and ramose bryozoans, and corals. Tabulate coral and fenestrate bryozoan colonies range up to about 20 cm long. Rarely tabulate corals, fenestrate bryozoans, and crinoid holdfasts are in apparent growth position.

Crinoid calyces and coral colonies probably could not have withstood much bed load transport without substantial amounts of disarticulation or fragmentation. The abundance of these fossils and the low degree of abrasion and sorting suggest limited reworking of most fossils even though they occur within cross-bedded grainstones (particularly when compared with the grainstone facies in the Salem Limestone). Certainly the fossils in growth position were not transported at all, and they are the same species as less well-preserved fossils. However, the lack of articulated crinoid arms or blastoid brachioles suggests that these organisms, and probably most others, were subject to at least some decay and disarticulation prior to final burial. The wide range of preservational qualities of the fossils likely reflects a

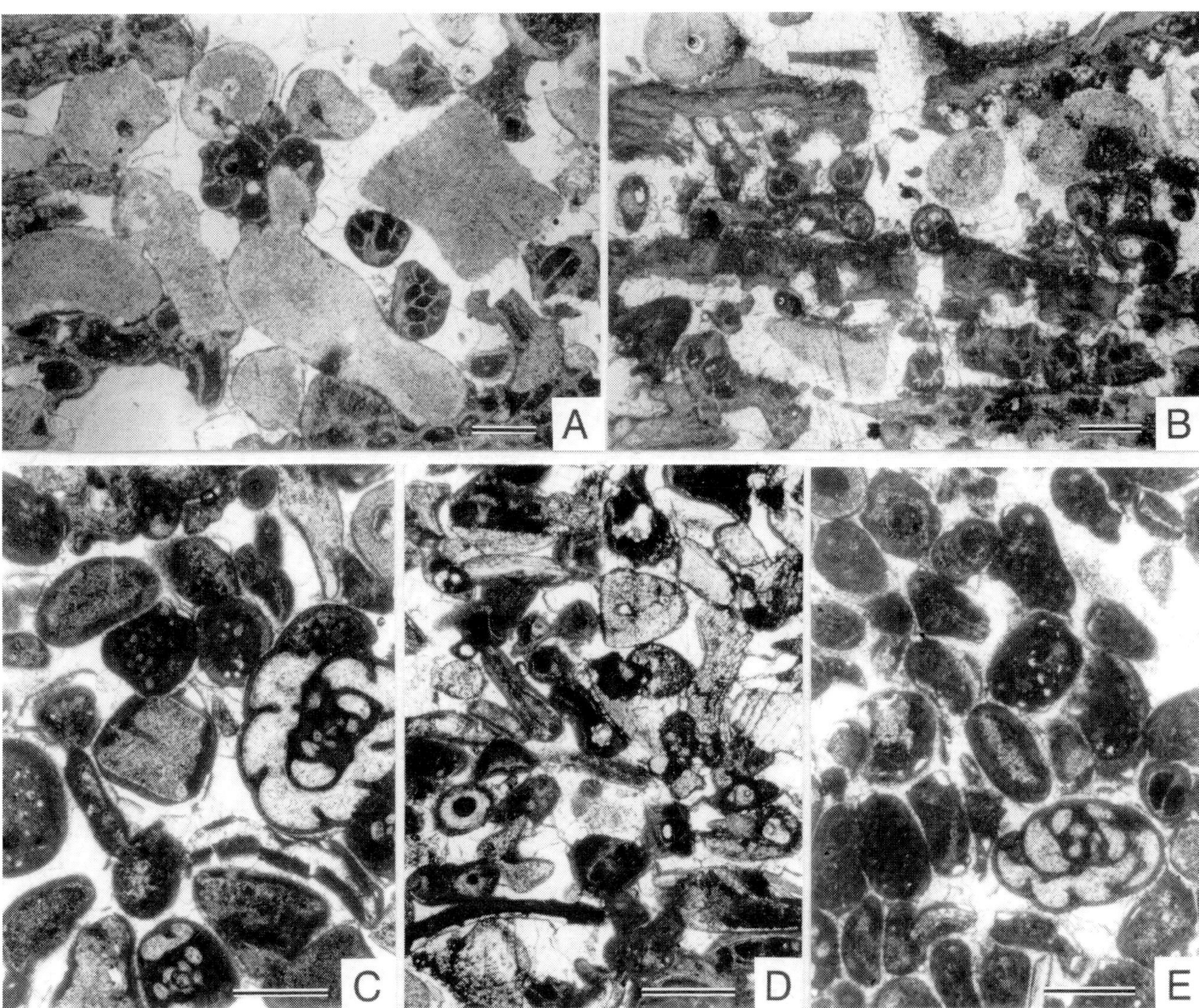

Figure 4. Thin sections of facies discussed in text. (A) Thin section of the upper shoreface facies dominated by echinoderm grains. (B) Thin section of upper shoreface facies dominated by bryozoans. (C) Thin section of the intershoal facies. (D) Thin section of the shoal facies. (E) Thin section of the sand-flat facies. Scale bars for all photomicrographs are 0.5 mm.

wide range of residence time on the sea floor and amount of reworking of skeletons prior to final burial. Overall, the preservational quality suggests that the biota was tumbled and disarticulated, but originated within the environment in which it was preserved.

The fossil assemblage from the upper shoreface facies is almost entirely composed of sessile benthos. Most of these organisms could not move or reorient themselves after settlement and metamorphosis of mobile larvae, and several of the species have morphologic features that would have specifically prevented locomotion within the sediment. An example is the rugose coral *Hapsiphyllum cassedayi* (Milne-Edwards) (Figure 6), which is one of the most abundant macrofossils in this facies. Juveniles of this species were initially cemented to a small bit of shell or other material. As individuals grew, they became effectively free-lying and probably sank into the loose sediment, leaving only the upper part of the organism exposed. Spines projecting from the lower part of the shell stabilized the organism on the sea floor. Most individuals grew 2–3 cm in length, but some reached 4 cm. Deposition of a few centimeters of sediment would have covered the coral, thus preventing it from feeding and subsequently resulting in death (modern coral polyps can remove a thin sediment covering, but are killed by thick sediment layers or repeated deposition). Erosion of the sediment would have removed the coral's support, causing it to drop to one side. Some rugose coral species were able to recover from such events and resume growing upward again, resulting in a sharp bend in their skeletons. Similar bends, however, were rarely

Figure 5. Trough cross-bedding in the upper shoreface facies, locality 2.

TABLE 2. Number of genera in benthic assemblages from the uppermost Harrodsburg and lowermost Salem limestones.

Taxon	US	SF	IS1	IS2
Sessile Benthos				
Tabulata	1	0	0	0
Rugosa	1	1	0	1
Brachiopoda	12	10	5	8
Bryozoa	11	4	5	5
Crinoidea	7	3	2	1
Blastoidea	3	0	2	1
TOTAL	35	18	14	16
Mobile Benthos				
Gastropoda	4	20	18	18
Bivalvia	1	4	6	7
Rostroconchia	0	0	1	1
Polyplacophora	0	0	2	2
Scaphopoda	0	1	1	
Trilobita	1	0	1	1
TOTAL	6	25	29	30

US, upper shoreface facies, data represent compilations from several localities and Cumings et al. (1906); SF, sand-flat facies, data from locality 3; IS1, IS2, intershoal facies, data from localities 11 and 4 respectively.

observed in *Hapsiphyllum*, demonstrating that the coral did not frequently survive (or were not subjected to) erosional events. Deposition of even 1 cm of sediment following such an erosional event would have covered and killed a toppled coral. Therefore, the sediment must have remained stable during the life span of most corals.

The size of *Hapsiphyllum* corals is related to the length of time during which the substrate was stable; thus, knowledge of the coral's growth rate would allow estimation of the minimum length of time between reworking events. Estimation of growth rates of extinct organisms is difficult at best; nevertheless, analyses of presumably daily growth rings on some species of rugose corals have been used to estimate growth rates ranging from about 0.5 cm to a few centimeters per year (see Hill [1981] for a review). These rates are similar to growth rates measured on modern scleractinian corals. Based on these comparisons, and assuming *Hapsiphyllum* had a growth rate typical of rugose corals, most specimens of *Hapsiphyllum* from the Harrodsburg Limestone were at least several months old and possibly several years old at the time of death. Thus, the sediment surface must have remained stable for at least this long.

Figure 6. *Hapsiphyllum cassedayi* (Milne-Edwards) from the Harrodsburg Limestone. Note that the apex is attached to a crinoid or blastoid columnal.

Arguments similar to those concerning growth rates for corals can be made for the many brachiopods in the upper shoreface facies. Most Paleozoic brachiopods had pedicles attached to the substrate, analogous to most living pedunculate brachiopods (Richardson, 1981). Free-lying brachiopods from the uppermost Harrodsburg Limestone typically had numerous rigid, fixed spines and were apparently incapable of movement. Other sessile benthos that are abundant in this facies and would have been killed by deposition of thin layers of sediment include the common colonial tabulate coral *Syringopora monroensis* Beede and encrusting bryozoans. The largest *Syringopora* colonies are several tens of centimeters in diameter, and individual coralites are commonly several centimeters tall, suggesting that there was little to no deposition for considerable periods of time relative to coral life spans. *Fistulipora*, as well as other bryozoans, encrusted shell surfaces and grew as loose sheets over the sediment surface. These organisms would have required prolonged periods of sediment stability, probably several months at least, in order to grow.

Stemmed echinoderms could have potentially survived burial of lower parts of the column as long the arms or brachioles were free for feeding. However, blastoid and crinoid columns did not possess muscles and could not have wiggled out from under a covering of sediment if buried.

Truly mobile benthos are rare in this facies. Trilobites are consistently present only in low abundances. Mollusks, particularly gastropods, also are present, but the most common gastropods are platyceratids that were coprophagous, living on crinoid calyces. Other benthic mollusks are rare.

Environmental Interpretation

The vast majority of preserved, autochthonous benthos in the upper shoreface facies required long periods of sediment stability relative to their life spans. The dominance in this facies of sessile benthos unable to tolerate shifting sediment suggests a predominantly quiet, normal-marine environment. If there were long periods of sediment stability, then the cross-bedding was produced during rare episodes of sediment reworking. The lack of micrite throughout most of the upper shoreface facies, however, suggests turbulent conditions that seem to contradict the paleoecologic evidence. A possible resolution is that currents were too weak to transport sand-sized clasts but were strong enough to prevent deposition of carbonate mud. The dominance of rheophilic (current-loving) crinoids in the upper shoreface facies (Feldman, 1987; Kammer and Ausich, 1987) further supports an interpretation of active and not stagnant conditions.

INTERSHOAL FACIES

Facies Description

The intershoal facies is stratigraphically adjacent to the shoal facies, occurring above the upper shoreface facies and below the micritic lagoon facies. Between Bedford and northern Monroe County (Figure 3) the intershoal facies is confined to two belts (approximately 5 and 16 km wide) between adjacent shoals (Brown, 1990). South of Bedford, the intershoal facies dominates the Salem Limestone. The contact with the underlying Harrodsburg Limestone (upper shoreface facies) is gradational. An abrupt change from grainstone to packstone marks the upper contact with lagoonal facies.

The facies consists of cross-bedded, medium- to coarse-grained skeletal grainstone. The sediment is well-sorted, and most grains are abraded (Figure 4C). Grains larger than a few millimeters are rare. The two dominant allochems are abraded echinoderm grains and the foraminifer *Globoendothyra baileyi* (Hall). High concentrations of *G. baileyi* are unique to this facies. Other allochems include bryozoans, brachiopods, calcispheres, codiacean algae, mollusks, and peloids (primarily micritized skeletal grains). Most grains have micritic envelopes or superficial oolitic coatings.

Trough cross-bedding is prominent in the intershoal facies. Channels up to tens of meters wide and about 5 m deep are common in some exposures. A rose diagram constructed from 90 paleocurrent measurements from trough cross-beds reveals a pronounced bimodality (Brown et al., 1990) (Figure 7A).

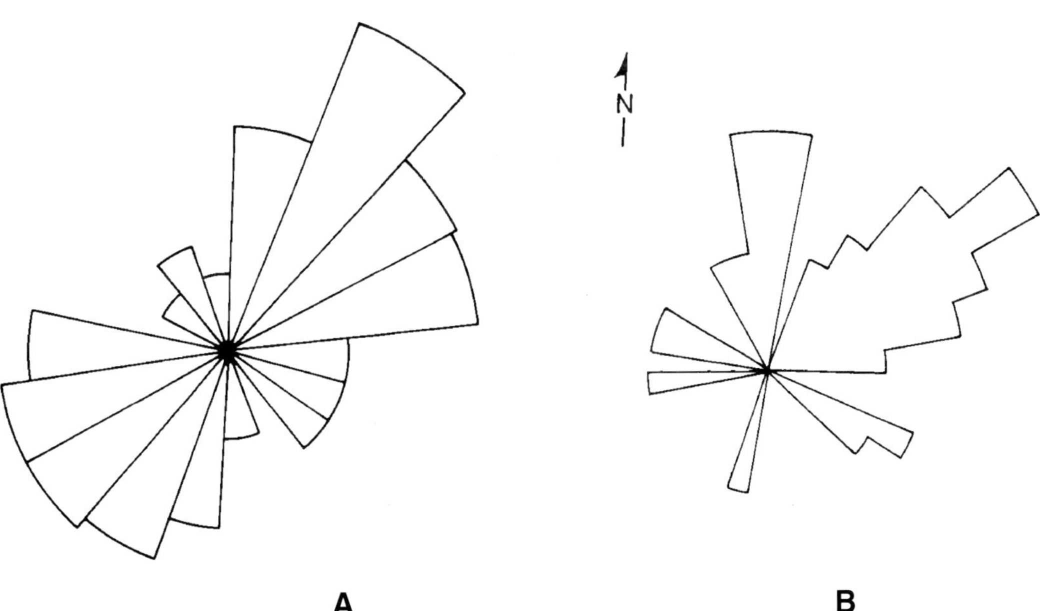

Figure 7. (A) Rose diagram of paleocurrent orientations from the intershoal environment based on 90 measurements of trough cross-bed sets from exposures at and near locality 4 (modified from Brown et al., 1990). (B) Rose diagram of paleocurrent orientations from a single sand wave in the shoal facies based on 40 measurements from locality 7 (modified from Thompson et al., 1990). Arrow indicates north.

Cross-bed orientations are predominantly to the northeast and southwest, perpendicular to depositional strike.

Composition and Interpretation of Fossil Assemblage

The fossil assemblage can be divided into autochthonous and allochthonous components based on preservational characteristics. The majority of allochems are shell fragments of sessile benthos with few whole shells of these species. *Pentremites* calyces are consistently articulated in coarser beds, but the calyces are mostly abraded. Most crinoid and blastoid plates are severely abraded. Abraded specimens of *Hapsiphyllum* are also present, but well-preserved specimens are invariably less than 5 mm long. Articulated juvenile pedunculate brachiopods are present albeit rare. Bryozoans are represented mostly by small fragments of fenestrate bryozoans. Encrusting bryozoans are rare, and none were observed on shell surfaces.

In contrast, the mobile benthos is well-preserved, diverse, and is represented by many whole and articulated fossils (Tables 2, 3). The most diverse and abundant whole macrofossils are gastropods (more than 25 species of gastropods are present). The largest and most abundant gastropod is the nearly planispiral *Straporolus*, which attains diameters of up to 1.5 cm. Most other gastropods are small, high-spired forms. Platyceratids, the most common gastropod of the upper shoreface facies, are rare in the intershoal facies. Other mobile benthos include bivalves, rostroconchs, chitons, scaphopods, and trilobites. The contrasting quality of preservation of mobile vs. sessile benthos suggests that the sessile benthos are allochthonous and the mobile benthos are autochthonous. The mobile benthos were mostly infaunal or semi-infaunal and were at least tolerant of, or perhaps preferred, frequently shifting sediment.

TABLE 3. Percent of specimens of solitary macrofossils of mobile and sessile benthos from the autochthonous benthic assemblages in Salem grainstone facies.

Taxon	SF	IS1	IS2
Sessile Benthos			
Rugosa	8	1	1
Brachiopoda	34	10	11
Crinoidea	2	<1	<1
Blastoidea	0	<1	<1
Total	44	12	12
Mobile Benthos			
Gastropoda	51	79	72
Bivalvia	4	6	10
Rostroconchia	0	<1	<1
Polyplacophora	0	0	1
Scaphopoda	1	3	3
Trilobita	0	<1	1
TOTAL	56	88	87
n	257	252	541

Locality abbreviations same as in Table 2. n, number of specimens.

Environmental Interpretation

The intershoal facies was deposited in a shallow, normal-marine environment. Abundance of oolitic coatings and abraded grains, and the degree of sorting all indicate multiple episodes of sediment reworking. Much of this facies was likely deposited within tidal inlets (Brown, 1990). Primary evidence for a tidal inlet environment is the bimodal cross-bedding oriented along depositional dip (Figure 7A) and restriction of this lithology to narrow belts between shoals.

Within the shifting substrate thrived a diverse assemblage of mobile benthos dominated by small gastropods. Lack of abundant autochthonous sessile benthos indicates that periods of substrate stability must have been less than the months or weeks necessary to allow development of a diverse sessile benthos assemblage. Some shells may have been transported to the intershoal facies from the upper shoreface facies, resulting in the deposition of rare and abraded corals, blastoids, and other sessile benthos. Tidal currents likely reworked sediment to produce abraded and coated grains.

SHOAL FACIES

Facies Description

The shoal facies lies above the upper shoreface facies and below the sand-flat facies and is stratigraphically adjacent to the intershoal facies. Upper and lower contacts of the shoal facies are gradational. The shoal facies is characterized by thick-bedded, cross-stratified, medium- to coarse-grained, very well-sorted, fossiliferous grainstone (Figure 4D). Most grains are abraded, and superficial ooid coatings are common, particularly toward the top of the facies, where it merges with the sand-flat facies. The composition of allochems is similar to the upper shoreface facies. Bryozoan and echinoderm fragments compose about 70% of the allochems. Other allochems in order of decreasing abundance are peloids, ooids, codiacean algae, brachiopods, mollusks, foraminifers, and ostracodes. This facies has the highest abundance of true ooids of any Salem facies. The ooids are generally concentrated in small lenses.

The shoal facies is the primary lithology quarried for dimension stone in Indiana. The quarries expose large continuous sections that generally are not observable in other exposures. Sedimentary structures in this facies are observable in two quarries (localities 1 and 7). In both localities, the facies contains a mixture of large-scale planar tabular cross-beds (Figure 8) and large-scale trough cross-beds. The planar tabular cross-beds are generally more com-

Figure 8. Planar tabular cross-bedding in the shoal facies, locality 7. Arrows point to top and base of a prominent set.

mon, although either may dominate a particular quarry face. Maximum set thickness is about 1.5 m and averages 0.5–1.0 m. Set thickness decreases upward to small-scale planar cross-beds in the sand-flat facies above. The majority of cross-bed sets in both quarries are oriented to the northeast, contrasting with the bimodal cross-bedding of the intershoal facies. One set of measurements from a small part of locality 7 illustrates the dominance of northwest-oriented cross-bedding (Thompson et al., 1990) (Figure 7B).

The tops of some sets contain hardgrounds. Hardgrounds are typically continuous throughout the length of the outcrop and have local relief of up to 0.5 m (Feldman and Archer, 1990).

Composition and Interpretation of Fossil Assemblage

There are few whole macrofossils and little evidence of in-situ sessile benthos in the shoal facies. The sessile benthos assemblage is represented almost exclusively by sand-sized, well-sorted, skeletal debris, although echinoderm calyces, particularly *Pentremites*, are also present. The skeletal debris is less abraded and has less well-developed oolitic coatings than in the intershoal facies. This evidence suggests significant winnowing and sorting of the sediment, but fewer cycles of reworking than in the intershoal facies, where oolitic coatings are well developed. The grain composition in the shoal facies is very similar to the coarser grainstones of the upper shoreface facies. The upper shoreface facies was deposited immediately basinward of the shoals and may have been the source of much of the fine shoal facies sediment. Paleocurrent directions in shoal deposits are predominantly to the northeast, supporting a southwesterly source for the sediment.

Bioturbation is patchy in the shoal facies (Archer and Brown, 1990). The uppermost portion of thick, tabular cross-bed sets may be bioturbated, and locally this bioturbation is extensive. The trace-fossil assemblage suggests an environment in which high rates of sedimentation were followed by brief periods of stabilization. Many of the traces, such as *Eione* and *Macaronichnus*, are similar to forms that comprise the *Curvolithus* ichnofacies identified in some Pennsylvanian sandstones of Colorado (Lockley et al., 1987; Maples et al., 1989). In these sandstones, the *Curvolithus* ichnofacies occurs in marine deltaic environments characterized by high sedimentation rates. Similarities of trace-fossil assemblages within the Pennsylvanian quartzose sands and shoal facies suggest similar depositional and biological environments. The *Curvolithus*-assemblage style of bioturbation in the Salem grainstones appears to be related to short-term stabilization of grainstone shoals following rapid deposition. High depositional rates probably were associated with subaqueous dune migration. High-angle foresets that developed during progradation of dunes rarely contain trace fossils except for vertical escape burrows (Archer and Brown, 1988). These can extend for tens of centimeters and do not cross set boundaries. The organisms that made the escape structures apparently were buried by the migrating dune and then burrowed up to its surface. After progradation, the well-washed dune surfaces served as a suitable substrate for a diverse suite of bioturbating organisms (see Archer, 1984).

The virtual lack of sessile or even mobile benthos suggests that periods of stabilization were not long enough for shelled organisms to become established. The most common whole sessile benthos are concentrations of bellerophontid gastropods, a fossil also present in the Pennsylvanian sandstones studied by Lockley et al. (1987). These organisms may be characteristic of unstable sand substrates. The only whole sessile benthos fossils are rare concentrations of crinoid holdfasts and calyces, bryozoans, brachiopods, and corals that are associated with hardgrounds and apparently became established during periods of stability preceding hardground development (Feldman and Archer, 1990). These concentrations are not common and do not characterize the shoal facies.

Environmental Interpretation

The shoal facies was deposited in linear shoals occurring between coarser grainstones of the upper shoreface facies to the southwest (basinward) and micritic lagoonal deposits to the northeast (Figure 2). Planar tabular cross-bedding likely records the migration of submarine dunes to the northeast, toward the lagoon environment. The majority of fossil grains are allochthonous and likely were derived from the upper shoreface facies. Decrease of bedding thickness and increase of concentration of ooids near the top of some shoal deposits indicate that shoal crests had reached the zone of maximum water turbulence. Throughout shoal development, deposition was rapid and the substrate was rarely stable.

SAND-FLAT FACIES

Facies Description

The sand-flat facies is about 2–4 m thick and found only on top of the shoal facies. The lower contact with shoal deposits is gradational, but the upper contact with micritic lagoonal deposits is abrupt. The sand-flat facies consists of medium- to coarse-grained, moderately to well-sorted, cross-bedded, fossiliferous grainstone (Figure 4E). Grain types and abundances are similar to the intershoal facies, but with less foraminifers. Peloids and micrite-coated grains are common. Many peloids were probably originally shell fragments. Oolitic coatings are common on echinoderm and bryozoan grains. Micrite, as matrix, is more abundant in this facies than in any other Salem grainstone facies, composing about 4% of the lithology.

Small-scale tabular cross-bedding is prominent on weathered exposures. Most cross-bed sets are less than 0.5 m thick. Cross-bed orientations were not measured, but appear to be oriented both to the northeast and southwest.

Composition and Interpretation of Fossil Assemblage

Fossils were collected from one exposure (locality 3) of the sand-flat facies. Macrofossils primarily consist of a mobile benthos association similar to the intershoal facies (Table 2), including a diverse assemblage dominated by gastropods. This assemblage is easily observed on partially weathered outcrops, but cannot be collected in most exposures. Some exposures contain abundant sessile benthos, particularly brachiopods (10 species), the rugose coral *Hapsiphyllum*, the bryozoan *Cystodictya*, and ramose bryozoans. The coral *Syringopora* occurs in apparent life position at several localities, suggesting periods of sediment stability. Other sessile benthos such as encrusting bryozoans and crinoids are rare and are commonly fragmented or disarticulated. The preservational quality of sessile benthos fossils is similar to that of the mollusks in that they occur as both abraded fragments and well-preserved articulated shells. The sessile benthos probably were not transported in from other facies, because some corals are in life position, large fossils are common, and there is no nearby preserved facies that could have been the source. Shoreward of the sand flat were lagoonal facies that contain abundant *Hapsiphyllum* and *Cystodictya* but few brachiopods and tabulate corals. Seaward of the sand flat was the shoal facies with few macrofossils other than bellerophontid gastropods, which are not common in this mixed assemblage of sessile and mobile benthos.

Environmental Interpretation

The sand-flat facies was deposited in an environment transitional between shoal and lagoon environments. The abundance of mollusks and coated grains suggests high rates of reworking associated with shallow water at the tops of shoals, but some areas were stable long enough to support sessile benthos. The sand-flat facies contains grains apparently derived from both shoal and lagoonal environments. Echinoderm and bryozoan fragments are abraded, and some have oolitic coatings as in the shoal environment where they probably originated. Foraminifers, calcispheres, and peloids typically have micrite envelopes, but not oolitic coatings, as in the lagoonal deposits where these grains types are particularly abundant and commonly micritized.

The sand-flat environment was transitional in position and shares characteristics of both the shoals and lagoon. Unlike the shoals, shelly benthos are abundant and apparently thrived on the sand flat. This may have resulted partly from protection of the sand flat from the main force of waves or currents by the shoals. The sand-flat facies may represent a mosaic of environments that are no longer resolvable. Sediment was probably more mobile in some areas and more stable in others.

DISCUSSION

Transitions from sessile benthos in the upper shoreface environment to mobile benthos in the intershoal environment and then to a *Curvolithus* ichnoassemblage in the shoal environment likely reflect decreasing sediment stability (Figure 9). Sessile benthos thrived under conditions of relatively stable sediment where animals that were anchored to the substrate could survive. Mobile shelly benthos thrived in higher-energy, tide-dominated environments because

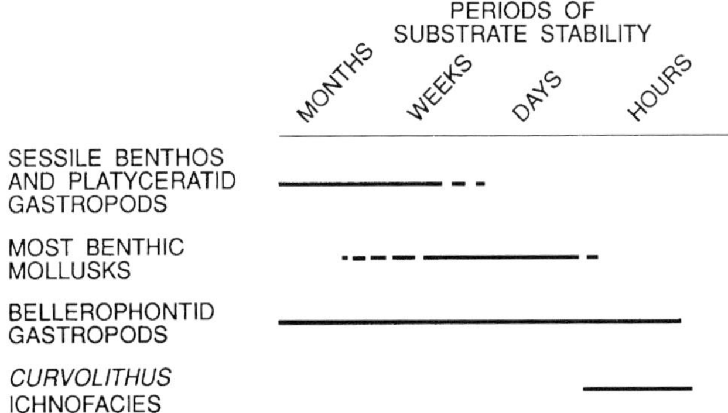

Figure 9. Relationship between benthic assemblage composition and average length of time of substrate stability between mobilization events proposed for Carboniferous grainstones.

they could escape shallow burial or burrow into the sediment after erosion. Very high rates of deposition prevented most shelly biota from becoming established within shoals, except on stabilized surfaces associated with hardground formation. The proposed increase in sediment mobility identified from faunal trends corresponds to increased sorting, decreased grain size, and increased development of oolitic coatings. Low abundance of oolitically coated grains in the lower and middle portions of the shoal grainstones may reflect high rates of deposition but relatively few cycles of reworking.

The relationship between benthos and sediment stability postulated for the Harrodsburg and Salem grainstones is somewhat similar to relationships described between modern benthic communities and sediment stability in Bahamian oolitic shoals as summarized by Purdy (1961, 1964) and Newell et al. (1959). The analogy is not perfect, however, because nearly all modern soft-substrate benthic animal communities are dominated by mobile infauna, mostly mollusks. Few sessile benthos species occur on coarse-grained carbonate substrates (primarily corals and bryozoans). Functional analogs to sessile benthos are represented, however, by some plants such as the algae *Halimeda*, *Penicillus*, and *Udotea* and the sea grass *Thalassia*. All these plants are rooted in the sediment and require stable sediment surfaces, yet thrive on coarse-grained substrates.

A transect across a modern oolite shoal can be used to illustrate the relationship between sediment stability and biota (Purdy, 1964, p. 254–259) (Figure 10). Skeletal sand in the open shelf and lagoon both support diverse faunal and floral assemblages, but not the oolite shoal. In fact, only one species, a gastropod, inhabits the crest of the shoal. The same relationship between sediment stability and abundance and type of sessile benthos appears to exist today as it did on the Mississippian sea floor. The main difference is the general replacement of brachiopod-crinoid shelly assemblages in the Carboniferous deposits with the mollusk-dominated assemblages of today.

The relationships between lithology and fossil assemblage described for the Harrodsburg and Salem limestones extend to other Carboniferous grainstones throughout the Midwest. Cross-bedded skeletal grainstones dominated by sessile benthos are well known from the Burlington, Keokuk, and Salem limestones in the Mississippi Valley and the Monteagle Limestone in Alabama. All these grainstones contain fossil assemblages composed of crinoids, brachiopods, bryozoans, and other organisms that would have required long periods of sediment stability to sustain growth.

Mollusk-dominated assemblages in oolitic grainstones, particularly lithologies dominated by superficial oolites, are also widespread, though less well known. Examples include the Gilmore City Oolite (Osagian, Iowa) (Harper, 1977), Ste. Genevieve Limestone (Chesterian, Illinois) (Weller, 1916), Marigold Oolite (Chesterian, Illinois) (Thein and Nitecki, 1974), and Drum Limestone (Missourian, Kansas) (Sayre, 1930; Stone, 1984).

Trace-fossil assemblages similar to the Salem shoal ichnofacies are not known from other limestones, but fine-grained, well-sorted oolites commonly are devoid of well-preserved macrofossils and represent high rates of sediment reworking.

If the relationship between fossil composition and sediment stability is consistent, as suggested above, this relationship could provide sedimentologists with a method of semiquantitatively estimating energy regimes independent of sedimentary structures and sediment parameters such as grain size and abrasion. We suggest the following approximate limits for the length of time between periods of sediment movement required for the establishment of the faunal assemblages preserved in grainstones. These limits are based on interpretations of functional morphology and assume that growth rates of fossil organisms are broadly similar to their modern relatives (Figure 9). The sessile benthos assemblage (crinoids, corals, bryozoans, etc.) in the upper shoreface facies required at least months to develop. As the period of time between depositional events decreased, those organisms most susceptible to death by burial would have been prevented from living in the environment. This includes organisms such as corals, encrusting bryozoans, and brachiopods that did not extend far above the sediment surface. Other organisms that extended some distance above the sediment surface, such as crinoids, blastoids, and upright bryozoans, may have been able to survive deposition of a few centimeters of sediment.

In carbonate sand that was mobilized more frequently than once every few weeks, sessile benthos could not survive and a mollusk-dominated assemblage could dominate, such as in the intershoal facies. Highly unstable substrates, such as the shoal facies, were inhabited by few shelly benthos because of the high amount of energy required to keep up with the shifting sediment surface. Modern sediments mobilized by every tide also are biologically barren. Carboniferous mollusks probably were unable to tol-

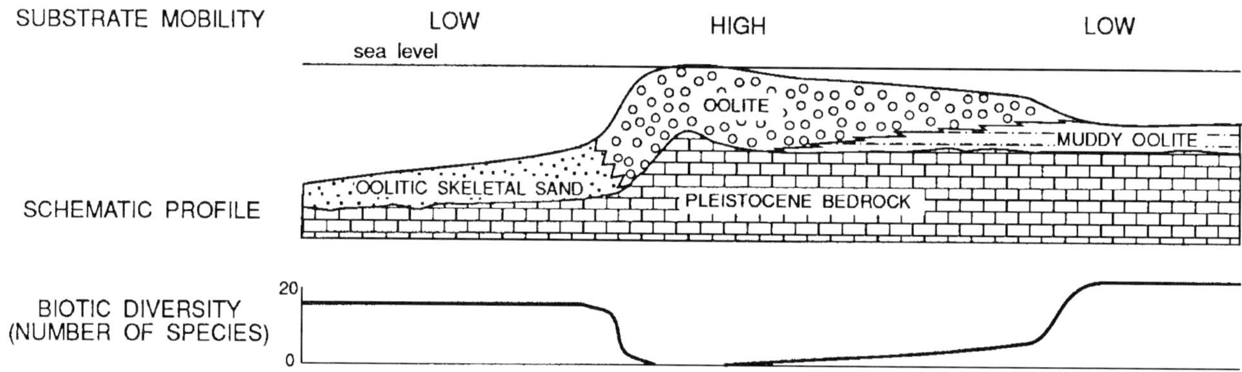

Figure 10. Schematic cross section of a Bahamian oolite shoal showing the relationships among substrate stability, sediment composition, and biotic diversity (modified from Purdy, 1964).

erate greater substrate mobility than their modern relatives. We suggest that twice-daily sediment reworking is sufficient to prevent most shelly benthos from inhabiting the substrate both today and in the Carboniferous.

Cross-bedded grainstones, such as the uppermost Harrodsburg Limestone, commonly have been interpreted as having been deposited in shallow water, high-energy environments (e.g., Carbonate Petrology Seminar, 1987). This oft-repeated concept conjures up images of frequently moving sediment, perhaps including sediment mobilization with each passing wave. The analysis presented here suggests a broad range of sediment stability within facies typically interpreted as having been deposited above normal wave base. We interpret the uppermost Harrodsburg Limestone (the upper shoreface facies) as having been deposited in shallow turbulent water, but potentially with minor sediment movement for extended periods of time. Most of the cross-bedding likely was formed during occasional storms that are preserved as amalgamated tempestites. Other grainstones in the intershoal and shoal facies likely were deposited in more turbulent environments where wave or tidal currents served to continually rework the sediment.

CONCLUSIONS

The uppermost Harrodsburg and lowermost Salem limestones consist of a series of grainstone facies, each with a unique autochthonous fossil assemblage that represents the preserved portions of once-living communities. Autochthonous benthic assemblages are represented only by complete to nearly complete fossils because small fragments probably were easily transported between environments. The distribution of benthic assemblages probably was a direct result of sediment stability. Assemblages dominated by sessile benthos occur in coarse, poorly sorted, cross-bedded, skeletal grainstones that were deposited in environments where turbulence rarely was sufficient to mobilize the coarse sediment. Cross-bedding was produced by short-duration, high-energy events, most likely storms. Assemblages dominated by mobile benthos (mostly gastropods) in coated-grain facies were deposited in shallow water where currents frequently reworked the sediment. The *Curvolithus* ichnoassemblage occurs in well-sorted grainstone where rapid deposition prevented establishment of shelly organisms. Oolitic grainstones lacking well-preserved macrofossils (not common in the Salem Limestone, but widespread in other formations) were deposited in very turbulent water where constantly shifting substrates prevented the establishment of benthic communities. The association between shelly benthos and facies described in this study also is present in many other Carboniferous grainstones; thus the faunal-facies model may be widely applicable. The *Curvolithus* ichnoassemblage is not known from other carbonate deposits but has been documented from several siliciclastic facies characterized by rapid deposition (Lockley et al., 1987). We suggest that macrofossil assemblages can provide a powerful guide for estimating sediment stability when used in conjunction with analyses of sedimentary structures and sediment textures.

ACKNOWLEDGMENTS

The authors are grateful to E. Franseen (Kansas Geological Survey), B. Keith, C. Zuppann (Indiana Geological Survey), and J. R. Dodd (Indiana University) for critical reviews of this manuscript. Drafting was done by R. Hensiek and B. Forgey of the Kansas Geological Survey. Acknowledgment is made to the Donors of The Petroleum Research Fund, administered by the American Chemical Society, for support of this research.

REFERENCES CITED

Ager, D. V., 1981, The nature of the stratigraphic record, 2nd ed.: New York, John Wiley and Sons, 122 p.

Archer, A. W., 1984, Preservational control of trace-fossil assemblages: middle Mississippian carbonates of south-central Indiana: Journal of Paleontology, v. 54, p. 285-297.

Archer, A. W., and M. A. Brown, 1988, Analysis of bioturbation and associated trace fossils within the Salem Limestone (Mississippian), Indiana: Geological Society of America, Abstracts with Programs, v. 20, p. 332.

Archer, A. W., and M. A. Brown, 1990, Bioturbation and associated trace fossils within the Salem Limestone (Mississippian), Indiana, *in* T. A. Thompson, ed., Architectural elements and paleoecology of carbonate shoal and intershoal deposits in the Salem Limestone (Mississippian) in south-central Indiana, Guidebook 14, State of Indiana Department of Natural Resources, Bloomington, Indiana, p. 21-26.

Brett, C. E., and G. C. Baird, 1986, Comparative taphonomy: A key to paleoenvironmental interpretation based on fossil preservation: Palaios, v. 1, p. 207-227.

Brett, C. E., S. E. Speyer, and G. C. Baird, 1986, Storm-generated sedimentary units: Tempestite proximality and event stratification in the Middle Devonian Hamilton Group of New York, *in* C. E. Brett, ed., Dynamic Stratigraphy and Depositional Environments of the Hamilton Group (Middle Devonian) in New York State, Pt. 1: New York State Museum Bull. 457, p 129-156.

Brown, M. A., 1987, Depositional and diagenetic history of a Middle Mississippian shoal and related facies, Salem Limestone, Lawrence County, Indiana: unpubl. M.S. thesis: Indiana University, 164 p.

Brown, M. A., 1990, Regional lithofacies and depositional environments of Salem Limestone (Valmeyeran, Mississippian), south-central Indiana, *in* T. A. Thompson, ed., Architectural elements and paleoecology of carbonate shoal and intershoal deposits in the Salem Limestone (Mississippian) in south-central Indiana, Guidebook 14, State of Indiana Department of Natural Resources, Bloomington, Indiana, p. 1-12.

Brown, M. A., A. W. Archer, and E. P. Kvale, 1990, Neap-spring tidal cyclicity in laminated carbonate channel-fill deposits and its implications: Salem Limestone (Mississippian), south-central Indiana: Journal of Sedimentary Petrology, v. 60, p. 152-159.

Carbonate Petrology Seminar, 1987, Ramp Creek and Harrodsburg Limestones: A shoaling-upward sequence with storm-produced features in southern Indiana, U.S.A.: Sedimentary Geology, v. 52, p. 207-226.

Cumings, E. R., J. W. Beede, E. B. Branson, and E. A. Smith, 1906, Fauna of the Salem Limestone: 30th Annual Report, Indiana Department of Natural Resources, v. 30, p. 1187-1486.

Dodd, J. R., 1987, Introduction to the Valmeyeran carbonate rocks of southern Indiana, *in* J. R. Dodd, ed., Valmeyeran (Middle Mississippian) Carbonate Rocks of Southern Indiana, A Guidebook for Annual Field Trip of the Great Lakes Section, SEPM: Indiana University, Bloomington, Indiana, p. 3-20.

Dott, R.H. Jr., 1983, SEPM presidential address: Episodic sedimentation—How normal is average? How rare is rare? Does it matter?: Journal of Sedimentary Petrology, v. 53, p. 5-23.

Feldman, H. R., 1987, Facies and faunas of the Salem Limestone (Mississippian) in southern Indiana and central Kentucky: Southeastern Geology, v. 27, p. 171-183.

Feldman, H. R., and A. W. Archer, 1990, Hardgrounds in the Salem Limestone, Mississippian, Indiana: Geological Society of America, Abstracts with Programs, v. 22(2), p. 10.

Feldman, H. R., and C. G. Maples, 1989, Sedimentologic implications of encrusting organisms from the phylloid-algal mound in the Sniabar Limestone near Uniontown, *in* W. L. Watney, J. A. French, and E. K. Franseen, eds., Guidebook for a Field Conference on Sequence Stratigraphic Interpretations and Modeling of Cyclothems: Kansas Geological Survey, Lawrence, Kansas, p. 173-177.

Harper, J. A., 1977, Gastropods of the Gilmore City Limestone (Lower Mississippian) of northern central Iowa: Ph.D. dissertation, University of Pittsburgh, Pittsburgh, Pennsylvania.

Hill, D., 1981, Coelenterata: Anthozoa, Subclass Rugosa, Tabulata, Treatise of Invertebrate Paleontology, Part F, Supplement 1: Boulder, Colorado, and Lawrence, Kansas, The Geological Society of America and the University of Kansas, 762 p.

Kammer, T. W., and W. I. Ausich, 1987, Aerosol suspension feeding and current velocities: distributional controls for Late Osagian crinoids: Paleobiology, v. 13, p. 379-395.

Lockley, M. G., A. K. Rindsberg, and R. M. Zeiler, 1987, The paleoenvironmental significance of the nearshore *Curvolithus* ichnofacies: Palaios, v. 2, p. 255-262.

Maples, C. G., A. W. Archer, L. J. Suttner, M. A. Brown, and H. R. Feldman, 1989, Recurring ichnoassemblages and sedimentary processes: Comparison of Carboniferous siliciclastic- and carbonate-dominated units from North America: Abstracts, 28th International Geological Congress, v. 2, p. 2-336–2-3667.

Newell, N. D., J. Imbrie, E. G. Purdy, and D. L. Thurber, 1959, Organism communities and bottom facies, Great Bahama Bank: Bulletin of the American Museum of Natural History, v. 116, art. 4, p. 181-228.

Purdy, E. G., 1961, Bahamian oolite shoals, *in* J. A. Peterson, J. C. Osmond, eds., Geometry of Sandstone Bodies: AAPG, Special Volume, p. 53-62.

Purdy, E. G., 1964, Sediments as substrates, *in* J. Imbrie, and N. Newell, eds., Approaches to Palaeoecology: New York and London, John Wiley and Sons, Inc., p. 238-271.

Richardson, J. R., 1981, Brachiopods and pedicles: Paleobiology, v. 7, p. 87-95.

Sayre, A. N., 1930, The fauna of the Drum Limestone of Kansas and western Missouri: Bulletin of the

University of Kansas, Bulletin 17, v. 31, p. 75-203.

Stanton, R. J., Jr., 1976, Relationship of fossil communities to original communities of living organisms, *in* R. W. Scott, and R. R. West, eds., Structure and Classification of Paleocommunities, Dowden, Hutchinson and Ross, Stroudsburg, Pa., p. 107-142.

Stone, W. P. Jr., 1984, Origin and evolution of oolite in the Drum Limestone (Pennsylvanian, Missourian), Montgomery, Kansas, *in* N. J. Hyne, Limestones of the Mid-Continent: Tulsa Geological Society Special Publication No. 2, p. 51-74.

Thein, M. L., and M. H. Nitecki, 1974, Chesterian (Upper Mississippian) gastropoda of the Illinois Basin: Fieldiana Geology, v. 34, 238 p.

Thompson, T. A., E. P. Kvale, and M. A. Brown, 1990, Physical sedimentary structures in the Salem Shoals, *in* T. A. Thompson, ed., Architectural elements and paleoecology of carbonate shoal and intershoal deposits in the Salem Limestone (Mississippian) in south-central Indiana, Guidebook 14, State of Indiana Department of Natural Resources, Bloomington, Indiana, p. 27-36.

Weller, S., 1916, Description of a Ste. Genevieve fauna from Monroe County, Illinois: Contributions to the Walker Museum, University of Chicago, v. 10, p. 243-265.

Chapter 9

Depositional Aspects of Golconda Group (Chesterian) Oolite Bodies, Southwestern Illinois Basin

Clayton D. Harris
University of Southern Indiana
Evansville, Indiana, USA

G. H. Fraunfelter
Southern Illinois University
Carbondale, Illinois, USA

ABSTRACT

The Golconda Group was deposited during the early to middle Chesterian, a time of transition in the Mississippian from when conditions favored ooid formation (Valmeyeran) to when conditions were unfavorable (late Chesterian). As a result, the Golconda is one of the youngest Chesterian carbonate units in the Illinois basin containing abundant ooids. Oolitic limestone is common along the western, southern, and eastern margins of the basin. Elsewhere, oolitic deposits are more sporadic and thin. The presence of a positive area (i.e., the Ozark dome) in Missouri induced conditions appropriate to ooid formation in the southwestern part of the basin. Antecedent topography formed by bioaccumulation also favored oolite deposition. Terrigenous influx and storm activity tended to discourage ooid formation. Although there are similarities between Valmeyeran and Chesterian oolite characteristics, notable differences exist.

INTRODUCTION

The middle and Upper Mississippian System of the Illinois basin (Figure 1) comprises formations that increase in lithologic complexity upward in the section. The Valmeyeran Series is an overall shallowing-upward depositional sequence with abundant carbonates, including oolitic limestones. The Chesterian Series consists of a variety of siliciclastic and carbonate facies that exhibit complex lateral and vertical relationships.

There is a general decline in the abundance of oolitic limestone upward in the Chesterian Series (Atherton et al., 1975). The Golconda Group, a mixed carbonate-siliciclastic unit that is early to middle Chesterian in age, is one of the youngest Chesterian units in which oolitic limestone is common. Unlike Valmeyeran oolite bodies, those of the Chesterian are generally less extensive and thinner. Golconda oolites are more abundant along the eastern, western, and southern margins of the basin. This chapter provides an overview of some depositional aspects of these oolite bodies in the southwestern basin area. A brief comparison of the nature and distribution of selected Chesterian oolites bodies with Valmeyeran-age examples is included as well.

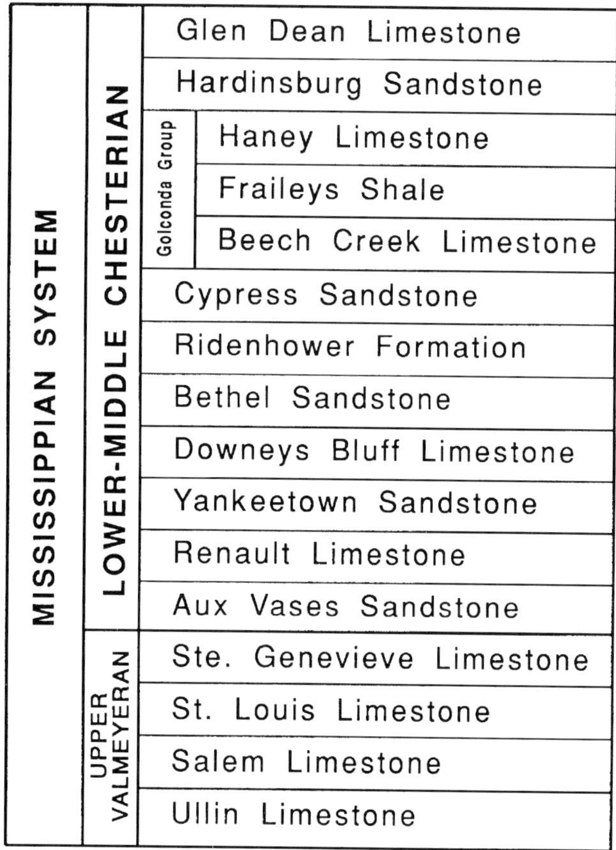

Figure 1. Chart showing the stratigraphic nomenclature of the lower and middle Chesterian and upper Valmeyeran in the southwestern Illinois basin.

REGIONAL SETTING

The study area in the southwestern Illinois basin includes 16 field localities (Figure 2) described in an earlier report (Harris, 1987). It is situated between the western and southern shelves of the Illinois basin (Figure 3) and lies to the east of the Ozark dome.

Treworgy (1988) has shown that the Illinois basin acted as a tidally influenced ramp, open to the south, during early and middle Chesterian time. Siliciclastics were delivered to the basin from the north or northeast (Figure 3), thereby influencing lithofacies patterns, especially carbonate deposition (Swann, 1963, 1964; Droste and Keller, 1989).

Swann (1963) proposed a model for Chesterian sedimentation in which a single river, the Michigan River, was responsible for delivery of siliciclastic sediments from the Canadian shield area to the Illinois basin. Repeated migrations of delta lobes within this system, along with numerous transgressive-regressive cycles, resulted in complex facies distribution patterns. Northwest-directed longshore currents also contributed to facies complexity, according to Swann.

Some aspects of Swann's model have been revised or are being challenged. Recent research involving lithofacies patterns suggests several river systems were active at this time (Droste and Keller, 1989). Repeated, large-scale transgressive-regressive cycles are considered unlikely (Gray, 1979). In addition, the Ozark dome may have been a secondary sediment source, especially for the southwestern area of the basin (Sable, 1979; Harris, 1987).

Within the Golconda Group, coarse siliciclastics are most abundant toward the east and northeast. Shale is concentrated in the center of the basin and to the northwest (Eldridge, 1961). This distribution suggests that at this time the delta was situated on the eastern margin of the basin, possibly at its southeasternmost position (Sable, 1979; Treworgy, 1988). Carbonates are most abundant toward the southeast, southwest, and west (Figure 3), probably due to reduced terrigenous input (Harris, 1987, Treworgy, 1988).

STRATIGRAPHY AND LITHOLOGIC CHARACTER

In Illinois, the Golconda Group comprises three formations: the Beech Creek Limestone, a coarsening-upward carbonate unit; the Fraileys Shale, a shale-dominated sequence with sporadic skeletal limestone lenses; and the Haney Limestone, a carbonate-dominated unit with variable amounts of limestone and shale. The Golconda Group is underlain by the Cypress Sandstone, and is overlain by the Hardinsburg Sandstone (Figure 1). In the study area, the thickness of the Golconda Group ranges from 90 to 130 ft (27–40 m).

The Beech Creek Limestone in the study area is composed of two units: a lower fine-grained wackestone and packstone unit and an upper unit of coarse-grained packstone and grainstone. The lower unit, characteristically 1–4 ft (0.3–1.2 m) thick, is composed primarily of skeletal material, and rarely contains ooids. The upper unit, which ranges in thickness from 8 to 20 ft (2.4–6 m), contains a wide variety of carbonate lithofacies but is predominantly pelmatozoan-bryozoan limestone with lesser amounts of oolitic limestone.

The Fraileys Shale, ranging in thickness from 20 to 40 ft (6–12 m) in the study area, contains fossiliferous calcareous shale at many localities, being especially noted for its fenestellid and *Archimedes* bryozoans and well-preserved brachiopods. The lenticular limestones common in the Fraileys consist of pelmatozoan-bryozoan packstones and grainstones, and intraclastic rudstones interpreted to be storm deposits. Oolitic limestone is rarely present.

The Haney Limestone, which ranges from 30 to 50 ft (9–15 m) thick in the study area, generally consists of equal parts limestone (coarse skeletal, biomicritic, and oolitic) and calcareous shale. Among the formations of the Golconda Group, the Haney contains the largest amount of oolitic limestone. As with the Fraileys, calcareous shale in the Haney often contains abundant fossils (Somasekhara, 1970).

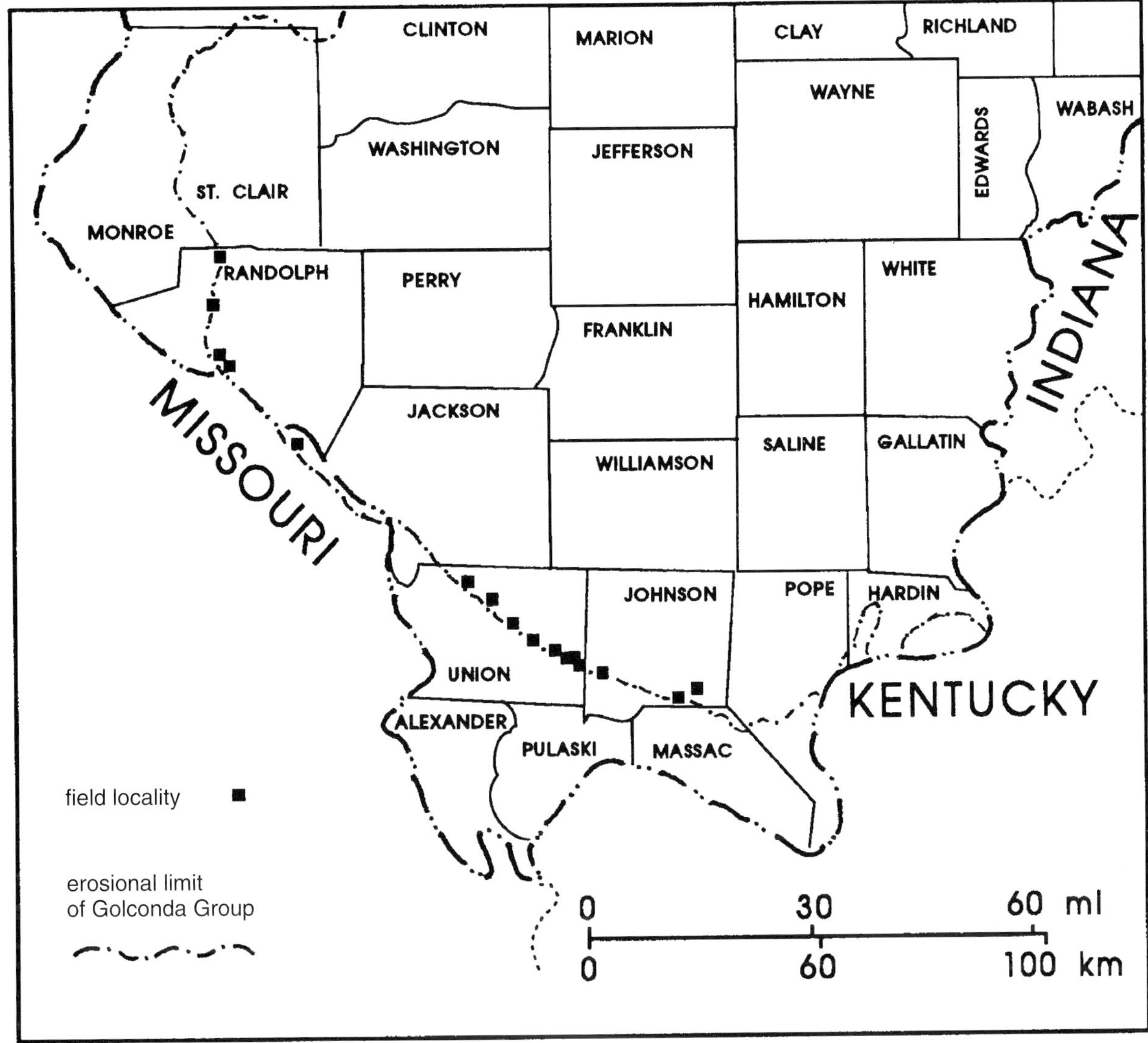

Figure 2. Location map of the 16 field localities used in this study.

GENERAL GOLCONDA DEPOSITIONAL MODEL

An approach similar to the classic X-Y-Z carbonate zone depositional model of Irwin (1965) was utilized as an overall model for Golconda Group deposition. Terrigenous input was, however, an additional influence on carbonate deposition in the Golconda; Irwin's model was modified slightly to emphasize this fact.

In the Golconda model, the nearshore zone (Irwin's Z-zone) is a locus of deltaic sedimentation or at least an area of significant terrigenous influx; it is termed Zone I herein (Figure 5). The zone of maximum shoaling (Irwin's Y-zone) has essentially the same characteristics as in Irwin's model; it is here referred to as Zone II. Finally, in the offshore zone (Irwin's X-zone), deposits consist of mixed fine-grained carbonate and siliciclastic sediments, and coarser grained skeletal accumulations; this zone is termed Zone III.

OOLITIC SHOAL DEPOSITIONAL MODEL

A depositional model specific to Golconda oolites was developed based on the character and relationships of oolitic shoal facies (Figures 4, 5, 6). Several factors were considered: association of oolites with bioclastic deposits (especially the pelmatozoan-bryozoan facies), storm deposits (intraclastic rudstone), and calcareous shale; lateral discontinuity of oolitic beds; and scarcity of thick, well-developed oolitic sequences. These factors are discussed further below.

Controls on Ooid Shoal Development

Areas of antecedent topography are known to promote ooid formation in modern carbonate environments (Harris, 1979). Relatively thick, well-developed oolite bodies in the Golconda are typically underlain by pelmatozoan-bryozoan shoal deposits. The coars-

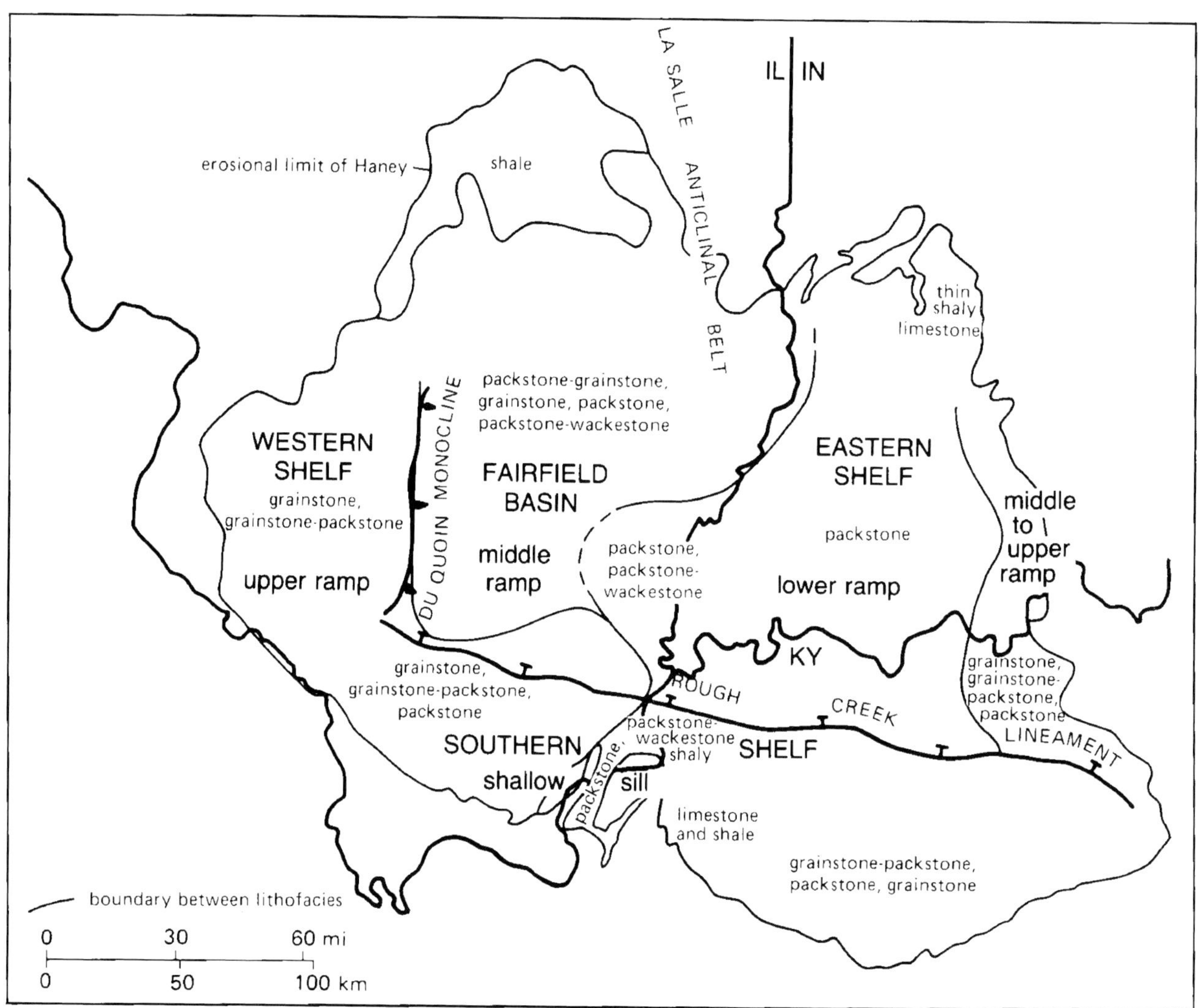

Figure 3. Lithofacies distribution, depositional setting, and structural features of the Illinois basin during Haney Limestone deposition, the time of maximum ooid formation during the Golconda. Lithofacies are presented in order of abundance in each area (from Treworgy, 1988).

ening-upward texture of these bioclastic deposits is interpreted to indicate that they progressively built up into shallower depths where higher energy conditions prevailed. These topographic highs concentrated current and wave activity, promoting conditions favorable to ooid formation (Ball, 1967; Harris, 1979). In addition to providing shoal areas, bioclastic deposits may have been a more favorable substrate for ooid formation than the open sea floor.

Other factors were important in limiting the extent and duration of ooid shoal development. Influx of siliciclastic material from the Michigan River delta to the northeast and the Ozark dome to the west (Sable, 1979; Harris, 1987) (Figure 3) limited the thickness of nearshore oolitic deposits by halting ooid production, if only temporarily. The lateral extent of ooid shoals was similarly diminished. Siliciclastic influx also discouraged bioclastic shoal growth in nearshore areas, thereby deterring the formation of topographic highs on which ooid production might commence.

Siliciclastic sedimentation prohibited oolitic shoal formation in offshore areas. Bioclastic shoals were recognized in offshore deposits; however, their size and number were greatly diminished as compared to nearshore shoals. The association of abundant siliciclastics interbedded with and overlying these diminutive bioclastic shoals is interpreted to indicate that siliciclastic sedimentation was a limiting factor in the size and longevity of offshore bioclastic shoals. This scarcity of bioclastic shoals resulted, for reasons discussed above, in the absence of ooid shoals in offshore areas.

Storm events were an important influence on the migration and termination of ooid-producing subenvironments. Ooid shoal sediments were transported into lower energy, bioclastic-dominated subenviron-

FACIES	COMPOSITION	COLOR AND TEXTURE	BEDDING AND GEOMETRY	DIMENSIONS W x L (ml^2)	Thick (ft)	DEPOSITIONAL ENVIRONMENT	DEPOSITIONAL PROCESS
Calcareous shale	Non-fossilliferous to moderately fossilliferous, minor slit	Light gray or greenish-gray	Thinly laminated or massive; tabular	10s	1–10s	Protected shelf, backshoal, open shelf	Suspension settling
Mixed-grain wackestone-packstone	Skeletal grains, intraclasts, ooids, and pellets	Light to medium gray; medium grained, poorly sorted	Thin bedded; tabular	1–10s	1–10	Backshoal, intershoal and foreshoal	In-situ accumulation, local transport, suspension settling
Skeletal packstone-grainstone	Skeletal grains, especially bryozoan, pelecypod, and brachiopod	Light to medium gray; medium cross grained, moderately well sorted	Medium to thick bedded, planar cross-bedded; lenticular or tabular	1–10	1–10	Shoal flanks and core	In-situ accumulation, local transport, suspension settling
Pelmatazoan-bryozoan packstone-grainstone	Crinoids and blastoids, fenestrate bryozoans	Light to dark gray; medium cross grained, moderately well sorted	Thin to medium bedded, trough cross-bedded; lenticular to tabular	1–10	1–10	Shoal flanks and core	In-situ accumulation, local transport, suspension settling
Ooid grainstone	True and superficial ooids, minor skeletal grains, and ooids	Light brown or gray; well to very well sorted	Medium to thick bedded, planar and trough cross-beds; lenticular	1–3	1–10	Ooid cap	In-situ accumulation, local transport
Intraclastic grainstone-rudstone	Intraclasts, skeletal grains, and ooids	Light to medium gray; fining upwards, moderately well sorted	Thin to medium bedded; lenticular	< 1–3	< 1–2	Throughout shoal complex, rare in ooid cap	Storm sedimentation

Figure 4. General character of Golconda ooid shoal facies.

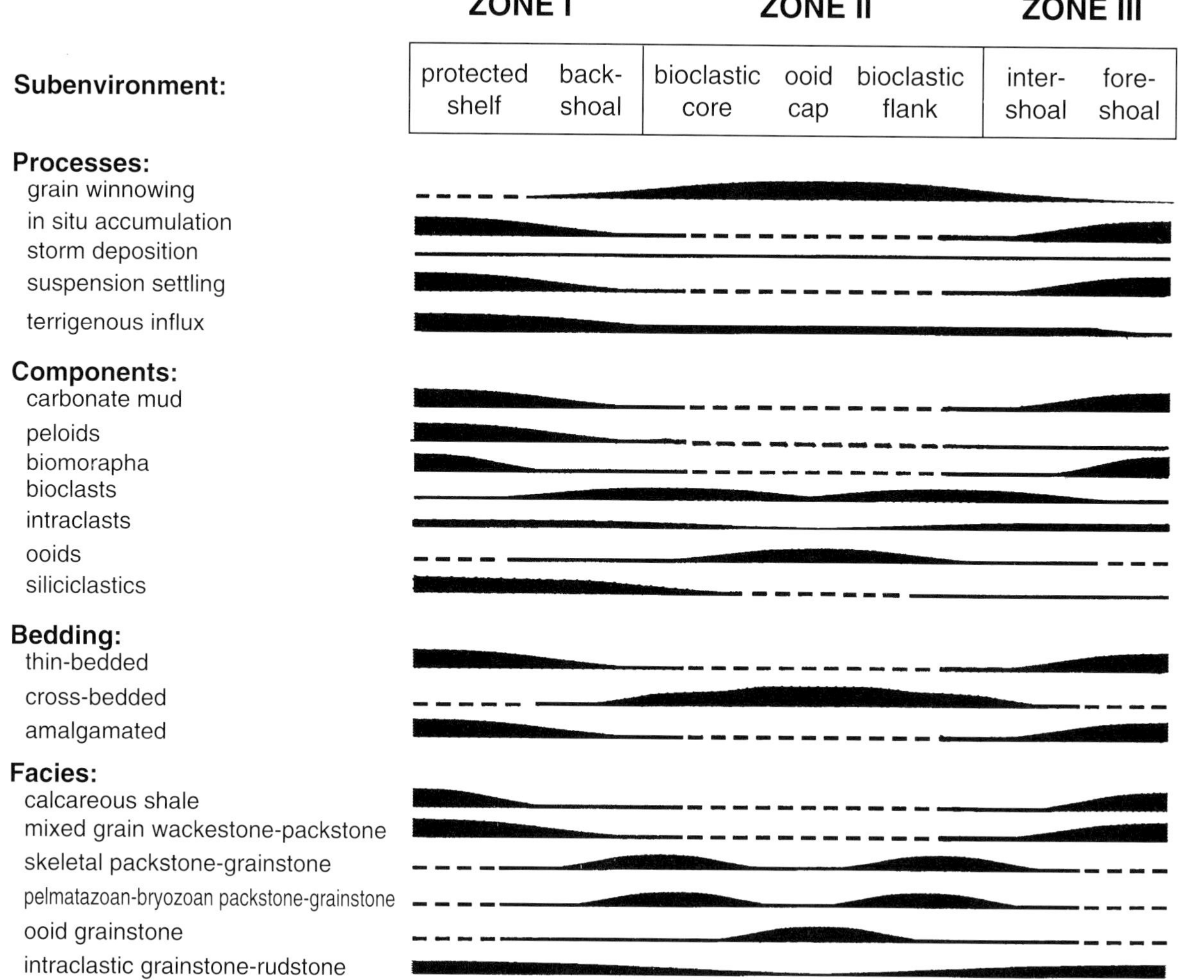

Figure 5. Characteristics of Golconda facies zones within the study area (after Irwin, 1965).

ments as a result of storm-forced waves and currents. Major storm events changed the bathymetry of shoals, halting ooid production in some areas (i.e., causing shoal abandonment) while promoting production elsewhere.

Storms also transported terrigenous sediment into shoal areas. This storm deposition resulted in the siliciclastic facies that are interbedded with oolitic and bioclastic shoal deposits (as discussed above). In extreme cases, terrigenous influx caused cessation of ooid production and eventual return to bioclastic accumulation.

Of the 16 localities studied, two contain abnormally thick oolitic zones relative to other Golconda oolites (20–40 ft [6–12 m] vs. less than 5 ft [1.5 m]). These thicker deposits developed because of the prolonged existence of hydrodynamic and sedimentologic conditions appropriate to ooid formation. Oolitic deposition at these sites, however, was not uninterrupted, as evidenced by the interbedding of oolitic limestone with bioclastic limestone and calcareous shale. These were deposited in what was probably an area of regular current activity, where conditions rarely deviated from those necessary for ooid production. Their uniqueness indicates the ephemeral nature of these conditions elsewhere in the study area.

Ooid Shoal Subenvironments

The character and distribution of Golconda facies (Figures 4, 5, 6) provide many insights into the Golconda Group oolitic shoal setting. These shoals consisted of several subenvironments with a predictable sequence and distribution.

Zone I Subenvironments

Protected shelf: This is a low-energy area, shoreward of the shoal, where calcareous shales were common. During times of reduced terrigenous influx, mixed-grain wackestones and packstones accumulated. Bioclastic, oolitic, and intraclastic grains from adjacent bioclastic and oolitic shoals were sometimes carried into this environment by storm currents or waves. After intense bioturbation, this resulted in a highly diverse assemblage of grain types in a mud-rich matrix. Slightly reworked or unreworked storm

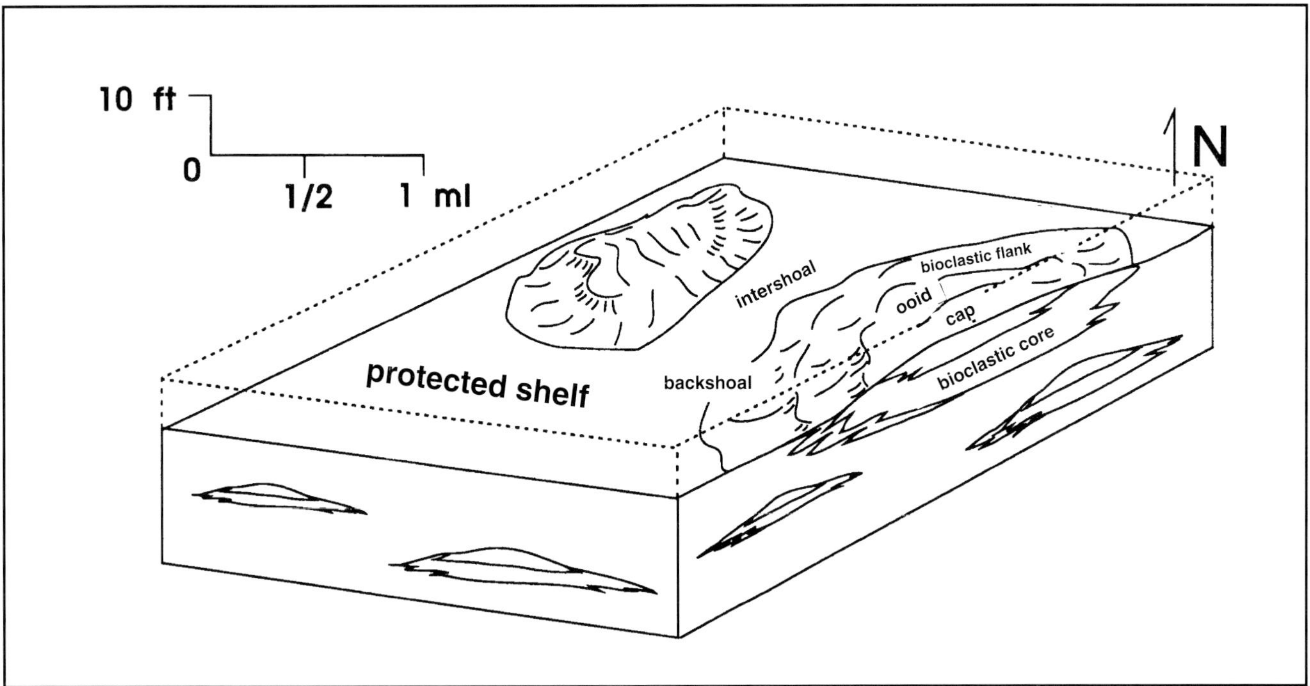

Figure 6. Schematic block diagram of the depositional setting for Golconda ooid shoals. Relict shoals are shown in cross-section.

deposits were also noted (intraclastic grainstone-rudstone).

The protected shelf environment often prograded over abandoned oolitic shoals and was in turn prograded over by active shoals. This area was less than a mile to several miles in width and length.

Transitional Zone I/II Subenvironments

Backshoal: This environment, produced by washover from the shoal, was a transition zone between the protected shelf and shoal environments. As such, it comprises an assortment of low-energy, moderate-energy, and high-energy deposits. Mixed-grain wackestone and packstone and skeletal packstone and grainstone resulted from background sedimentation and contain locally derived skeletal material. Intraclastic grainstone-rudstone was storm-generated and contains grain types from adjacent environments. This area was of limited extent, probably no more than a mile or so in maximum dimension.

Zone II Environments

Bioclastic flanks and core: The bulk of shoal material was deposited on the shoal flanks, a nearly continuous band of bioclastic deposits that surrounded the ooid cap (discussed below). The core consists of the bioclastic deposits that preceded ooid production and underlie the ooid cap. Mechanical energy levels were low to moderate, but almost constant water movement prohibited significant fine-grained sediment deposition. Pelmatozoan and bryozoan remains dominate these deposits, indicating that the flanks and adjacent areas were populated by dense colonies of these organisms. Fair-weather deposits are typically skeletal and pelmatozoan-bryozoan packstones and grainstones. Storm deposits (intraclastic grainstone-rudstone) contain a wide variety of grain types from this and adjacent subenvironments.

The protected shelf and intershoal areas (discussed below) were important sources for the skeletal grains that make up the shoal flanks. Skeletal material was probably being swept onto the shoal flanks by both fair weather and storm waves and currents. The shoal flanks were probably not more than a mile or so wide and up to a few miles in length.

Ooid cap: This was the shallow, high-energy zone of ooid production situated on the upcurrent or windward side of the shoal complex. Well-developed ooids formed where conditions were particularly suited to ooid production; elsewhere, grains transported from adjacent subenvironments were thinly coated to produce superficial ooids (i.e., with a single oolitic laminae).

Storms were responsible, in part, for lateral migration of the shoal over adjacent areas. Changes in shoal morphology caused by storm currents or waves resulted in a shift in the locus of ooid formation and episodes of abandonment in some areas. Storm deposits (intraclastic grainstone-rudstone) from this subenvironment were probably deposited in areas less susceptible to reworking, either along the outer margins of the cap or in subsequently abandoned areas.

Because of the often rapid abandonment of the shoal, ooid cap deposits (ooid grainstone) are usually thin (2–3 ft, 0.6–0.9 m) and of limited extent (less than a mile to a few miles). Based on field observations, it is likely that the long dimension of the cap was rarely more than a mile or two.

Transitional Zone II/III Subenvironments

Intershoal: In areas where shoals were abundant and closely spaced, the intershoal consisted mainly of bioclastic flank deposits, with minor ooid content. In areas where shoals were widely spaced, the intershoal was essentially an open, low- to moderate-energy shelf environment where mixed-grain and skeletal wackestones and packstones were accumulating. Storm deposits (intraclastic grainstone- rudstone) are commonly well-developed. Storms were probably responsible for the bulk of the oolitically coated grains found in intershoal deposits.

This subenvironment was of variable extent. Between closely spaced shoals it was essentially absent, whereas in some areas it was several miles or more in width and length.

Foreshoal: This area is a nearshore, shallow shelf environment. It served as a grain source for shoal deposits. Limestones from the foreshoal environment (mixed-grain wackestone-packstone, skeletal and pelmatozoan-bryozoan packstone-grainstone) are generally muddy and thin, and calcareous shale is common. Proximal to the shoal environment, however, grainstones are more common and shales less so. Storm deposits (intraclastic grainstone-rudstone) are well-developed. Foreshoal deposits extend outward into and are indistinguishable from Zone III shallow shelf deposits.

SUMMARY OF GOLCONDA OOLITE SHOAL ORIGIN AND CHARACTER

In southwestern Illinois and southeastern Missouri, the Golconda Group contains a significant number of oolitic limestone bodies of limited thickness (average 2–3 ft, 0.6–0.9 m) and lateral extent. These appear to have formed where terrigenous influx was minor and conditions were otherwise favorable for ooid development.

Lateral and vertical relationships of oolitic and nonoolitic facies are complex, suggesting variable settings of short duration. Development of a bioclastic substrate provided relief above the sea floor and apparently favored oolite shoal formation. Terrigenous influx and structurally induced relief were also important controls on oolite shoal development. No simple model can adequately account for all these factors; however, both storm and tidal effects are recognized, helping to explain the complexities.

Bioclastic shoals containing packstones and grainstones are found subjacent to oolitic deposits at several localities. Interbedded with or overlying the oolitic facies (ooid grainstone) are: bioclastic facies (mixed-grain wackestone-packstone, skeletal and pelmatozoan-bryozoan packstone-grainstone), which signal a return to more moderate hydrodynamic conditions; calcareous shale, which suggests minimal mechanical energy and increased siliciclastic sediment supply; and non-calcareous shales or sandstones, which resulted from greatly increased terrigenous sedimentation. The entire oolite shoal complex averages 5–10 ft (1.5–3 m) in thickness, a few miles in length, and somewhat less in width.

Some oolitic bodies contain sporadic occurrences of storm deposits (intraclastic grainstone-rudstone). These are often located at or near the upper surface of oolitic shoals, suggesting that storm deposition was sometimes responsible for terminating ooid formation.

The oolite bodies formed in shoal settings with a predictable association of subenvironments (Figure 6): the protected shelf, situated well behind the main shoal body; backshoal, an area of transition between the protected shelf and the shoal with traits from both; the bioclastic core and flanks, which, respectively, underlay and encircled the central shoal area and consisted mainly of reworked pelmatozoan and bryozoan debris; the central shoal area or ooid cap, where ooid production was favored; the intershoal, an area of mixed shoal and low-energy character; and the foreshoal, analogous to the open shelf, where low-energy deposits predominated.

Based on these facies associations and environmental characteristics, an idealized model for oolitic limestone production is proposed. This model suggests a storm-influenced, environmentally variable setting in which lateral migration of oolitic shoals, and dispersal of ooids into adjacent low-energy environments, outpaced vertical accretion. Bioclastic shoal development preceded and often succeeded ooid deposition. Interruption of ooid production by sporadic, possibly storm-generated, siliciclastic influx was common in the nearshore environment. In the offshore setting, fair-weather siliciclastic deposition was sufficient to prohibit ooid production.

COMPARISON OF VALMEYERAN AND CHESTERIAN OOLITES

Oolitic limestone is a common facies in middle Mississippian rocks. The Salem and Ste. Genevieve limestones (Figure 1) are known for their extensive and thick shoal deposits; however, notable changes in oolite character occurred during the Valmeyeran. Significant changes also can be seen when comparing Valmeyeran oolites to those of the Chesterian.

Salem Limestone Oolites

The Salem Limestone is composed of one or more laterally continuous grainstone intervals produced by migration of ooid shoals over open marine wackestones and packstones. Lagoonal wackestones and packstones commonly overlie shoal grainstones (Cluff, 1984). Salem grainstones usually contain some oolitically coated grains; however, these coatings are commonly thin, suggesting rapid deposition and limited reworking (Zeng et al., 1983). Ooid grainstones composed of thickly coated ooids are not abundant (Cluff, 1984).

Oolite bodies in the Salem Limestone are thickest in the center of the basin. These thin toward the Illinois basin margins, with lagoonal deposits common to the north and open shelf facies to the south (Cluff, 1984). In the Indiana subsurface, oolitic shoals 0.5–2 mi (0.8–3.2 km) wide and 2–5 mi (3.2–8 km) long with thicknesses of 20 ft (6 m) or more (Figure 7) are recognized (Keller and Becker, 1980). Thompson and Brown (1989) investigated an 8-ft-thick (2.4-m-thick) sand wave in an Indiana quarry. Stacking of sand waves of this scale could easily account for typical oolite thicknesses reported in the subsurface.

Salem oolitic and skeletal grainstones form widespread blankets or sand flats that are only broadly linear, trending approximately northwest–southeast, perpendicular to regional paleoslope (Cluff, 1984). Smaller scale, tidally influenced bars and channels, oriented parallel with paleoslope (Figure 7), are sometimes superimposed on these larger scale sand bodies (Keller and Becker, 1980). As in modern shoal settings (Ball, 1967), storm events were an important mechanism in Salem shoal migration (Thompson and Brown, 1989).

Multiple, thick, shoaling-upward cycles contributed to Salem shoal morphology, continuity, and distribution. Shoals formed on a broad, shallow, tidally influenced, carbonate ramp with gentle slope and stable depositional conditions (Cluff, 1984).

Ste. Genevieve Oolites

Ste. Genevieve ooid shoals are commonly composed of two major lithologies: oolitic bar and fine-grained interbar facies (Cluff, 1984). Many minor lithologies are also present but have limited lateral extents (Cluff, 1981; Zuppann, this volume). Shales and minor sandstones are interbedded with and overlie oolitic deposits (Choquette and Steinen, 1980; Zinn, 1983). Typically, terrigenous material is more common in the upper half of the formation (Atherton and Palmer, 1979). The Ste. Genevieve is known for its abundant ooid grainstones (Carozzi, 1989) and well-developed true ooids (i.e., with multiple laminae; Figure 7). Long periods of agitation or periodic reworking of ooids were needed to produce the thick ooid cortexes (Ball, 1967; Zeng et al., 1983).

Individual Ste. Genevieve oolite bodies are widespread, but generally smaller and more discontinuous than those of the Salem Limestone (Figure 7). Cluff (1986, p. 5) reported that Ste. Genevieve oolite bars in Illinois are usually "1/2 mile to several miles in length, 1/8th to 1/4 mile in width." Other studies provide similar or only slightly greater dimensions (Carr, 1973; Choquette and Steinen, 1980). The Ste. Genevieve Limestone, like the oolitic zones it contains, thickens toward the center of the basin (Carr et al., 1986; Atherton et al., 1975). A typical thickness range for its oolitic zones is 10–20 ft (3–6 m) (Cluff, 1986; Choquette and Steinen, 1980) although excep-

	Golconda Group	Ste. Genevieve Limestone	Salem Limestone
Dominant ooid type	superficial	true	superficial
Abundance of oolitic facies	uncommon to common	abundant	abundant
Thickness of individual ooid grainstone bodies (ft)	< 5	10–20	> 2
Orientation of oolite bodies relative to SW paleoslope	parallel	parallel	parallel
Area of maximum oolite shoal abundance	basin margins	basin-wide	basin-centered
Oolite body dimensions (9 miles):			
width	< 0.5	< 1.5	0.5–2
length	0.5–2	1–3	2–5
Oolite body morphology	irregular shoals and tidal bars	tidal bars	linear banks with superimposed tidal bars

Figure 7. Comparison of Golconda Group oolite characteristics with Valmeyeran oolites. Sources: Kissling, 1967; Keller and Becker, 1980; Cluff, 1981, 1984; Harris, 1987; and Treworgy, 1988.

tions are known (McIntire and Pryor, this volume). Lower Ste. Genevieve oolites are commonly somewhat thicker than those in the upper Ste. Genevieve (Cluff, 1984; Zinn, 1983).

Ste. Genevieve shoals tend to be elongate, lenticular bodies, parallel with paleoslope (approximately southwest; Figure 7). They have been interpreted as tidal sand bars (Choquette and Steinen, 1980; Cluff, 1984, 1986). When clustered together, Ste. Genevieve oolite bars tend to form poorly defined belts normal to paleoslope (Choquette and Steinen, 1980; Zinn, 1983).

Shoaling-upward cycles in the Ste. Genevieve are thinner and less easily recognized than those in the Salem Limestone (Cluff, 1984). These smaller scale cycles were a response to several factors, including local sedimentation rates, variations in current or wave activity, or sea floor topography. Few Ste. Genevieve oolites occur as well-developed marine sand belts (Cluff, 1986), but rather as scattered, isolated tidal bars. This may be due to the low paleoslope of the Illinois basin during this time (Cluff, 1984).

Golconda Oolites

The character and distribution of Golconda oolites was discussed above. These demonstrate the sporadic and ephemeral nature of Golconda ooid production. Outcrop studies suggest that Golconda oolites, as well as other Chesterian oolite bodies, display sedimentologic and morphologic traits similar to Ste. Genevieve oolites (Kissling, 1967; Balthaser, 1969; Vincent, 1975; Treworgy, 1985; Harris, 1987). Very little subsurface data has been collected on Chesterian oolites, probably because they are not known to contain economic petroleum reservoirs.

Three factors—reduced turbulence, increased turbidity, and storm influence—are probable causes for the notable decrease in oolitic limestones from late Valmeyeran to the end of Chesterian time. Thick shale sequences commonly associated with Chesterian carbonates indicate quiet and turbid water conditions were prevalent through much of Golconda deposition.

By lower to middle Chesterian time, a shallow, widespread shoal area had developed along the southern shelf (Treworgy, 1985). This probably decreased water movement between the flat, shallow ramp of the Illinois basin and the epicontinental sea to the south. The resulting drop in hydrodynamic energy promoted fine-grained sedimentation and further diminished ooid production. The relative abundance of oolites in the southwestern and western Illinois basin suggests that unrestricted circulation with the open sea was maintained in these areas.

In modern depositional settings, when conditions favorable to ooid production exist, storm-generated waves and currents may aid shoal migration (Ball, 1967). However, storms may have adversely affected ooid formation in the Chesterian by dispersing ooids to low-energy areas surrounding the shoal and by blanketing shoals with terrigenous sediments. The abundance of storm beds in many lower and middle Chesterian formations (Visher, 1980; Kissling, 1967; Harris, 1987; Foster, 1990) attests to the prevalence of storm deposition in the Illinois basin at this time.

Small-scale shoaling-upward carbonate cycles are present in the Golconda and elsewhere in Chesterian rocks. Yet many of these cycles appear to have been overwhelmed by other sedimentary processes, such as fair-weather siliciclastic deposition, storm deposition, and energy restriction. Active ooid shoals were short-lived, probably as a result of these same processes.

SUMMARY AND CONCLUSIONS

Golconda Group oolite deposits are similar to those of the Ste. Genevieve and Salem Limestones. Important differences do exist, however, related to rates of ooid production, shoal migration, terrigenous influx, and environmental stability. Overall the amount of ooid production decreased from the Valmeyeran until the end of the Chesterian.

By the middle Mississippian, tectonism on the eastern flank of the North American continent had resulted in upwarp and consequent infilling of the Illinois basin (Ettensohn, this volume; Frazier and Schwimmer, 1987). This led to the formation of a warm, shallow water platform favorable to ooid production. Thick, shoaling-upward cycles associated with Salem oolites are evidence of these favorable depositional conditions.

Near the Valmeyeran–Chesterian transition, sedimentation conditions developed that were less favorable to the formation of thick, extensive oolitic sequences. Ste. Genevieve ooid shoals were therefore smaller and the individual oolitic bodies somewhat thinner than those of the Salem Limestone.

Tectonism was widespread during Chesterian time (Frazier and Schwimmer, 1987). Shallowing, resulting from uplift on the margins of the Illinois basin, discouraged ooid formation in some areas, but aided ooid formation elsewhere. Increased terrigenous sedimentation due to uplift was another important deterrent to oolite production. Oolites were mainly restricted to those areas along the basin margins where terrigenous input was low, and sufficient mechanical energy was present to keep mud suspended. Typically, only thin, isolated oolite bodies are present in the Golconda.

ACKNOWLEDGMENTS

Discussions with J. B. Droste were instrumental in improving the focus of this paper, and reviews by J. B. Droste, B. Keith, J. Treworgy, and C. Zuppann were also valuable. Their help is greatly appreciated. The senior author would like to thank J. Utgaard for allocating research funds from the Department of Geology at Southern Illinois University for the preparation of thin sections. Responsibility for any shortcomings found in this paper lie squarely on the shoulders of the senior author.

REFERENCES CITED

Atherton, E., and Palmer, J. E., 1979, The Mississippian and Pennsylvanian (Carboniferous) Systems in the United States—Illinois: USGS Prof. Paper 1110, p. L1-L42.

Atherton, E., Collinson, C., and Lineback, J. A., 1975, Mississippian System, *in* Willman, H. B., Atherton, E., Buschbach, T. C., Collinson, C., Frye, J. C., Hopkins, M. E., Lineback, J. A., and Simon, J. A., eds., Handbook of Illinois Stratigraphy: Illinois State Geological Survey, Bulletin 95, p. 123-163.

Ball, M. M., 1967, Carbonate Sand Bodies of Florida and the Bahamas: Journal of Sedimentary Petrology, v. 37, p. 556-591.

Balthaser, L. H., 1969, Petrology and paleoecology of middle Chester (Miss.) rocks of the S.W. Indiana outcrop: Ph.D. thesis, Indiana University, Bloomington, 257 p.

Carozzi, A. V., 1989, Carbonate rock depositional models—A microfacies approach: N.J., Prentice Hall, 604 p.

Carr, D. D., Rexroad, C. B., and Gray, H. H., 1986, Ste. Genevieve Limestone; *in* Shaver, R.H. et al.; (eds.), Compendium of Paleozoic rock-unit stratigraphy in Indiana—A revision: Indiana Geological Survey Bull., v. 59, p. 128-130.

Carr, D. D., 1973, Geometry and origin of oolite bodies in the St. Genevieve Limestone (Mississippian) in the Illinois Basin: Indiana Geological Survey Bulletin, v. 48, 81 p.

Choquette, P. W., and Steinen, R. P., 1980, Mississippian non-supratidal dolomite, Ste. Genevieve Limestone, Illinois Basin—evidence for mixed-water dolomitization: SEPM Spec. Publ., v. 28, p. 163-196.

Cluff, R. M., 1981, Upper Valmeyeran deposition, *in* Cluff, R. M., and Lineback, J. A., eds., Middle Mississippian carbonates of the Illinois Basin: a seminar and core workshop: Mt. Vernon, Illinois Geological Society, p. 46-85.

Cluff, R. M., 1984, Carbonate sand shoals in the Middle Mississippian (Valmeyeran) Salem-St. Genevieve Limestones, Illinois Basin, *in* Harris, P. M., ed., Carbonate Sands—A Core Workshop: SEPM Core Workshop 5, p. 94-135.

Cluff, R. M., 1986, Application of modern carbonate sand models to oil and gas exploration, Mississippian Ste. Genevieve Limestone, Illinois Basin, *in* Seyler, B., ed., Aux Vases and Ste. Genevieve Formations: a core workshop and field trip guidebook: Mt. Vernon, Illinois Geological Society, p. 5-7.

Droste, J. B., and Keller, S. J., 1989, Development of the Mississippian-Pennsylvanian unconformity in Indiana: Indiana Geological Survey Occasional Paper, v. 55, 11 p.

Eldridge, W. F., 1961, The Golconda Formation in the Illinois Basin: An integrated facies study: M.S. thesis, University of Illinois, Urbana, 74 p.

Foster, Z. A., 1990, Depositional and diagenetic history of the Haney Limestone (Upper Mississippian) Sulphur, Indiana: M.S. thesis, Indiana University, Bloomington, 157 p.

Frazier, W. J., and Schwimmer, D. R., 1987, Regional Stratigraphy of North America: N.Y., Plenum Press, 719 p.

Gray, H. H., 1979, The Mississippian and Pennsylvanian (Carboniferous) Systems in the United States—Indiana: USGS Prof. Paper 1110, p. K1-K20.

Harris, C. D., 1987, Depositional and diagenetic environments of the Golconda Group, southwestern Illinois and southeastern Missouri: M.S. thesis, Southern Illinois University, Carbondale, 177 p.

Harris, P. M., 1979, Facies anatomy and diagenesis of a Bahamian ooid shoal—Sedimenta VII: Miami, University of Miami Comparative Sedimentology Laboratory, 163 p.

Irwin, M. L., 1965, General theory of epeiric clear water sedimentation: AAPG Bull., v. 49, p. 445-459.

Keller, S. J., and Becker, L. E., 1980, Subsurface stratigraphy and oil fields in the Salem Limestone and associated rocks in Indiana: Indiana Geological Survey Occasional Paper 30, 63 p.

Kissling, D. L., 1967, Environmental history of Lower Chesterian rocks in southwestern Indiana: Ph.D. thesis, Indiana University, Bloomington, 366 p.

Sable, E. G., 1979, Eastern Interior basin Region, *in* Craig, L. C. et al., ed., Paleotectonic investigations of the Mississippian System in the United States—Part I, Introduction and regional analyses of the Mississippian System: USGS Prof. Paper 1010E, p. 59-106.

Somasekhara, K., 1970, Megafauna and paleoecology of two shale units in the Haney and Menard formations (U. Miss.) in southern Illinois: M.S. thesis, Southern Illinois University, Carbondale, 87 p.

Swann, D. H., 1963, Classification of Genevievian and Chesterian (Late Mississippian) rocks of Illinois: Illinois State Geological Survey Report of Investigations, v. 216, 91 p.

Swann, D. H., 1964, Late Mississippian rhythmic sediments of Mississippi Valley: AAPG Bull., v. 48, p. 637-658.

Thompson, T. A., and Brown, M. A., 1989, Inhomogeneities, paleomorphology, and architectural elements in shoal deposits of Salem Limestone (Mississippian) in southern Indiana (Abstract): AAPG Bull., v. 73, p. 1287.

Treworgy, J. D., 1985, Stratigraphy and depositional settings of the Chesterian (Miss.) Fraileys/Big Clifty and Haney Formations in the Illinois Basin: Ph.D. thesis, University of Illinois, Urbana, 202 p.

Treworgy, J. D., 1988, The Illinois Basin—a tidally and tectonically influenced ramp during mid-Chesterian time: Illinois State Geological Survey Circular, v. 544, 20 p.

Vincent, J. W., 1975, Lithofacies and biofacies of the Haney Limestone (Miss.), Illinois, Indiana, and Kentucky: Kentucky Geological Survey, Series X, Thesis Series, v. 4, 64 p.

Visher, P. M., 1980, Sedimentology and three dimensional facies relations within a tidally-influenced Carboniferous delta: The Big Clifty Formation,

Sulphur, Indiana: M.S. thesis, Indiana University, Bloomington, 156 p.
Zeng, Y. F., Lee, N. H., and Huang, Y. Z., 1983, Sedimentary Characteristics of Oolitic Carbonates from the Jialing-Jian Formation (Lower Triassic), South Sichuan Basin, China, *in* Peryt, T., ed., Coated Grains: Berlin, Springer-Verlag, p. 176-187.
Zinn, L. A., 1983, Subsurface stratigraphy of the Ste. Genevieve Limestone (Meramecian) and relations to underlying Silurian reefs, Green County, Indiana: M.S. thesis, Indiana University, Bloomington, 187 p.

Chapter 10

The Drowning of Ooid Shoals: Mississippian Greenbrier Limestone Near the West Virginia Dome

Cindy Carney
Department of Geological Sciences
Wright State University
Dayton, Ohio, USA

ABSTRACT

The Greenbrier Limestone was deposited in a subsiding basin in southeastern West Virginia and adjacent Virginia and across a broad, shallow shelf located to the north and northwest. Oolitic limestones are thick and extensive in southern West Virginia, but most thin and change in character to the north. An area of uplift, the West Virginia dome was active along the zone of change in north-central West Virginia. The dome was exposed during early Greenbrier deposition and remained as a submarine high throughout deposition of these predominantly oolitic, Mississippian-aged limestones.

The transition from oolitic grainstones that contain well-developed ooids through grainstones that comprise less well-developed ooids mixed with detrital quartz sand to poorly washed, quartz-sandy, oolitic packstones is interpreted to indicate flooding of the shoals around the West Virginia dome.

INTRODUCTION

The Mississippian Greenbrier Limestone is a relatively thick and extensive unit in West Virginia. Because it produces oil and gas, it has been studied at numerous localities around the state, both at surface exposures and in the subsurface (Rittenhouse, 1949; Wells, 1950; Flowers, 1956; Overbey et al., 1962; Youse, 1964; Overbey, 1967; Leonard, 1968; Wray and Smosna, 1982; Yeilding and Dennison, 1986; Sullivan and Textoris, 1988; Carney and Smosna, 1989; Smosna and Koehler, this volume; Kelleher and Smosna, this volume).

The Greenbrier Limestone of southern West Virginia comprises alternating sequences of oolitic grainstone, skeletal packstone and wackestone, and silty laminated mudstone reaching a maximum total thickness of 550 m (Leonard, 1968). These sequences are thought to represent upbuilding in a discontinuously subsiding basin (Leonard, 1968). The skeletal and oolitic grainstones represent deposition on shoals that formed across topographic highs on an underlying erosional surface and along shorelines formed around partially emerged hills (Youse, 1964). Skeletal wackestones accumulated in low-energy environments below wave base in somewhat restricted areas, possibly lagoons behind shoals. The silty laminated mudstones represent slow sedimentation in low-energy environments such as tidal flats. The interbedded nature of these sediments suggests rapid change as the high-energy shoals migrated across the lagoons and tidal flats (Leonard, 1968).

The Greenbrier Limestone changes in character and thins to the north, where only 30–100 m of limestone is present. These carbonates typically consist of thin-bedded to massive packstone and wackestone that can be argillaceous and sandy. The Greenbrier Limestone of northern West Virginia records a change from extremely shallow-water to open-marine conditions produced during transgression of the sea into the Appalachian basin (Carney and Smosna, 1989).

These changes over such a short distance suggest the presence of a hinge-line zone (Figure 1) (Arkle et al., 1979). Along the hinge zone an area of uplift, the West Virginia dome, developed as a result of broad-scale arching of the crust or movement along a basement fault with a roughly east–west trend (Yeilding and Dennison, 1986). The West Virginia dome of Kammer and Bjerstedt (1986) is the equivalent of the Catskill Island of Dally (1956), West Virginia positive area of Craig and Connor (1979), Pocono dome of Donaldson and Shumaker (1981), and Beverly Uplift of Yeilding and Dennison (1986). The northernmost occurrence of well-developed oolitic limestones in West Virginia is found near this feature. The present study describes depositional conditions and the resulting facies sequence developed during the establishment and subsequent drowning of ooid shoals around the West Virginia dome.

PALEOGEOGRAPHY

At the time of deposition of the Greenbrier Limestone, an embayment or gulf bounded on the northwest by lowlands and on the east and southeast by delta plains and low hills extended into the central Appalachian basin (Figure 1) (Craig and Varnes, 1979; deWitt and McGrew, 1979). The depositional framework for the Greenbrier Limestone consisted of a shallow shelf in the north and west separated from a southern subsiding basin (Craig and Varnes, 1979). The hinge line is thought to run northeast to southwest through West Virginia, dividing the two depositional settings (Arkle et al., 1979). Carbonate deposition initiated first in the basin, spreading gradually up onto the shelf during continued transgression of shallow seas.

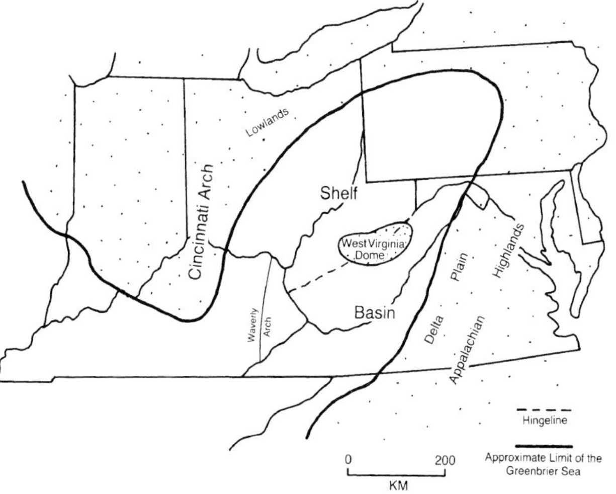

Figure 1. Paleogeographic reconstruction of the Greenbrier sea. Note location of the West Virginia dome. Hinge-line after Donaldson (1974), deWitt and McGrew (1979), and Kammer and Bjerstedt (1986).

STRATIGRAPHY

Limestones are exposed all along the Mississippian outcrop belt in West Virginia. A nearly complete section of the Greenbrier Limestone is present 16 km south of Davis, West Virginia, on WV Route 32, just south of the entrance to Canaan Valley State Park (Figure 2). Approximately 90 m of section (Figure 3) is exposed along the highway and in an abandoned quarry at the study locality.

The Greenbrier Limestone is middle to late Meramecian and early Chesterian in age (Horowitz and Rexroad, 1972). In southern West Virginia the Greenbrier Limestone is given group status and is divided into seven formations (Figure 4). Leonard (1968) recognized the Denmar through Alderson formations at the study locality (Figure 3). The lowermost unit of the Greenbrier Group, the Hillsdale Limestone, pinches out to the south (Figure 5) (Arkle et al., 1979). The Greenville Shale was not recognized at the Canaan Valley locality. Greenbrier Group is the formal designation only in southern West Virginia; therefore, the informal term Greenbrier Limestone will be used for convenience throughout the paper. The Greenbrier Limestone is underlain by the Maccrady Formation, a red and purple arenaceous shale and siltstone whose variable thickness indicates an upper erosional surface (Flowers, 1956; Youse, 1964). The Greenbrier Limestone is overlain by the Mauch Chunk Group, composed of red siltstone and shale with thin limestone beds.

The Denmar Formation is the oldest Mississippian unit present at the Canaan Valley locality (see Figure 3). The Denmar consists of 19 m of thin- to medium-bedded, often cross-bedded, oolitic grainstone and skeletal packstone and wackestone. Seven meters of red shale interbedded with quartz-sandy, oolitic grainstone make up the Taggard Formation at this locality. The Pickaway Limestone is represented by a thick sequence (35 m) of cross-bedded, quartz-sandy, oolitic grainstone with interbedded red shale and sandstone near the top of the unit. The Union Limestone is 21 m thick at Canaan. This unit comprises quartz-sandy, oolitic grainstone, skeletal packstone and wackestone and thin beds of dolomitic mudstone. The Alderson Formation consists of 8 m of quartz-sandy, peloidal and oolitic packstone and wackestone.

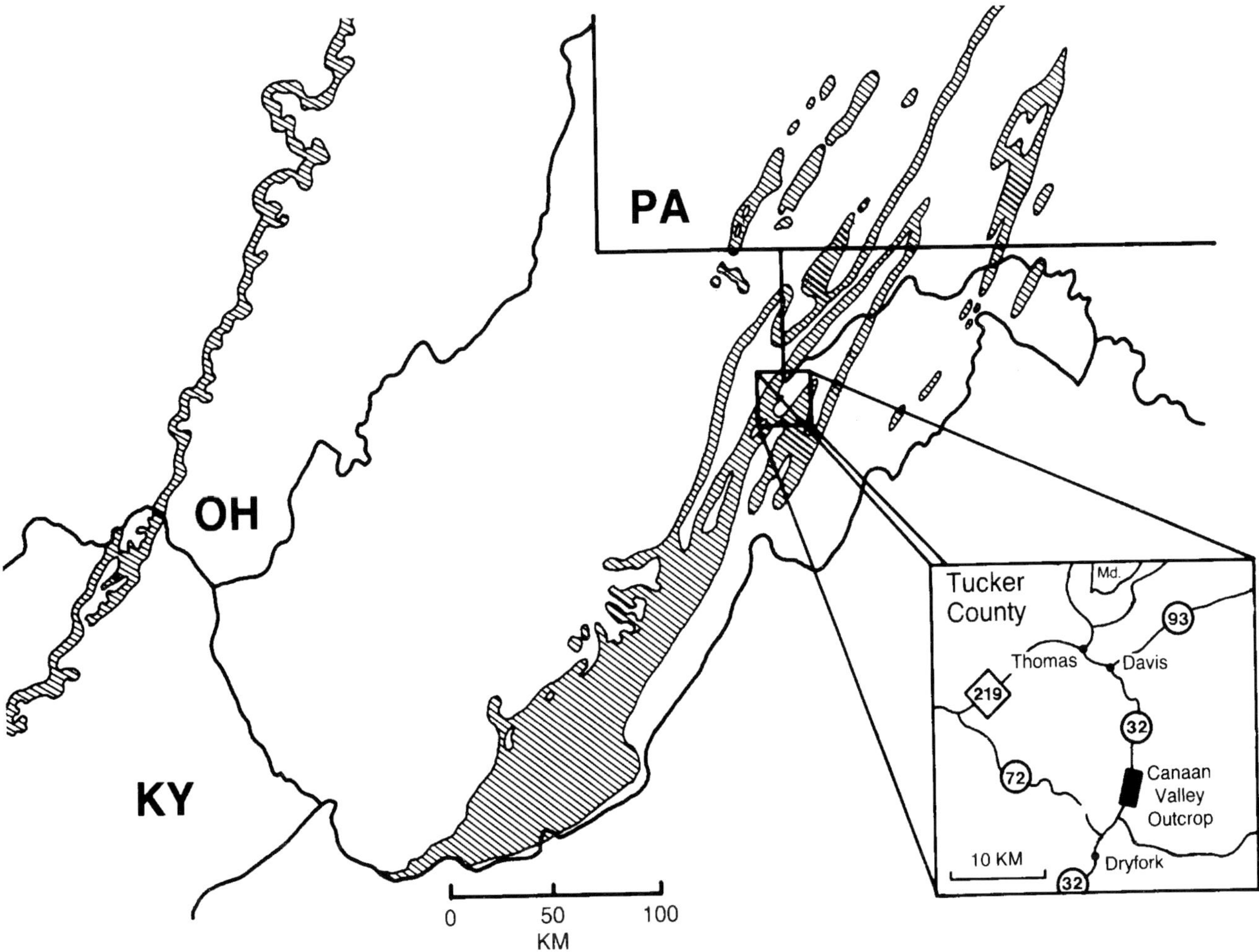

Figure 2. Study locality near Canaan Valley State Park, West Virginia. Outcrop pattern of Mississippian rocks is shown with a hatchured pattern.

OOLITE DEPOSITION AROUND THE WEST VIRGINIA DOME

At the Canaan Valley study locality, the Greenbrier Limestone is composed of light gray to reddish gray limestones and interbedded shales. The limestones are occasionally cross-bedded. Cross-bed sets vary from several centimeters to 1.0 m thick. Four units are oolitic at Canaan Valley: the Denmar, Pickaway, Union, and Alderson limestones (see Figure 3).

These limestones represent deposition on and around ooid shoals that formed along the flanks of the West Virginia dome. With some variation, ooid shoals were present in this general area throughout the time of Greenbrier deposition. The rounding of grains, oolitic coatings, cross-bedding, and near absence of micrite all indicate, in general, sedimentation on a well-agitated bottom in shallow water.

Before and during deposition of the earliest portion of the Greenbrier Limestone, the West Virginia dome was exposed as an island along the hinge-line zone (Figure 5) (deWitt and McGrew, 1979; Yeilding and Dennison, 1986). During the remainder of Greenbrier sedimentation, the West Virginia dome was a submarine high separating the shelf from the subsiding basin to the south (Leonard, 1968; Yeilding and Dennison, 1986). Greenbrier carbonates record a marine transgression across the uplift.

Ooid shoals around the dome varied through time as a result of changing conditions produced during the transgression. Three different shoal types were recognized: (1) carbonate-dominated shoals of the Denmar Formation, (2) clastic-influenced shoals of the Pickaway and Union limestones, and (3) flooded shoals of the Alderson Limestone. Figure 6 shows the variation in composition among these three shoal types.

Denmar Formation

Around the West Virginia dome, several different facies of the Denmar Formation were developed. Yeilding (1984) described micritic and dolomitic beds characteristic of a tidal-flat complex over the uplift itself. Cross-bedded oolitic grainstones formed on relatively high-energy ooid shoals around the flanks of

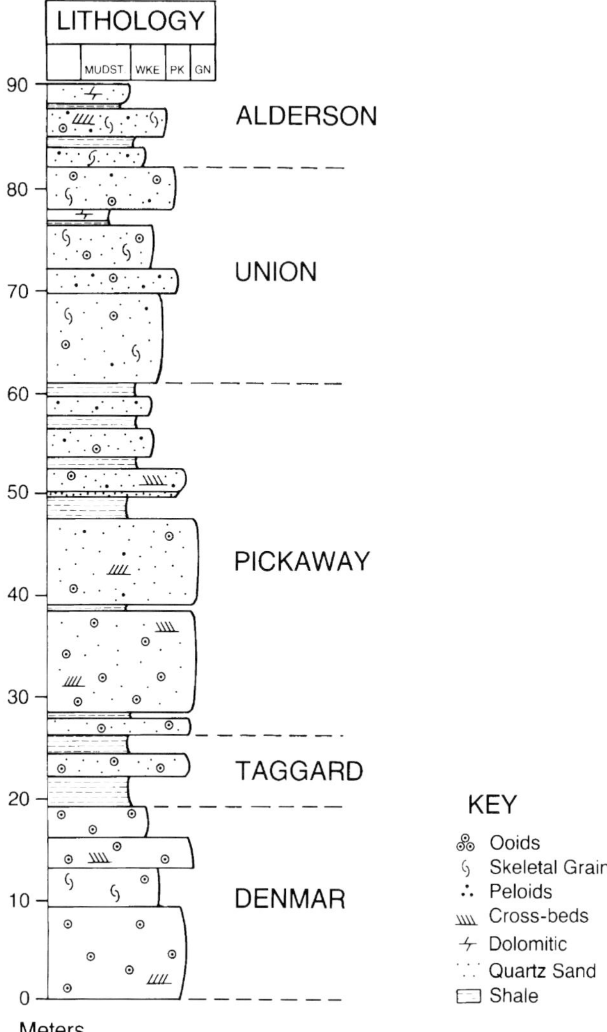

Figure 3. Stratigraphic section at Canaan Valley, West Virginia. The entire section is Greenbrier Limestone. Formation names follow Leonard (1968).

SYSTEM	SERIES	SOUTHEAST WEST VIRGINIA		
PENN.	MORROWAN	POTTSVILLE GP.		
MISSISSIPPIAN	CHESTERIAN	MAUCH CHUNK GROUP	BLUESTONE FM.	
			PRINCETON SS.	
			HINTON FM.	
			BLUEFIELD FM.	
		GREENBRIER GROUP	ALDERSON LS.	← Flooded Shoals
			GREENVILLE SH.	
			UNION LS.	← Clastic-Influenced Shoals
			PICKAWAY LS.	← Clastic-Influenced Shoals
	MERAMECIAN		TAGGARD FM.	
			DENMAR FM.	← Carbonate-Dominated Shoals
			HILLSDALE LS.	
		MACCRADY FM.		

Figure 4. Stratigraphy of the Greenbrier Group in southeastern West Virginia, after Arkle et al. (1979). The units that are mostly oolitic at the Canaan Valley outcrop are indicated by arrows.

the uplift. Ooid grainstones (Figure 7A) of the Denmar contain an average of 74% ooids (maximum of 86%) of the total grain content. Ooids are typically medium-sand sized or slightly larger. Of the ooids present, superficial ooids with only one or two coatings make up an average of 13%. Nine percent of the ooids are partially to completely micritized. A variety of grains serve as nuclei, including peloids, intraclasts, and skeletal fragments, especially pelmatozoan columnals and forams. Uncoated skeletal grains (Figure 7B) are an important constituent, averaging 25% of total grains. Most are broken and show evidence of abrasion. Peloids and intraclasts are less common, averaging under 2%.

A few thin skeletal wackestones are interbedded with the oolitic grainstones. These muddier deposits may have accumulated in subtidal areas, perhaps in lagoons behind shoals. Periodically the oolitic shoals migrated over the subtidal deposits, producing interbedded ooid grainstone and wackestone. Migration of ooid shoals over lagoonal sediments has been noted in modern carbonate-forming environments during storms (Hine, 1977).

Taggard Shale

Source lands near the northeastern margin of the embayment were intermittently elevated and eroded, producing pulses of terrigenous material that at times were sufficient to temporarily interrupt carbonate deposition. Alternatively, sea level may have dropped, producing the same sedimentary response at the Canaan Valley outcrop. During one of these episodes, the Taggard Shale was deposited in southern and central West Virginia. This unit consists of grayish red silty shale interbedded with oolitic and skeletal grainstone and represents deposition in quiet, tidal-flat or shallow lagoonal environments crossed by higher-energy channels (Yeilding, 1984).

Pickaway and Union Limestones

During continued transgression, ooid shoals were reestablished around the West Virginia dome, and the Pickaway and Union limestones were deposited. Recently a tidal origin for the Pickaway and Union limestones has been documented (Kelleher and Smosna, this volume; Smosna and Koehler, this volume).

Grainstones of the Pickaway Limestone are sometimes cross-bedded, and those of the Union Limestone contain a higher percentage of uncoated

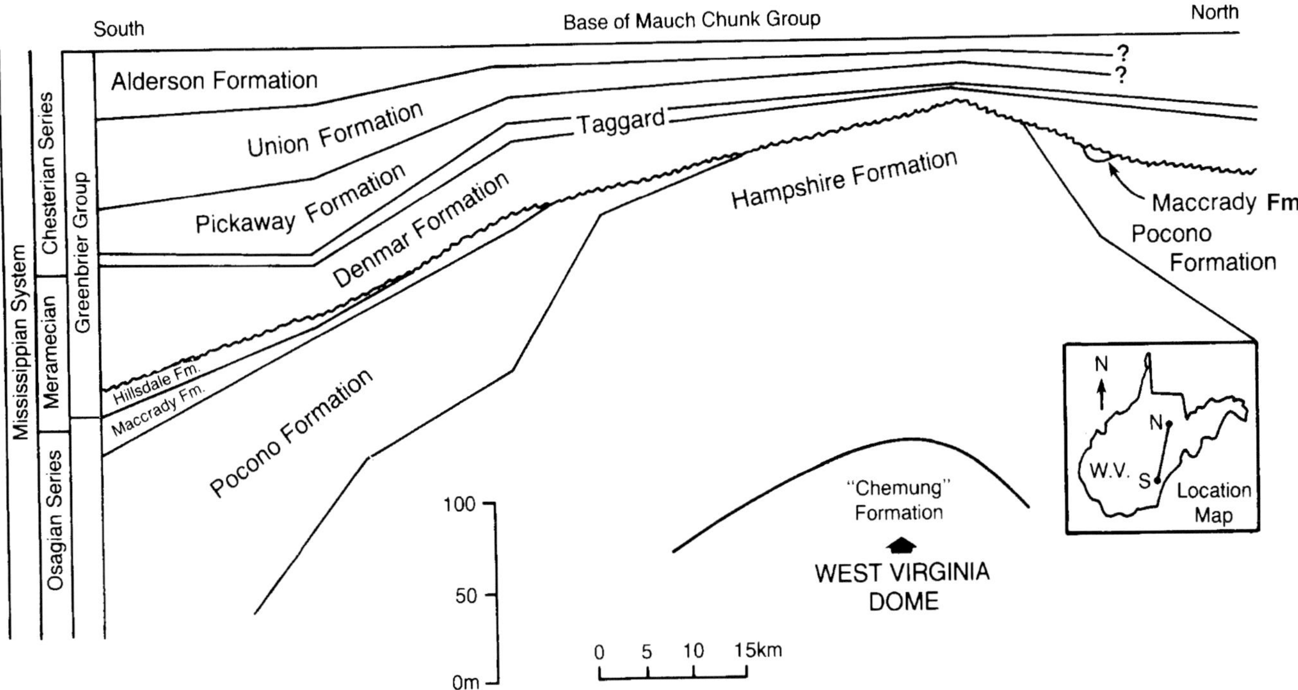

Figure 5. Stratigraphic cross section of the Greenbrier Limestone and underlying units. After Yeilding and Dennison (1986), Leonard (1968), and Dally (1956).

skeletal grains. However, at Canaan Valley other characteristics are so similar that for the purposes of this discussion, interpretations will be combined.

Terrigenous influx continued to some degree throughout deposition of the remainder of the Greenbrier Limestone. Fine- to medium-grained quartz sand averages 39% of total grains in the Pickaway and Union limestones (Figure 7C). The average ooid concentration falls to 40%, with a maximum of only 55%. Superficial ooids are more common in these later developed shoals (averaging 57% of total ooids), as are micritized ooids (28% of total ooids) (Figure 7D). Conditions for ooid production must have been less ideal than during Denmar deposition, considering the rather low percentage of ooids with thick cortexes. Modern shoals often consist of oolitic deposits composed of greater than 80% nonskeletal sand grains (Purdy, 1963; Harris, 1979). The presence of abundant superficial and micritized ooids suggests that currents were not always energetic enough to keep ooids in suspension. In modern ooid shoals such as those on the Bahamas platform, ooids are mobile only at the crest of shoals (Dravis, 1977). Most of the grains remain motionless on the sea floor for an undetermined amount of time, are covered with an organic film, and show evidence of boring by algae. Ooids in the Greenbrier shoals probably acted similarly, resulting in abundant micritized grains. The quartz grains are not oolitically coated. Perhaps the influx was too rapid for oolitic laminae to form, or the grains were too coarse to be moved by currents.

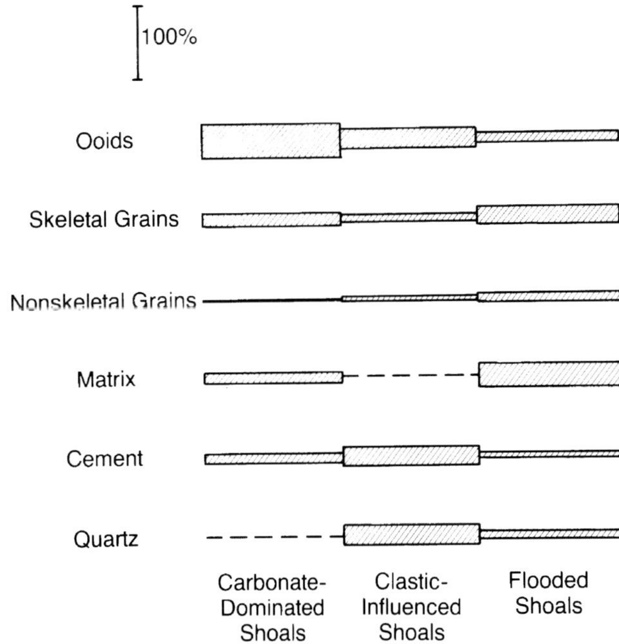

Figure 6. Diagram showing the relative proportions of major constituents in the three types of ooid shoals recognized in the Greenbrier Limestone at Canaan Valley. Carbonate-dominated shoals are characterized by relatively high percentages of ooids. Clastic-influenced shoals contain fewer ooids and abundant quartz sand. Flooded shoals have even fewer ooids and are muddy.

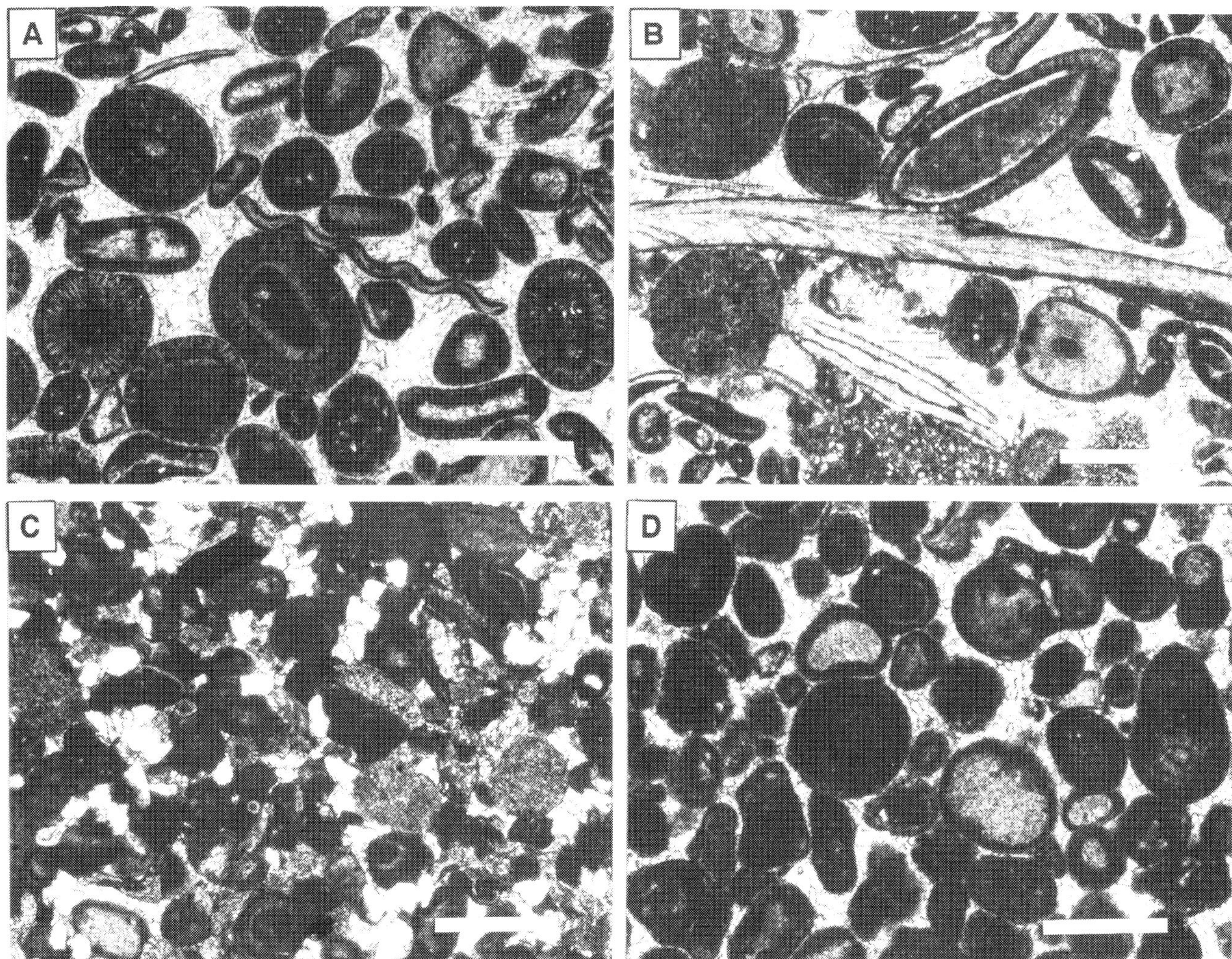

Figure 7. (A) Photomicrograph of ooid grainstone of the Denmar Formation. Scale bar is 0.5 mm. (B) Photomicrograph of uncoated skeletal grains from the Denmar Formation. Scale bar is 0.5 mm. (C) Photomicrograph of quartz-sandy peloidal and oolitic grainstone of the Pickaway Limestone. Fine- to medium-grained detrital quartz averages 30% of total grains. Scale bar is 0.5 mm. (D) Photomicrograph of oolitic grainstones in the Union Limestone. Ooids are found in all stages of micritization; many are superficial. Scale bar is 0.5 mm.

The clastic-influenced shoals of the Pickaway and Union limestones may have differed from their predecessors in that water depth was slightly greater as a result of continued transgression; consequently wave action was less effective, and conditions were not as favorable for development of thick oolitic coatings.

Alderson Limestone

The Alderson Limestone formed as a result of flooding of ooid shoals during continued transgression of the epicontinental sea. The limestones comprise poorly washed peloidal and oolitic packstones that contain a higher percentage of skeletal fragments and a muddier matrix than previously deposited limestones (Figure 8A). Skeletal fragments are less abraded than in earlier ooid shoal deposits. Ooids (as much as 52%, but averaging 20% of total grains) are not as abundant or as well developed in the flooded shoals. Many are superficial with skeletal fragments as nuclei (Figure 8B). Grapestones and peloids are abundant, averaging 19% of total grains. Some peloids may represent ooids micritized to the extent that they can no longer be recognized. Detrital quartz is somewhat less common than in the Pickaway and Union shoals, averaging less than 20%. Abundant (38% of total grains) unabraded skeletal grains indicate that organisms were living on the flooded shoal or in nearby environments. Current energy was not strong or consistent enough to prevent accumulation of mud between grains.

Conditions were far from ideal for ooid production in the flooded shoals. Water depth continued to increase, and the more stable substrate allowed colonization by a variety of organisms. Modern examples of flooded shoals such as sandy areas near Lily Bank, Bahamas, are now in 3–5 m of water and have been extensively colonized by sea grasses and green algae

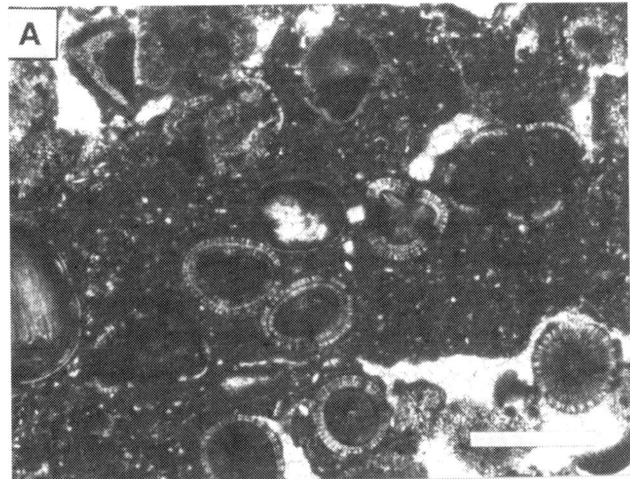

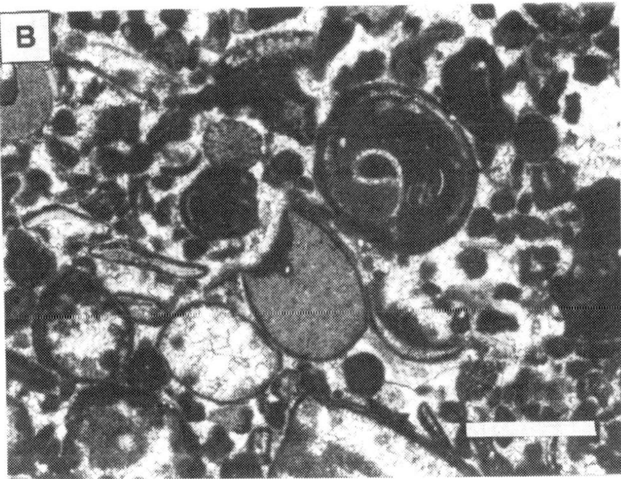

Figure 8. (A) Photomicrograph of poorly washed, oolitic packstones of the Alderson Limestone (flooded shoal). Scale bar is 0.5 mm. (B) Photomicrograph of skeletal grains and superficially coated ooids from the Alderson Formation. Scale bar is 0.5 mm.

(Hine, 1977). Relict bed forms such as sand ridges and sand waves are preserved, and bottoms are disturbed only during very large storms. Aggregate grains (grapestones) and micritized grains (peloids) are common constituents of these flooded shoals.

CONCLUSIONS

A record of transgression of shallow seas into the central Appalachian basin is preserved in oolitic units of the Mississippian Greenbrier Limestone in central West Virginia. Soon after initial transgression, cross-bedded, oolitic grainstones formed on ooid shoals over and along the flanks of the West Virginia dome. Skeletal wackestones and packstones were deposited in subtidal areas away from the uplift. Later, ooid deposition was temporarily interrupted by a pulse of terrigenous sedimentation. During continued transgression, ooid shoals were reestablished but differed from their predecessors. Conditions were not as favorable for development of thick coatings on grains, and terrigenous material (quartz sand) was periodically deposited along with carbonate grains. Eventually, the epicontinental sea flooded the ooid shoals around the dome. As substrates became more stable, the shoals were colonized by a variety of benthic organisms, and skeletal-rich, poorly washed, peloidal and oolitic packstones were deposited.

ACKNOWLEDGMENTS

The author wishes to thank John M. Dennison, Brian D. Keith and Charles W. Zuppann for reviews of this paper and Danielle Foye for photographic reproductions.

REFERENCES CITED

Arkle, T., Jr., D.R. Beissel, L.E. Larese, E.B. Nuhfer, D.G. Patchen, R.A. Smosna, W.H. Gillespie, R. Lund, C.W. Norton, and H.W. Pfefferkorm, 1979, West Virginia and Maryland, *in* J.W. Shekan, D.P. Murray, J.C. Hepburn, M.P. Billings, P.C. Lyons, and R.G. Doyle, eds., The Mississippian and Pennsylvanian (Carboniferous) Systems in the United States: USGS Professional Paper 1110-A-L, p. D1-D35.

Carney, C., and R. Smosna, 1989, Carbonate deposition in a shallow marine gulf, the Mississippian Greenbrier Limestone of the central Appalachian basin: Southeastern Geology, v. 30, p. 25-48.

Craig, L.C., and C.W. Connor, eds., 1979, Paleotectonic Investigations of the Mississippian System in the United States—Part II: USGS Professional Paper 1010, 559 p.

Craig, L.C., and K.L. Varnes, 1979, History of the Mississippian System—an interpretative summary, *in* Craig, L.C. and C.W. Connor, eds., Paleotectonic Investigations of the Mississippian System in the United States—Part II: USGS Professional Paper 1010, p. 371-406.

Dally, J.L., 1956, The stratigraphy and paleontology of the Pocono Group in West Virginia: Ph.D. dissertation, Columbia University, New York, 248 p.

deWitt, W., Jr., and L.W. McGrew, 1979, The Appalachian basin region, *in* L.C. Craig, and C.W. Connor, eds., Paleotectonic Investigations of the Mississippian System in the United States: USGS Professional Paper 1010, p. 13-48.

Donaldson, A.C., 1974, Pennsylvanian sedimentation of Central Appalachians, *in* G. Briggs, ed., Carboniferous of the Southeastern United States: GSA Special Paper 148, p. 47-78.

Donaldson, A.C., and R.C. Shumaker, 1981, Late Paleozoic molasse of the central Appalachians, *in* A.D. Miall, ed., Sedimentation and Tectonics in Alluvial Basins: Geological Association of Canada Special Paper 23, p. 100-123.

Dravis, J.J., 1977, Holocene sedimentary depositional environments on Eleuthera Bank, Bahamas: M.S.

thesis, University of Miami, Coral Gables, Florida, 386 p.

Flowers, R.R., 1956, A subsurface study of the Greenbrier Limestone in West Virginia: West Virginia Geological Survey Report of Investigations 15, 17 p.

Harris, P.M., 1979, Facies anatomy and diagenesis of a Bahamian ooid shoal: Sedimenta VII, Comparative Sedimentology Laboratory, University of Miami, Florida, 163 p.

Hine, A.C., 1977, Lily Bank, Bahamas: History of an active oolite sand shoal: Journal of Sedimentary Petrology, v. 47, p. 1554-1581.

Horowitz, A.S., and C.D. Rexroad, 1972, Conodont biostratigraphy of some United States Mississippian sites: Journal of Paleontology, v. 46, p. 884-891.

Kammer, T.W., and T.W. Bjerstedt, 1986, Stratigraphic framework of the Price Formation (Upper Devonian-Lower Mississippian) in West Virginia: Southeastern Geology, v. 27, p. 13-33.

Leonard, A.D., 1968, The petrology and stratigraphy of Upper Mississippian Greenbrier limestones of eastern West Virginia: Ph.D. dissertation, West Virginia University, Morgantown, West Virginia, 219 p.

Overbey, W.K., Jr., 1967, Lithologies, environments, and reservoirs of the Middle Mississippian Greenbrier Group in West Virginia: Producers Monthly, v. 31, p. 25-31.

Overbey, W.K., Jr., R.C. Tucker, and E.E. Ruley, 1962, Recent developments in the Big Injun horizon in West Virginia: Producers Monthly, v. 27, p. 18-22.

Purdy, E.G., 1963, Recent calcium carbonate facies of the Great Bahama Bank, part 2, sedimentary facies: Journal of Geology, v. 71, p. 472-497.

Rittenhouse, G., 1949, Petrology and paleogeography of Greenbrier Formation: AAPG Bulletin, v. 33, p. 1704-1730.

Sullivan, E.M., and D.A. Textoris, 1988, Microfacies and paleoenvironments of the Missippian Denmar Formation, eastern West Virginia: Southeastern Geology, v. 28, p. 133-152.

Wells, D., 1950, Lower–Middle Mississippian of southeastern West Virginia: AAPG Bulletin, v. 34, p. 882-922.

Wray, L.L., and R. Smosna, 1982, Sedimentology of a carbonate-red bed association, Mississippian Greenbrier Group, eastern West Virginia: Southeastern Geology, v. 23, p. 99-108.

Yeilding, C.A., 1984, The stratigraphy and sedimentary tectonics of the Upper Mississippian Greenbrier Group of eastern West Virginia: M.S. thesis, University of North Carolina, Chapel Hill, North Carolina, 117 p.

Yeilding, C.A., and J.M. Dennison, 1986, Sedimentary response to Mississippian tectonic activity at the east end of the 38th Parallel fracture zone: Geology, v. 14, p. 621-624.

Youse, A.C., 1964, Gas producing zones of Greenbrier (Mississippian) Limestone, southern West Virginia and eastern Kentucky: AAPG Bulletin, v. 48, p. 465-486.

Chapter 11

Tidal Origin of a Mississippian Oolite on the West Virginia Dome

Richard Smosna
Department of Geology & Geography
West Virginia University
Morgantown, West Virginia, USA

Bryan Koehler
The Advent Group, Inc.
Brentwood, Tennessee, USA

ABSTRACT

Petrographic descriptions, sedimentary structures, and facies analysis document the predominantly tidal control on deposition of the Pickaway Limestone. Cross-bedded ooid-peloid grainstones accumulated over the West Virginia dome as decimeter-sized intertidal sand waves and meter-sized subtidal sand waves. Evidence of tidal sedimentation includes bidirectional foresets, tidal bedding, mud drapes, reverse grading, wavy bedding, reactivation surfaces, and tidal bundles. Sand holes, desiccation cracks, mud chips, and interbeds of red shale and finely crystalline dolomite formed during periods of subaerial exposure, whereas ripple bedding hints of occasional wave activity. Storms passing over the sand waves produced thin sets of trough cross-beds and plane beds. Major storms also led to the migration of the subtidal sand waves over other facies. In the quiet offshore environment, pelletal-skeletal packstones and wackestones were deposited, although advancing spillover lobes or tempestites produced interbedded sets of ooid grainstone.

INTRODUCTION

Many, if not most, modern ooid shoals develop under the primary influence of tides, with secondary influences from waves and storms. Tide-dominated ooid sand bodies include, among others: tidal-bar belts at Schooner Cay, Tongue of the Ocean, and Joulters Cays, Bahamas, and along the coast of Abu Dhabi; marine sand belts at South Cat Cay-Browns Cay and Lily Bank, Bahamas; and tidal deltas off the Trucial Coast (summarized in Halley et al., 1983). In contrast, tidal processes that might have been effective in forming ancient oolites are generally assumed but not confirmed. Two notable exceptions in which tidal activity has been documented are the Mississippian Ste. Genevieve Limestone of the Illinois basin (Knewston and Hubert, 1969; Choquette and Steinen, 1980) and the Jurassic oolites of southern England (Klein, 1965; Wilson, 1968).

The purposes of this paper are twofold: to verify the tidal origin for a Mississippian Greenbrier oolite in the Appalachian basin and, by extension, to provide a set of criteria for the recognition of tidal facies and depositional patterns in other carbonate rocks. This synthesis includes a collection of petrographic data, sedimentary structures, and facies associations. Although no single feature is by itself evidence for tidal sedimentation, the assemblage of features (including bidirectional foresets, tidal bedding, mud drapes, reverse grading, wavy bedding, reactivation surfaces, and tidal bundles) collectively demonstrates a tidal control.

The Mississippian Greenbrier Group outcrops in eastern West Virginia, and four exposures of the lower Pickaway Limestone (Figure 1) are the focus of this study: (1) Canaan Valley (road cut at intersection of State Routes 32 and 72), (2) Elkins (J.F. Allen quarry, 4 km west on U.S. 250), (3) Huttonsville (abandoned quarry 10 km south on U.S. 250), and (4) Mingo (road cut 12 km south on U.S. 219). The Pickaway Limestone in this part of the state averages 40 m thick and has been informally divided into two members. The lower member (6–24 m thick) is dominated by ooid grainstones and is the subject of this report. The upper member is dominated by red and yellow shale. Field descriptions concentrated on bed configurations, dimensions, and stratigraphic relations. Cross-beds exposed in three dimensions were measured for strike and dip, which were subsequently corrected for regional structure. Petrographic analysis is based on thin sections prepared from 80 samples, and 300 points were counted on each thin section.

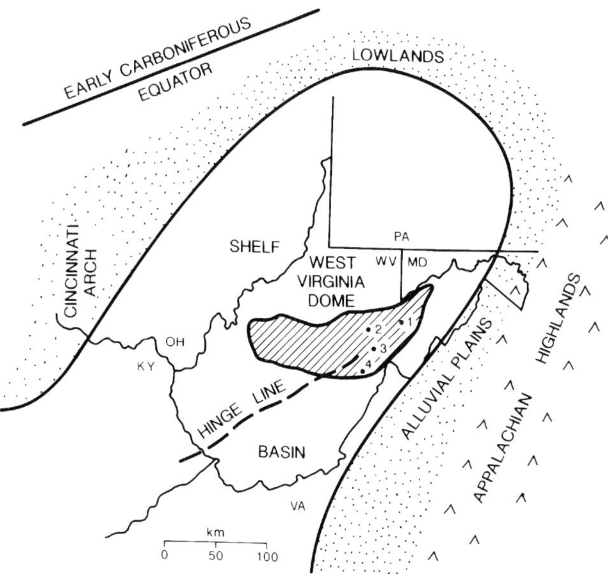

Figure 1. Paleographic map of Greenbrier gulf (unpatterned) in the central Appalachian basin, surrounded on three sides by low-lying land masses (stippled pattern). The West Virginia dome (diagonally lined) separated the southern basin center from a northern shelf. Four outcrops over the dome were studied for this report: (1) Canaan Valley, (2) Elkins, (3) Huttonsville, and (4) Mingo.

GEOLOGICAL SETTING

The Meramecian-Chesterian Greenbrier Group consists chiefly of limestones deposited in a shallow gulf (Carney and Smosna, 1989). An arm of the Central Interior sea extended into the Appalachian region, and land masses surrounded this gulf on three sides (Figure 1). The Appalachian highlands and alluvial plains lay to the east, lowlands to the north, and the Cincinnati arch to the west. Paleomagnetic data place the region within the tropics and just a few degrees south of the equator (Scotese et al., 1979).

The Greenbrier sea was at first confined to the southern basin, where a thick sequence of shallow to deep marine limestones accumulated (Arkle et al., 1979). These initial deposits include the Hillsdale Limestone and the Denmar Formation (Figure 2). The sea eventually flooded the northern shelf, extending across northern West Virginia and into neighboring Ohio, Pennsylvania, and Maryland (Carney and Smosna, 1989). At this time shallow water lime sediments were laid down throughout the area—the Pickaway, Union, and Alderson limestones.

Nonmarine clastic sediments were occasionally introduced into the basin from the east and north, temporarily interrupting carbonate sedimentation with thin units of red and green shale, siltstone, and sandstone (Taggard Formation and the informal upper Pickaway member). Carbonate deposition of the Greenbrier was finally brought to a close when siliciclastics of the overlying Mauch Chunk Group overwhelmed the basin. Uplift and erosion of the eastern highlands produced voluminous quantities of alluvial-plain sediment, which prograded across the basin (Donaldson and Shumaker, 1981).

Situated between the basin and shelf was the West Virginia dome, a topographically positive area centered in Randolph County (Figure 1). Stratigraphic evidence documents the presence of the dome: the Lower Mississippian Price Formation is absent in this area or consists of a thin shoreface sandstone (Bjerstedt and Kammer, 1988), the underlying Maccrady red beds and Hillsdale Limestone pinch out here (Figure 2), and the Denmar and overlying Greenbrier limestones display local facies changes and anomalous thinning (Yeilding and Dennison, 1986). Dally (1956) believed the dome originated as a deltaic lobe of the Upper Devonian Hampshire Formation that stood above sea level, remaining high in later times because of minimal subsidence. Donaldson and Shumaker (1981) postulated that tilting of a basement fault block, triggered by Early Mississippian thrusting, uplifted the dome. Yeilding and Dennison (1986) explained the dome's origin by uplift along a major basement fault.

The dome was subaerially exposed in Early Mississippian time, and erosion locally removed or

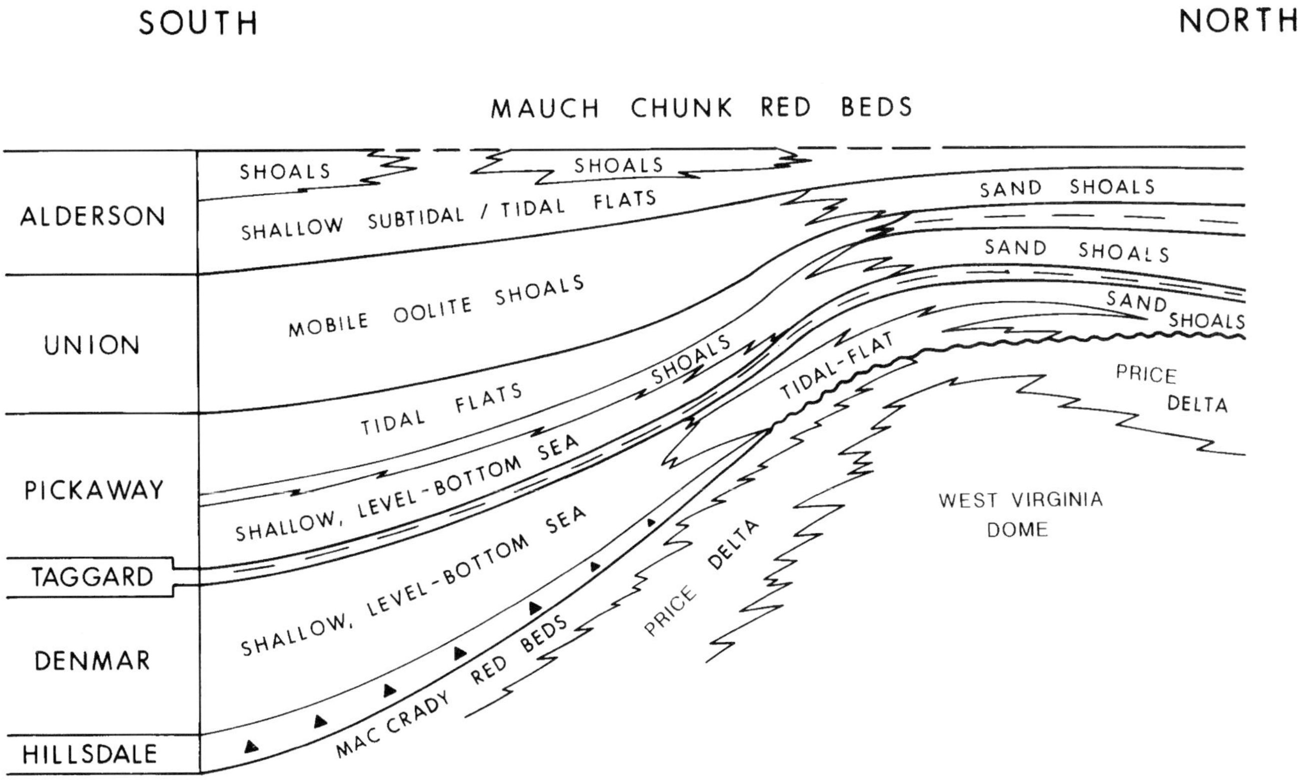

Figure 2. Formations of the Greenbrier Group (left column) and reconstruction of environmental facies in the outcrop belt of eastern West Virginia. The subject of this study is the lower Pickaway member over the West Virginia dome.

thinned underlying clastic units. This erosion surface was then overlapped by Greenbrier Group limestones. During the Greenbrier transgression the dome was flooded, but water depth remained shallow. Ooid sand bars, muddy tidal flats, and shallow lagoons developed atop this and other, smaller paleotopographic highs in West Virginia (Youse, 1964; Yeilding and Dennison, 1986; Sullivan and Textoris, 1988). In the lower Pickaway member of this study, we have identified three environmental facies—subtidal tide-dominated grainstones, intertidal grainstones, and offshore packstones and wackestones—which generally reflect the shallow, turbulent conditions over the West Virginia dome.

SUBTIDAL TIDE-DOMINATED GRAINSTONES

Petrography

Grainstones of this facies consist of varying proportions of peloids (average 22%), quartz (15%), ooids (average 12%, range up to 50%), fossils (12%), and calcite cement (25%). Micritization of carbonate grains has been extensive, and judging by their texture, most peloids were probably ooids originally. Peloids and ooids are fine- to medium-sand-sized and generally well sorted. Oncolites are present, along with minor amounts of fecal pellets, erosional intraclasts of mud and fossils, and aggregate grains of peloids and ooids. Skeletal fragments include crinoids, echinoids, brachiopods, bryozoans, gastropods, ostracodes, trilobites, rugose corals, and foraminifera. Very fine to fine quartz sand is often concentrated along quartz-rich laminae.

Sedimentary Characteristics

Bed Configuration

The basic bed configuration of these grainstones consists of numerous sets of wedge-shaped cross-strata (Figure 3). Nonparallel planar surfaces bound the large-scale sets, which range in thickness from 20 cm to 2 m and average near 50 cm. Cross-bed sets represent the internal foresets of meter-sized sand waves that advanced across the sea floor.

Although the bounding surfaces truncate underlying cross-beds, the planar shape of the set boundaries and an absence of deep, uneven scour demonstrate that the sand waves had straight, rather than sinuous, crests (compare with Harms et al., 1975). The height of these truncated sand waves must have been somewhat greater than their preserved thicknesses, and water depth obviously exceeded their height. In general, subtidal sand waves attain a height of up to half the water depth (Swift, 1985), suggesting that the

Figure 3. Outcrop photograph illustrating sets of wedge-shaped cross-strata with planar bounding surfaces. Hammer near center for scale.

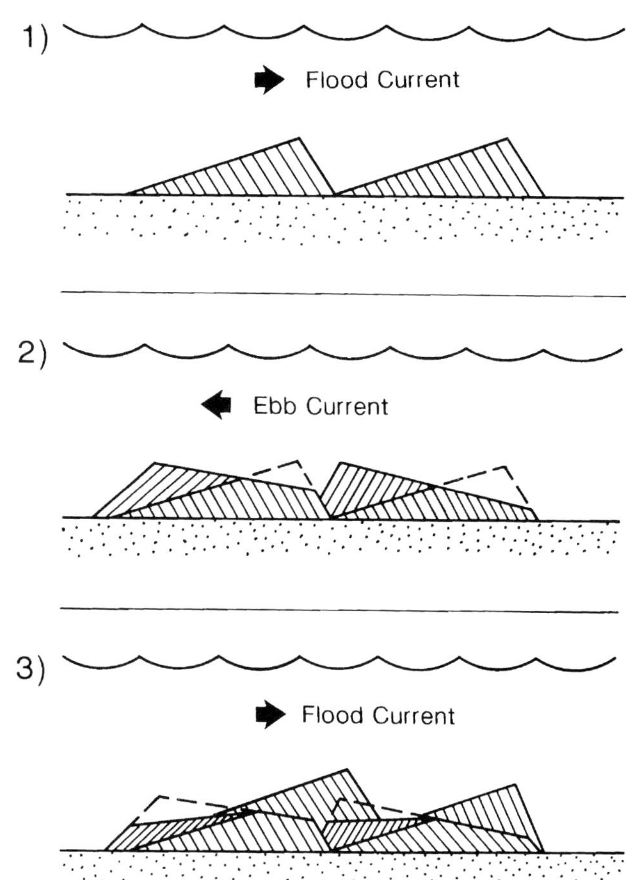

Figure 4. Diagram of wedge-shaped cross-strata with bidirectional foresets that formed when sand waves migrated under the influence of reversing tidal currents.

large Pickaway bed forms developed in water up to 5 m deep. Modern sand waves typically form on fine to medium sands as a transitional bed phase between ripples and dunes, and mean flow velocities of approximately 50 cm/sec are required for their formation (Harms et al., 1975; Hine, 1977). Sets of cross-strata may thicken in either of two directions, and the erosional bounding surfaces resulted from one sand wave overriding and partially destroying another under the influence of reversing currents (Figure 4).

Individual cross-beds of these large wedge-shaped sets are mostly accretion cross-beds. Foreset dip angles near 15–16° formed where the current had barely separated from the slip face, and dip angles near 6°, where a strong downslope component of the current existed (Imbrie and Buchanan, 1965; Walker, 1985). Currents apparently carried much of the carbonate sand as suspended load, which rained out of the water column, evidenced by the tangential lower bounding surfaces of the accretion cross-beds (Swift, 1985). Moreover, the alternation of grain sizes from one lamina to the next suggests that the bed load sediment moved intermittently (Blatt et al., 1980), as may be expected from fluctuating tidal currents. In this manner transport by traction or saltation (coarser grains) alternated with transport by suspension (finer grains).

Less common are avalanche cross-beds, having formed under lower flow velocities. For these, the current separated from the bed surface, and avalanching grains produced a slip face near the angle of repose (mode at 25–27°). Coarse grains tended to become concentrated along the base of these slip faces.

Bidirectional Foresets

Reversing tidal currents of nearly equal velocity can produce cross-bedding with a bipolar distribution of foresets. This bed form has long been accepted as a criterion for sediment transport by tidal process (Klein, 1977). Ideally the bipolar orientations are 180° apart. In the Pickaway Limestone, however, cross-bedding in the sand waves is distinctly bidirectional but not exactly bipolar (Figure 5). At Huttonsville, for example, beds dip in two dominant directions separated by 132°; at Canaan Valley, by 130°; and at

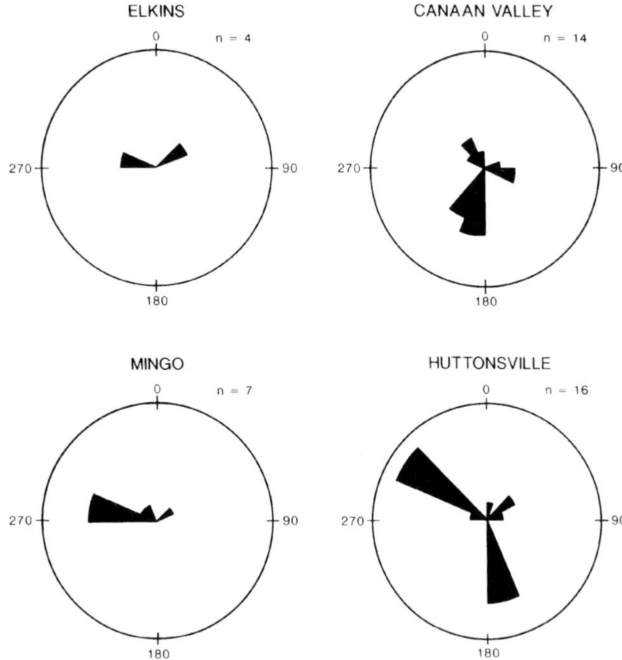

Figure 5. Rose diagrams of cross-bed dip directions at four outcrops, corrected for regional structure. Roses are divided into classes of 22.5°, and the length of each rose petal is proportional to number of observations (n).

Elkins, by 137°. (Cross-bedding at Mingo is essentially unidirectional, to be discussed in a following section.) Bidirectional foresets separated at angles other than 180° illustrate that ebb and flood currents followed somewhat different transport paths around the sand waves (Houbolt, 1968; Klein, 1970). Scatter does exist in the two dominant directions recorded at each outcrop, comparable to the scatter in dip directions of cross-bedding on Bahamian ooid sand bodies (Ball, 1967). Deviation from the mean cross-bed dip direction is most likely due to local differences in bottom topography and water depth plus the action of waves.

Reactivation Surfaces

Inclined discontinuity surfaces, separating sets of cross-beds with a similar orientation, are termed reactivation surfaces (Klein, 1970). They are inclined at a lower angle than the cross-beds themselves, truncate underlying cross-beds, and are commonly curved. These surfaces reflect time-velocity asymmetry of tidal currents. The foresets advance during a constructional phase, recording the direction of the dominant tidal current. The weaker reverse current partly erodes the sand wave's lee side, and deposition is temporarily interrupted (Klein's destructional phase). With the next dominant current, migration of the sand wave resumes, burying and preserving the reactivation surface.

In the Pickaway Limestone, reactivation surfaces (Figure 6A) are present but rare. Bidirectional foresets

Figure 6. Outcrop photographs of subtidal tide-dominated grainstones. (A) Reactivation surface (between arrows) separates two sets of cross-beds with similar orientations. 14-cm pencil near center for scale. (B) Tidal bundles: thin foresets of neap tides (between arrows) alternate with thicker foresets of spring tides. All foreset bedding planes appear dark due to thin shale partings (mud drapes), deposited during slack tides. Pencil is 14 cm long. (C) Set of cross-beds (to immediate left of 10-cm knife) with dark shaly partings on foresets. Several mud drapes (arrows) extend only a short distance up the slip face from the toe.

reveal that the currents were generally of equal strength: the degree of time-velocity asymmetry was small. In this environment (no dominant tidal direction and two constructional phases), cross-beds with opposing dips prevailed over reactivation surfaces. Moreover, the consistently low angle between reactivation surfaces and underlying foreset beds implies that erosion by destructional tidal phases was minor (compare with de Mowbray and Visser, 1984).

Tidal Bundles

One small example of tidal bundles is observed in the Pickaway (Figure 6B). Bundles constitute alternating groups of cross-beds: spring tides transport a greater volume of sediment and produce thicker foresets of mostly sand, whereas neap tides transport a minimal volume of sediment and produce thin foresets with numerous laminae of fines (Visser, 1980; Allen and Homewood, 1984). Foreset bundles of the Pickaway, interpreted to have been deposited by spring tides, have a thickness of nearly 2.0 cm, whereas those deposited by neap tides, 0.2 cm (Figure 7). This one example hints of lunar-month cyclicity (Allen and Homewood, 1984); the period of one complete Pickaway cycle is 32 bundles.

Mud Drapes

Terrigenous clays were introduced to the Greenbrier gulf from the eastern alluvial plain, and green shale partings occasionally draped the foresets of Pickaway sand waves (Figure 6C). Allen (1981), de Mowbray and Visser (1984), and Walker (1985) attributed mud drapes to the deposition of fines from suspension during either (1) a single slack tide or (2) a series of slack waters around neap tide, when sand deposition is minimal.

Some Pickaway mud drapes extend only a short distance up the slip face from the toe, rather than reaching up the entire length of the slip face. Short mud drapes such as these reflect variations in flow strength of the tidal currents. A dominant tidal current generated the foreset beds, shale-mud drapes were laid down at slack tide, and during reverse flow the upper part of the mud drape was removed by erosion (de Mowbray and Visser, 1984).

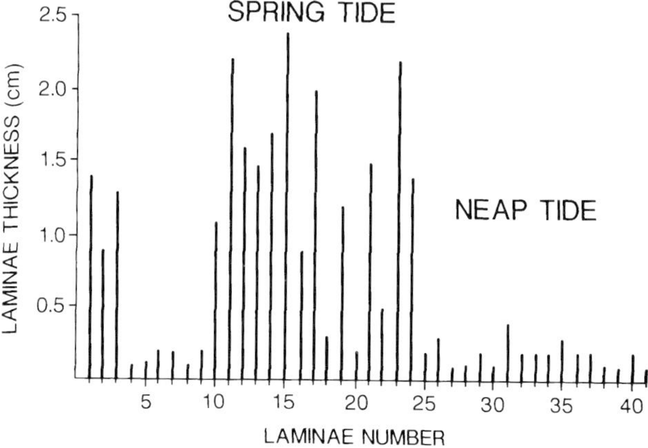

Figure 7. Histogram illustrating laminae thicknesses within a small set of Pickaway tidal bundles. Thickness of interpreted spring bundles is near 2.0 cm, whereas thickness of neap bundles is only 0.2 cm. One complete tidal cycle is represented by laminae 10 through 41.

Trough Cross-Bedding

Medium-scale trough cross-bedding is rare in the subtidal grainstones (Figure 8A). Troughs are 15–25 cm deep; the filling cross-beds are slightly asymmetric and tangential to the underlying erosion surface. These structures may have been generated by major storms with higher than normal velocities: hurricanes passing over modern ooid shoals can give rise to cuspate megaripples (Hine, 1977), which form similar trough cross-beds as they migrate.

Cross-beds in one trough were oversteepened by excessive deposition at the brink point, leading to penecontemporaneous deformation. The beds moved as a unit as they slumped downward and deformed into simple folds.

Plane Beds

Sets of plane beds, averaging around 30 cm thick, alternate with the cosets of wedge-shaped cross-strata. These sets denote deposition in the lower part of the upper flow regime and, like the trough cross-bedding, could have formed during the passing of hurricanes (compare with Hine, 1977). At these times, sand waves would have been destroyed by rapid sand transport in the upper flow regime, only to be reformed with the return of normal (tidal) conditions. A second interpretation is that plane beds developed during periodic increases in the tidal-current velocity.

Erosion Surfaces

Planar erosion surfaces are present at the base of several sand-wave units in the Pickaway (Figure 8B). At the Huttonsville outcrop, for example, one surface with a relief of 0.5 m can be traced laterally for a distance of over 30 m. A thin lag deposit of gravel-sized intraclasts, shells, and ooids generally covers these surfaces.

Johnson (1978) and Klein (1977) interpreted similar surfaces in tide-dominated sandstones as the product of scour by tidal currents augmented by powerful storms. Their position in the Greenbrier—at the base of a sand-wave unit and scouring into intertidal sand or offshore sediment—indicates that these subtidal sand waves migrated laterally over neighboring facies. Migration was preceded by intense erosion of the sea floor from tidal and storm currents. Under normal tidal conditions the Pickaway sand waves probably experienced little or no net movement, as evidenced by the bidirectional foresets; thus, wholesale facies migration apparently took place only in conjunction with powerful storm-generated flows.

Burrows

During temporary lulls in the movement of sand, animals infrequently burrowed the sediment. The

Figure 8. Outcrop photographs of subtidal tide-dominated grainstones. (A) Trough cross-bedding (lower half of photo) overlain by planar, wedge-shaped set of cross-bedding. 14-cm pencil in trough for scale. (B) Erosion surface at base of sand-wave unit, with thin lag deposit. Hammer to left for scale.

rare burrows are small (diameter less than 1 cm) and mostly vertical. Like the deposition of mud drapes, burrowing probably occurred at times of slack or low tide.

INTERTIDAL GRAINSTONES

Petrography

Petrographically, grainstones of this facies differ little from those of the subtidal tide-dominated grainstones. Peloids account for an average of 25%, quartz 19%, fossils 12%, ooids 5% (range up to 10%), and calcite cement 22%. Micrite envelopes commonly coat the carbonate grains, and most peloids (spherical shape, fine to medium sand, well sorted) are interpreted to be micritized ooids. The same kinds of fossil fragments occur in these rocks as in the subtidal grainstones, as well as small amounts of fecal pellets, erosional intraclasts, and oncolites. Mean grain size, however, is slightly finer (up to medium sand) in this facies than in the subtidal tide-dominated grainstones (up to coarse sand). Very fine quartz sand and silt is most abundant in these intertidal rocks (maximum of 45%), and in many places it is concentrated as quartz-rich laminae along with other terrigenous grains (zircon, tourmaline, chert, and mica). Overall sorting is poor because of a bimodal grain-size distribution; coarser carbonate grains intermix with finer quartz sand and silt.

Sedimentary Characteristics

Bed Configuration

The dominant bed configuration is a composite set, consisting of plane beds and cosets of cross-beds (Figure 9A). Cross-bed sets range in thickness from 10 to 50 cm and are bounded by planar, parallel surfaces (not wedge-shaped). The foresets display a variety of inclinations: dips of 20–30° reflect avalanche bed forms; dips of 10–20°, weak separation of flow; and dips of less than 10°, accretion bed forms. Foresets are also strongly bidirectional. These structures represent decimeter-sized sand waves that developed within the intertidal zone under the control of tidal currents. They apparently had straight crests, and their heights typically did not exceed 0.5 m. Most likely, foresets advanced at times around high tide. With exposure at low tide, sand movement must have ceased.

Plane beds and laminae, in sets of less than 10 cm, alternate with the cross-beds. Like plane beds in the subtidal grainstones, this structure formed under conditions of upper flow regime, such as at times of midtide when flow strength peaked or during seasonal increases in current velocity. They may also have formed with the passing of storms, when the sand waves would have been temporarily destroyed.

Though rare, reactivation surfaces mark the erosion of intertidal sand waves by subordinate reverse currents.

Tidal Bedding

Alternate waxing and waning of currents within the intertidal zone produces a couplet of sand and mud, termed tidal bedding (Klein, 1977). In the Pickaway this structure is common, consisting of very thin beds of grainstone (grains transported by flood and ebb currents) alternating with beds or laminae of green shale, red shale, or silty, finely crystalline dolomite (fines that settled during slack tides). In many places thin but continuous layers of green shale drape over ripple cross-beds of grainstone, forming a wavy bedding (Figure 9B).

Subaerial exposure of the red shale and dolomite created desiccation cracks and curled mud chips (Figure 9C). Mud cracks outline 3- to 13-cm polygons, and the mud chips frequently became reworked (1- to 8-cm clasts) into overlying grainstone layers. The red color of these shales comes from exposure and oxidation of iron-bearing terrigenous clay minerals. Lime muds of the intertidal zone may have been in contact with both marine and meteoric waters, which could

Figure 9. Outcrop photographs of intertidal grainstones. (A) Composite set consists of cross-beds below and planar beds above. 14-cm pen at bottom for scale. (B) Ripple cross-beds draped by thin green shales (which weather as recesses on outcrop face) form a wavy bedding. Pen is 14 cm long. (C) Curled desiccation chips of red shale in thin unit immediately below 10-cm knife. (D) Ripple bedding. Ripples to right of pen have a height of 1 cm and wave length of 6 cm.

have dolomitized the fine-grained carbonate sediment (compare with Badiozamani, 1973).

Sporadic and solitary reverse grading occurs in the dolomite layers, which contain dolomitized lime mud at the base and coarsen upward through fine carbonate pellets to quartz silt at the top (Figure 10A). These beds were produced by waxing current activity. The mud was deposited under conditions of high suspension load and low flow regime, before the arrival of peak flow strength, whereas the carbonate and siliciclastic silt was deposited at peak flow strength, either flood or ebb (Reineck and Singh, 1980).

Ripple Bedding

Small-scale cross-beds, as ripple form-sets, are superimposed on some of the intertidal sand waves. These form sets have a height of around 1 cm and a wave length of 4–6 cm (Figure 9D). The symmetric shape and low ripple index (4–6) distinguish them as wave ripples (Reineck and Singh, 1980). Their presence shows that waves sometimes disturbed the carbonate sand, most likely at slack tides when the influence of tidal currents was nil.

Sand Holes

Air can become trapped in foreshore sands where the surf is not too strong (Emery, 1945). With a rapidly incoming wave, air is trapped as bubbles in the upper part of the sand layer, which may be preserved as sand holes. In the Pickaway these vugs are circular in cross section, around 0.6 mm in diameter, and infilled by sparry calcite (Figure 10B); they document sedimentation in the intertidal environment.

Burrows

Intertidal Pickaway sediments were often excavated by unidentified infaunal organisms. Burrows are locally few in number to quite extensive (bioturbated), have both horizontal and vertical orientations, and range from less than 1 cm in diameter up to 5 cm. The burrowers may have worked the sediment at times when current velocity dropped below the

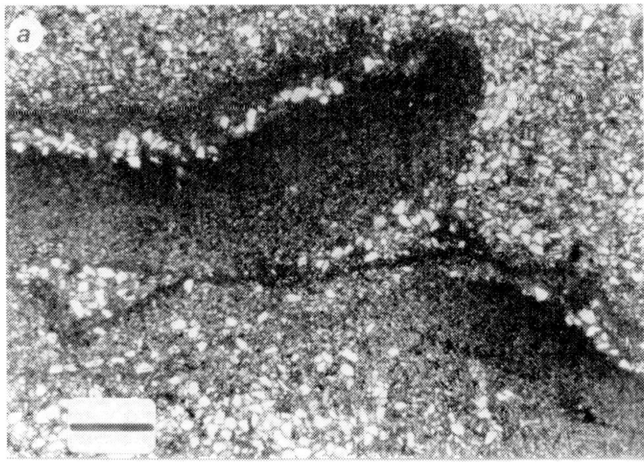

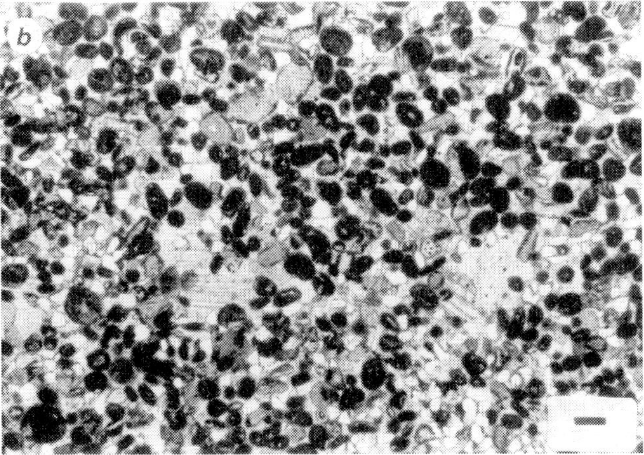

Figure 10. Photomicrographs of intertidal facies; bar scales are 0.400 mm. (A) Curled, dolomitized desiccation chips with reverse grading. Within each mud chip are: dolomicrite at the base, overlain by dark carbonate pellets, and white quartz silt at the top. Matrix is silty shale. (B) Two intertidal sand holes (lower half of photo) infilled by coarse calcite cement.

threshold for sand movement, or they may have been rapid excavators and thus tolerant of sediment movement. Burrows are infilled with the same kind of grainstones as the surrounding rock.

OFFSHORE PACKSTONES AND WACKESTONES

Petrography

The offshore facies consists of an interbedding of two rock types (Figure 11).

1. Packstones and wackestones contain fecal pellets (15%), fossils (12%), and quartz sand (4%) in an abundant micrite matrix (44%). Fossils include crinoids, echinoids, ostracodes, brachiopods, bryozoans, trilobites, gastropods, foraminifera, rugose corals, and calcareous *Girvanella* algae. Oncolites are locally common (up to 20%). The matrix of

Figure 11. Outcrop photograph of offshore facies. Cross-bedded grainstone (GS, interpreted as spillover lobe or tempestite) is sandwiched between burrowed packstone-wackestones (P-W, deeper subtidal facies). 10-cm knife at lower left for scale.

two samples (at Elkins) has been completely replaced by finely crystalline dolomite.

2. Ooid grainstones resemble those of the other two Pickaway facies with varying proportions of peloids (average 22%), ooids (average 22%, range from 1 to 43%), very fine quartz sand (11%), fossils (8%), and calcite cement (30%). The fossil assemblage is about the same as previously described grainstones, but intraclasts of ooid and muddy-skeletal sediment are somewhat more abundant.

Sedimentary Characteristics

Packstones-Wackestones

The characteristic feature of these muddy sands is extreme bioturbation. Infaunal organisms thoroughly churned the sediment, producing thickly bedded or homogeneous units, and biological activity obviously outstripped any sediment movement by traction. *Chondrites* burrowing is recognized in these beds, an ichnogenus usually attributed to deposit feeders in shallow, low-energy marine environments (Kennedy, 1975).

Green, gray, and yellow shale partings and beds are numerous, and the limestones themselves are argillaceous. Shales are generally less than 6 cm thick but range up to 60 cm. Terrigenous influxes must have been significant. The fines originated in source areas to the east and settled from suspension in the relatively low-energy, offshore environment.

Grainstones

The ooid sediment occurs in cosets of cross-beds and plane beds. Thickness of bed sets ranges from 20 to 160 cm. These structures apparently developed under the influence of currents in both the lower and the upper flow regimes.

The small amount of lime mud in these rocks displays an infiltration structure (Dunham, 1962), a geopetal infilling of interstitial pores by mud after deposition of the sand grains. With the waning of a tidal current, mud in suspension may have filtered down into the already deposited ooid sand. Alternatively, infiltration structure can be produced by deposition during the waning of a storm (Kreisa, 1981).

DEPOSITIONAL SETTING

The Greenbrier epeiric sea extended into the central Appalachians as an elongate gulf surrounded on three sides by low-lying land masses. And the West Virginia dome, a major topographic element in the gulf, was situated between the northern shelf and southern basin center (Figure 1). The dome's crest, taken from Dally's (1956) isopach map of the Lower Mississippian Price (Pocono) Formation, trended east–northeast through Randolph County, and the sloping sea floor provided a ramp profile.

Four depositional environments are interpreted from the Pickaway and underlying Taggard formations around the dome (Figure 12). Red shales of the upper Pickaway member and the Taggard Formation reflect sedimentation on a *coastal plain*. Ooid-peloid grainstones, which dominate the lower Pickaway member in this part of the state, accumulated in high-energy zones of the *onshore and shallow-nearshore* environments. Finally, north and south of the dome, grainstones passed downslope into *offshore* packstones, wackestones, and mudstones (Carney and Smosna, 1989).

Coastal Plain

Red shales of the underlying Taggard Formation represent a terrestrial, coastal-plain environment over which lower Pickaway limestones transgressed. The red color, polygonal desiccation cracks, rain-drop imprints, and absence of fossils collectively point to subaerial deposition, and thin interbeds of ooid limestone denote the proximity to a shoreline. Red and yellow shales of the upper Pickaway member mark a second progradation of terrigenous sediment and a return to a coastal-plain environment.

Carbonate Shore

Ooids, peloids, and other carbonate sediment of the intertidal-grainstone facies originated in offshore environments but were transported to the shoreline by tidal currents, waves, and storms. Quartz, on the other hand, originated in source areas to the east and north and was probably transported by winds or longshore currents to the shoreline around the West Virginia dome. The mixing of grains from different sources and of different transport agents accounts for the bimodal size distribution and the alternation of grain types (carbonate vs. clastic) from one lamina to the next.

Decimeter-sized sand waves existed along the shore near the level of low tide. Reversing currents oriented the foreset bedding in both flood and ebb directions, although there was little or no net movement of these sand waves. Near the times of slack tide, reduced current action allowed the deposition of lime and siliciclastic muds, producing tidal bedding. Frequent periods of exposure led to the formation of mud cracks, desiccation chips, and sand holes. Burrowing organisms were relatively active, presumably because the sediment was not being continuously moved by currents, and boring algae and fungi readily micritized carbonate grains (Bathurst, 1966).

The intertidal facies developed as follows. (1) With a rise of sea level following deposition of Taggard red shales, carbonate sediments of the Pickaway shore transgressed the coastal plain. Ooid-peloid grainstones (with interbeds of red shale and dolomite) of the intertidal environment directly overlie the Taggard; as the sea advanced and tidal activity intensified, the small sand waves formed. (2) Through time the subtidal sand waves built upward to sea level, and at two outcrops they are capped by intertidal deposits. Coarsening-upward sequences within the nearshore facies (generally from very fine sand at the base to medium sand near the top) provide evidence for sediment aggradation into shallower water. Moreover, the scale of subtidal bed forms can change vertically. At Huttonsville, for example, cross-bed sets are 30–40 cm thick at the base of a 6-m unit, 60 cm in the middle, and 200 cm at the top. Thickening of bed forms implies that tidal flow intensified, the sand supply increased, and sediment movement accelerated (Walker, 1985). When the aggrading sand waves intersected sea level, the subtidal facies gave way to the intertidal.

Nearshore Tide-Dominated Environment

Large subtidal sand waves formed under the influence of strong tidal currents. Individual Pickaway sand waves grew to 2 m in height, and water depth ranged up to about 5 m. With sediment aggradation and sea level rise, ooid-sand bodies 1–8 m thick accumulated.

The water in which these grainstones accumulated must have been supersaturated with $CaCO_3$ to facilitate the formation of ooids (accounting for up to 50% and possibly more of the total sediment) and aggregates. Currents in the depositional environment generally kept any mud in suspension, while sweeping sand-sized particles along the sea floor. Ooids with well-developed coatings and oncolites both illustrate that the sediment was frequently in motion. Storms occasionally passed by, ripping up intraclasts from adjacent carbonate facies and redepositing them on the sand waves. Composite ooids testify that many ooids proceeded through several episodes of growth before final deposition. Also, boring algae and fungi attacked grains on the bottom, creating the abundant micritized peloids (as much as 36% of the total rock volume). Widespread micritization indicates a slow

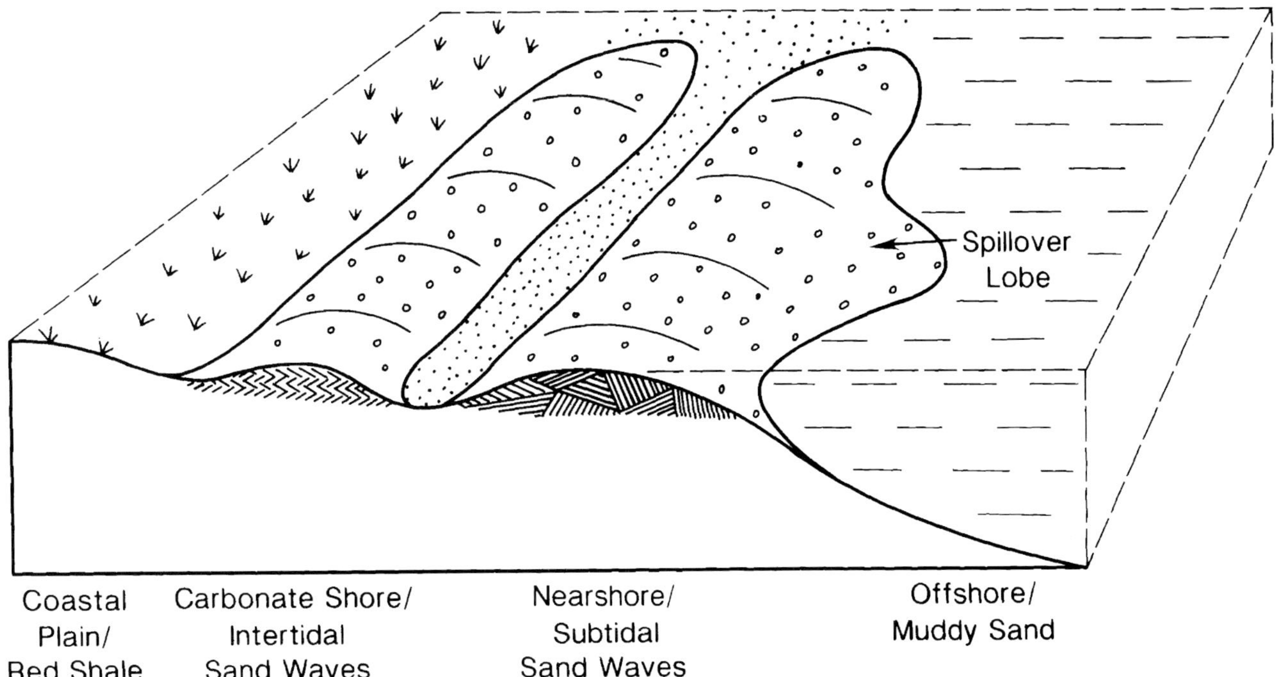

Figure 12. Depositional model of the Pickaway ramp. Lower Pickaway consists of small intertidal sand waves, large subtidal sand waves, offshore spillover lobes/tempestites, and offshore muddy sands. Coastal-plain facies occurs in Taggard Formation and upper Pickaway.

sedimentation rate, providing sufficient time for blue-green algae to alter the grains before final burial.

Paleoflow directions can be interpreted from an examination of cross-bed orientations (Figure 5). On the north side of the dome at Canaan Valley, foreset beds dip in one major direction, south (mean vector of 200°), with a secondary, almost bipolar direction to the northwest (330°). Paleoflow directions must have paralleled the dip of internal cross-beds; thus, flood currents flowed southward while ebb currents flowed to the northwest. Paleoflow data at Elkins show that flood currents flowed to the northeast (55°) and ebb currents, to the west (278°). On the south side of the dome at Huttonsville (Figure 5), flood currents were oriented to the northwest (302°) and ebb currents, to the south (170°). Paleoflow data at Mingo suggest only one tidal direction—flood currents to the northwest (290°). In summary, tidal currents at all four outcrops roughly trended perpendicular to the crest of the West Virginia dome (Figure 13). They were generally directed northwest and south. Where the dome's axis turned near Elkins, tidal currents likewise changed their direction, becoming more east and west.

Although bimodal sedimentary structures suggest that the reversing tidal currents were of nearly equal velocity, flood currents in fact may have been slightly stronger. Of 33 dip measurements interpreted to represent tidal activity, two-thirds are flood oriented. Flood tides, therefore, dominated the movement of Pickaway sediment. In addition, ebb foresets are lacking at Mingo, situated farthest from the dome's crest and in deepest water. Flood currents could move carbonate sand at this locality, but weaker ebb currents were ineffective. Last, the occurrence of reactivation surfaces, tidal bundles, and short mud drapes, although rare, denotes a small degree of time-velocity asymmetry for Greenbrier tidal currents.

Cross-bed dip directions at three outcrops actually identify a third mode (Figure 5). A small mode points eastward (100°) at Canaan Valley and northeastward (52°) at both Huttonsville and Mingo. This third mode aligns with the inferred shoreline of the dome and is attributed to wind-generated waves. Other evidence (ripple bedding, scatter of the cross-bed dip directions) likewise demonstrates an element of wave action superimposed on tidal action. Mississippian waves thus came from the west in Pickaway time (Chesterian) as they did earlier in the Meramecian (see Carney and Smosna, 1989).

The stratigraphic succession at Huttonsville, the most extensive and best exposed of the four outcrops, is illustrated in Figure 14. At the base of the ooid-sand wave is a low-relief erosion surface and lag gravel, which formed when the subtidal sand wave migrated over the intertidal facies (transgression). Thin sets of cross-beds occur near the base, but set thickness increases upward as sediment aggradation built the sand wave into shallower water. Grain size also increases upward from 0.300 to 0.600 mm. Intertidal carbonate sediments as small ooid-sand waves cap the sequence, laid down after the sand body emerged above sea level. Overlying the intertidal limestones are ooid grainstones, interpreted as

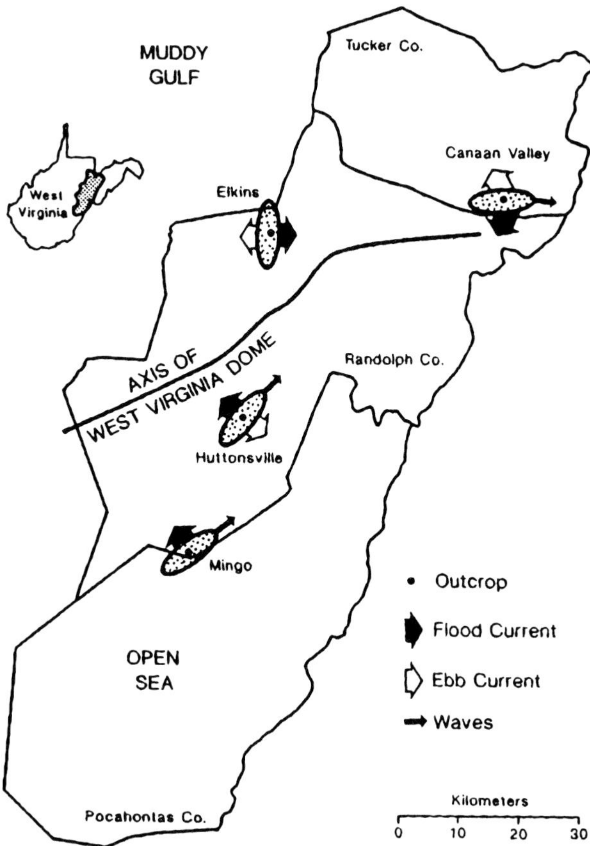

Figure 13. Paleogeographic elements around the axis or crest of the West Virginia dome. Flood and ebb tidal currents flowed perpendicular to the axis, and waves came from the west-southwest. The carbonate ramp sloped northward into a muddy gulf and southward into the open sea.

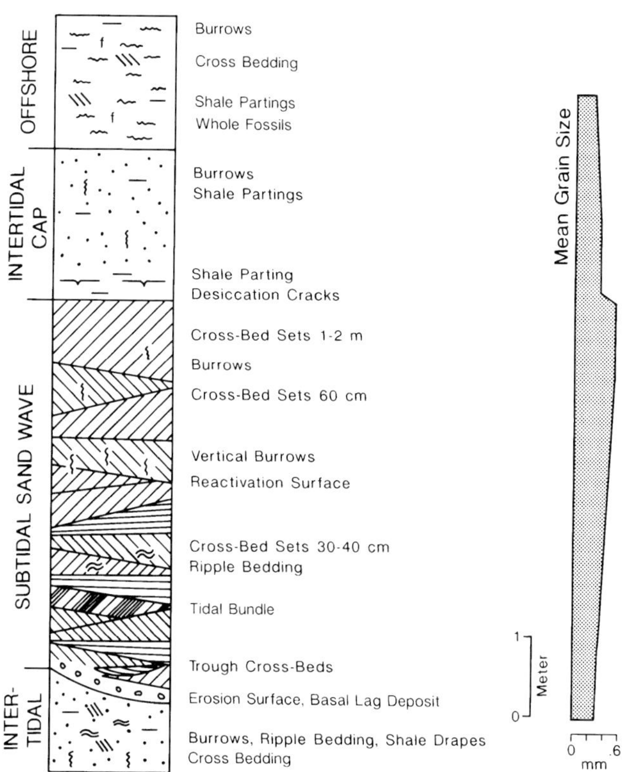

Figure 14. Stratigraphic succession of facies at Huttonsville outcrop. Note basal erosion surface and lag deposit, upward increases in thickness of cross-bed sets and in mean grain size, and capping facies of intertidal deposits.

tempestites/spillover lobes, and interbedded skeletal packstones and wackestones. The transgression had continued and offshore sediments were deposited.

Offshore

Muddy pelletal-skeletal sands represent relatively low-energy deposits that accumulated offshore from the subtidal sand waves (Figure 12). The sea floor was generally below the action of any currents and waves. However, the presence of oncolites indicates that currents at times rolled the sand-sized grains along the bottom (Logan et al., 1964), and a small volume of intergranular calcite cement (micrite/cement ratio of 6) hints of a slight washing of the sediment. Good circulation provided necessary nutrients to the invertebrate community, clear water allowed sufficient light penetration, and a stable substrate favored the benthonic organisms.

The well-washed, well-sorted grainstones formed where tidal currents or storms swept ooid sediment off the nearshore sand waves (compare with Ball, 1967). Offshore in deeper water, it was redeposited as spillover lobes or tempestites that became interbedded with the pelletal-skeletal packstones and wackestones. Common erosional intraclasts also denote allochthonous sedimentation; tidal and storm currents eroded and transported these clasts from both the adjacent sand waves and the surrounding open shelf. The infiltration structure of lime mud into the ooid sand was produced by waning tidal or storm currents. Afterward, normal quiet-water sedimentation resumed.

Offshore facies constitute 80% of the stratigraphic section at Mingo, and this is dominated by the low-energy packstones and wackestones. Situated 25 km south of the dome's axis, this locality occupied a position low on the carbonate ramp and in relatively deep water (Figure 13). By comparison, offshore facies constitute only 15% of the stratigraphic section at Canaan Valley, and this is mainly the spillover/tempestite grainstones. The Canaan Valley locality must have been situated very near to the crest of the West Virginia dome. Elkins and Huttonsville were positioned at intermediate depths on the Greenbrier ramp.

SUMMARY

Mississippian oolites of the lower Pickaway Limestone accumulated in shallow, turbulent water over the West Virginia dome. A succession of facies developed from onshore to offshore, and tidal currents were chiefly responsible for shaping the carbonate-sand bodies.

Red shales with subaerial-exposure structures represent deposition on a *coastal plain* (underlying Taggard Formation and overlying upper Pickaway member). The *shore environment* was marked by small intertidal sand waves. Though predominantly composed of peloids and ooids, shoreline grainstones contain abundant quartz. Sedimentary structures of this facies include composite sets of cross-beds and plane beds, bidirectional foresets, tidal bedding, desiccation features, reverse grading, wavy bedding, reactivation surfaces, sand holes, and ripple bedding. The *subtidal tide-dominated environment* contained large subtidal sand waves that developed directly adjacent to the shore at the top of the carbonate ramp. This facies is characterized by the following sedimentary structures: large wedge-shaped sets of cross-beds, bidirectional foresets, reactivation surfaces, mud drapes, tidal bundles, planar erosion surfaces, trough cross-bedding, and plane beds. The dip directions of foreset bedding illustrate that tidal currents flowed perpendicular to the dome's crest, that is, northwest and south, and flood currents might have been slightly stronger. Wind-generated waves, though of minor importance, came from the west. *Offshore* from the ooid-sand waves, sedimentation at depths normally below the influence of tidal currents produced muddy, bioturbated, pelletal-skeletal sand. Not infrequently, however, currents swept ooid-peloid sediment from shallower environments and redeposited it offshore as spillover lobes or tempestites. These grainstones contain cosets of cross-beds and plane beds, infiltration structure, and erosional intraclasts.

The ooid-sand waves appear to produce a distinctive vertical stratigraphic sequence, which is marked by a basal erosion surface and lag deposit (migration of the sand wave by tidal and storm currents), an upward increase in both cross-bed set thickness and grain size (sediment aggradation into shallower water), and a capping intertidal facies (building up above the level of low tide).

ACKNOWLEDGMENTS

This research was partly funded by a Chevron Corporation Field Grant. Special thanks go to Kathy Bruner for her help in the field. Allison Hanham drafted the illustrations. We also wish to thank Brian Keith, Erik Kvale, Cindy Carney, and Charles Zuppann, who reviewed the paper and made helpful suggestions for its improvement.

REFERENCES CITED

Allen, J.R.L., 1981, Lower Cretaceous tides revealed by cross-bedding with mud drapes: Nature, v. 289, p. 579-581.

Allen, P.A., and P. Homewood, 1984, Evolution and mechanics of a Miocene tidal sand wave: Sedimentology, v. 31, p. 63-81.

Arkle, T., and others, 1979, West Virginia and Maryland, *in* The Mississippian and Pennsylvanian (Carboniferous) Systems in the United States: United States Geological Survey Professional Paper 1110-D, p. D1-D35.

Badiozamani, K., 1973, The Dorag dolomitization model—application to the Middle Ordovician of Wisconsin: Journal of Sedimentary Petrology, v. 43, p. 965-984.

Ball, M.M., 1967, Carbonate sand bodies of Florida and the Bahamas: Journal of Sedimentary Petrology, v. 37, p. 556-591.

Bathurst, R.G.C., 1966, Boring algae, micritic envelopes and lithification of molluscan biosparites: Geology Journal, v. 5, p. 15-32.

Bjerstedt, T.W., and T.W. Kammer, 1988, Genetic stratigraphy and depositional systems of the Upper Devonian-Lower Mississippian Price-Rockwell deltaic complex in the central Appalachians, U.S.A.: Sedimentary Geology, v. 54, p. 265-301.

Blatt, H., G.V. Middleton, and R.C. Murray, 1980, Origin of sedimentary rocks: Englewood Cliffs, NJ, Prentice-Hall, Second edition, 782 pp.

Carney, C., and R. Smosna, 1989, Carbonate deposition in a shallow marine gulf, the Mississippian Greenbrier Limestone of the central Appalachian Basin: Southeastern Geology, v. 30, p. 25-48.

Choquette, P.W., and R.P. Steinen, 1980, Mississippian non-supratidal dolomite, Ste. Genevieve Limestone, Illinois Basin: evidence for mixed-water dolomitization, *in* D.H. Zenger, J.B. Dunham, and R.L. Ethington, (eds.), Concepts and models of dolomitization: Society of Economic Paleontologists and Mineralogists Special Publication 28, p. 163-196.

Dally, J.L., 1956, The stratigraphy and paleontology of the Pocono Group in West Virginia: Columbia Univ., Ph.D. thesis, 238 pp.

de Mowbray, T., and M.J. Visser, 1984, Reactivation surfaces in subtidal channel deposits, Oosterschelde, southwest Netherlands: Journal of Sedimentary Petrology, v. 54, p. 811-824.

Donaldson A.C., and R.L. Shumaker, 1981, Late Paleozoic mollasse of central Appalachians, *in* A.D. Miall, (ed.), Sedimentation and tectonics in alluvial basins: Geological Association of Canada Special Paper 23, p. 99-124.

Dunham, R.J., 1962, Classification of carbonate rocks according to depositional texture, *in* W.E. Ham, (ed.), Classification of carbonate rocks: American Association of Petroleum Geologists Memoir 1, p. 108-121.

Emery, K.O., 1945, Entrapment of air in beach sand: Journal of Sedimentary Petrology, v. 15, p. 39-49.

Halley, R.B., P.M. Harris, and A.C. Hine, 1983, Bank margin, *in* P.A. Scholle, D.G. Bebout, and C.H. Moore, (eds.), Carbonate depositional environments: American Association of Petroleum Geologists Memoir 33, p. 463-506.

Harms, J.C., J.B. Southard, D.R. Spearing, and R.G. Walker, 1975, Depositional environments as interpreted from primary sedimentary structures and stratification sequences: Society of Economic Paleontologists and Mineralogists Short Course Notes 2, 161 pp.

Hine, A.C., 1977, Lily Bank, Bahamas: history of an active oolite sand body: Journal of Sedimentary Petrology, v. 47, p. 1554-1581.

Houbolt, J.J.H.C., 1968, Recent sediments in the southern bight of the North Sea: Geologie en Mijnbouw, v. 47, p. 245-273.

Imbrie, J., and H. Buchanan, 1965, Sedimentary structures in modern carbonate sands of the Bahamas, *in* G.V. Middleton, (ed.), Primary sedimentary structures and their hydrodynamic interpretation: Society of Economic Paleontologists and Mineralogists Special Publication 12, p. 149-172.

Johnson, H.D., 1978, Shallow siliclastic seas, *in* H.G. Reading, (ed.), Sedimentary environments and facies: New York, Elsevier, p. 207-258.

Kennedy, W.J., 1975, Trace fossils in carbonate rocks, *in* R.W. Frey, (ed.), The study of trace fossils: New York, Springer-Verlag, p. 377-398.

Klein, G.d., 1965, Dynamic significance of primary structures in the Middle Jurassic Great Oolite Series, southern England, *in* G.V. Middleton, (ed.), Primary sedimentary structures and their hydrodynamic interpretation: Society of Economic Paleontologists and Mineralogists Special Publication 12, p. 173-191.

Klein, G.d., 1970, Depositional and dispersal dynamics of intertidal sand bars: Journal of Sedimentary Petrology, v. 40, p. 1095-1127.

Klein, G.d., 1977, Clastic tidal facies: Champaign, Illinois, Continuing Education Publication Co., 149 pp.

Knewston, S.L., and J.F. Hubert, 1969, Dispersal patterns and diagenesis of oolitic calcarenites in the Ste. Genevieve Limestone (Mississippian), Missouri: Journal of Sedimentary Petrology, v. 39, p. 954-968.

Kreisa, R.D., 1981, Storm-generated sedimentary structures in subtidal marine facies with examples from the Middle and Upper Ordovician of southwestern Virginia: Journal of Sedimentary Petrology, v. 51, p. 823-848.

Logan, B.W., R. Rezak, and R.N. Ginsburg, 1964, Classification and environmental significance of algal stromatolites: Journal of Geology, v. 72, p. 68-83.

Reineck, H.-E., and I.B. Singh, 1980, Depositional sedimentary environments: Berlin, Springer-Verlag, Second edition, 549 pp.

Scotese, C.R., R.K. Bambach, C. Barton, R. Van der Voo, and A.M. Ziegler, 1979, Paleozoic base maps: Journal of Geology, v. 87, p. 217-277.

Sullivan, E.M., and D.A. Textoris, 1988, Microfacies and paleoenvironments of the Mississippian Denmar Formation, eastern West Virginia: Southeastern Geology, v. 28, p. 133-152.

Swift, D.J.P., 1985, Response of the shelf floor to flow, *in* R.W. Tillman, D.J.P. Swift, and R.G. Walker, (eds.), Shelf sands and sandstones: Society of Economic Paleontologists and Mineralogists Short Course Notes 13, p. 135-241.

Visser, M.J., 1980, Neap-spring cycles reflected in Holocene subtidal large-scale bedform deposits: a preliminary note: Geology, v. 8, p. 543-546.

Walker, R.G., 1985, Ancient examples of tidal sand bodies formed in open, shallow seas, *in* R.W. Tillman, D.J.P. Swift, and R.G. Walker, (eds.), Shelf sands and sandstones: Society of Economic Paleontologists and Mineralogists Short Course Notes 13, p. 303-341.

Wilson, R.C.L., 1968, Carbonate facies variation within the Osmington Oolite Series in southern England: Palaeogeography, Palaeoclimatology, Palaeoecology, v. 4, p. 89-123.

Yeilding, C.A., and J.M. Dennison, 1986, Sedimentary response to Mississippian tectonic activity at the east end of the 38th Parallel fracture zone: Geology, v. 14, p. 621-624.

Youse, A.C., 1964, Gas producing zones of Greenbrier (Mississippian) Limestone southern West Virginia and eastern Kentucky: American Association of Petroleum Geologists Bulletin v. 48, p. 465-486.

Chapter 12

Oolitic Tidal-Bar Reservoirs in the Mississippian Greenbrier Group of West Virginia

Gregory T. Kelleher
CNG Producing Company
New Orleans, Louisiana, USA

Richard Smosna
Department of Geology & Geography
West Virginia University
Morgantown, West Virginia, USA

ABSTRACT

In the Rhodell field of southern West Virginia, initial potentials of oolitic reservoirs in the Greenbrier Group commonly exceed 2.0 MMCFGD of natural gas. The major reservoir, an oolitic member of the Union Limestone, stands out as distinct northwest-trending thicks, comparable in size, shape, and location to modern tidal bars. They formed along the hinge line that separated a rapidly subsiding basin from a stable shelf. Individual bars have 15–30 ft (4.6–9.1 m) of net thickness with greater than 6% porosity, and the producing zones are ooid grainstones with a mean log porosity of 9.5% and permeability of up to 0.15 md. Ooids were generally altered to microrhombic low-magnesium calcite during meteoric diagenesis and developed considerable intragranular microporosity. Reservoir permeability is greatly enhanced where ooids have undergone chemical compaction, which produced long contacts between neighboring microporous grains.

INTRODUCTION

The Greenbrier Group of Meramecian–Chesterian age is one of the more prolific, shallow petroleum-producing reservoirs in West Virginia. A significant volume of oil has been produced from this unit, and Ruley (1970) calculated reserves of 2400–6500 bbl/acre in the north-central portion of the state. In southwestern counties, reserves of natural gas commonly reach 1 bcf per well. In our study area of southern West Virginia, reserve data are difficult to calculate because natural gas from the Greenbrier is typically commingled with gas from other reservoirs. Recent drilling suggests that probable reserves can exceed 1 bcf per well in this area.

Economic analyses indicate that the Greenbrier Group may have the best rate of return for any shallow petroleum reservoir in the state. But because of a lack of subsurface information, this unit has not been a popular exploration target. In 1988, for example, just 19 Greenbrier wells were drilled and 22 permits issued, accounting for only 2% of the state's drilling activity (Avary et al., 1989). Our studies, though, suggest that Greenbrier carbonates in the southern counties have a good potential for large undiscovered reserves.

Production is generally from oolitic zones, representing several different depositional and diagenetic environments. In north-central West Virginia (Figure 1), production comes from a basal sandy, oolitic facies, which was deposited in a coastal environment. These limestones have been extensively dolomitized, and porosity is mainly intercrystalline in dolomite (Overbey, 1967). In contrast, oolites of southwestern West Virginia developed offshore and over topographic highs of the underlying Maccrady Formation (Youse, 1964). Porosity there is mainly interparticle.

Previously published interpretations, however, do not explain the stratigraphic position, orientation, or porosity distribution of oolites within the Rhodell field of southern West Virginia. The purpose of this study, therefore, is to present (1) a depositional model that resolves the geometry and stratigraphy of the two main producing zones at Rhodell field and (2) a diagenetic model that addresses reservoir quality in terms of porosity and permeability.

The Rhodell field is located in the area of McDowell, Mercer, Raleigh, and Wyoming counties, West Virginia (Figure 1). The field has production from several Mississippian sandstone members of the younger Mauch Chunk Group and older Pocono (Price) Formation (Figure 1), including the Ravencliff, Maxton, and Weir, in addition to the Greenbrier carbonates. To date, approximately 500 wells have been drilled in this area. Sidewall cores show that the Greenbrier reservoirs are oolitic, and detailed isopach maps of the productive members delineate a northwest–southeast trend. The depth to Greenbrier reservoirs is approximately 3000 ft (915 m).

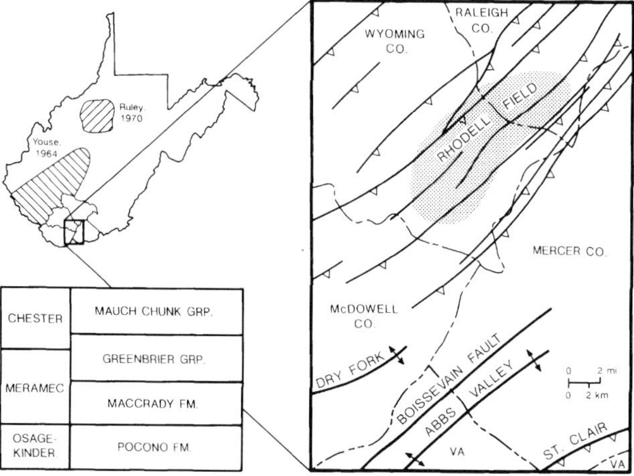

Figure 1. Map of the study area, showing major (heavy lines) and minor structural features and the location of Rhodell field (stippled pattern) in southern West Virginia. Inset map indicates two other important areas of petroleum production from the Greenbrier Group, as described by Youse (1964) and Ruley (1970). The stratigraphic section of Mississippian rocks in the study area is also included.

DATA BASE

Gamma-ray, density, and induction logs from 355 wells were used to divide the Greenbrier Group into stratigraphic units and to correlate these units across the study area. Density logs and production data identify the porous reservoir zones. Initial test gauges are available for the vast majority of wells in the study area, as are cumulative production data for all Consolidated Natural Gas wells. Sidewall cores were taken from two of the wells (Figure 2), and samples from these cores were analyzed by petrographic microscope and SEM. Most of the data, including geophysical logs, drafted base maps, and cores, were generously provided by CNG Development Company.

GEOLOGIC SETTING

Limestones of the Greenbrier Group were deposited within a broad gulf that extended into the central Appalachians from the southwest (Figure 3) (Carney and Smosna, 1989). This shallow seaway was surrounded by land masses on three sides: (1) highlands to the east (a remnant of the Acadian Mountains) and the adjacent alluvial coastal plain, (2) a low-lying land mass to the north, and (3) the exposed Cincinnati arch to the west. The eastern highlands provided small volumes of clastic material to the gulf, producing the red shales characteristically interbedded with Greenbrier limestones. Erosion of the northern land mass introduced small volumes of quartz sand to the nearby shoreline facies, but the Cincinnati arch apparently shed no sediments into the basin. Paleomagnetic evidence (Scotese et al., 1979) places the gulf a few degrees south of the equator. The climate was somewhat dry, perhaps due to a rain-shadow effect behind the Acadian highlands (Cecil, 1990; Ettensohn, 1985).

The Greenbrier Group unconformably overlies siliciclastic marine, deltaic, and alluvial sediments of the Mississippian Maccrady formation (Figure 1). When the major influx of these siliciclastics ceased, carbonate sedimentation began. Lime sediments were initially confined to a depocenter or basin in southern West Virginia and neighboring Virginia (Figure 3), an area of substantial downwarping where more than 1600 ft (500 m) of rock accumulated. Subsidence occurred during a relaxation stage of lithospheric deformation following the Acadian orogeny (Quinlan and Beaumont, 1984). The adjacent shelf of northern West Virginia, eastern Ohio, and southwestern Pennsylvania at first lay exposed; however, with continued subsidence this shelf, too, was eventually trans-

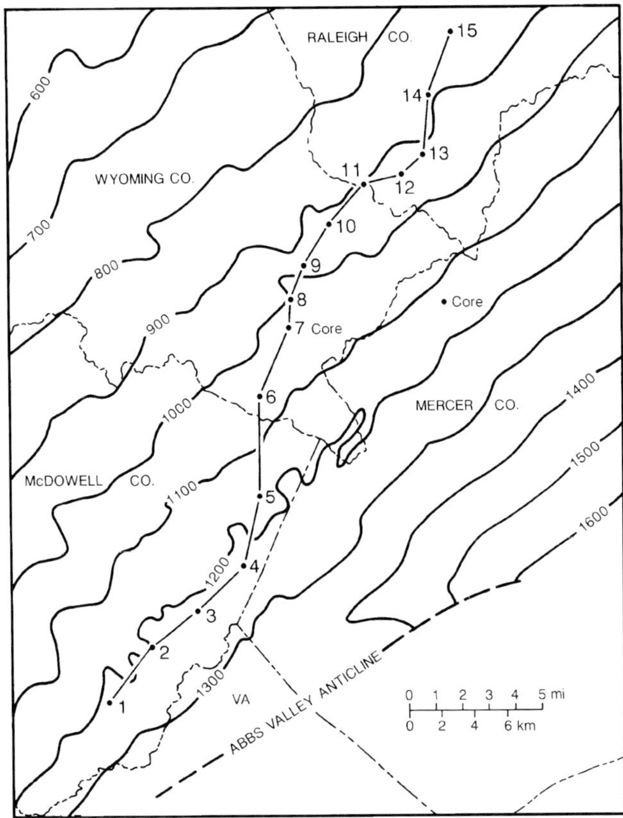

Figure 2. Isopach map of the total interval from the base of the Greenbrier Group to the base of the Reynolds Limestone Member of the overlying Bluefield Formation (see Figure 5). Depositional strike was northeast, and dip to the southeast. Location of the stratigraphic cross section in Figure 5 is shown as are the locations of two cored wells used for petrological analyses. Contour interval is 100 ft.

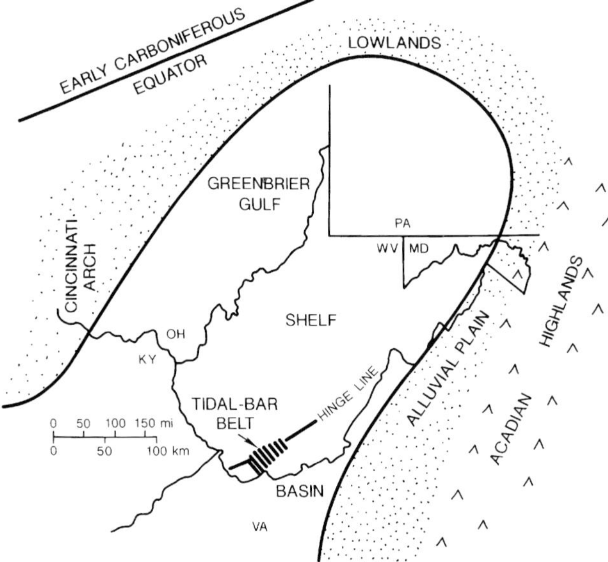

Figure 3. Paleogeographic reconstruction of the Greenbrier gulf in central Appalachian basin (modified from Carney and Smosna, 1989; Scotese et al., 1979). A belt of tidal bars in the Union oolite member, which produces natural gas in the Rhodell field, is positioned along a hinge line that separated the depocenter or basin on the south from the shelf to the north.

gressed. Over most of the shelf, the Greenbrier is less than 300 ft (90 m) thick.

The basal formation of the Greenbrier Group (Figure 4) is the Hillsdale Limestone, which contains cherty skeletal wackestones interbedded with packstones and lime mudstones. The overlying Denmar Formation is composed of skeletal wackestones that are lithologically similar to the Hillsdale, but with less chert. These lower two units reflect low-energy, relatively deep-water deposition in the rapidly subsiding basin. The overlying Taggard Formation consists of red silty shales interbedded with ooid and skeletal grainstones, deposited in a coastal-plain/shoreline setting during a minor progradation of terrigenous sediment. The Meramecian-Chesterian boundary is placed at the base of the Taggard (Carney and Smosna, 1989), and its top is used as a datum for stratigraphic cross sections of this study.

The overlying Pickaway and Union limestones are quite similar. Both contain ooid grainstones and skeletal packstones and wackestones. These oolites (informally designated here as members) are the principal Greenbrier reservoirs in the southern part of West Virginia. The Alderson Limestone, above the Greenville Shale, marks the top of the Greenbrier Group. Ooid and skeletal grainstones and skeletal packstones and wackestones of the Alderson, like those of the Pickaway and Union below, were deposited on mobile sand shoals and in associated lagoons.

Overlying the Greenbrier is a sequence of interbedded shales and limestones of the Bluefield Formation of the Upper Mississippian Mauch Chunk Group. Included in this formation are the black Lillydale Shale Member, called the Pencil Cave shale by drillers, and the overlying Reynolds Limestone Member, consisting of nodular, argillaceous, skeletal wackestones and packstones. The Reynolds (the drillers' Little Lime) is areally extensive and uniform in thickness; it also has a distinct gamma-ray signature. It therefore serves as an excellent marker bed for well correlations.

A number of prominent faults and folds, which formed during the Permian-Pennsylvanian Allegheny orogeny, occur within the study area (Figure 1). In the northern region of the study area, 13 southeast-dipping thrust faults are identified by repeated stratigraphic sections on the geophysical logs. Displacement decreases consistently from the southeast (222 ft or 68 m) to the northwest (18 ft or 5 m). Faults transect the productive trends with no effect, and the

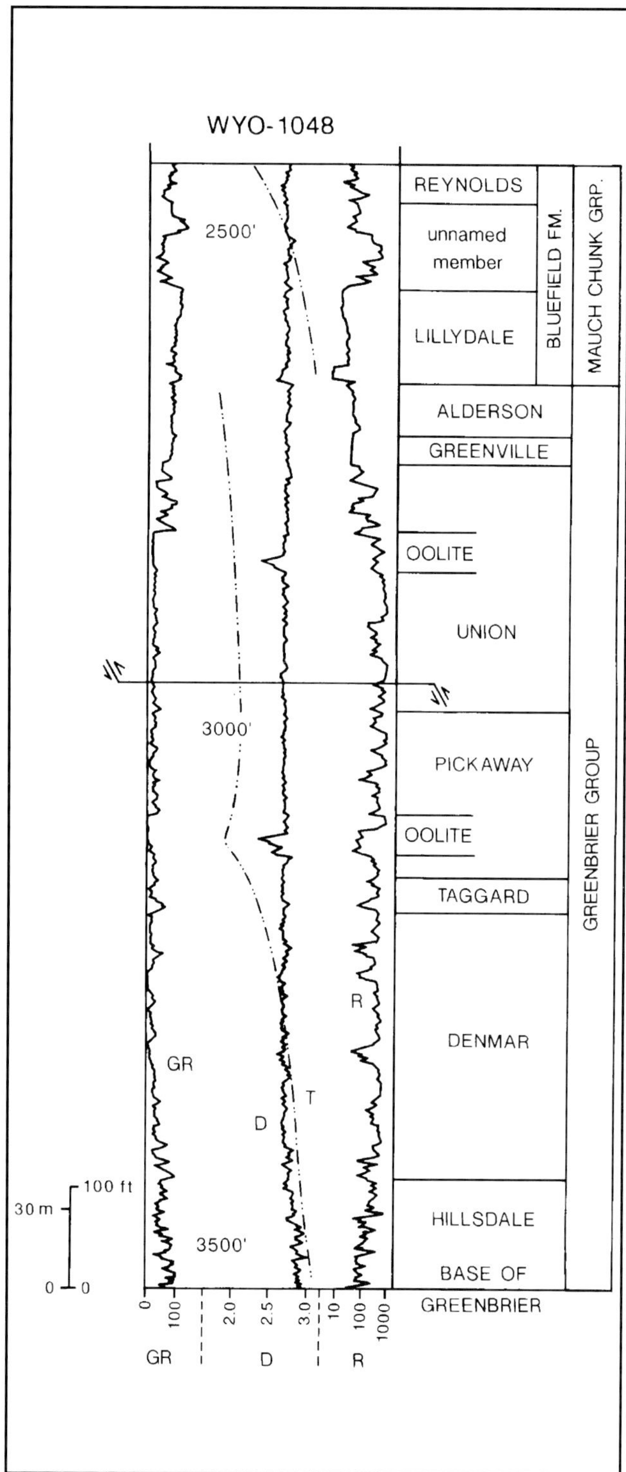

Figure 4. Plot of geophysical-log curves for Wyo-1048 well includes gamma-ray (GR; scale at bottom is in API units), bulk density (D; scale in g/cc), resistivity (R; scale in ohms-m), and temperature (T; no scale). Production comes mainly from the oolite member in the lower Pickaway Limestone, as indicated by the inflection of the temperature curve to the left. The Union Limestone is cut by a thrust fault that repeats 52 ft (16 m) of section. Wyo-1048 is identified as well #7 on the cross section of Figures 2 and 5.

trapping mechanism for Greenbrier reservoirs of the Rhodell field is purely stratigraphic.

DEPOSITIONAL MODEL

The stratigraphic interval from the base of the Greenbrier to the top of the Reynolds thickens from 600 ft (180 m) in the northwest part of the study area to more than 1600 ft (500 m) in the southeast (Figure 2). Thickening is related to subsidence of the Appalachian basin which resulted from relaxation of the viscous lithosphere (Quinlan and Beaumont, 1984). Relaxation and subsidence followed loading of the crust by overthrust sheets of the Acadian orogeny. The eastern thrust load, estimated by Quinlan and Beaumont to have been 10 km (6 mi) thick, remained in place for an extended length of time following its emplacement in the Late Devonian and perhaps earliest Mississippian. As the underlying lithosphere relaxed, a progressively deeper foreland basin was created. Subsidence in the Meramecian Epoch outpaced carbonate sedimentation, leading to deposition of the deeper-water Hillsdale and Denmar formations.

Depositional dip was to the southeast into the foreland basin. On the total-interval isopach map (Figure 2), a change in the rate of thickening occurs near the 900-ft contour line. Northwest of this line, the unit thickens into the basin at an average rate of 34 ft/mi (6.5 m/km), but to the southeast it is almost 47 ft/mi (8.9 m/km). This rate change marks a local hinge line that separated a rapidly subsiding basin from the adjoining stable shelf.

Basin subsidence slowed by early Chesterian time, and shallow water deposits of the Pickaway, Union, and Alderson limestones accumulated across the basin. Cross sections show that the middle formations (Taggard, Pickaway, and Union) are relatively uniform across the study area, except for thickness variations in the Union and Pickaway oolite members (Figure 5).

Union Limestone

The Union oolite member produces gas in 32 of 106 wells within the Rhodell field. Porous intervals of the Greenbrier Group produce only dry gas; formation water has not been encountered regardless of structural position. Log porosity in these producing wells averages 9.5%, but permeability is generally less than 0.15 md (based on 14 sidewall-core samples). Initial potential tests frequently exceed 2.0 MMCFGD. The gamma-ray character clearly defines this member: the unit has a clean, blocky signature, and the top is marked by an overlying shale (Figure 4).

Thin-section analyses of sidewall cores show the productive member to be an ooid grainstone containing 48–72% medium to coarse ooids (Figure 6). Minor components include fossils (fragments of brachiopods, bryozoans, echinoderms, gastropods, forams, ostracodes, oncolites, and dasyclad algae), muddy and fossiliferous intraclasts, peloids, and ooid

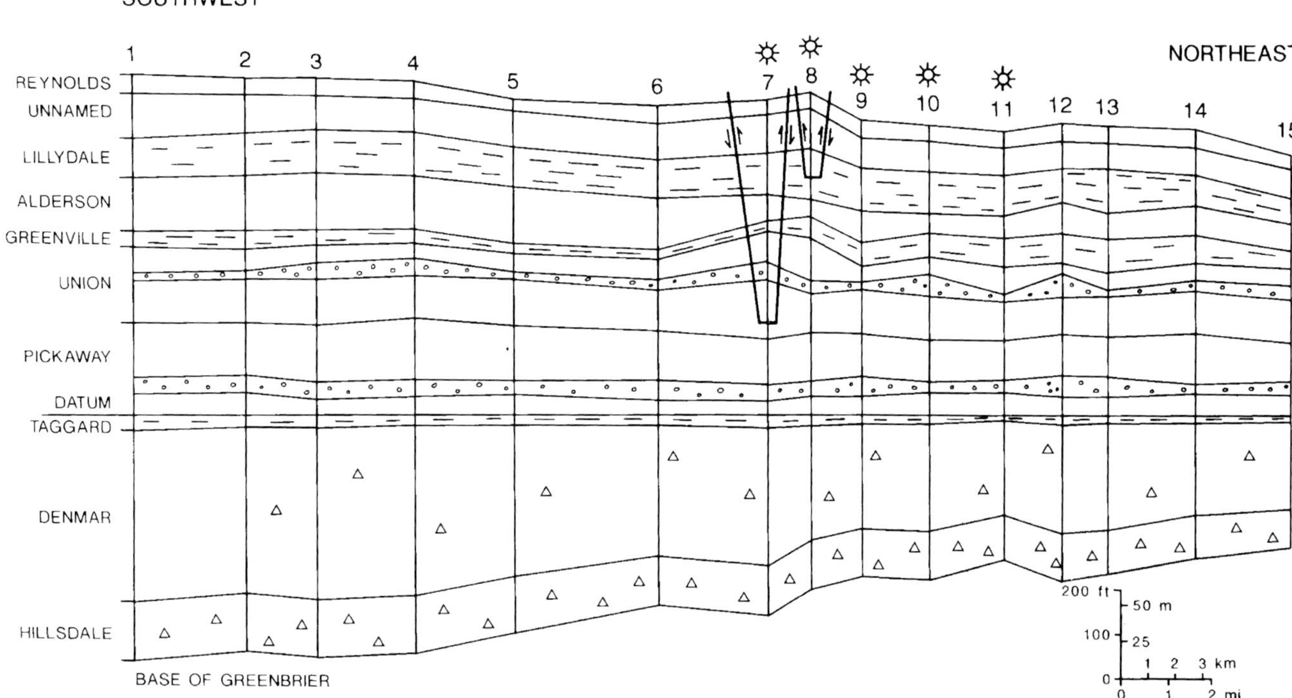

Figure 5. Northeast–southwest stratigraphic cross section through the study area using the top of the Taggard Formation as a datum. Oolitic members of the Union and Pickaway limestones are indicated by a pattern of circles; shales in the Taggard, Greenville, and Lillydale by dashes; and chert in the Hillsdale and Denmar formations by triangles. Wells #7 and #8 are cut by thrust faults, and productive wells #7– #11 lie within the Rhodell field. The location of this section is shown in Figure 2.

aggregates. Sorting is generally poor at the base but becomes very good toward the top of the member.

The isopach map of the oolite member of the Union Limestone (Figure 7) differs considerably from that of the entire Greenbrier (Figure 2). Thick dip-oriented accumulations appear in this member. Both the orientation and the geometry of these isopach thicks suggest that the member consists of a number of ooid sand bars. Individual bars in the northern part of the study area average 40 ft (12 m) in thickness and approximately 4500 ft (1.4 km) in width. They extend beyond the limits of the study area, but additional subsurface control indicates that their overall length is 20 mi (32 km). Bars are spaced 1.5 mi (2.4 km) apart. These sand bars are interpreted to have had a tidal origin, based on a comparison with modern tidal bars. They closely resemble in size those on modern tide-dominated shelves of the Bahamas, and modern tidal bars are similarly oriented perpendicular to regional strike (Halley et al., 1983).

The bars collectively constitute a belt that trends northeast–southwest across the study area, and the location of the Union tidal-bar belt coincides with the postulated hinge line (Figure 3). In modern tidal-bar belts of the Bahamas, currents are unusually strong because of a resonance between tidal waves and the natural frequency of the water mass on the shelf (Ball, 1967). At peak tide (velocities greater than 50–100 cm/sec), the incoming current becomes unstable and separates into a number of approximately equally spaced and sized digits. Channelized flow is then responsible for creating a tidal-bar geometry perpendicular to the shelf edge. Sediment transport and perhaps erosion occur in the digits or channels with a high-energy flow, whereas ooid deposition takes place in slower moving currents over the rising shoal. Daily tidal currents keep crestal grains in near-constant motion, thus promoting rapid ooid development on top of the bars. By analogy, tidal bars of the Greenbrier are assumed to have formed under similar hydrodynamic conditions.

The tidal-bar belt of the Union Limestone stretches for a distance of almost 20 mi (32 km) parallel to the Greenbrier hinge line. Future exploration for productive oolite sand bodies should be concentrated along the hinge line (900-ft contour line on the isopach map) to the northeast and southwest of the study area. If comparable in size to modern belts in the Bahamas, the Greenbrier tidal-bar belt may extend for 25–60 mi (45–100 km) across portions of Raleigh, Summers, Fayette, and Greenbrier counties (northeast) and McDowell County and adjacent Virginia (southwest).

The linear trend of tidal-bar belts is poorly defined in the southwestern region of the study area (Figure 7). Bars there display a greater variability: some are short and wide, whereas others have variable orientations. Intervening channels, too, are much wider (3.0 mi or 4.8 km) and notably irregular. These changes in geometry may indicate that water depth increased slightly (just a few meters) to the southwest; hence,

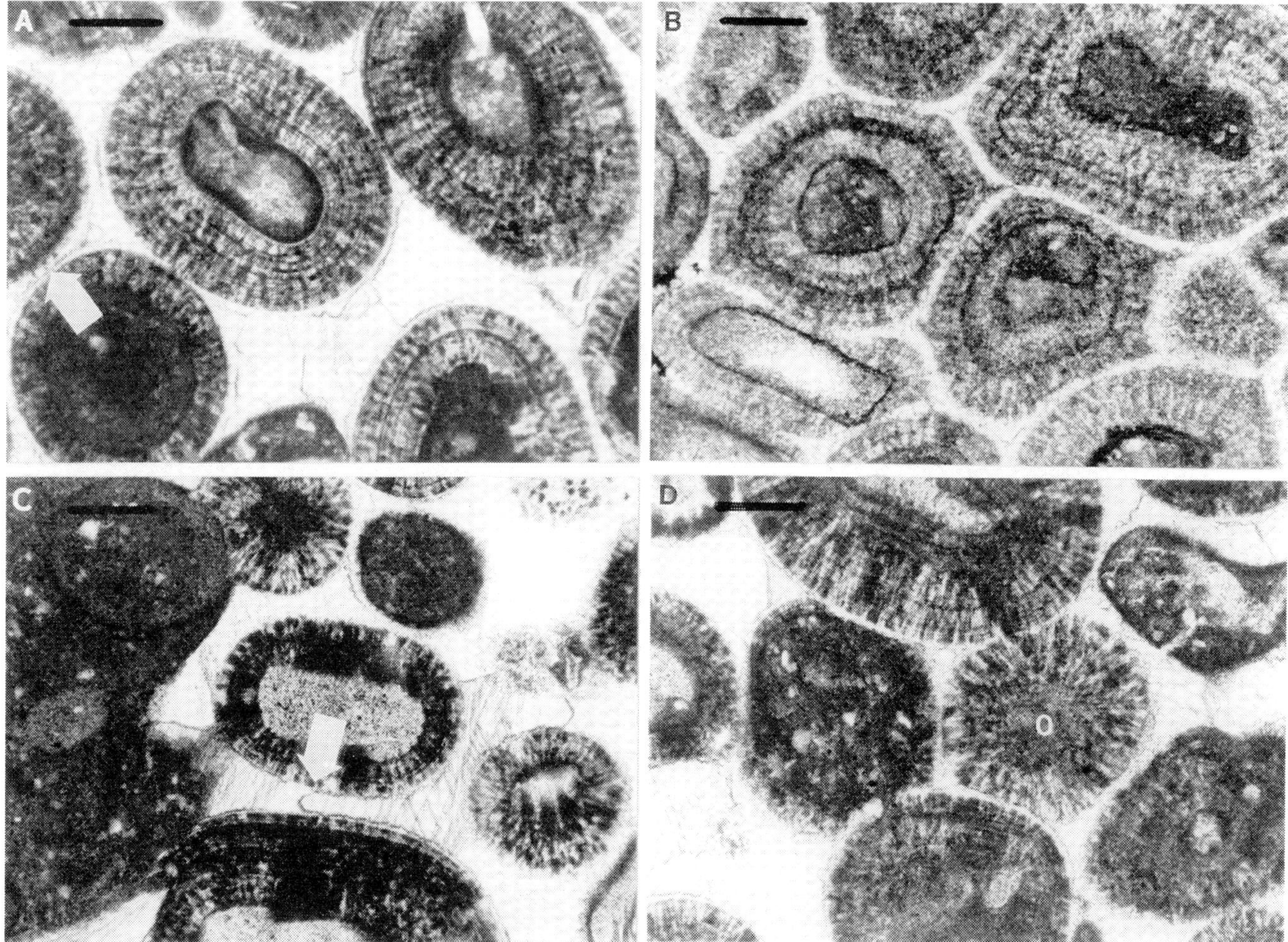

Figure 6. Thin-section photomicrographs of Greenbrier oolites. Bar scales in upper left of each photo are 0.200 mm. (A) Ooid grainstone of the Union Limestone displays loose packing and point-to-point contacts. Thin isopachous rim of calcite around most grains constitutes the first-generation cement. Rarely these rims meet along polygonal boundaries (arrow at left). Coarse, second-generation cement filled the remaining intergranular pore space. (B) Fitted fabric of ooid grainstone in the Pickaway Limestone produces long contacts between neighboring grains. Fringe of first-generation cement is absent or discontinuous on several grains. (C) Meniscus cement (arrow) formed after the thin isopachous rim cement and before the coarse void-filling calcite. (D) Chemical compaction of an ooid grainstone. Center grain (O) is in contact (long and concave-convex) with five adjacent grains in the plane of thin section.

tidal energy would have been lower. Thus, the bar-and-channel topography was less well-developed in this direction. A similar relationship exists between sand-body geometry and water depth at Joulters Cay, Bahamas (Harris, 1979).

Pickaway Limestone

Significant production in eight wells of the study area has also been established from the Pickaway oolite member. Reservoir parameters, including porosity and permeability, are similar to those of the Union producing zone. However, lateral facies changes occur in the Pickaway that make regional mapping of the oolite bodies difficult. The net-pay isopach map of the Pickaway oolite member (Figure 8), using a porosity cutoff value of 6%, illustrates that porosity occurs in both the northern and southern areas. Natural gas is produced in each of the northern isopach bodies, but insufficient permeability renders the southern trends unproductive.

Based on thin-section analyses of sidewall cores, the productive zone is an ooid grainstone, consisting of 47–66% medium-sand-sized ooids. Other nonskeletal grains are intraclasts of muddy skeletal sediment, peloids (totally micritized particles), fecal pellets, and ooid aggregates. Fossils include fragments of echinoderms, brachiopods, bryozoans, ostracodes, gastropods, and oncolites. Toward the top of the oolite member, the percentage of ooids increases and sorting improves—denoting that the ooid shoal grew upward into shallower, more turbulent water.

On the net-pay isopach map (Figure 8), porous oolitic intervals of the Pickaway range in thickness

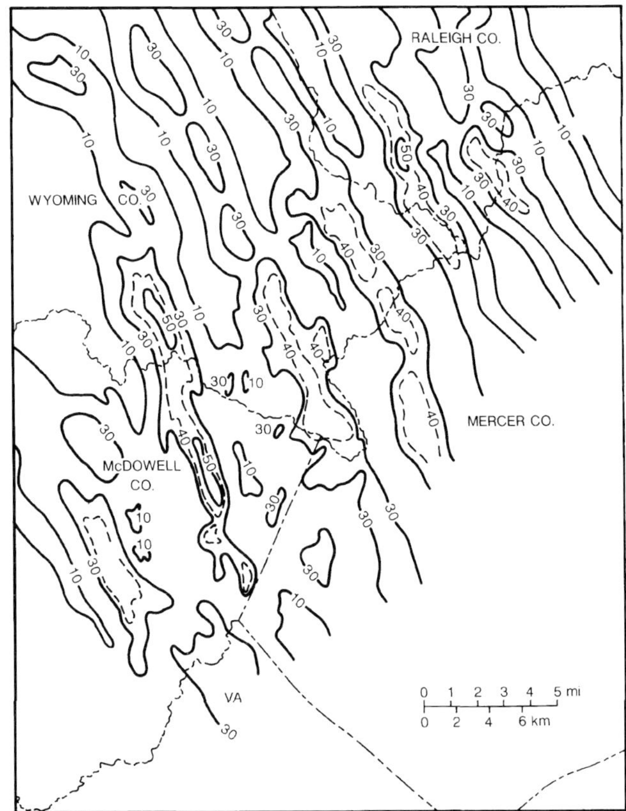

Figure 7. Isopach map of the Union oolite member (informal). The oolite thicks are interpreted as northwest-trending tidal bars. They average 40 ft (12 m) in thickness, 4500 ft (1.4 km) in width, and 20 mi (32 km) in length. Contour interval is 20 ft, with 40-ft contour lines dashed within the ooid bars. The member thins to 0 between some tidal bars.

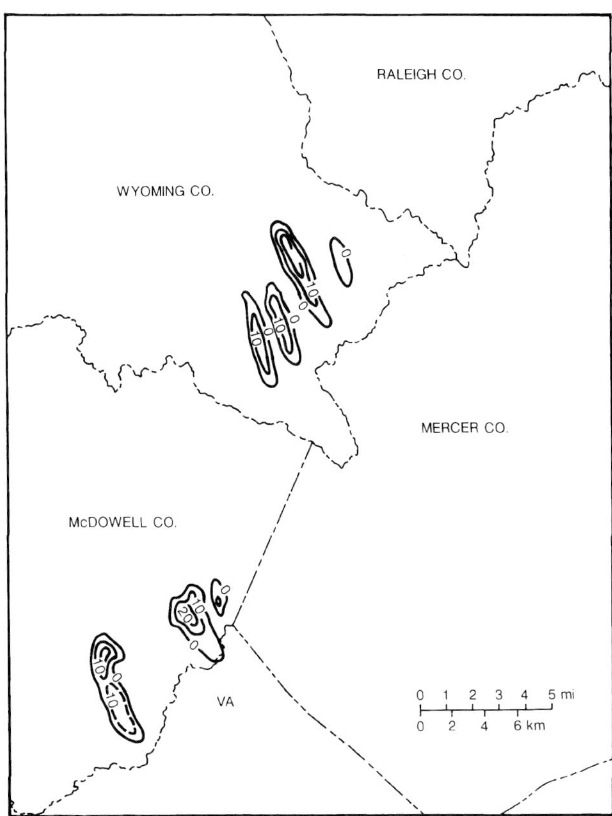

Figure 8. Isopach map of net pay in the Pickaway oolite member, calculated as the stratigraphic interval with a bulk density equal to or less than 2.6 g/cc (equivalent to a porosity of 6% or more). Contour interval is 10 ft (3.0 m).

from 2 to 22 ft (1–7 m), but stratigraphic cross sections depict the entire oolite member to be 20–25 ft (6–8 m) thick. The northern bars are at least 3600 ft (1.1 km) wide, 3.2 mi (5.1 km) long, and have a 2400-ft (730-m) spacing. Thus, in comparison to those of the overlying Union Limestone, Pickaway ooid bars are significantly smaller and not as areally extensive. Pickaway oolites in outcrops of east-central West Virginia, 160 km (100 mi) to the north, contain sedimentary structures, textures, and bed geometries interpreted as having been produced by tidal currents (Smosna and Koehler, this volume), supporting indirectly a tidal origin of ooid sand shoals in the Rhodell field.

DIAGENETIC MODEL

The net-pay isopach map of the Union oolite (Figure 9) is constructed using a porosity cutoff of 6%. Significant porosity exists along the axes of the northern tidal bars, where thicknesses exceed 30 ft (9 m) (compare with Figure 5). On flanks of these bars and in areas between bars, however, the oolite member is generally nonporous. This distribution indicates that porosity developed only on the exposed crests of the oolite shoals, when relative sea level fell and meteoric water entered the sediments (compare with Ahr, 1989).

Only the northern group of Union ooid bars is productive. Tidal bars in the southwestern region of our study area are nonporous and nonproductive and may not have been subjected to meteoric diagenesis (although no cores are available from these wells for petrological examination). They formed in somewhat deeper water, as determined by their poorly defined geometry, and perhaps remained in the marine environment when relative sea level dropped slightly. This portion of the tidal-bar belt thus never underwent the freshwater diagenesis necessary for porosity development as did the productive (and shallower) northern region.

To explain the origin of porosity and permeability, a closer examination of the rocks' diagenetic history is needed. The density log for well Wyo-1048 (Figure 4), from which 10 sidewall cores were taken, indicates that porosity occurs within both the Union and Pickaway oolite members. The temperature curve,

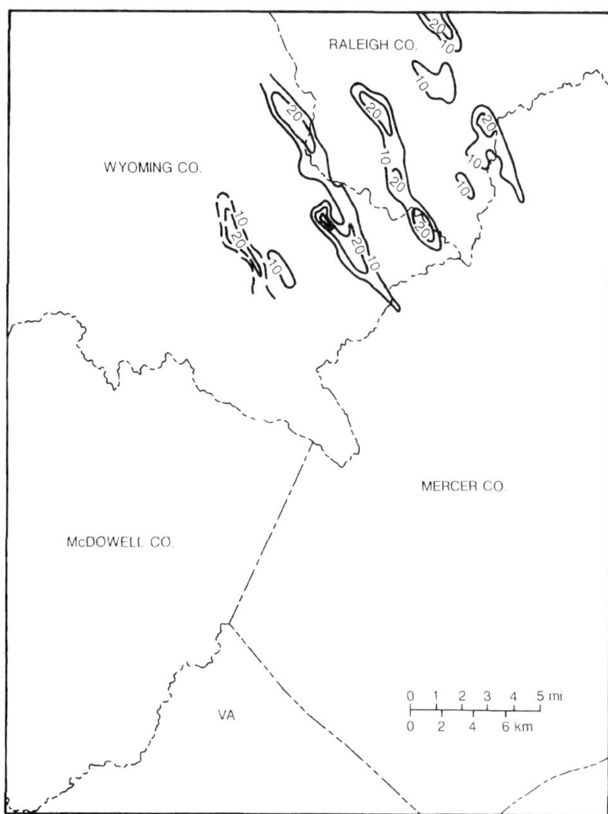

Figure 9. Isopach map of net pay in the Union oolite member, calculated as the stratigraphic interval with a bulk density equal to or less than 2.6 g/cc (equivalent to a porosity of 6% or more). Contour interval is 10 ft (3.0 m).

though, shows that gas is entering the borehole from the Pickaway oolite member; thus, in this well natural permeability is notably greater in the lower unit.

Samples from the Union porous zone are ooid grainstones, but interparticle pore space has been filled by two stages of calcite cement (Figure 6A). Intraooid porosity, however, does exist. Ooids are now composed of microrhombic calcite, and micropores within the ooids range from 1 to 5 μm in diameter. Minor porosity also occurs in leached fossils and between ooids. Union ooids have small, point-to-point contacts, and connections between the intraparticle pore spaces of neighboring grains are few in number. Because the pathway for gas migration is highly tortuous, this unit has low permeability (generally less than 0.01 md).

Samples from the Pickaway porous zone of the same well are similar ooid grainstones. They too have little interparticle porosity but significant intraparticle porosity. Furthermore, inside a single ooid the micropores have relatively large throats and numerous connections to adjacent pores (Figure 10A). This unit differs from the Union in one important aspect: Pickaway ooids here have suffered chemical compaction that has increased the contact area between individual ooids (Figure 6B). This close-packing arrangement, termed fitted fabric, greatly enhances the rock's permeability (up to 0.15 md) by increasing the available pathways for gas migration between ooids.

Both oolite members were perforated in well Wyo-1048, and the initial potential was 4.2 MMCFGD. It is still too early to perform a decline-curve analysis with accuracy, but during the first six months the well produced 126.5 MMCFG with little decline. This early production history suggests that the well may ultimately produce more than 2.0 bcf.

Cementation

Ooids were locally cemented (first stage) by the precipitation of an isopachous cement shortly after deposition (Figures 6A, 10B). This cement occurs as blades or equant crystals 8–30 μm in size that fringe the ooids and other grains; rarely the rims meet along polygonal boundaries. It is interpreted to have formed in the marine environment similar to modern isopachous cements of high-magnesium calcite described by Shinn (1975) and Harris (1979). Aggregates of ooids were also produced by cementation in this diagenetic environment. Where it is relatively abundant, this first-stage marine cement occluded 13–17% of the interparticle pore space and effectively inhibited later chemical compaction in the ooid zones.

Thin-section analyses from Wyo-1048 samples show that marine cementation was generally more common in the Union ooid grainstones than in Pickaway grainstones. Moreover, its distribution in the Pickaway is spotty; cement fringes are discontinuous or absent around many grains. Previous studies have demonstrated that the degree of marine cementation can vary directly with the saturation level of the water for $CaCO_3$ and sediment permeability and indirectly with sedimentation rate and water agitation (Bathurst, 1975). Union and Pickaway samples with significant volumes of marine cement (2–4% of the total rock volume) also possess comparatively high percentages of oolitically coated grains, hinting that the distribution of marine cement was controlled by water-saturation level.

At some later time a second generation of cement precipitated in the Greenbrier ooid grainstones. Calcite overgrowths on crinoid fragments and meniscus (Figure 6C) and equant cements precipitated to fill all remaining interparticle pore space. The morphologies of these cements suggest that they formed in meteoric-vadose and meteoric-phreatic environments (Harris, 1979; Halley and Harris, 1979).

Mineral Stabilization

In both oolite members diagenesis by meteoric water resulted in the dissolution of aragonitic gastropods (now calcite-filled molds) and the mineral stabilization of ooids to low-magnesium calcite. The (probable) precursor aragonite needles of the ooids

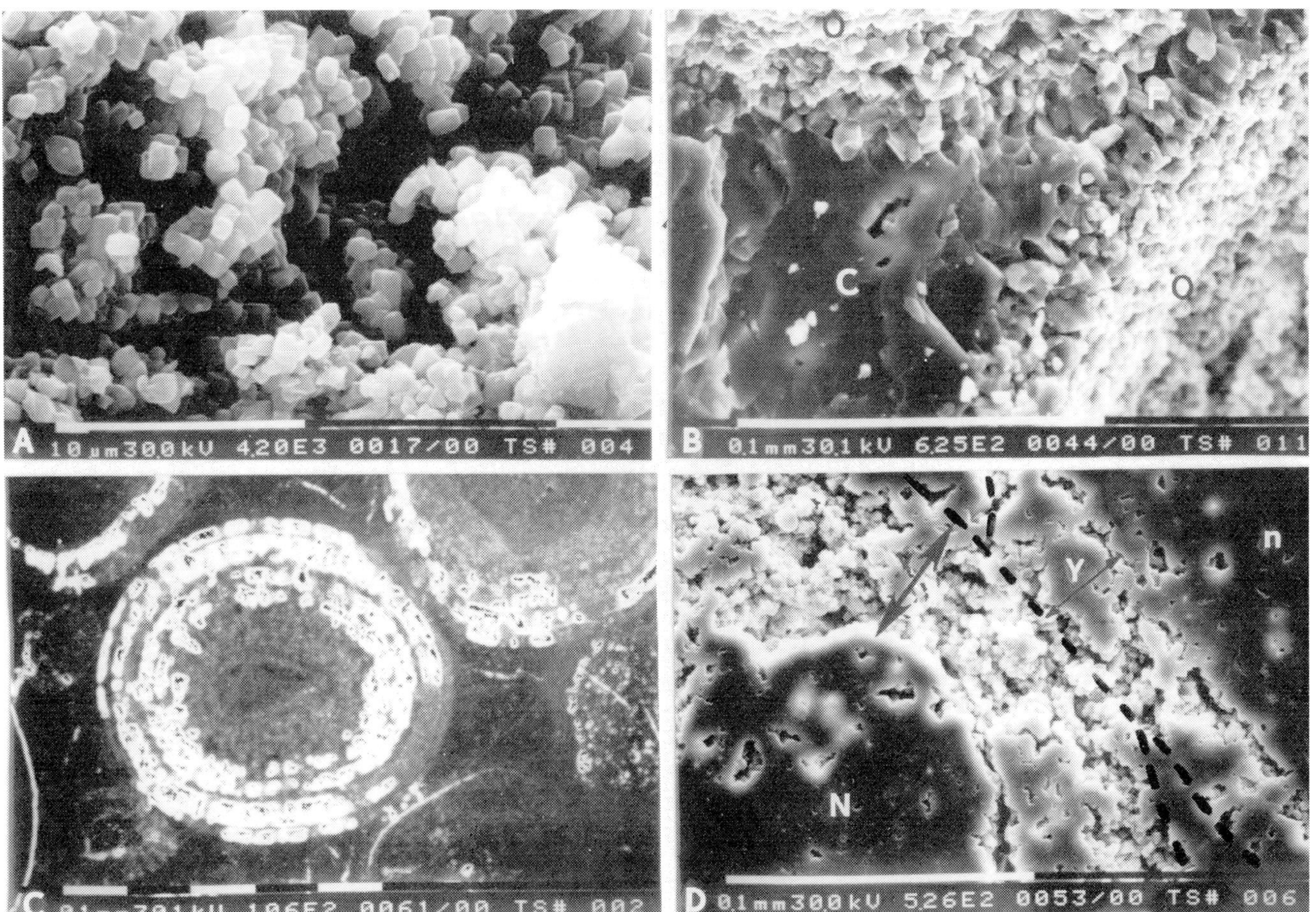

Figure 10. SEM photomicrographs of Greenbrier ooid grainstones. White bar scales are at bottom of pictures (10 µm or 0.1 mm). (A) Microrhombic calcite that now composes the ooids. Intraooid microporosity is quite high, and the pores have relatively large throats and numerous interconnections. (B) Parts of two ooids (O) composed of microrhombic calcite. The first-generation cement is an isopachous fringe (F, upper right), which was followed by coarse void-filling calcite cement (C). (C) Microporosity within the ooids (white areas) occurs along dissolved laminae of the cortex. However, the small point-to-point contacts between these porous grains yield a low permeability. Polished sample. (D) Illustration of the long interconnection created by a fitted fabric between pores of one ooid and those of the next. Arrow X indicates the porous cortex of one ooid (nucleus in lower left marked by N), and arrow Y, the porous cortex of its neighbor (nucleus in upper right marked by n). Boundaries of the two grains are marked by dashes. The permeability pathway here is 0.085 mm wide. Polished sample.

inverted to calcite with a microrhombic crystal habit (1–5 µm). This process of inversion is assumed to have taken place by reaction with fresh waters, the same waters that leached aragonitic fossils and precipitated the second-generation cements. Comparable interpretations for the recrystallization of porous ooids in other limestone reservoirs have been offered by Keith and Pittman (1983) and Ahr (1989).

Chemical analysis by SEM energy-dispersive analyzer shows that Greenbrier ooids have a very low and uniform distribution of the trace elements strontium, iron, and magnesium. One might expect the distribution and abundance of these elements to reflect the original mineralogy of the ooids. However, mineral stabilization during meteoric diagenesis was apparently complete, and these elements must have been flushed from the system.

Intraparticle microporosity in Greenbrier ooids resulted from the consumption of aragonite needles to provide material for the equant calcite rhombs, which are similar in length but much fatter than the original aragonite crystals (Lasemi and Sandberg, 1984; Moshier, 1989). Porosity is therefore intercrystalline—between the calcite microrhombs. Moreover, some aragonite laminae were completely dissolved (Figure 10C). Selective leaching of layers within an ooid cortex led to the formation of collapsed and spalled ooids and to the development of missing laminae (additional intraparticle microporosity). These dissolution textures in ooids are attributed to meteoric diagenesis (Halley and Harris, 1979; Wilkinson et al., 1984).

Chemical Compaction

After mineral stabilization the rocks were subjected to eogenetic chemical compaction. Concave-convex contacts between ooids are rather common in the Pickaway oolite member of well Wyo-1048, and grains commonly exhibit a fitted fabric; that is, they are in contiguity with neighboring grains all along their margins (Figure 6B). Fitted fabric results from the percolation of undersaturated meteoric-vadose water, resulting in dissolution at grain contacts and a closer packing of the grains (Clark, 1979; Hird and Tucker, 1988). In addition, chemical compaction led to grain-to-grain pressure solution and ooids with concave-convex contacts. Closer packing increases the natural permeability of the Pickaway oolite by juxtaposing and interconnecting the microporosity of adjacent ooids at long grain contacts (Figures 6B, 10D). In this way a continuous pathway for gas migration is formed.

Neighboring grains with a fitted fabric are separated in places by a thin fringe of first-stage cement, a porous 10-µm lamina of aphanocrystalline calcite (Figures 6B, 10B). Involvement in chemical compaction illustrates that this cement pre-dated development of the fitted fabric. Because the fringe is itself porous, permeability pathways from one ooid to another are not adversely affected.

Chemical compaction occurred more commonly in the Pickaway member of well Wyo-1048 because this oolite had not been extensively cemented previously in the eogenetic marine zone. Marine cement is relatively rare in grainstones with fitted fabric (accounting for an average of 0.8% of the total rock volume), discontinuous around the carbonate grains, and irregularly distributed throughout the rock. In contrast, the Union oolite was subjected to significant early marine cementation and had the strength to resist later chemical compaction when flushed with undersaturated meteoric-vadose water. Marine cements are more common and uniformly distributed around the grains. Consequently, Union ooids in well Wyo-1048 have only small point contacts and low natural permeabilities (Figure 10C).

In an attempt to quantify the degree of chemical compaction in the oolite members of this one well, thin sections were point counted to determine (1) the number of grain contacts per ooid and (2) the different types of grain contacts. Pickaway ooids, which underwent significant compaction, exhibit 1–7 contacts per grain, with an average value of 3.9, and two-thirds of the contacts are fitted fabric. Union ooids, on the other hand, have an average value of 3.0 contacts per grain (23% fewer contacts), and more than half of these are point-to-point. Other, less common grain contacts are straight or long, concave-convex, and sutured (compare with Taylor, 1950).

Additional point-count data illustrate that compaction is related to the presence or absence of first-generation cement (Figure 11). Grain fraction (GF) is defined as the volume of constituent grains in a rock as a percent of total rock volume (the complement of intergranular cement plus porosity), and this value is a function of the amount of compaction (Smosna, 1989). In particular, grain fraction in the 10 Greenbrier samples from sidewall cores increases linearly as the volume of first-generation cement decreases. Those ooid grainstones with the least cement (0–2%) suffered the greatest amount of chemical compaction (GF = 82–88%). Conversely, grainstones with 3–4% first-generation cement—which strengthened the rock and retarded compaction—have grain fractions of 74–80%. By way of comparison, newly deposited ooids with no compaction whatever have an initial grain fraction of near 65% (Coogan, 1970).

CONCLUSIONS

Prominent dip-oriented thicks on the isopach map of the Union oolite member suggest the presence of a major tidal-bar belt. Individual sand bars are approximately 40 ft (12 m) thick, 4500 ft (1.4 km) wide, and 20 mi (32 km) long. Bars of the underlying Pickaway are somewhat smaller. These two tidal-bar belts developed along a hinge line that separated the basin to the southeast from a shelf on the northwest. The belts stretch for almost 20 mi (32 km) parallel with the hinge line, and future exploration should be concentrated where this line extends to the northeast and southwest.

Ooid grainstones of the tidal bars underwent several stages of diagenesis that strongly influenced the rocks' final porosity and permeability. Most importantly, during meteoric diagenesis aragonite needles of the ooids stabilized to microrhombs of low-magnesium calcite with considerable intragranular microporosity. Permeability exists, however, only where

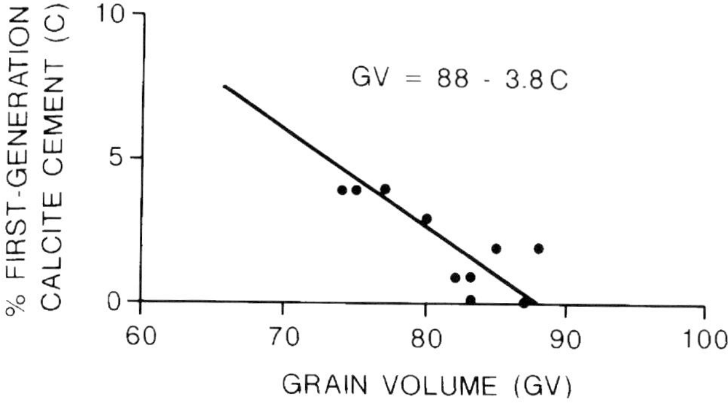

Figure 11. Graph of grain volume (GV) vs. percent first-generation calcite cement (C) for 10 samples from sidewall cores in well Wyo-1048. Grain volume in these ooid grainstones, as an indicator of the degree of chemical compaction, is inversely related to the amount of early cement. (Coefficient of deter-

meteoric-vadose chemical compaction produced a fitted fabric between adjacent ooids. The fitted fabric establishes long grain-to-grain contacts and enhances the flow of natural gas from one microporous ooid to the next. Ooid shoals in the southern part of the study area may have been deposited in slightly deeper water and perhaps spent little time exposed to meteoric water; these bars are consequently nonporous and nonproductive.

ACKNOWLEDGMENTS

We wish to thank Brian Keith, Stephen Moshier, Charles Zuppann, and Lee Avary, who reviewed the paper and made helpful suggestions for its improvement. Allison Hanham drafted the illustrations.

REFERENCES CITED

Ahr, W.M., 1989, Early diagenetic microporosity in the Cotton Valley Limestone of east Texas: Sedimentary Geology, v. 63, p. 275-292.

Avary, K.L., M.C. Behling, R.F. Kleinschmidt, D.G. Patchen, and N.L. Simcoe, 1989, Preliminary oil and gas drilling statistics for West Virginia during 1988: Twentieth Annual Appalachian Petroleum Geology Symposium, West Virginia Geological and Economic Survey, Section 3, 10 pp.

Ball, M.M., 1967, Carbonate sand bodies of Florida and the Bahamas: Journal of Sedimentary Petrology, v. 37, p. 556-591.

Bathurst, R.G.C., 1975, Carbonate sediments and their diagenesis: Amsterdam, Elsevier Publishing Company, Second edition, 658 p.

Carney, C., and R. Smosna, 1989, Carbonate deposition in a shallow marine gulf, the Mississippian Greenbrier Limestone of the central Appalachian basin: Southeastern Geology, v. 30, p. 25-48.

Cecil, C.B., 1990, Paleoclimate controls on stratigraphic repetition of chemical and siliclastic rocks: Geology, v. 18, p. 533-536.

Clark, D.N., 1979, Patterns of porosity and cement in ooid reservoirs in Dogger (Middle Jurassic) of France: discussion: American Association of Petroleum Geologists Bulletin, v. 63, p. 676-677.

Coogan, A.H., 1970, Measurements of compaction in oolitic grainstone: Journal of Sedimentary Petrology, v. 40, p. 921-929.

Ettensohn, F.R., 1985, Controls on development of Catskill delta complex basin-facies, *in* D.L. Woodrow, and W.D. Sevon, (eds.), The Catskill delta: Geological Society of America Special Paper 201, p. 65-75.

Halley, R.B., and P.M. Harris, 1979, Fresh-water cementation of a 1000-year-old oolite: Journal of Sedimentary Petrology, v. 49, p. 969-988.

Halley, R.B., P.M. Harris, and A.C. Hine, 1983, Bank margin, *in* P.A. Scholle, D.G. Bebout, and C.H. Moore, (eds.), Carbonate depositional environments: American Association of Petroleum Geologists Memoir 33, p. 463-506.

Harris, P.M., 1979, Facies anatomy and diagenesis of a Bahamian ooid shoal: Sedimenta VII, University of Miami, Florida, 163 p.

Hird, K., and M.E. Tucker, 1988, Contrasting diagenesis of two Carboniferous oolites from South Wales: a tale of climatic influence: Sedimentology, v. 35, p. 587-602.

Keith, B.D., and E.D. Pittman, 1983, Bimodal porosity in oolitic reservoir—effect on productivity and log response, Rodessa Limestone (Lower Cretaceous), East Texas basin: American Association of Petroleum Geologists Bulletin, v. 67, p. 1391-1399.

Lasemi, Z., and P.A. Sandberg, 1984, Transformation of aragonite-dominated lime muds to microcrystalline limestones: Geology, v. 12, p. 420-423.

Moshier, S.O., 1989, Microporosity in micritic limestones: a review: Sedimentary Geology, v. 63, p. 191-213.

Overbey, W.K., 1967, Lithologies, environments, and reservoirs of the Middle Mississippian Greenbrier Group in West Virginia: Producers Monthly, v. 31, p. 25-32.

Quinlan, G.M., and C. Beaumont, 1984, Appalachian thrusting, lithospheric flexure, and the Paleozoic stratigraphy of the eastern interior of North America: Canadian Journal of Earth Science, v. 21, p. 973-996.

Ruley, E.E., 1970, "Big Injun" oil and gas production in north-central West Virginia: American Association of Petroleum Geologists Bulletin, v. 54, p. 758-782.

Scotese, C.R., R.K. Bambach, C. Barton, R. Van der Voo, and A.M. Ziegler, 1979, Paleozoic base maps: Journal of Geology, v. 87, p. 217-277.

Shinn, E.A., 1975, Polygonal cement sutures from the Holocene: a clue to recognition of submarine diagenesis (abs.): American Association of Petroleum Geologists and Society of Economic Paleontologists and Mineralogists, Annual Meetings Abstracts, v. 2, p. 68.

Smosna, R., 1989, Compaction law for Cretaceous sandstones of Alaska's North Slope: Journal of Sedimentary Petrology, v. 59, p. 572-584.

Taylor, J.M., 1950, Pore-space reduction in sandstone: American Association of Petroleum Geologists Bulletin, v. 34, p. 701-716.

Wilkinson, B.H., C. Buczynski, and R.M. Owen, 1984, Chemical controls of carbonate phases: implications from Upper Pennsylvanian calcite-aragonite ooids of southeastern Kansas: Journal of Sedimentary Petrology, v. 54, p. 932-947.

Youse, A.C., 1964, Gas producing zones of Greenbrier (Mississippian) Limestone, southern West Virginia and eastern Kentucky: American Association of Petroleum Geologists Bulletin, v. 48, p. 465-486.

Chapter 13

Ooid Mineralogy and Diagenesis of the Pitkin Formation, North-Central Arkansas

Ezat Heydari
Department of Geology and Geophysics
Louisiana State University
Baton Rouge, Louisiana, USA

Ronald D. Snelling
Basin Research Institute
Louisiana State University
Baton Rouge, Louisiana, USA

William C. Dawson
Texaco Inc., E & P Technology Division
Houston, Texas, USA

Maria L. Machain
Instituto De Ciencias Del Mar Y Liminol
Mexico City, Mexico

ABSTRACT

The Pitkin Formation is a Late Mississippian (Chesterian) high-energy marine oolitic-bioclastic limestone. The original mineralogy of Pitkin ooids varied across the Pitkin carbonate shelf. In the eastern part of the study area, originally aragonite ooids deposited in shoal and shoreface environments. These ooids are presently replaced by calcite and exhibit high Sr and low Mg concentrations. Originally calcite ooids in overlying lagoonal facies are composed of well-preserved, radial fabric. These ooids contain low Sr and high Mg values. Similar characteristics indicate that ooids in the central and western part of the study area were all originally calcite.

An early marine nonferroan low-Mg calcite cement occurs as fibrous-to-bladed circumgranular crust, equant mosaic pore fill, and syntaxial overgrowth on echinoderm bioclasts. An extensive early dissolution event related to meteoric diagenesis dissolved aragonite and high- and low-Mg calcite components and created moldic and vuggy porosity. A late burial ferroan calcite occurs as mosaic and poikilotopic cements filling intergranular and moldic pores. Early marine and late burial calcite cementation eliminated virtually all porosity within Pitkin grainstones.

INTRODUCTION

It is demonstrated that original mineralogy of ancient ooids and marine cements has changed cyclically since the Phanerozoic time (Sandberg, 1983; Wilkinson et al., 1985). Mineralogical changes of marine abiotic carbonates have implications for changes in the chemistry of oceans and atmosphere through time. In addition, because calcite behaves differently from aragonite, original mineralogy of a carbonate grainstone has significant influence on reservoir potential of carbonate rocks (Moore, 1989).

Sedimentology and stratigraphy of the Pitkin Limestone are well documented in north-central Arkansas (Jehn and Young, 1976; Handford, 1986). However, diagenetic studies of this unit have not previously been reported. The purpose of this study is to document ooid mineralogy, cementation history, and porosity evolution of the Pitkin Limestone in Stone and Searcy counties in north-central Arkansas (Figure 1). It has been demonstrated that ooid mineralogy has had an important role in the porosity evolution of the Late Jurassic Smackover Formation (Moore and Druckman, 1981; Moore et al., 1986; Swirydczuk, 1988; Moore, 1989). Ooid mineralogy and diagenetic study of Pitkin grainstones should enhance our understanding of processes that lead to porosity preservation, generation, and destruction.

METHODS

Petrographic observations were made on standard polished petrographic thin sections, stained with alizarin red-s and K-ferricyanide. Samples for scanning electron microscopy (SEM) were polished and etched with formic acid for 10 sec (Lasemi and Sandberg, 1984).

Samples for chemical analyses were powdered and x-rayed from 24° to 32° 2Θ to detect 100% intensity peaks of calcite, dolomite, anhydrite, celestite, and siderite. Powdered samples (100 mg) were washed in distilled and deionized water five times, dissolved in 0.5 N HCl, and analyzed by Inductively Coupled Plasma (ICP) for Sr and Mg. Instrument detection limits for both elements are below 0.1 ppm. The relative precision of the analyses is 2.5% for Mg and 0.8% for Sr.

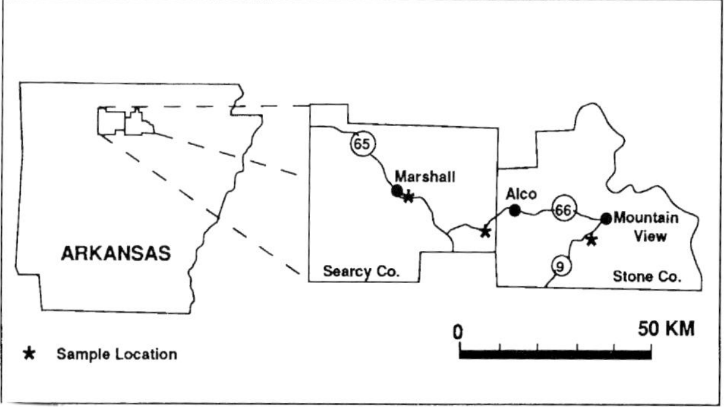

Figure 1. Map of the study area. Sample locations are marked with star.

STRATIGRAPHY AND DEPOSITIONAL ENVIRONMENT

The Pitkin Limestone and the underlying Fayetteville Shale deposited on a shallow, stable carbonate shelf in northern Arkansas during Late Mississippian (Chesterian) (Glick, 1979; Handford, 1986). The Pitkin Limestone was measured, described, and sampled in three road cuts in Stone and Searcy counties, Arkansas, one south of the town of Mountain View along Arkansas Highway 5, one west of the town of Alco along Arkansas Highway 66, and one near the town of Marshall along U.S. Highway 65 (Figure 1). These are referred to as the Mountain View, Alco, and Marshall sections. Stratigraphic sections presented here are modified from those reported by Handford (1986) from the same outcrops.

At the Mountain View section, the Pitkin Limestone overlies the Fayetteville Shale with a gradational contact. The lower part of the Pitkin Limestone consists of ooid grainstone with plane laminations thought to represent shoal and shoreface facies (Handford, 1986). This facies grades upward to ooid grainstones with gently dipping cross-laminations interpreted as foreshore facies. These units are followed by a spillover ooid grainstone containing bidirectional cross-beds, algal mound, shales, and finally by another ooid grainstone (Figure 2).

At the Alco section, the Pitkin Limestone is composed of, from base to top, crinoidal grainstone with gently dipping laminations, ooid-crinoid-bryozoan grainstone containing bidirectional cross-beds, ooid-bryozoan packstone, and a thick crinoid-bryozoan-brachiopod packstone (Figure 2). This sequence is interpreted as ooid shoal at the base and bryozoan mound at the top.

At the Marshall section, the Pitkin Limestone consists of ooid-crinoid grainstones that display large, unidirectional planar cross-beds followed by ooid-crinoid-bryozoan-brachiopod grainstone with bidirectional cross-beds. The lower part of this section is interpreted as ooid shoal and the upper part as a tidal inlet deposit.

The Pitkin and Fayetteville formations represent a single shoaling-upward progradational system (Jehn and Young, 1976; Handford, 1986). The lower shale member of the Fayetteville Formation deposited on a deep shelf, and its upper limestone member on a storm-dominated, muddy shelf (Handford, 1986). Ooid grainstones of the lower Pitkin Limestone deposited in a lower shoreface environment that prograded over the Fayetteville Shale. Upper Pitkin strata deposited in lagoonal and marsh environments and contain leaf molds and fossilized wood (Jehn and Young, 1976).

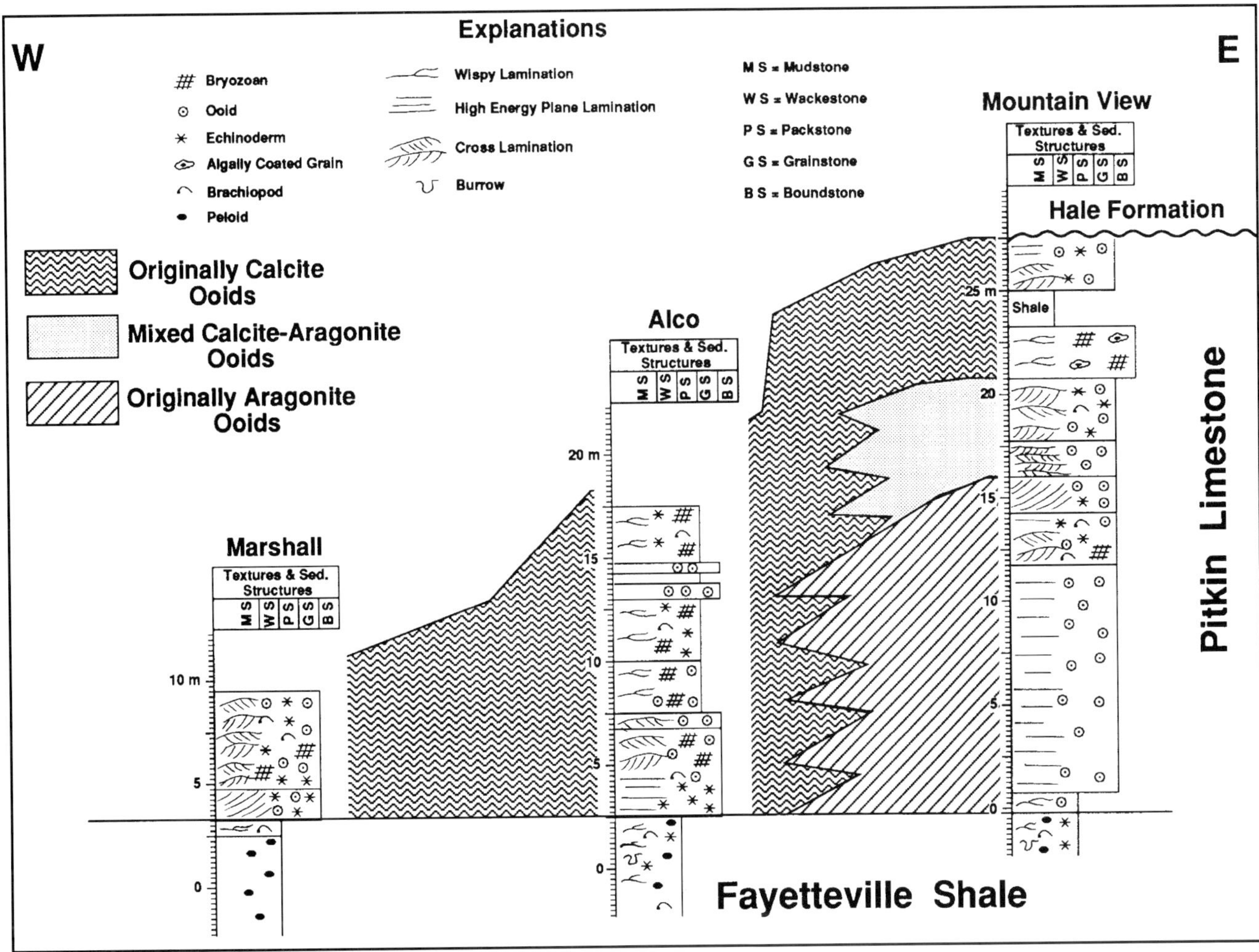

Figure 2. Composite measured stratigraphic sections of the Pitkin and Fayetteville formations at Mountain View, Marshall, and Alco localities. Distribution of aragonite and calcite ooids are shown in space and time.

OOID MINERALOGY AND FABRIC

The fabrics of modern ooids are composed of tangentially or radially oriented crystals of either aragonite, high-Mg calcite, or a combination of both mineralogies. The more common fabric is concentric growth of tangentially oriented aragonite crystals. Ooids with tangential fabric are described from Baffin Bay, Texas (Land et al., 1979), the Bahama Bank (Bathurst, 1975), the Florida shelf (Major et al., 1988), the Persian Gulf (Loreau and Purser, 1973), and the Yucatan shelf, Mexico (Bathurst, 1975).

Ooids with concentric growth of radially oriented crystals are less common. Although cortexes of all modern tangential ooids are aragonite, those of radial ooids may be composed of either aragonite or high-Mg calcite or both. High-Mg calcite radial ooids are reported from the Great Barrier Reef, Australia (Marshall and Davies, 1975), the Amazon shelf, Brazil (Milliman and Barretto, 1975), Baffin Bay, Texas (Land et al., 1979), and the Florida shelf (Major et al., 1988). Aragonite radial ooids are described from the Lizard Island, Australia (Davies and Martin, 1976), the Great Salt Lake, Utah (Sandberg, 1975a), the Persian Gulf (Loreau and Purser, 1973), and the Red Sea (Friedman et al., 1973).

Stabilization of Carbonate Minerals

Studies of low-Mg calcite fauna (e.g., brachiopods) indicate that this mineral generally retains its original microfabrics and chemistry during diagenesis (Al-Aasm and Veizer, 1982; 1986). Therefore, it may be inferred that ooids with preserved fabric were originally calcite (Sandberg, 1975a; Wilkinson et al., 1984, 1985).

The microtextures of originally aragonite fauna (e.g., gastropods and corals) are lost by dissolution of metastable aragonite and reprecipitation of calcite during diagenesis (Sandberg et al., 1973). It may, therefore, be generally inferred that ooids that have lost their primary microtextures were originally aragonite (Sandberg, 1975a; 1983; Wilkinson et al., 1984,

1985). Diagenesis of originally aragonite ooids may create moldic porosity (Moore and Druckman, 1981; Moore et al., 1986; Budd, 1988), calcite mosaic texture (Sandberg, 1983, 1975a; Wilkinson et al., 1984, 1985), or bricklike calcite fabric (Assereto and Folk, 1976; Tucker, 1985).

Stabilization textures of fauna and flora composed of high-Mg calcite are complex. Although when viewed through a petrographic microscope the microtextures of high-Mg calcite allochems appear to be preserved, SEM observations indicate that microstructural details of these fauna and flora have, in fact, been modified (Sandberg, 1975b; Oti and Muller, 1985). Bryozoans (4–5 mole % $MgCO_3$) lose their original laminated cell walls and are typically replaced by elongated calcite crystals (Sandberg, 1975b). Modern red algae (14 mole % $MgCO_3$) contain micron-sized crystals aligned normal and parallel to the cell wall (Oti and Muller, 1985). During diagenesis these crystals are dissolved and replaced by equant micron-sized crystals of low-Mg calcite (Oti and Muller, 1985). Foraminifera (15 mole % $MgCO_3$) apparently retain their original wall microtexture during stabilization (Towe and Hemleben, 1976). Transformation of echinoderm fragments (10–15 mole % $MgCO_3$) to low-Mg calcite may be accompanied by formation of microdolomite (Lohmann and Meyers, 1977). Observations made during this study indicate that limited generation of moldic porosity has occurred by dissolution of high-Mg calcite. Because experimental studies by Plummer and MacKenzie (1974) and Walter (1985) suggest that high-Mg calcite with >12 mole % $MgCO_3$ dissolves at least as rapidly as aragonite, it is not entirely clear why stabilization of high-Mg calcite fauna does not seem to produce moldic porosity as commonly as the dissolution of aragonite fauna.

Because of the variability in stabilization textures of high-Mg calcite fauna, it is not clear whether high-Mg calcite ooids will assume different textures when transformed to low-Mg calcite. Some ooids with well-preserved radial texture exhibit strong luminescence and contain high concentrations of Fe and Mn. This implies that the ooids were originally unstable and acquired Fe and Mn during stabilization in Fe- and Mn-rich waters (Richter and Fuchtbauer, 1978). However, luminescent ooids with preserved original fabrics have been reported from the Florida shelf by Major et al. (1988), indicating that this criterion may not always be reliable.

Ancient Ooids

It has been suggested that mineralogy and possibly fabric of marine ooids have changed cyclically over geologic time (Sandberg, 1983; Wilkinson et al., 1985; Wilkinson and Given, 1986). Secular variation in original ooid mineralogy corresponds roughly with major eustatic sea level changes (Sandberg, 1983; Wilkinson et al., 1985). Calcite ooids dominated during sea level highs; aragonite ooids during sea level lows. Sandberg (1983) related secular variation of abiotic carbonates to climatic cycles as proposed by Fischer (1981). By this interpretation, the mineralogy of marine abiotic carbonate changed in response to variations in partial pressure of CO_2 and possibly variation of Mg/Ca in marine water (MacKenzie and Pigott, 1981; Sandberg, 1983; Wilkinson et al., 1985).

MINERALOGY AND FABRIC OF PITKIN OOIDS

Mountain View Section

Petrographic observations indicate that Pitkin ooids were composed originally of two mineralogies in the Mountain View section. Ooid cortexes in the lower part of the section exhibit mosaic calcite and bricklike calcite textures (Figure 3). These ooids are interpreted to have been originally aragonite. Ooids with similar textures have been reported from the Mississippian of Georgia by Rich (1982). Original aragonite fabric of these ooids is uncertain.

Microtextural details of ooids in the upper part of the Mountain View section are well preserved (Figure 4). These ooids are interpreted as having been originally calcite. Most cortexes have radial concentric fabric, and some are composed of equigranular micron-sized calcite crystals (Figure 5). A transition zone containing both originally calcite and aragonite ooids is present in the middle of the Mountain View section (Figure 2).

Calcite ooids in the Mountain View section are generally larger than aragonite ooids (mean diameter 0.7 mm and 0.1 mm, respectively) (Figure 6). Reasons for this size difference between the two ooid types are not apparent. The problems associated with the size of marine oolites are discussed by Carozzi (1972) and Bathurst (1975). The size of abiotic coated grains depends upon rate of growth, rate of sedimentation, and probably the energy of the environment. The originally aragonite ooids formed in the shoreface environment of Pitkin ooid shoals where high sedimentation rates prevailed. These ooids may not have remained on the sea floor long enough to grow large. Calcite ooids formed in lagoons, behind the shoals, with slower sedimentation rate but constant agitation.

Alco and Marshall Sections

All ooids in the Alco and Marshall sections have well-preserved, radial fabrics (Figure 2). These ooids are interpreted to have been originally calcite. However, cortexes of some ooids in the Marshall section are composed of loosely packed micron-sized calcite crystals (i.e., chalky). The chalky ooids are soft and friable. The chalky texture is similar to that developed during stabilization of red algae (Oti and Muller, 1985) and some high-Mg calcite mudstones (Moshier, 1989). It is not clear whether the original fabric of ooids with chalky textures in radial fabric of high-Mg calcite that was subsequently altered to microcrystalline calcite or whether they were com-

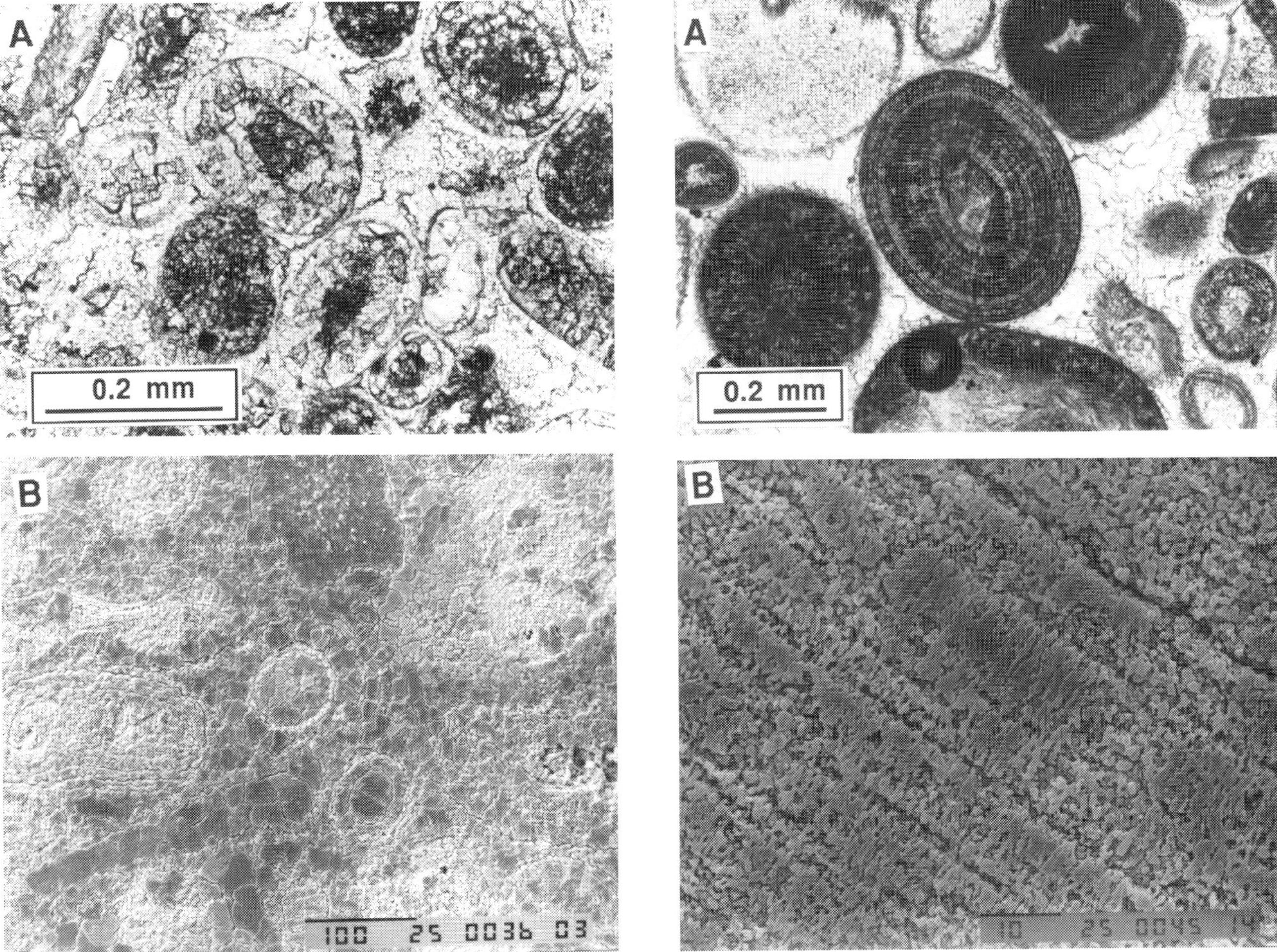

Figure 3. Originally aragonitic ooids (Mountain View locality). Photomicrograph (A) and SEM image of polished and etched sample (B) (100 μm scale bar in the data strip).

Figure 4. Originally calcite ooids (Mountain View locality). Photomicrograph (A) and SEM image of polished and etched sample (B) (10 μm scale bar in the data strip).

posed originally of high-Mg calcite micrite that was transformed to the present chalky appearance.

Trace Element Geochemistry

Interpretations of original calcite and aragonite compositions for Pitkin ooids are verified by Sr and Mg contents of the two ooid types. Abiotic aragonite precipitated from sea water contains about 10,000 ppm Sr, while low-Mg calcite and high-Mg calcite contain only 1000 ppm and 2000 ppm, respectively (Kinsman, 1969; Morrow and Mayers, 1978; Land et al., 1979). Some loss of original trace element content occurs during stabilization (Brand and Veizer, 1980). Still, diagenetically stabilized carbonates that had an aragonite precursor tend to contain more Sr than stabilized low- and high-Mg calcite (Veizer and Demovic, 1974).

The Sr composition of the inferred aragonite Pitkin ooids ranges from 1400 ppm to 2700 ppm (mean = 1855 ppm). Sr values of originally calcite ooids range from 740 ppm to 1400 ppm (mean = 1044 ppm). Mg composition of originally aragonite ooids ranges from 1000 ppm to 3600 ppm (mean = 1787 ppm). The Mg content of originally calcite ooids ranges from 2100 ppm to 5700 ppm (mean = 4878 ppm). Trace element data (Figure 7) support the interpretations made by petrographic observations.

Original mineralogies of the Pitkin ooids varied both vertically and laterally in the sections studied. This indicates that calcite and aragonite ooids precipitated simultaneously in different locations along the Pitkin carbonate shelf. Carbonate shelves containing ooids of different mineralogies have been described from the Holocene (Land et al., 1979), the Pleistocene (Major et al., 1988), the Jurassic (Moore et al., 1986; Swirydczuk, 1988), the Pennsylvanian (Wilkinson et al., 1984), the Late Cambrian (Chow and James, 1987), the late Precambrian (Singh, 1987), and the middle Proterozoic (Tucker, 1984). The question is: what controls mineralogy during precipitation of ooids on a given carbonate platform?

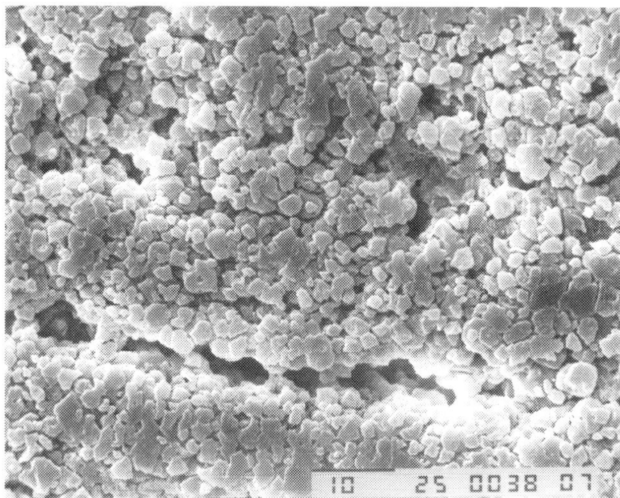

Figure 5. SEM photograph of chalky ooids from the Mountain View locality (10 μm scale bar in the data stip).

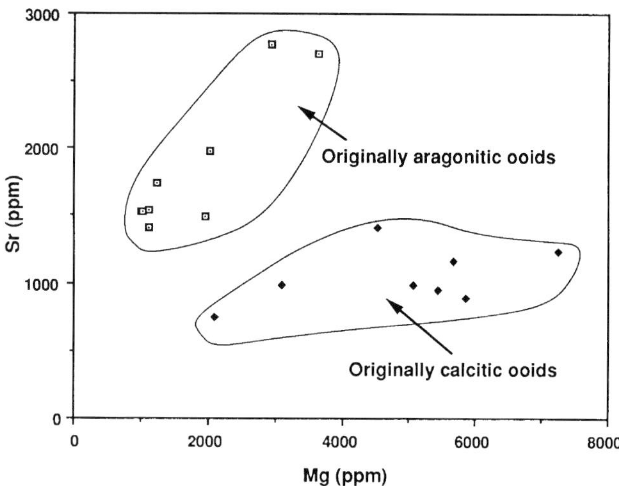

Figure 7. Crossplot of Sr and Mg content of originally calcitic and aragonitic ooids.

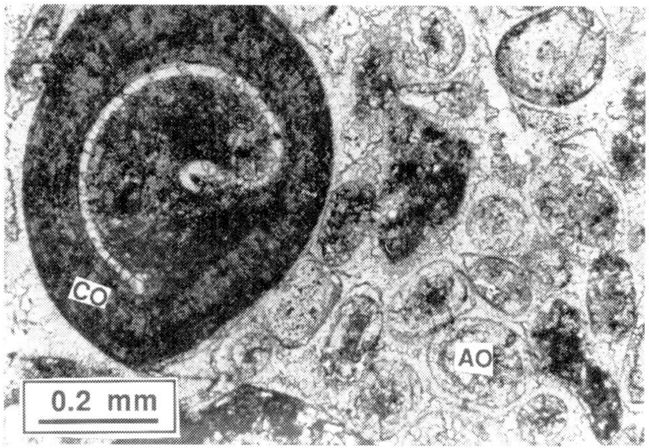

Figure 6. Representative size differences between originally calcite ooids (CO) and aragonite ooids (AO). Plane light photomicrograph.

FACTORS CONTROLLING OOID MINERALOGY AND FABRIC

Previously cited factors that influence mineralogy of abiotic carbonate minerals are: (1) Mg/Ca ratio (Folk, 1974; Berner, 1975); (2) partial pressure of CO_2 (MacKenzie and Pigott, 1981); (3) temperature (Burton and Walter, 1987); (4) rate of CO_3^{-2} supply (Lahann, 1978; Given and Wilkinson, 1985); (5) organic compounds (Kitano and Kanamori, 1966; Suess and Futterer, 1972); and (6) salinity. It is unlikely, however, that organic compounds influence ooid mineralogy in a normal marine setting because concentrations of organic compounds used in experimental studies are significantly higher than those present in sea water. Effect of salinity on ooid mineralogy and texture is not satisfactorily investigated. Because high temperature favors aragonite precipitation (Burton and Walter, 1987), water temperature variations across a carbonate shelf, particularly in an arid environment, could conceivably affect the mineralogy of coated grains. The most commonly accepted factors controlling the mineralogy of abiotic carbonates are partial pressure of CO_2, rate of CO_3^{-2} supply, and Mg/Ca ratio.

However, experimental studies (Berner, 1975; Burton and Walter, 1987) indicate that neither Mg/Ca ratio nor the rate of CO_3^{-2} supply can totally inhibit aragonite precipitation when sea water is supersaturated relative to both calcite and aragonite. At the present time, no one kinetic variable seems to explain the observed distributions of ooid mineralogy of the Pitkin Formation. It is possible that local variations in saturation state of late Mississippian sea water and/or combination of kinetic factors resulted in the distribution of calcite and aragonite ooids. Certainly, two paleoenvironments of ooid formation existed side by side for a considerable period of time, one producing only calcite and the other only aragonite ooids. In addition, a transition zone resulted from physical mixing of calcite and aragonite ooids from the separate paleoenvironments. This is supported by laminae of originally calcite ooids alternating with originally aragonite ooids. Alternatively, calcite and aragonite ooids in the transition zone could have formed in place by changing the environmental conditions causing alternate formation of calcite or aragonite ooids. Individual bimineralic ooids (ooids containing both aragonite and calcite in their cortexes) are present in the Mountain View section, but they are not very common.

Calcite ooids of the Pitkin Limestone are luminescent, in contrast to brachiopod fragments, which are non-luminescent. This indicates that the originally calcite ooids probably contained some $MgCO_3$ and acquired luminescence during diagenesis. Chalky ooids may have had higher Mg contents than well-preserved radial ooids. This is because of the similarity of the texture of the chalky ooids to that of the red algae.

DIAGENESIS OF THE PITKIN LIMESTONE

Marine, meteoric, and burial diagenesis can play a significant role in porosity evolution of carbonate rocks (James and Choquette, 1984; Choquette and James, 1987; Moore, 1989). In addition, diagenetic processes can cause near total destruction of hydrocarbons in a reservoir (Heydari and Moore, 1989). In the Mountain View section, the Pitkin Limestone has been affected by three major diagenetic events: an episode of early nonferroan calcite cementation, a period of pervasive dissolution, and an episode of late ferroan calcite cementation.

Early nonferroan low-Mg calcite occurs as fibrous-to-bladed circumgranular crusts, equant mosaic cement, and as syntaxial overgrowths on echinoderm fragments (Figure 8). It stains pink with alizarin red-s and K-ferricyanide and is non-luminescent. These characteristics suggest that nonferroan calcite cement precipitated either in well-oxygenated conditions (during marine diagenesis) or in reducing conditions with Mn absent (meteoric diagenesis). Nonferroan calcite cement was extensively dissolved during subsequent meteoric dissolution (Figure 8), suggesting that this cement precipitated during marine diagenesis.

A Late Mississippian sea level drop (Vail et al., 1977) exposed the Pitkin Limestone and resulted in an episode of meteoric diagenesis. Aragonite grains were totally dissolved. High-Mg calcite such as echinoderm fragments and even nonferroan circumgranular and equant marine calcite cements were partly to completely dissolved (Figure 9).

The pores created by dissolution during the meteoric diagenesis are filled by a ferroan calcite cement (Figures 9, 10). This cement occurs as mosaic and poikilotopic porefill, stains purple with alizarin red-s and K-ferricyanide, and luminesces dark brown. Ferroan calcite is inferred to have been precipitated from an Fe-rich pore water in a reducing environment and postdates grain fracturing (Figure 10). This cement must have precipitated after the meteoric dissolution event and continued through burial diagenesis. Minor precipitation of saddle dolomite is the only other burial diagenetic event that affected the Pitkin Limestone.

In the Mountain View section, grain density is low in most ooid grainstones, and grain-to-grain pressure solution is uncommon. This indicates that massive early cementation of Pitkin grainstones inhibited physical and chemical compaction during progressive burial. Most intergranular porosity in the grainstones was destroyed by early marine cementation. Porosity generated during early meteoric dissolution was subsequently cemented by ferroan calcite cement. Pitkin grainstones are presently nonporous and nonpermeable, except for minor microporosity in chalky ooids.

The effects of early meteoric diagenesis on porosity generation and destruction of Pitkin oolitic grainstones are the reverse of those in the oolitic grainstones in the Late Jurassic Smackover Formation (Moore and Druckman, 1981; Swirydczuk, 1988; Moore, 1989). Early diagenesis affecting aragonite ooids in the northern part of the Smackover carbonate shelf resulted in significant generation of moldic porosity. Calcite ooids in the southern part of the Smackover shelf resulted in the preservation of intergranular porosity despite slight early diagenesis.

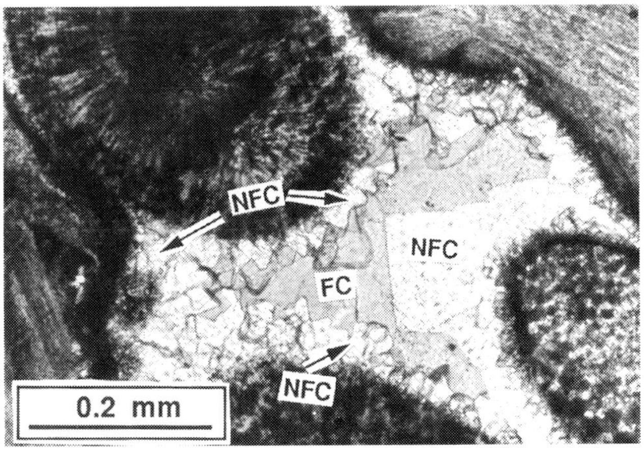

Figure 8. Early, non-ferroan calcite cement (NFC) and late, ferroan calcite cement (FC). Note corroded edges of ferroan calcite cement. Plane light photomicrograph of stained thin section.

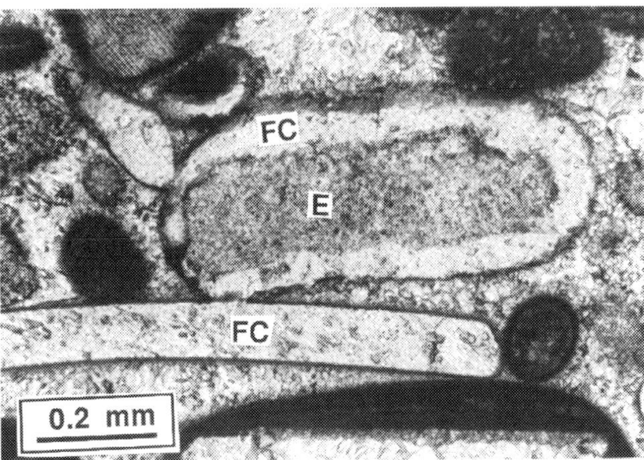

Figure 9. Dissolution of originally aragonitic and high Mg-calcite echinoderm fragments (E). FC: Ferroan calcite cement. Plane light photomicrograph of stained thin section.

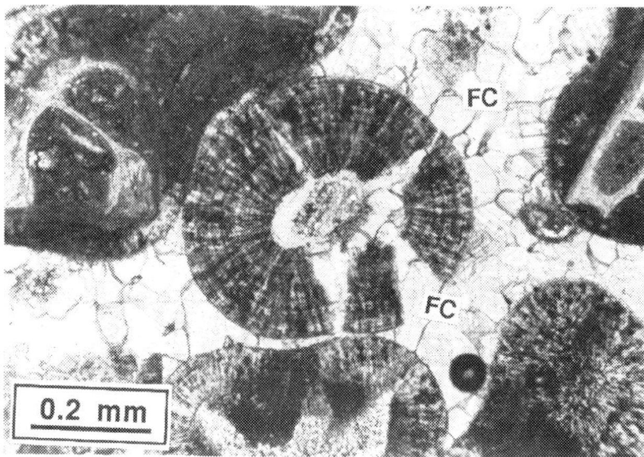

Figure 10. Ferroan calcite cement (FC) postdating grain fracturing. Plane light photomicrograph of stained thin section.

These differences in Pitkin and Smackover ooid grainstones demonstrate the complexities of meteoric diagenesis and thereby indicate that subaerial exposure should not uncritically be assumed to have resulted in significant porosity generation.

CONCLUSIONS

Petrographic and geochemical analyses of the Pitkin Limestone in three locations (Mountain View, Alco, and Marshall) in north-central Arkansas indicate that ooids were originally of two mineralogies: aragonite and calcite. Ooid mineralogy apparently varied both normal and parallel to shoreline along the Pitkin shelf.

Pitkin grainstones were affected by three main diagenetic events: (1) precipitation of nonferroan calcite occurring as fibrous-to-bladed circumgranular crusts, equant mosaic cements, and syntaxial overgrowths on echinoderm fragments during marine diagenesis; (2) dissolution of aragonite, high-Mg calcite grains, and ferroan low-Mg calcite cement during intense meteoric diagenesis; and (3) precipitation of late ferroan calcite cement during burial diagenesis.

ACKNOWLEDGMENTS

We thank Robert Handford for his assistance in geologic and sedimentologic interpretations of the area. Discussions with Brian Carter and Clyde H. Moore were helpful. Thanks to Wanda LeBlanc for x-ray diffraction analyses and Kerry Lyle and Dana Maxwell for photography. Reviews by Robert Handford, Brian Keith, Brenda Kirkland-George, Clyde H. Moore, Bruce H. Wilkinson, Bill Wade, and Charles W. Zuppann significantly improved the manuscript. This research was supported by grants from Chevron Petroleum Company, Applied Carbonate Research Program, and the Department of Geology and Geophysics of the Louisiana State University. We thank Texaco, Inc. for granting permission to publish this manuscript.

REFERENCES CITED

Al-Aasm, I. S., and J. Veizer, 1982, Chemical stabilization of low-Mg calcite: an example of brachiopods: Journal of Sedimentary Petrology, v. 52, p. 1101-1109.

Al-Aasm, I. S., and J. Veizer, 1986, Diagenetic stabilization of aragonite and low-Mg calcite, I. Trace elements in rudists: Journal of Sedimentary Petrology, v. 56, p. 138-152.

Assereto, R., and R. L. Folk, 1976, Brick-like texture and radial rays in Triassic pisolites of Lombardy, Italy: A clue to distinguish ancient aragonitic pisolites: Sedimentary Geology, v. 16, p. 205-222.

Bathurst, R. G. C., 1975, Carbonate sediments and their diagenesis: New York, Elsevier Scientific Company, 658 p.

Berner, R. A., 1975, The role of magnesium in the crystal growth of calcite and aragonite from seawater: Geochimica et Cosmochimica Acta, v. 39, p. 489-504.

Brand, U., and J. Veizer, 1980, Chemical diagenesis of a multicomponent carbonate system—1: Trace elements: Journal of Sedimentary Petrology, v. 50, p. 1219-1236.

Budd, D. A., 1988, Petrographic products of freshwater diagenesis in Holocene ooid sands, Schooner Cays, Bahamas: Carbonates and Evaporites, v. 3, p. 143-163.

Burton, E. A., and L. M. Walter, 1987, Relative precipitation rates of aragonite and Mg calcite from sea water: Temperature or carbonate ion control: Geology, v. 15, p. 111-114.

Carozzi, A. V., 1972, Microscopic sedimentary petrography: Kringer Publishing, Huntington, New York, 485 p.

Choquette, P. W., and N. P. James, 1987, Diagenesis #12: diagenesis in limestone—3. The deep burial environment: Geoscience Canada, v. 14, p. 3-35.

Chow, N., and N. P. James, 1987, Facies-specific, calcitic and bimineralic ooids from middle and Upper Cambrian platform carbonates, western Newfoundland, Canada: Journal of Sedimentary Petrology, v. 57, p. 907-921.

Davies, P. J., and K. Martin, 1976, Radial aragonite ooids, Lizard Island, Great Barrier Reef, Queensland, Australia: Geology, v. 4, p. 120-122.

Fischer, A. G., 1981, Climatic oscillations in the biosphere, *in* M. Nitecki, ed., Biotic crises in geological and evolutionary time: New York, Academic Press, p. 103-131.

Folk, R. L., 1974, The natural history of crystalline calcium carbonate: Effect of magnesium content and salinity: Journal of Sedimentary Petrology, v. 44, p. 40-53.

Friedman, G. M., A. J. Amiel, M. Braun, and D. S. Miller, 1973, Generation of carbonate particles and laminites in algal mats—Example from sea-marginal hypersaline pool, Gulf of Aqaba, Red Sea: AAPG Bulletin, v. 57, p. 541-557.

Given, R. K., and B. H., Wilkinson, 1985, Kinetic control of morphology, composition, and mineralogy of abiotic sedimentary carbonates: Journal of Sedimentary Petrology, v. 55, p. 109-119.

Glick, E. E., 1979, Arkansas: U. S. Geological Survey Professional Paper 1010, p. 125-145.

Handford, C. R., 1986, Facies and bedding sequences in shelf-storm-deposited carbonates—Fayetteville Shale and Pitkin Limestone (Mississippian), Arkansas: Journal of Sedimentary Petrology, v. 56, p. 123-137.

Heydari, E., and C. H. Moore, 1989, Burial diagenesis and thermochemical sulfate reduction, Smackover Formation, southeastern Mississippi salt basin: Geology, v. 17, p. 1080-1084.

James, N. P., and P. W. Choquette, 1984, Diagenesis 9. Limestone—the meteoric diagenetic environment: Geoscience Canada, v. 11, p. 161-194.

Jehn, P. J., and L. M. Young, 1976, Depositional environments of the Pitkin Formation, northern Arkansas: Journal of Sedimentary Petrology, v. 46, p. 377-386.

Kinsman, D. J. J., 1969, Interpretation of Sr^{+2} concentrations in carbonate minerals and rocks: Journal of Sedimentary Petrology, v. 39, p. 486-508.

Kitano, Y., and N. Kanamori, 1966, Synthesis of magnesian calcite at low temperatures and pressures: Geochemical Journal, v. 1, p. 1-10.

Land, L. S., E. W. Behrens, and S. A. Frishman, 1979, The ooids of Baffin Bay, Texas: Journal of Sedimentary Petrology, v. 49, p. 1269-1278.

Lahann, R. W., 1978, A chemical model for calcite crystal growth and morphology control: Journal of Sedimentary Petrology, v. 48, p. 337-344.

Lasemi, Z., and P. A. Sandberg, 1984, Transformation of aragonite-dominated lime muds to microcrystalline limestones: Geology, v. 12, p. 420-423.

Lohmann, K. C., and W. J. Meyers, 1977, Microdolomite inclusions in cloudy prismatic calcites: a proposed criterion for former high-magnesium calcites: Journal of Sedimentary Petrology, v. 47, p. 1078-1088.

Loreau, J., and B. H. Purser, 1973, Distribution of Holocene ooids in the Persian Gulf, *in* B. H. Purser, ed., The Persian Gulf: New York, Springer-Verlag, p. 279-328.

MacKenzie, F. T., and J. D. Pigott, 1981, Tectonic controls of Phanerozoic sedimentary rock cycling: Geological Society of London Journal, v. 138, p. 183-196.

Major, R. P., R. B. Halley, and K. J. Lukas, 1988, Cathodoluminescent bimineralic ooids from the Pleistocene of the Florida continental shelf: Sedimentology, v. 35, p. 843-855.

Marshall, J. F., and P. J. Davies, 1975, High-magnesium calcite ooids from the Great Barrier Reef: Journal of Sedimentary Petrology, v. 45, p. 285-291.

Milliman, J. D., and H. T. Barretto, 1975, Relict magnesian calcite oolite and subsidence of the Amazon shelf: Sedimentology, v. 22, p. 137-145.

Moore, C. H., 1989, Carbonate diagenesis and porosity: New York, Elsevier, 338 p.

Moore, C. H., and Y. Druckman, 1981, Burial diagenesis and porosity evolution, Upper Jurassic Smackover, Arkansas and Louisiana: AAPG Bulletin, v. 65, p. 597-628.

Moore, C. H., A. Chowdhury, and E. Heydari, 1986, Variation of ooid mineralogy in Jurassic Smackover limestones as control of ultimate diagenetic potential (abstract): AAPG Bulletin, v. 70, p. 622.

Morrow, D. W., and I. R. Mayers, 1978, Simulation of limestone diagenesis—a model based on strontium depletion: Canadian Journal of Earth Science, v. 15, p. 376-396.

Moshier, S. O., 1989, Microporosity in micritic limestones: a review: Sedimentary Geology, v. 63, p. 191-213.

Oti, M. and G. Muller, 1985, Textural and mineralogical changes in coralline algae during meteoric diagenesis: An experimental approach: Neues Jahrbuch Miner. Abh., v. 151, p. 163-195.

Plummer, L. N., and F. T. MacKenzie, 1974, Predicting mineral solubility from rate data: Application to magnesian calcite: American Journal of Science, v. 274, p. 61-283.

Rich, M., 1982, Ooid cortices composed of neomorphic pseudospar: possible evidence for ancient originally aragonitic ooids: Journal of Sedimentary Petrology, v. 52, p. 843-847.

Richter, D. K., and H. Fuchtbauer, 1978, Ferroan calcite replacement indicates former magnesian calcite skeletons: Sedimentology, v. 25, p. 843-860.

Sandberg, P. A., 1975a, New interpretations of Great Salt Lake ooids and of ancient non-skeletal carbonate mineralogy: Sedimentology, v. 22, p. 497-537.

Sandberg, P. A., 1975b, Bryozoan diagenesis: Bearing on the nature of the original skeleton of rugose corals: Journal of Paleontology, v. 49, p. 587-606.

Sandberg, P. A., 1983, An oscillating trend in Phanerozoic non-skeletal carbonate mineralogy: Nature, v. 305, p. 19-22.

Sandberg, P. A., Schneidermann, N., and Wunder, S. J., 1973, Aragonitic ultrastructural relics in calcite-replaced Pleistocene skeletons: Nature Physical Science, v. 245, p. 133–134.

Singh, U., 1987, Ooids and cements from the Late Precambrian of the Flinders Ranges, South Australia: Journal of Sedimentary Petrology, v. 57, p. 117-127.

Suess, E., and D. Futterer, 1972, Aragonitic ooids: experimental precipitation from seawater in the presence of humic acid: Sedimentology, v. 19, p. 129-139.

Swirydczuk, K., 1988, Mineralogical control on porosity type in upper Jurassic Smackover ooid grainstones, southern Arkansas and northern Louisiana: Journal of Sedimentary Petrology, v. 58, p. 339-347.

Towe, K. M., and C. Hemleben, 1976, Diagenesis of magnesian calcite: Evidence from miliolacean foraminifera: Geology, v. 4, p. 337-339.

Tucker, M. E., 1984, Calcitic, aragonitic and mixed calcitic-aragonitic ooids from the mid-Proterozoic Belt Supergroup, Montana: Sedimentology, v. 31, p. 627–644.

Tucker, M. E., 1985, Calcitized aragonitic ooids and cements from the late Precambrian Biri Formation of southern Norway: Sedimentary Geology, v. 43, p. 67-84.

Vail, P. R., R. M. Mitchum, and S. Thompson, III, 1977, Seismic stratigraphy and global changes of sea level, part 3: Relative changes of sea level from coastal onlap, *in* C. E. Payton, ed., Seismic stratigraphy—application to hydrocarbon exploration: AAPG Memoir 26, p. 63-97.

Veizer, J. and R. Demovic, 1974, Strontium as a tool in facies analysis: Journal of Sedimentary Petrology, v. 44, p. 93-115.

Walter, L. M., 1985, Relative reactivity of skeletal carbonates during dissolution: Implications for diagenesis, *in* N. Schneidermann and P. M. Harris eds., Carbonate cements: SEPM Special Publication 36, p. 3-16.

Wilkinson, B. H., C. Buczynski, and R. M. Owen, 1984, Chemical control of carbonate phases: Implications from upper Pennsylvanian calcite-aragonite ooids of southeastern Kansas: Journal of Sedimentary Petrology, v. 54, p. 932-947.

Wilkinson, B. H., and R. K. Given, 1986, Secular variation in abiotic marine carbonates: Constraints on Phanerozoic carbon dioxide content and oceanic Mg/Ca ratios: Journal of Geology, v. 94, p. 321-333.

Wilkinson, B. H., R. M. Owen, and A. R. Carroll, 1985, Submarine hydrothermal weathering, global eustasy, and carbonate polymorphism in Phanerozoic marine oolites: Journal of Sedimentary Petrology, v. 55, p. 171-183.

Chapter 14

Oolite Shoals of the Mississippian St. Louis Formation, Gray County, Kansas: A Guide for Oil and Gas Exploration

Kerry D. Parham
Kansas Geological Survey
Lawrence, Kansas, USA

Peter G. Sutterlin
Wichita State University, Wichita
Kansas, USA

ABSTRACT

The productive ooid shoals of the St. Louis Formation (Mississippian, Meramecian) in southwestern Kansas have been interpreted as representing linear ramp barrier-type deposits that developed southwest of, and parallel to, a southeasterly trending shoreline. However, examination of available cores and interpretation of petrophysical log data from the Ingalls field in Gray County suggest that production is from an oolite shoal situated on the leeward side of a small "island" mudflat. Positive magnetic and gravity anomalies associated with the Ingalls field imply deep structures that might have resulted in subtle perturbations on the Mississippian sea floor that in turn provided loci for ooid shoal formation. Classified lithologically as an oolitic grainstone, this principal productive facies exhibits primary intergranular porosity with evidence of only minor diagenetic porosity enhancement. A possible implication for future exploration is that St. Louis Formation oolitic buildups may be present in pairs, occurring on both the leeward and basinward sides of coastal plain "islands" that might have formed an archipelago extending into the Hugoton embayment of the Anadarko basin.

INTRODUCTION

This subsurface investigation encompasses a 12×24 mi (19×39 km) area in northwestern Gray County, Kansas, that lies along the eastern flank of the Hugoton embayment, a northern extension of the Anadarko basin (Figure 1). Encountered at a depth of 4850 ft (1480 m), the B zone of the Upper Mississippian St. Louis Formation (Meramecian) is the main oil-producing interval in the area (Figure 2).

Many St. Louis oil fields in Kansas produce from oolitic grainstone reservoirs that lie along apparent southeasterly trends (Figure 3) aligned somewhat with the eastern flank of the Hugoton embayment

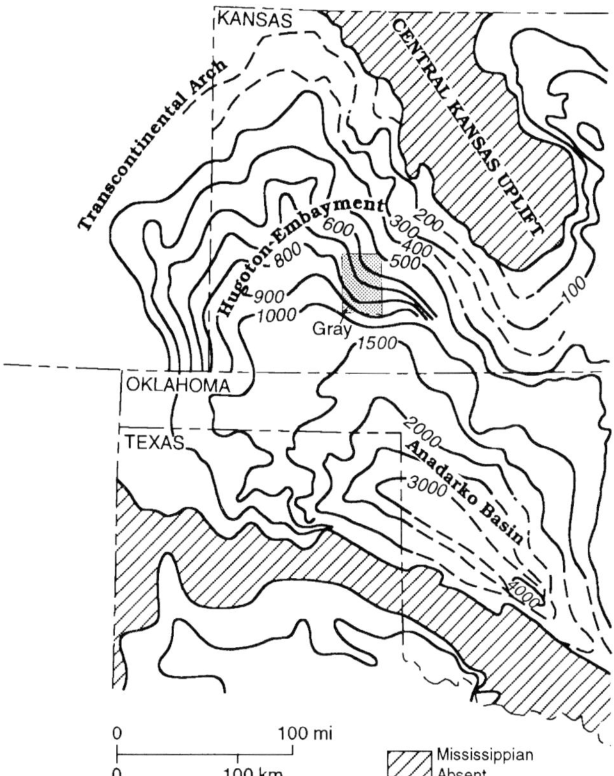

Figure 1. Isopach map (in ft) of Mississippian rocks in the Hugoton Embayment and Anadarko Basin; parts of Kansas, Oklahoma, Texas, and eastern Colorado (modified from Huffman, 1959).

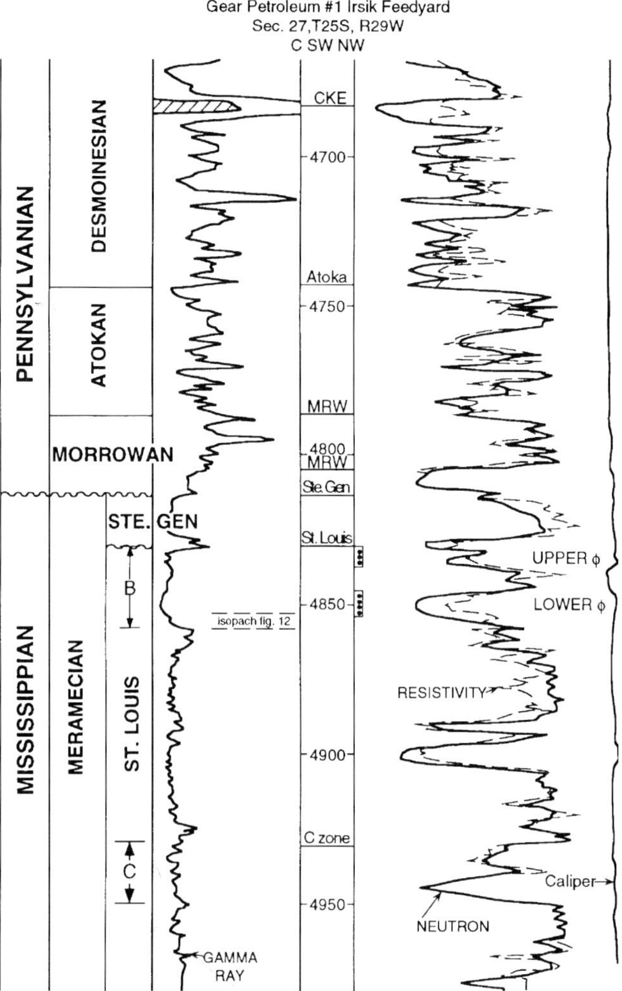

Figure 2. Ingalls field type log (see Figure 3 for location of field). Open-hole radiation-guard petrophysical log from Gear Petroleum #1 Irsik Feed-Yard, C SW NW, Sec. 27, T25S, R29W. Completed February 1980 from the St. Louis upper and lower B zone intervals with an initial potential pumping of 131 BOPD (measurements are in ft).

(Figure 1). Therefore, in southwestern Kansas, these deposits have been interpreted as linear ramp barrier-type shoals, which developed southwest of, and parallel to, a southeasterly trending shoreline (c.f. Irwin, 1965; Handford, 1988). However, on the basis of this study, the Ingalls field in Gray County has been interpreted as an oolite shoal situated on the leeward side of a small "island" mudflat.

The stratigraphically trapped 24-well Ingalls field, the largest oil field in Gray County and the focus of this study, has produced more than 1.5 million barrels of primary recovered oil from the B zone and is now in the early stages of a waterflood. The Ingalls field has two larger and well-studied analogs located in neighboring Finney County—the Pleasant Prairie field (Ball, 1966) and the Damme field (Handford, 1988) (Figure 3)—to which comparisons have been made. Similarities also are noted with Modern Bahamian oolites.

Data for this study were derived primarily from petrophysical (wireline) logs and geological reports from wells that penetrated Mississippian strata in the study area. Fourteen stratigraphic cross sections, in both dip and strike directions, were constructed throughout the area. Additional primary sources of data were samples of drill cuttings and 4-in. (10-cm) cores from two producing wells and two dry holes. From these cores, 14 thin sections were made and examined. Nine photomicrographs of thin sections from well cuttings of seven other wells in the area were available (Figure 4).

GENERAL STRUCTURE AND STRATIGRAPHY

Mississippian strata in southwestern Kansas are truncated progressively to the north by an angular unconformity (Merriam, 1963; Thompson and Goebel, 1968). In Gray County, the Mississippian St. Louis Formation dips generally southeasterly at a rate of about 50 ft/mi (9.5 m/km). Subtle, sinuous folding has resulted in a series of small, 1-mi-wide (1.6-km-wide) anticlines, gently plunging to the southeast and usually 4–5 mi (6.5–8 km) apart. An interpretation of proprietary Vibroseis data, collected east–west across

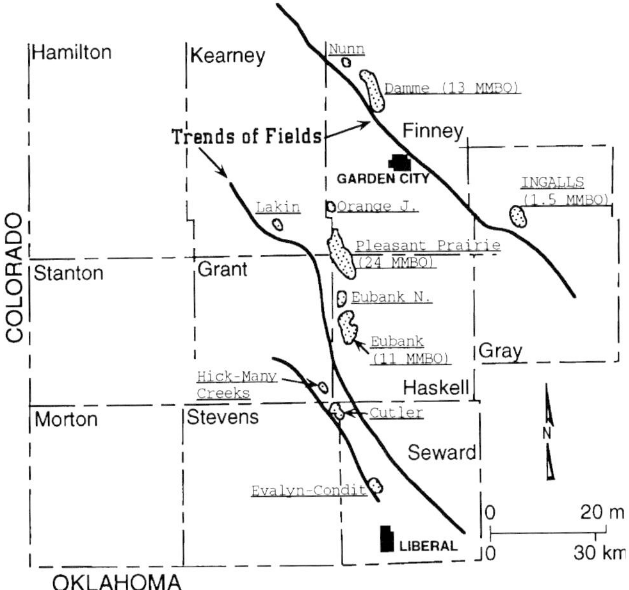

Figure 3. Mississippian oil fields and their apparent trends in the Hugoton embayment in southwestern Kansas. Cumulative production figures for larger fields also shown (modified from Handford, 1988).

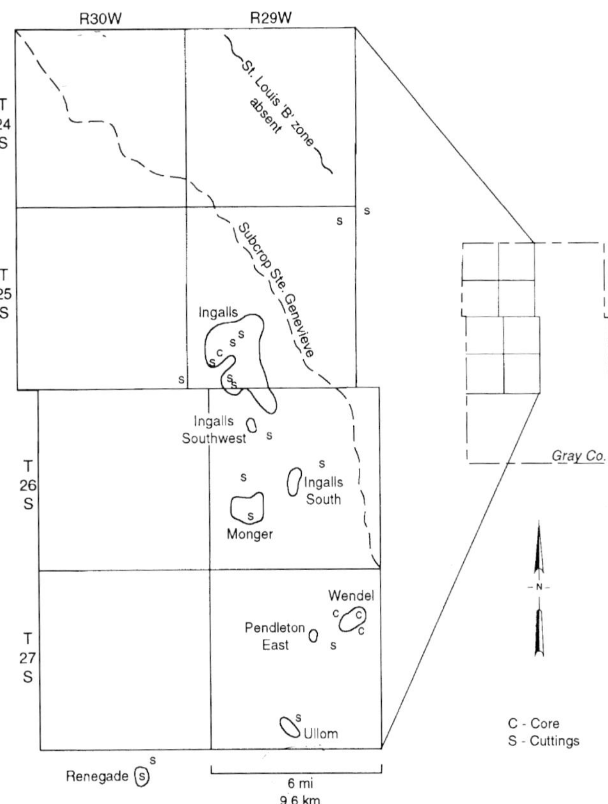

Figure 4. Location of Gray County oil fields that produce principally from the Mississippian St. Louis Formation. The B zone is generally absent, as a result of erosion, northeast of Ste. Genevieve subcrop line.

the center of Gray County, suggests the existence of northwesterly trending pre-Pennsylvanian normal faults with displacements of 20–50 ft (6–15 m) that are 2–6 mi (3–10 km) apart.

Lithology and Stratigraphy of the St. Louis Formation

The overall thickness of the St. Louis Formation in southwestern Kansas ranges from 50 to 200 ft (15–61 m). In the Ingalls field area, the St. Louis Formation is underlain by the Salem Formation (Meramecian) and overlain (where not absent because of erosion) by the St. Genevieve Formation (Chesterian) (Zeller, 1968) (Figures 2, 4). From the base, a complete St. Louis section consists of:

1. sucrosic dolostone beds
2. lime wackestones, packstones, brecciated limestone and anhydrite beds
3. discontinuous coarse oolitic grainstone beds, designated the C zone
4. lime wackestone and packstone beds, in part siliciclastic
5. oolitic grainstones, designated the B zone, which are divided into zones of "upper" and "lower" porosity by a non-porous chert or dense limestone
6. a thin discontinuous shale bed "cap"

The regional stratigraphy of the St. Louis Formation and local stratigraphic relationships in the Ingalls field have been determined in large part from petrophysical logs. Correlation along the northwesterly strike (rather than in a dip direction) is generally easier because there is less variation in thickness and greater lateral persistence of facies.

There are typically three to five closely spaced porous dolostone beds, each generally 5–10 ft (2–3 m) thick, that occur near the base of the St. Louis Formation (uppermost bed is shown at base of Figure 2) (Goebel, 1968). They are readily recognizable on electric logs by their high apparent porosity and low-resistivity signatures. In general, these dolomitic beds cannot be correlated from well to well and locally may be absent. Where present, they are separated from the overlying C zone by 25–50 ft (8–15 m) of non-porous lime wackestones, packstones, brecciated mudstones, and anhydrite.

Where present, the C zone is a somewhat laterally continuous marker bed within the lower St. Louis (Figure 2). It is characteristically widespread and has a coarse oolitic texture and a slightly higher porosity and lower resistivity than the dense crystalline limestones and fine-grained wackestones and mudstones that are above and below. The C zone thickness ranges to a maximum of 25 ft (8 m). Although the C zone is one of the principal producing strata in the Pleasant Prairie field (Ball, 1966), approximately 30 mi (48 km) to the west, it has not been productive in the Ingalls field. The interval thickness from the top of the C zone to the top of the St. Louis Formation (within the Ste. Genevieve subcrop) ranges from 85 to 110 ft (26–33 m).

The oil-producing upper part of the St. Louis Formation, referred to as the St. Louis B zone (where not eroded) may contain both an upper and a lower porous oolite facies, generally separated by a dense limestone or chert (Figure 2). The overall B zone thickness ranges from 15 to 55 ft (5–17 m). The two porous intervals within the B zone are aptly termed the upper B zone and the lower B zone. Many wells penetrate a nonproductive 2- to 10-ft-thick (0.67- to 3-m-thick) bed of red-to-orange fractured chert between the two porous intervals of the B zone. The top of the porous upper B zone in the Ingalls field occurs within 2–3 ft (0.67–1 m) of the St. Louis-Ste. Genevieve disconformity. A dense, green, waxy, highly fossiliferous, fissile, 2- to 5-ft-thick (0.67- to 1.5-m-thick), discontinuous shale (basal Ste. Genevieve?) usually caps the St. Louis.

Facies and Porosity Occurrence

The St. Louis oil fields of southwestern Kansas are characterized by oolitic grainstones with preserved primary interparticle porosity (Ball, 1966). These "carbonate sand" shoals in the better oil-producing fields commonly are poorly cemented and may almost totally disaggregate during drilling. The porous oolitic grainstone facies of the St. Louis B zone in Gray County, Kansas, is typified by an oil-stained core recovered from the Pendleton No. 2 Miller well, Wendell field (Figures 4, 5). Coarse, well-sorted ooids have thick cortexes of alternating concentric and radial fabrics. Ooid nuclei are primarily skeletal; however, some syntaxially overgrown quartz nuclei are present (Figure 6). Although average log-measured neutron-density porosity is less than 10%, air-tested average permeability of this core is 43 md.

Within St. Louis reservoirs, loss of porosity is almost always accompanied by a facies change to skeletal wackestones or packstones (Ball, 1966; Handford, 1988). This provides evidence that, although there may be some secondary porosity enhancement of St. Louis oolite reservoirs, the majority of porosity present was controlled principally by deposition.

Figure 5. Core slab of coarse ooid grainstone (e-log corrected depth 5071 ft) from Pendleton #2 Miller, Wendell field. Rock is oil-stained and has in part excellent interparticle porosity. Airtested average permeability of core is 43 md.

PLEASANT PRAIRIE FIELD, FINNEY COUNTY, KANSAS

Ball (1966) used the studies of Newell et al. (1960) and Purdy (1963) in comparing the 24-million-barrel Pleasant Prairie oil field (Figure 3) in Kansas to the South Cat Cay-Brown's Cay areas, Bahamas. Utilizing extensive cores and thin sections from both the B and C zones of the St. Louis Formation, Ball recognized three productive grainstone facies, which he termed *foreshoal, shoal,* and *backshoal*, coupled with two bounding nonproductive lithofacies (Figure 7).

According to Ball (1966), mud content of various grainstone facies is the primary controlling factor of permeability in the St. Louis. The ooid-skeletal grainstone facies at Pleasant Prairie field, although not the coarsest or best size-sorted, is the most lime–mud-free and therefore has the greatest intergranular porosity and permeability of any of the facies present. This is the result of deposition along the shoal axes where the ooid-skeletal grainstone facies was subjected to the greatest degree of winnowing.

Examination of well cuttings and nine grain-mounted thin sections, from both productive and dry wells in and near the Ingalls field, indicates general correspondence with Ball's observations. Based on grain-mounted thin sections from the Ingalls field area, poorly sorted, coarse, skeletal-ooid grainstones (presumably foreshoal), in which some very coarse skeletal grains lack oolitic coating and are generally unabraded, most commonly are associated with dry or marginally productive wells. Similarly, fine-grained, well-sorted ooid-grainstones (backshoal), in which nearly all grains exhibit both excellent size sorting and thick oolitic coatings, also most commonly are associated with dry or marginally productive

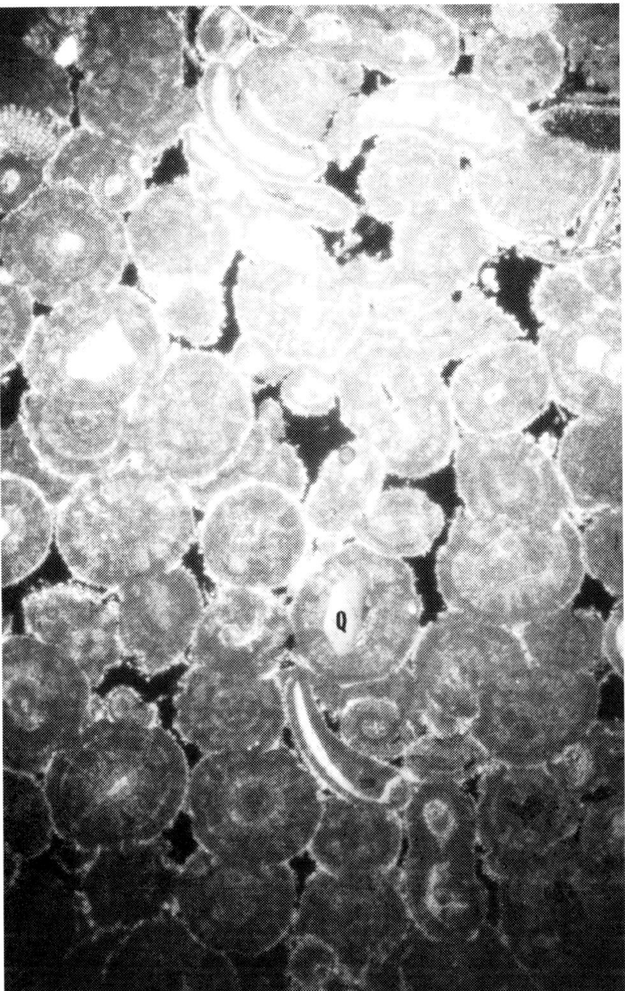

Figure 6. Photomicrograph from core slab in Figure 5 (crossed nicols, 28×). Coarse ooids are generally well-sorted, and have a thick coating that exhibits both concentric and radial fabric. Ooids are surrounded by a thin white isopachous calcite cement that was present prior to compaction. Evidence of compaction includes concave-convex grain contacts and some grain suturing. An euhedral quartz nucleus (Q, near center) has syntaxially overgrown at the expense of the ooid cortex.

wells. However, moderately sorted, coarse, ooid-skeletal grainstones, in which almost all grains have some oolitic coating, and numerous coarse, recognizable bryozoan, brachiopod, and echinoderm fragments are present, invariably are associated with the better oil-producing wells.

DAMME FIELD, FINNEY COUNTY, KANSAS

The Damme field (Figure 3) is similar in size and shape to the Bahamian shoals studied by Ball (1967). This field has produced about 13 million barrels of oil (about one-half of the amount of cumulative production of the Pleasant Prairie field), primarily from the St. Louis B zone. Unlike the smaller stratigraphically trapped Ingalls field, the Damme field (like the Pleasant Prairie field) (Roby, 1959) is a combination structural/stratigraphic trap that has approximately 50 ft of structural closure, but is limited areally by absence of reservoir rock (Schmidlapp and Hartman, 1959). The Damme field has a "typical" oolite shoal geometry, elongate and parallel to depositional strike, with a series of pod-like "thicks" separated by cross-cutting narrow storm channels. Large-scale cross-bedding oriented in a bankward direction is present, and there is an overall convex-upward shape (Handford, 1988). The Ingalls field is also elongate and parallel to depositional strike with an overall convex-upward shape, but it seems to lack the presence of cross-cutting storm channels in the porous lower B zone interval.

Upper St. Louis Lithofacies

On the basis of six cores from the north end of the Damme field, Handford (1988) recognized two general lithofacies in the upper St. Louis: "skeletal wackestones and packstones" and "porous skeletal-ooid grainstones," of which descriptions and geometries seem applicable to the Ingalls field. However, since no reservoir facies were cored in the Ingalls field, only generalized petrographic comparisons are possible.

Wackestones and packstones of the Damme field show evidence of bioturbation and are argillaceous and cherty, and cores have numerous discontinuous stylolites. Fossil assemblages include large, nearly whole and unabraded bryozoans, echinoderms, brachiopods, rugose corals, and gastropods. Handford (1988) concluded deposition was on a relatively shallow, open shelf inhabited by numerous benthonic organisms and occasionally swept by waves. The upper boundary of wackestone/packstones with overlying grainstones is gradational.

The three porous shoal facies recognized by Ball (1966) in the Pleasant Prairie field were not differentiated by Handford (1988) in the Damme field study, but were grouped together into the "porous skeletal-ooid grainstone" facies. These rocks exhibit a coarsening-upward texture, and an upward-increasing proportion of oolitically coated grains, indicating a relatively stable environment of deposition over a muddy open-marine sea floor in which energy levels were gradually increasing (Handford, 1988). Handford notes that this coarse-grained lithofacies is easily recognized in core by its large oil-stained interparticle pores and on a neutron-density log by a distinctive increase in porosity.

INGALLS FIELD AREA

The porous lower B zone interval is the most laterally continuous and productive oil zone in Gray County. A few excellent producing wells in the

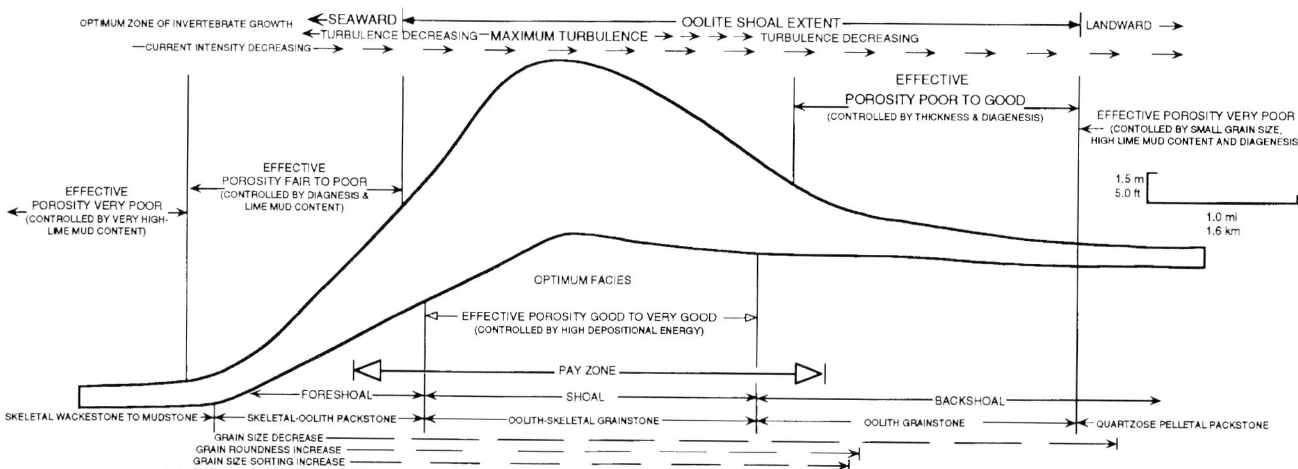

Figure 7. Transverse cross section illustrating geometry, attributes, and facies relationships of St. Louis oolite shoal, Pleasant Prairie field (modified from Ball, 1966). Ball recognized five principal changes that occur in the facies progressively from foreshoal toward backshoal, as follows: (1) An increase in percentage of oolitically coated grains, (2) an increase in grain roundness, (3) an increase in grain sorting, (4) a relative grain size decrease, and (5) an increase in the percentage of lime mud in both directions (from shoal axes). Location of the "backshoal" quartzose pelletal packstone facies corresponds with a brecciated fenestral mudstone facies identified in the Ingalls field (see text and Figures 10, 11, 13).

Ingalls field have recorded peak porosities of 28% and thicknesses up to 20 ft (6 m) (average porosity greater than 8%) in the lower B zone. Initial potentials pumping (IPP) of 150 barrels of oil per day (BOPD), with cumulative production of over 100,000 barrels of oil per well (100 MBO/well), are relatively common for wells that produce from the porous lower B zone interval.

The upper B zone ranges from absent to 20 ft (0–6 m) thick and has as high as 25% porosity in some wells. Commercial B zone production is possible from ooid grainstone reservoirs with average neutron-density log porosity as low as 8%. The porous upper B zone is less laterally continuous than the porous interval of the lower B zone. Although the IPP may be more than 100 BOPD from the upper B zone, cumulative production rarely reaches 100 MBO/well.

B Zone Geometry

Because of the direct relationship of porosity occurrence to lithology (Ball, 1966; Handford, 1988), an individual isopach map (porosity greater than 8%) of genetic porous B zone intervals in the Ingalls field area also serves as a general lithofacies map. However, modification must be made to delete values for porous fenestral mudstones that contain over 8% porosity. Thus, a modified isopach map of porous lower B zone oolites illustrates the general reservoir geometry of the main ooid shoal of the Ingalls field (Figure 8A). Location of the proposed "island" mudflat shown in Figure 8A is based on core data. Unfortunately, the fenestral mudstone facies is not commonly determinable from well cuttings, and where nonporous fenestral mudstones occur, there is no apparent difference on petrophysical logs from dense wackestone/packstones. The thickest portion of the Ingalls field porous lower B zone ooid shoal, in general, corresponds to the best producing wells. An isopach map of the upper B zone (porosity greater than 8%) shows the smaller areal extent and overlap of the porous upper B zone (Figure 8B).

Discussion

It has been proposed that ooid shoals in the area of the Ingalls field were deposited southwest of, and parallel to, a southeasterly trending shoreline (cf. Irwin, 1965; Handford, 1988). However, the following stratigraphic evidence does not seem to support this interpretation for the Ingalls field.

The depositional center of the Hugoton embayment is approximately 30 mi southwest of Gray County (Merriam, 1963) (Figure 1). Proceeding southwest from the southern flank of the Central Kansas uplift across Gray County and toward the center of the Hugoton embayment, there is an overall progressive thickening of Pennsylvanian and Mississippian strata. Therefore, one would anticipate a thickening of the St. Louis Formation from northeast to southwest across the Ingalls field area. However, just the reverse is true when considering the interval from the top of the St. Louis to the St. Louis C zone, as demonstrated by stratigraphic cross section A–A′ (Figures 8A, 9). (Note that this interval thins, rather than thickens, to the southwest from well no. 7 to well no. 2).

Based on core data from well no. 3 (Figures 8A, 9), the producing St. Louis lower B zone lithology is not oolitic grainstone, as in other Ingalls field wells, but rather is a brecciated fenestral lime mudstone (Fig-

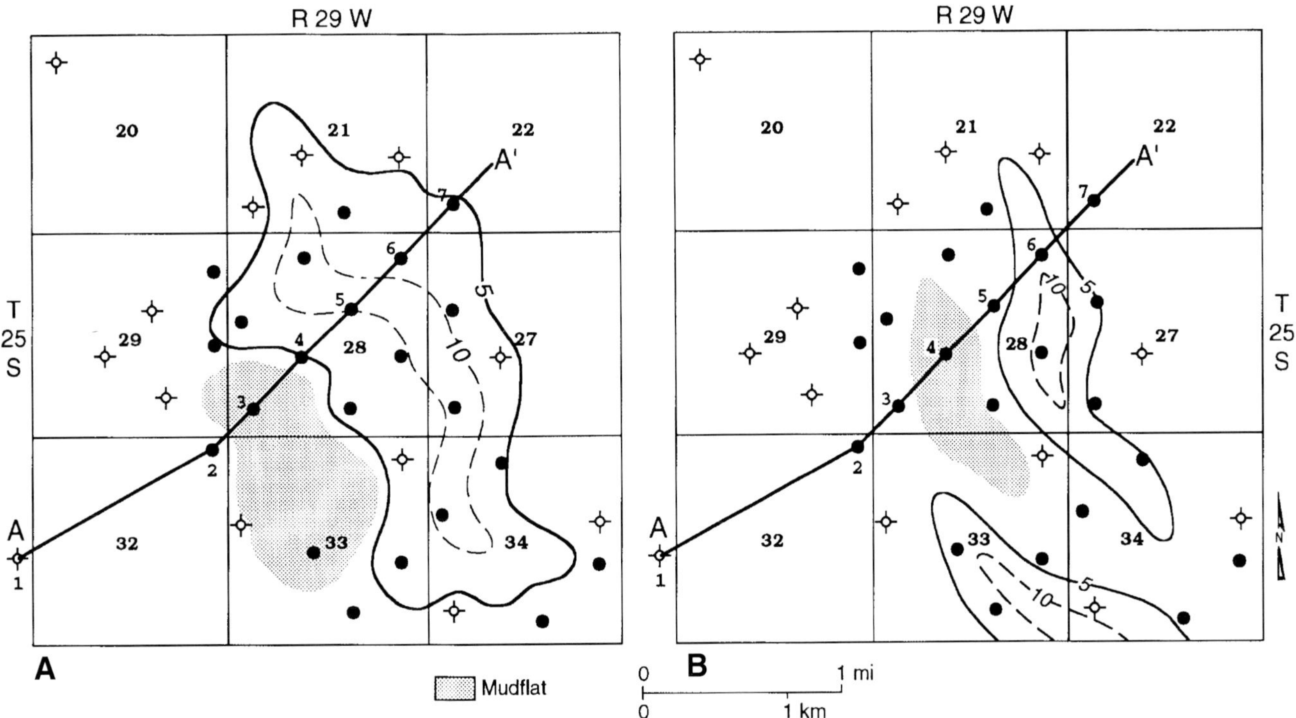

Figure 8. Isopach map of Mississippian St. Louis lower B zone ooid shoal (A), and upper B zone ooid shoal (B) and corresponding "island" mudflat (Ingalls field, Gray County, Kansas). The 5-ft isopach line (porosity greater than 8%) shows general net reservoir outline. Isopach based on petrophysical log and sample data. Note northeasterly (landward?) shift of island mudflat from lower to upper zones. Cross section shown in Figures 9 and 13.

ures 10, 11), indicative of an intertidal algal mudflat deposit (Shinn, 1968). This intertidal facies is not the lithology one would expect to find basinward of a marine bar shoal (Ball, 1967; Handford, 1988). It seems evident that the primary reservoir at Ingalls field is an oolite shoal that developed on the leeward side of a small island mudflat. This conclusion is substantiated by the isopach of the interval from the base of the porous lower B zone down to the base of the overall B zone interval (Figure 12) (isopach interval is illustrated on Figure 2).

To construct the isopach map shown in Figure 12, the upper datum point was selected using the neutron-porosity curve. This is accomplished by moving upward to the approximate depth where the neutron-porosity value first increases to 8% within the B zone interval. The lower point is selected using the gamma-ray curve, by moving downward within the B zone to the location where a distinctive increase in gamma-ray intensity at the base of the B zone occurs. This is the point where cleaner and coarser-grained lower B zone wackestones/packstones contact low-energy carbonates of the lower St. Louis. It is suggested that this formation contact between the B zone interval and underlying St. Louis carbonates is essentially a time/stratigraphic horizon.

Assuming that ooid genesis within lower B zone sediments required wave energy, and that the porous grainstone facies of the B zone were formed at or above the average wave base, the base of the porous lower B zone interval marks a generalized point of common water depth and therefore a near horizontal planar datum. Accordingly, an isopach of the interval from the base of the porous ooid-grainstone to the base of the B zone reveals the approximate depth from the average wave base datum to the sea floor. The net result is a generalized paleotopographic map of the sea floor prior to oolite deposition, which reveals a roughly circular sea floor perturbation (Ingalls Island) approximately 3 mi (5 km) across, on which a mudflat developed near the center. Obviously, lateral transport of porous grainstone facies and erosion could create spurious results when constructing maps using this method.

Given this island mudflat depositional model, it is assumed that a shoal similar to the Ingalls field might have been deposited on the southwest side (basinward) of the island. Evidence for this shoal on the other side of the island is substantiated by well no. 2 (Figures 9, 12), which produces from an oolitic grainstone correlative with the lower B zone, but on the basinward side of the postulated island mudflat. Based on this assumption, well no. 1 was drilled further southwest of well no. 2 to test the theory. This well encountered the lower B zone as well as the intermediate chert bed, confirming the existence of

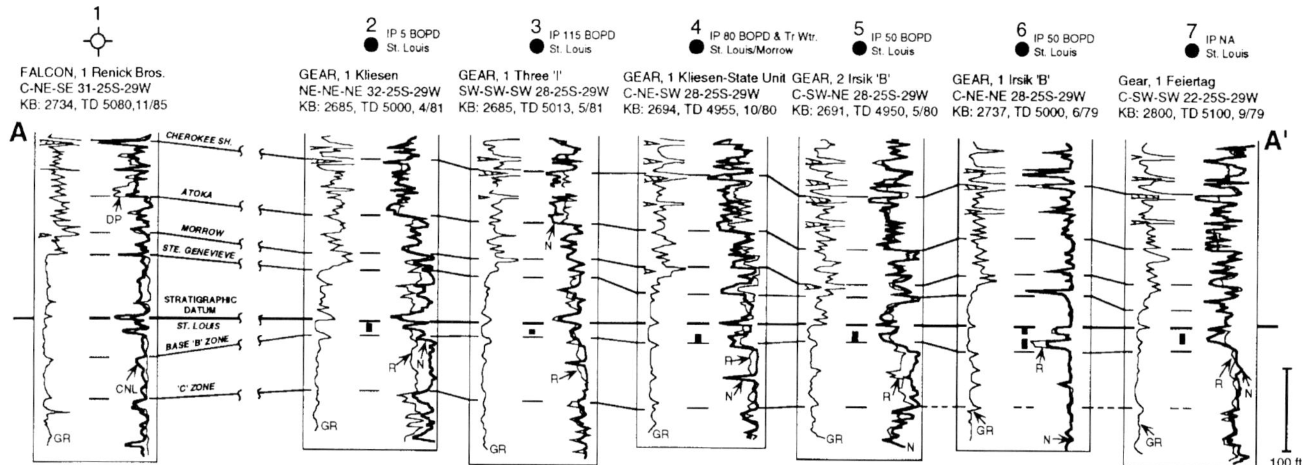

Figure 9. Electric log cross section across Ingalls field. Note apparent thinning to the southwest when considering the interval from the top of the St. Louis (stratigraphic datum) to the St. Louis C zone, between wells no. 7 and no. 2. Thinning was not anticipated, as Mississippian rocks generally thicken to the southwest. Only well no. 6 encountered both upper and lower porous B zone intervals. Line of section shown in Figures 8A, 8B, and 12 (same as line of section in Figure 13).

the shoal on the southwest side of the island mudflat. (Unfortunately, the porous lower B zone was encountered at a lower structural datum than the Ingalls field and recovered only oil and gas-cut water when drill-stem tested).

Schematic cross section A–A′ of the Mississippian St. Louis B zone through Ingalls field (Figure 13; same as well log cross section, Figure 9) demonstrates the depositional relationship of the oolite shoal to the associated sea floor perturbation. This highly vertically exaggerated cross section was constructed by integrating petrophysical logs with sample and core data. Where sample data were not available, lithology was implied based upon presence/absence of porosity. It was assumed that the fenestral lime mudstone contained in the core from well no. 3 was at the structurally highest position during deposition of the porous lower B zone. Additionally, it was assumed that the point where oolites contact the base of the B zone was initially the shallowest water depth point. Based on these fundamental assumptions, various angles of dip were tested until a logical dip relationship on the base of the B zone porosity was established between wells 3 and 5. Once this rate of dip was chosen, then the balance of the cross section was added, maintaining the same rate of dip on the base of the porous lower B zone (as was used between wells 3 and 5).

The concentration of oolite sands on the leeward side of an island is perplexing, as it is assumed that most agitation and wave energy would have come from open waters to the southwest, expending most energy on the basinward side of the mudflat. On the basis of the regional stratigraphic relationships, it is postulated that there might have been a distance of 50 mi (80 km) or more from the Ingalls field area to the actual shoreline during St. Louis lower B zone deposition. Because the Ste. Genevieve and St. Louis formations have been truncated by erosion just a few miles northeast of the Ingalls field (Figure 4), the actual distance to the shoreline and the distribution of other facies northeast of the Ingalls field cannot be known with certainty. This vast expanse of what was presumably open water may have permitted significant wave action and agitation in a southwesterly direction to supply the energy necessary to form the Ingalls field oolite deposits on the leeward side of the island mudflat.

Because of thickening to the northeast, indicated by the interval mapped in Figure 13, a scenario in which deposition occurred around an island mudflat that was located far from shore is favored. However, an alternative explanation for concentration of oolites on the leeward side of the island mudflat is offered. A proximal shoreline to the island mudflat could generate a "venturi" effect created by waters channeling between the island and shoreline, resulting in transportation and deposition of oolite sands on the leeward side of the "island" (G.B. Asquith, personal communication, 1989). This type of "channeled" oolite deposit would be elongated in the direction of current flow and would have less of a pod-like geometry than the Bahamian oolite shoals studied by Ball (1967), a configuration that matches the thickness pattern of the lower B zone shown in Figure 8A.

Transgression

Previous studies (Ball, 1966; Handford, 1988) have suggested the St. Louis strata reflect deposition in a transgressive sea. Local sedimentation or uplift in the Ingalls field area must have occurred at a greater rate than the marine transgression. Schematic cross section A–A′ (Figure 13) demonstrates the apparent

Figure 10. Core slab of "brecciated" fenestral lime mudstone (e-log corrected depth 4848 ft) from well no. 3, Figure 9. Core exhibits typical filled to partially filled (with clear to white sparry pore filling calcite cement) fenestral fabric, indicative of intertidal mudflat depositional environment. Subaerial exposure is implied; however, typical vadose fabrics such as meniscus and pendant cements are not present.

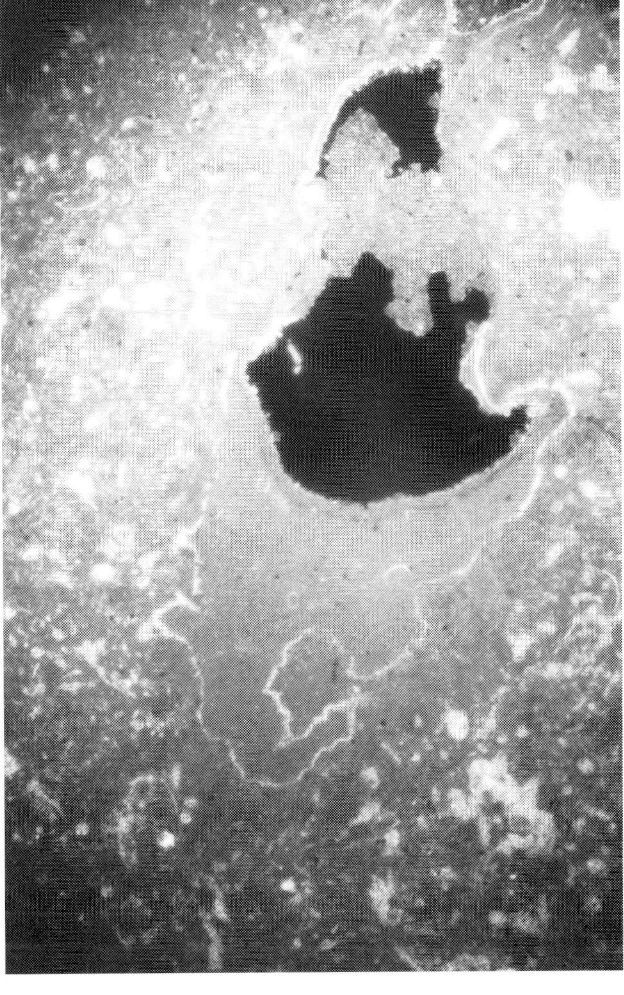

Figure 11. Photomicrograph from core slab in Figure 10 (cross nicols, 28×). Calcite lined vug is three-quarters full of geopetal sediment. Original vug shape is shown by thin isopachous cement lining. Quartz silt and sand content is less than 1%. Calcareous monaxon sponge spicules and small ostracodes are numerous.

overall transgressive nature of the B zone deposits. In the core from well no. 3 (Figures 9, 13), a poorly developed coarse oolite facies underlies a fenestral mudstone. This strongly suggests that water depths were first adequate for deposition of the oolite facies and then subsequently shallowed and became restricted enough to form and preserve intertidal deposits. The fenestral mudstone would underlie an oolite facies if, locally, transgression had exceeded deposition. Nevertheless, the overall transgressive nature of the deposit is seen by the northeasterly (landward?) shift of the fenestral mudstone facies from well no. 3 to well no. 4 (Figures 8A, 8B, and 13).

In both the Damme and Pleasant Prairie fields, deposition of the ooid-skeletal grainstone facies occurred in areas that were sufficiently shallow for ooid shoal deposition. However, because of sampling location or erosion or because deposition and subsidence equalled transgression, the intertidal mudstone facies (like that present in the Ingalls field) either has not been recognized, is not preserved, or was never formed.

Seismic Exploration Methods

At least three principal operators have conducted extensive seismic surveys in the study area. It seems many well locations drilled primarily on the basis of seismic data (presumably because of fast isotimes from a lower Pennsylvanian reflector to the Mississippian unconformity) have encountered abnormally thick Ste. Genevieve sections that correspond to

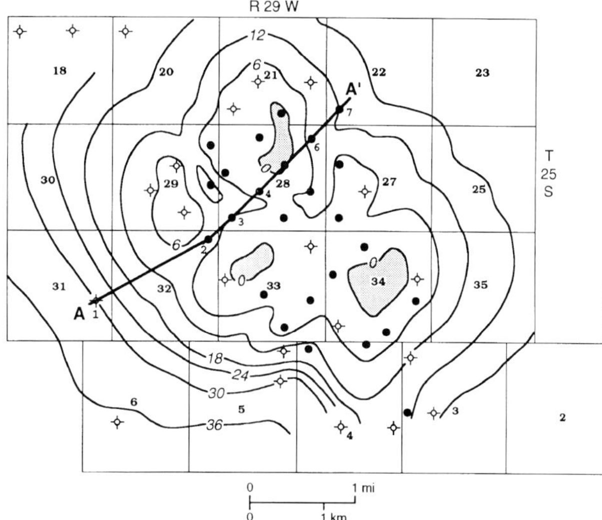

Figure 12. Isopach map of the interval from the base of the porous lower B zone interval (oolite shoal) to the base of the overall B zone interval (the point where there is a distinctive increase in gamma-ray intensity that is visible on most electric logs; see Figure 2). This map reveals the general sea floor configuration in the area of the Ingalls field prior to ooid deposition. Cross section shown in Figures 9 and 13. Isopach interval is also illustrated in Figure 13.

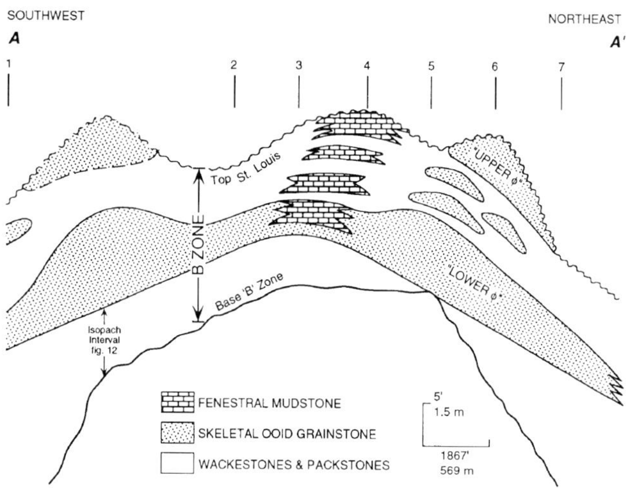

Figure 13. Schematic cross section of Mississippian St. Louis B zone through Ingalls field. Line of section shown in Figures 8A, 8B, and 12. Dip is estimated. Assumes that oolite shoal formed first at point where oolites are in contact with the base of the B zone and that the island mudflat was the topographically highest point. Overall transgressive nature is evident by northeasterly (landward?) shift of facies; however, deposits preserved are regressive. Note at well no. 3 that first there was adequate water depth for deposition of an oolitic grainstone and then later subsequent shallowing and formation of the brecciated lime mudstone.

remnant paleotopographic highs on the Ste. Genevieve unconformity. These paleotopographic high areas are apparently unrelated to the paleostructure of the underlying St. Louis Formation. There rarely seems to be any correlation between Mississippian erosional topography and buried shoal deposits in the St. Louis. It is speculated that seismic thins associated with structural features discernible from stratigraphic reflectors below the Mississippian unconformity may be more likely to correspond to St. Louis shoal deposits than the apparently unrelated topographic highs of the Mississippian unconformity.

Stratigraphic Thinning of the St. Louis and Occurrence of Oolites

Because of the thin nature of the B zone and a lack of velocity contrast between over- and underlying rocks, direct seismic mapping of, and between, beds of the upper and lower St. Louis is not possible. This is truly unfortunate, because in producing wells, the interval between the St. Louis B and C zones usually thins 10–50 ft (3–15 m) compared with nearby dry holes. We suggest this thinning indicates shallower water in areas where the B zone was relatively higher topographically and therefore in a higher energy regime in which formation and deposition of oolites shoals would be more likely. Moreover, the overall B zone interval is relatively 10–35 ft (3–11 m) thinner in producing wells, presumably the result of the relative structural position of the sea floor prior to ooid deposition.

Sea Floor Perturbations and Associated Magnetic and Gravity Anomalies

The Shelf Model of Handford (1988) assumes a nearly flat platform across which minor sea floor perturbations influenced deposition. The areal size of an oolite shoal thus is limited by the size of the associated sea floor elevation (i.e., small sea floor "bumps" result in small associated shoals, like the Ingalls field, while a larger perturbation is necessary for the formation of a large shoal, like Pleasant Prairie).

Previous investigations (Dietterich, 1986; Handford, 1988) in the area have led to the hypothesis that locations of St. Louis oolite shoals corresponded to minor sea floor perturbations that may be associated with subtle basement "highs." Airborne magnetic and gravity maps (Yarger et al. 1981, 1985) show that there are both obvious positive magnetic and positive gravity anomalies associated with the Ingalls field area (Figures 14, 15). The increase in magnetic field intensity and gravitational attraction is presumably caused in part by closer proximity to the surface of higher density, intermediate or basic composition basement rock. An alternative explanation for the existence of these anomalies could be enhanced mineralization of near-surface sediments because of microseepage of hydrocarbons (c.f.

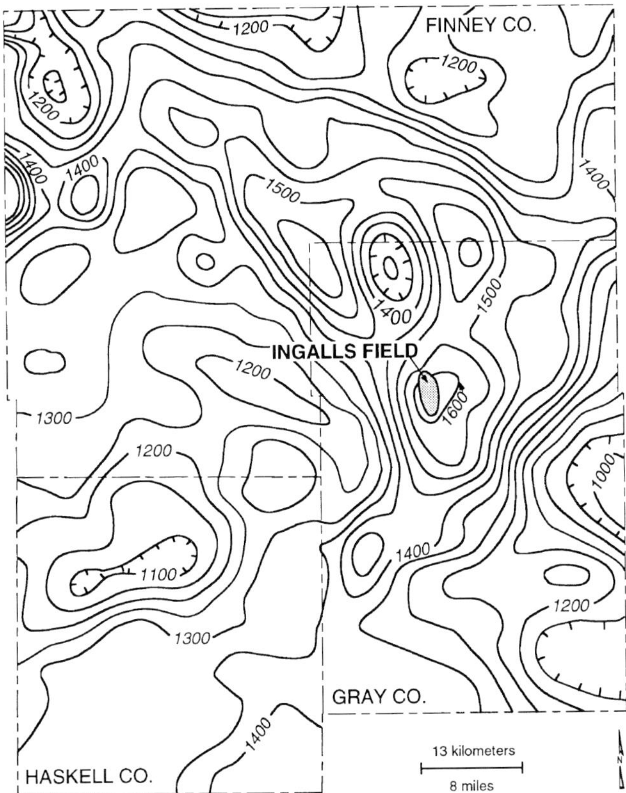

Figure 14. Aeromagnetic map of Finney, Haskell, and Gray counties, Kansas. Shaded area indicates location of Ingalls field in relation to positive magnetic anomaly. Other more prominent positive anomalies exist that are not associated with St. Louis oil fields; however, only the Ingalls field anomaly is also coincident with a pronounced gravity anomaly. Contour interval: 50 gammas (from Yarger et al., 1981). See also Figure 15.

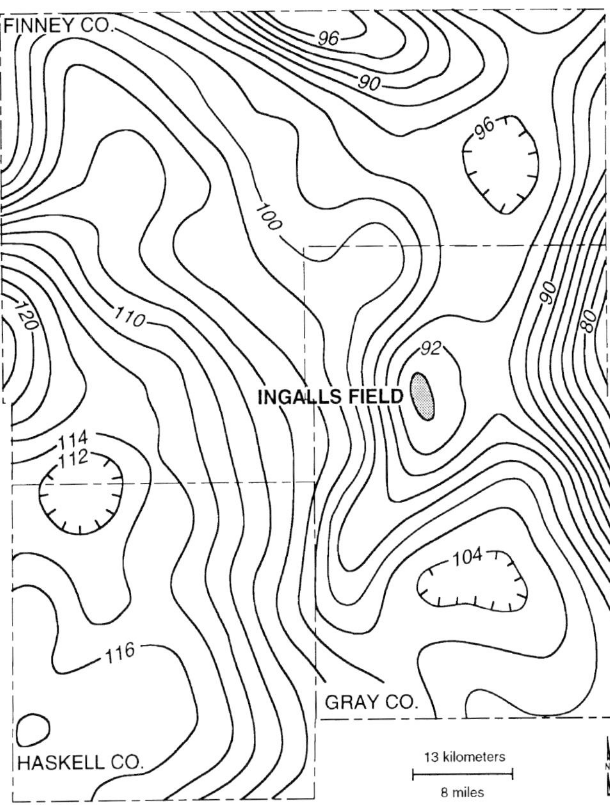

Figure 15. Bouguer gravity map of Finney, Haskell, and Gray counties, Kansas. Shaded area indicates location of Ingalls field in relation to positive gravity anomaly. Other positive gravity anomalies exist that are not associated with St. Louis production, but only the Ingalls field anomaly is also coincident with a magnetic anomaly. Contour interval: 2 mGal (from Yarger et al., 1985). See also Figure 14.

Duchscherer, 1981). Numerous sand-size spherules, in part containing oil and tentatively identified as magnetite, have been noted in well samples from this area by the authors (c.f. McCabe et al. 1987; Saunders et al., 1991; Machel and Burton, 1991). The coincidence of both positive gravity and positive magnetic anomalies appears to be significant, as there are numerous unrelated positive magnetic or gravity anomalies that are not associated with St. Louis oil production.

SUMMARY

Depositional History

In summary, the depositional history of the Ingalls field area is as follows:

1. As a result of uplift, possibly combined with a seaward regressive pulse, the area became sufficiently shallow to be within a high-energy marine environment.
2. Oolitic grainstone deposition began at the points of shallowest water depth, and skeletal wackestones and packstones were deposited in adjoining areas.
3. Uplift and deposition outpaced transgression, resulting in formation of the intertidal mudstone facies near the island center.
4. Transgression followed by a regressive pulse resulted in deposition of the areally smaller upper B zone shoal, which exhibits more of a pod-like geometry than the lower B zone shoal.

Mississippian St. Louis B zone oolite shoals seem to be associated with ancient sea floor perturbations where, because of slightly shallower water depths, there existed a higher energy environment where ooid-grainstones were more likely to form. The size of the perturbation controls the size of the associated overlying shoal, with small "bumps" influencing deposition of small shoals like the Ingalls field. The location of these shoals may be in part determinable from gravity and magnetic maps, with an apparent greater probability of occurrence at locations that exhibit both a positive magnetic and a positive gravity anomaly.

Subtle, sinuous folding of Mississippian strata has resulted in a series of small, 1-mi-wide (1.6-km-wide) anticlines, gently plunging to the southeast and generally 4–5 mi (6.5–8 km) apart. These gently folded Mississippian rocks are apparently intersected by northwesterly trending pre-Pennsylvanian normal faults with displacements of 20–50 ft (6–15 m), which are 2–6 mi (3–10 km) apart. It is speculated that the intersection of these folds and faults could correspond to the location of what may have been an ancient archipelago of coastal plain islands (mudflats) that may have extended either southwest or southeast from the Ingalls field.

Oil production from the St. Louis B zone is almost always associated with thinning of both the interval from the C zone to the B zone and the B zone overall. This thinning is thought to be the result of deposition in areas where the sea floor was relatively elevated to surrounding areas. Unfortunately, these intervals seem to be too thin and lack sufficient velocity contrast for direct seismic mapping. It would seem that greater success with seismic mapping could be achieved by searching for anomalies associated with mappable structures below the Mississippian unconformity. Fast seismic intervals from lower Pennsylvanian reflectors to the Mississippian unconformity generally indicate erosional topographic highs on the Ste. Genevieve that are nonproductive and are not related to buried oolite shoal deposits in the St. Louis.

In general, occurrence of upper B zone shoals are unpredictable, and they are limited in areal extent. The existence of lower B zone shoals is more predictable, and they are more likely to cover a larger area than shoals of the upper B zone. The productive facies relationships identified by Ball (1966) at the Pleasant Prairie field regarding lime-mud content, sorting, grain coatings, and grain size seem to be applicable in the Ingalls area. Likewise, many of the reservoir attributes and facies and porosity relationships noted by Handford (1988) also seem applicable in the Ingalls area. An important conclusion is that a St. Louis oolite shoal associated with an island mudflat could be one of a pair of northwesterly trending shoals that are separated by a fenestral mudstone facies. This study also illustrates the importance of core data. If it had not been for the "accidental" coring of a "poor" producing well, many of these observations would likely not have been made.

ACKNOWLEDGMENTS

We would like to recognize Mr. Jim Gowens and Mr. Floyd Wilson of the Hugoton Energy Corporation for their assistance and financial support during much of the original data collection. Also we wish to thank Mr. John Stone, Mr. Rene Stucky, and Gear Petroleum Corporation for the use of core, petrophysical log data, and grain-mount photomicrographs. In addition, we would like to acknowledge the assistance of Dr. W. Lynn Watney and Ms. Renate Hensiek of the Kansas Geological Survey and Mr. K. David Newell of the University of Kansas for their assistance in preparing and drafting the figures. Moreover, we would like to thank Drs. Sal J. Mazzullo and George B. Asquith, who critically reviewed the original paper.

REFERENCES CITED

Ball, M.M., 1967, Carbonate Sand Bodies of Florida and the Bahamas: Journal of Sedimentary Petrology, v. 37, p. 556-591.

Ball, S.M., 1966, Prediction of Effective Porosity in some Marine Oolite Deposits: Amoco Production Co., Liberal, Kansas District Exploration Report AMOCO-41, available as Kansas Geological Survey Open-File Report No. 66-6, 7 p., 3 plates.

Dietterich, R.J., 1986, Porosity and Permeability Analysis in Mississippian System, St. Louis Limestone Formation, Damme Field, Finney County, Kansas Utilizing Petrographic Image Analysis: M.S. Thesis, Wichita State University, 223 p., 1 plate.

Duchscherer, W. Jr., 1981, Nongasometric Geochemical Prospecting for Hydrocarbons with Case Histories: Oil and Gas Journal, Oct. 19th, p. 312-327.

Goebel, E.D., 1968, Mississippian Rocks in Western Kansas: AAPG Bulletin, v. 52, p. 1732-1778.

Handford, C. R., 1988, Review of Carbonate Sand-Belt Deposition of Ooid Grainstones and Application to Mississippian Reservoir, Damme Field, Southwestern Kansas: AAPG Bulletin, v. 72, p. 1184-1199.

Huffman, G.G., 1959, Pre-Desmoinesian Isopachous and Paleogeologic Studies in Central Midcontinent Region: AAPG Bulletin, v. 43, p. 2541-2574.

Irwin, M.L., 1965, General Theory of Epeiric Clear Water Sedimentation: AAPG Bulletin, v. 49, p. 445-459.

Machel, H.G., and E.A. Burton, 1991, Causes and Spatial Distribution of Anomalous Magnetization in Hydrocarbon Seepage Environments: AAPG Bulletin, v. 75, p. 1864-1876.

McCabe, C., R. Sassen, and B. Saffer, 1987, Occurrence of secondary magnetite within biodegraded oil: Geology, v. 15, p. 7-10.

Merriam, D.F., 1963, The Geologic History of Kansas: Kansas Geological Survey Bulletin, No. 162, 317 p.

Newell, N.D., E.G. Purdy, and J. Imbrie, 1960, Bahamian Oolitic Sand: Journal of Geology, v. 68, p. 481-497.

Purdy, E.G., 1963, Recent Calcium Carbonate Facies of the Great Bahama Bank. 2. Sedimentary Facies: Journal of Geology, v. 71, p. 472-497.

Roby, R.E., 1959, Pleasant Prairie field, *in* Kansas Oil and Gas Fields, Western Kansas: Kansas Geological Society, v. 2, p. 131-140.

Saunders, D.F., K.R. Burson, and C.K. Thompson, 1991, Observed Relation of Soil Magnetic Susceptibility and Soil Gas Hydrocarbon Analyses to Subsurface Hydrocarbon Accumulations: AAPG Bulletin, v. 75, p. 389-408.

Schmidlapp, R.L., and W.L. Hartman, 1959, Damme and Finnup fields, *in* Kansas Oil and Gas Fields, Western Kansas: Kansas Geological Society, v. 2, p. 8-12.

Shinn, E.A., 1968, Practical Significance of Birdseye Structures in Carbonate Rocks: Journal of Sedimentary Petrology, v. 38, p. 215-223.

Thompson, T.L., and E.D. Goebel, 1968, Conodonts and Stratigraphy of the Meramecian Stage (Upper Mississippian) in Kansas: Kansas Geological Survey Bulletin, No. 192, 56 p.

Yarger, H.L., R.R. Robertson, J.A. Martin, K.H. Ng, R.L. Sooby, and R.L. Wentland, 1981, Aeromagnetic Map of Kansas: Kansas Geological Survey Map M-16, Lawrence, Kansas.

Yarger, H.L., C. Lam, R.L. Sooby, C. Lee, J.A. Martin, K.H. Ng, R. Robertson, Y. Zhao, M.D. Adkins-Heljeson, and D.W. Steeples, 1985, Bouguer Gravity Map of Kansas: Kansas Geological Survey Open File Report 85-14, Lawrence, Kansas.

Zeller, D.E., 1968, The Stratigraphic Succession in Kansas: Kansas Geological Survey Bulletin, No. 189, 81 p.

Chapter 15

Comparison of Oolitic Sand Bodies Generated by Tidal vs. Wind-Wave Agitation

Harold R. Wanless
Department of Geological Sciences
University of Miami
Coral Gables, Florida, USA

Lenore P. Tedesco
Department of Geology
Indiana University/Purdue University at Indianapolis
Indianapolis, Indiana, USA

ABSTRACT

The trade-wind belt of the southeastern Bahamas provides a setting for wind-wave–generated oolitic sands and sand bodies that contrast dramatically with the tidal-generated oolitic sands and sand bodies of the northern Bahamas, which are restricted to shoals near the platform margins where tidal flow is strong.

In the southeastern Bahamas (Caicos Platform), waves from brisk prevailing trade winds and from open ocean swells provide intermittent bottom agitation for ooid formation. These mechanisms are producing thick and widespread oolitic grainstones of two types in four settings. *Regular concentric ooids* are forming on (a) wind-wave–agitated shallow subtidal and shoreface settings on the platform interior and (b) ocean-wave–agitated shorefaces facing the platform margin. These are forming stratified, prograding strand plains and are undergoing early meteoric and brine diagenesis. *Irregular ooids* are forming (c) broad banks and (d) widespread subtidal sheets in areas of intermittent wind-wave agitation across the platform interior.

INTRODUCTION

Carbonate ooids form in areas where five requirements are met: (1) presence of nuclei, (2) bottom agitation to move grains, (3) a source of supersaturated water, (4) a process for renewal of the water, and (5) minimal amount of grain degradation processes. For significant accumulation to occur, the area must be near a hydrodynamic balance with respect to on-bank and off-bank transport.

Most of the ooids forming in the Bahamas today contain concentric lamellae in which the fine aragonite needles are essentially tangentially oriented (Newell et al., 1960; Loreau, 1973). The morphology of the individual ooid grains and their resultant sediment body geometry, however, vary as the nature of the above requirements vary. Especially important are the type, intensity, and persistence of agitation and the nature of grain aggregation and degradation processes.

Bottom agitation can be from tidal currents or from wind-generated waves. Tidal-current agitation dominates in the northern Bahamas, where most of the early models of carbonate sedimentation were derived. By contrast, wind-generated wave agitation dominates in the southeastern Bahamas, including the Turks and Caicos islands. Recent research by the authors and others (Lloyd et al., 1987) on Caicos Platform has found oolitic sedimentation to be very different from that for the northern Bahamas. This difference includes differences in ooid microstructure; depositional structures, fabric, and texture; oolitic sediment body distribution, geometry, and orientation; and early diagenetic modifications.

The purpose of this chapter is to compare the ooids and oolitic sediment bodies forming in the northern Bahamas (by tidal agitation) with those forming on Caicos Platform (dominantly under wind-generated wave agitation). Oolitic sand bodies forming under the influence of wave agitation provide an important spectrum of alternative models for paleoenvironmental interpretation of oolitic facies. First, a comparison of the climatic/hydrographic setting of the northern and southeastern Bahamas is presented. Second, the characteristics of the tidal-generated oolitic sand bodies of the northern Bahamas are summarized. Third, the nature and distribution of wave-agitation–generated oolitic sand bodies on Caicos Platform is described. Finally, application to interpretation of ancient limestone sequences is discussed.

CLIMATIC/HYDROGRAPHIC SETTING

Eight conditions of the physical climatic/hydrographic environment of south Florida, the Bahamas, and Caicos control the character of carbonate sedimentation (Wanless et al., 1989): (A) prevailing easterly trade winds; (B) tides; (C) oceanic swells; (D) platform circulation; (E) rainfall; (F) evaporation; (G) winter cold-front intrusions; and (H) tropical storms and hurricanes. The first six of these conditions are critical to providing a setting for ooid generation. Three of these, trade winds, rainfall, and winter cold fronts, display striking gradients from south Florida and the northern Bahamas southward toward Caicos Platform.

(A) Prevailing Winds. Easterly trade winds dominate wind patterns throughout south Florida and the northern Bahamas from May through October (Figure 1). These areas are located near the northern range of the easterlies, so winds are gentle, averaging 5–10 knots (Traverse and Ginsburg, 1966). Daily convective winds may override regional wind flow near large land masses.

In the southeastern Bahamas and Caicos Platform, easterly trade winds are brisk and persistent through much of the year and average 10–15 knots (Milliman, 1966).

(B) Tides. Tide range is 1 m or less over most of south Florida, the Bahamas, and Caicos Platform (U.S. Dept. of Commerce, 1990). The strength of tidal currents at the platform margin is dependent on the presence or absence of emergent islands. Where islands block interchange with the platform interior, tidal currents are concentrated in breaks in the island chain and ooids form. Where islands block circulation, tidal-current agitation necessary for ooid formation is absent. Currents are moderate to strong toward the platform margins along those margins where there is open exchange to the platform interior. Margins of smaller platforms tend to have weaker tides than margins of larger platforms. Tidal flow may be accentuated across shoals or submerged ridges or through constrictions. Strong tidal currents penetrating 10–20 km onto the platform occur at the ends of basin reentrants adjacent to southern and eastern Great Bahama Bank (Ball, 1967).

(C) Oceanic Swells. Seas and waves from locally generated storms and hurricanes may be intense but are infrequent and of rather short wavelength and wave period. Platform margins adjacent to open ocean environments also receive large, long-period swells from distant Atlantic storms. Oceanic swells can cause persistent bottom agitation to exposed shallow bottoms and shorelines.

The eastern margins of the Bahamian platforms and the northern margin of Caicos Platform are exposed to the waves and swells of the open Atlantic Ocean (Figure 1). The interior platform margins of the Bahamas and the southeastern platform margin of south Florida are not exposed to open ocean conditions and receive only that wave energy generated within the adjacent basins and straits.

(D) Platform Circulation. Platform circulation is enhanced by wind, oceanic currents, and an absence of emergent barriers. Caicos Platform, situated within the regional west-flowing Antilles Current, subjected to brisk prevailing easterlies and having relatively open eastern and western margins (Figure 1), has effective east to west cross-bank circulation. Similarly, southern Great Bahama Bank has good cross-bank circulation because of prevailing easterlies and a lack of islands. In contrast, Andros Island effectively blocks bank circulation on central Great Bahama Bank (Smith, 1940; Traverse and Ginsburg, 1966).

(E-F) Rainfall and Evaporation. The climate of the northern Bahamas and south Florida is rainy (100–170 cm/yr) and humid, with a pronounced wet season from June through October and the higher rainfall associated with diurnal land-mass–induced convection (Thomas, 1974). Evaporation is about 165 cm/yr, producing accentuated salinities on platform interiors during the drier winter months (Smith, 1940; Traverse and Ginsburg, 1966).

The climate on Caicos Platform is tropical subarid with rainfall of 50–75 cm/yr (Milliman, 1966). Evaporation (also about 165 cm/yr) far exceeds rainfall, promoting evaporation of bank waters through

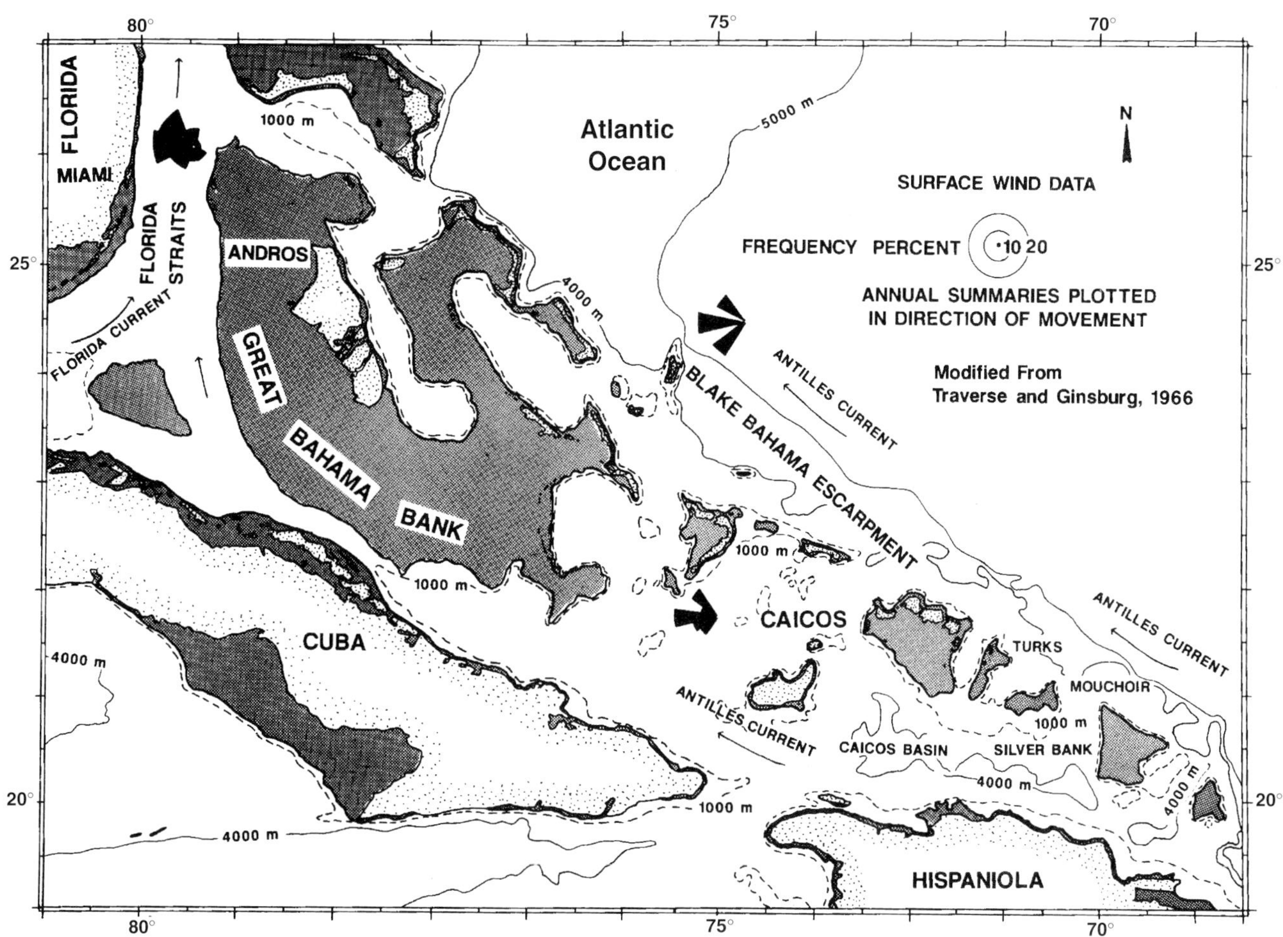

Figure 1. Map of south Florida and Bahamian platform chain showing land (stippled), platform areas less than 20 m in depth (shaded), the 1000-, 4000-, and 5000-m-depth contours, surface winds, and regional currents (from Wanless et al., 1989).

most of the year and formation of brines in coastal salinas and tidal flats (Leaver, 1985; Wanless et al., 1989; Browne et al., 1989). From the 1600s, solar salt production has been an important industry of the Turks and Caicos islands and the southeastern Bahamas.

(G-H) Winter Storms and Hurricanes. Thirty to 40 winter storms (arctic cold fronts) pass through south Florida and the northern Bahamas between November and May each year (Warzeski, 1976). These bring 1- to 3-day periods of stronger winds out of the northern quadrant following frontal passage and are an important influence on shallow marine sedimentation. Winter storms also affect Caicos Platform but are less frequent and generally less intense than in the northern Bahamas and south Florida (Milliman, 1966). South Florida, the Bahamas, and the Turks and Caicos islands lie directly in the hurricane belt. South Florida is affected by hurricane-force winds on the average of once every 7.5 yr; Caicos Platform, on the average of once every 5.5 yr (Neumann et al., 1978; see Wanless et al., 1988, 1989 for more detailed description).

OOLITIC SANDS OF THE NORTHERN BAHAMAS

Purdy (1963) provided the basis for what has become the classical Bahamian model of carbonate sedimentation. In this model, the high-energy platform margin is dominated by coralgal or oolitic grainstones, which reflect locally strong tidal or wave agitation. Toward the platform interior, water energy gradually decreases, giving rise first to grapestone aggregate sediment, then to hardened peloidal sediments, and finally to soft peloidal sediments in the most restricted interior portions of the platform. Later studies by Imbrie and Buchanan (1965), Dravis (1977, 1979), Harris (1977, 1979), Hine and Neumann (1977), Palmer (1979), and others showed that variations in sedimentary facies can occur along portions of platforms dominated by much stronger tidal current agitation. Those portions of platforms situated at the ends of deep-water embayments are dominated by much stronger tidal currents and exhibit a broader belt of platform-margin deposition (usually oolitic;

Ball, 1967) with significantly more attendant marine cementation (Dravis, 1977, 1979).

In the northern Bahamas, ooids are generating sand bodies near the platform margins because the ooids are forming under the influence of (a) persistent agitation by tidal currents, (b) on-bank transport of offshore waters, and (c) waters that become increasingly supersaturated because of warming and evaporation on the platform (Newell et al., 1960). The light prevailing winds of the northern Bahamas are not strong enough to cause bottom agitation, and the platform interiors are stabilized by algal scum mats and sea grasses (Newell et al., 1960; Purdy, 1963). Ooid generation is, thus, restricted to areas of bottom agitation by tidal currents. As a result, tidal-current–formed oolitic sediment bodies of the northern Bahamas occur in areas of stronger tidal agitation—near the platform margins.

These tidal-current–formed ooids occur in areas where there is flow between the platform margin and interior. Where emergent islands block this interchange, oolitic sands are generally absent. For example, Andros Island, on the eastern margin of Great Bahama Bank, blocks tidal exchange with the platform interior, and as a result, tidal currents are weak near the margin and modern oolitic sands are absent (Newell et al., 1960).

Today oolitic shoals are present only along those portions of the platform margin where there is effective tidal circulation. During different sea level stands of the late Pleistocene, the focus of oolitic sedimentation varied. During the Sangamon Interglacial, 120,000 years ago, sea level was about 7 m higher than present in the northern Bahamas (author's observations on Mores Island, Little Bahama Bank) and south Florida (Parker et al., 1955; and author's observations of Pleistocene oolitic beach deposits in Miami area). During this highstand, active tidal interchange and agitation formed the extensive oolitic sediment bodies that are now Andros Island, on the east side of Great Bahama Bank, and the oolitic portion of the Miami Limestone, southeastern Florida. Oolitic grainstone sediment bodies that formed during the Sangamon Interglacial, the Wisconsin Glacial, and the Holocene Interglacial, when considered together, form a nearly continuous oolitic rim to the margins of Great Bahama Bank, Little Bahama Bank, and south Florida (Figure 2).

It is these Holocene and Pleistocene oolitic sediment bodies in the northern Bahamas and south Florida that have provided the previous analogs for interpretation of ancient oolitic grainstones (Hoffmeister et al., 1967).

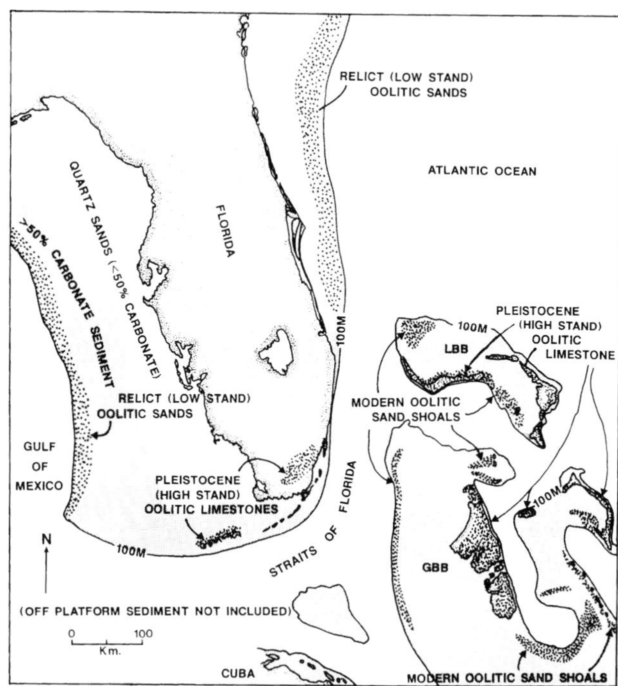

Figure 2. Map of Florida and the northern Bahamas showing the distribution of some of the surficial Holocene and late Pleistocene oolitic sand and limestone deposits (coarse stipple). The distribution of oolitic sand formation is spatially restricted at any one time. Ooids, however, form a nearly continuous rim to these platforms and peninsulas when considering an integration of the oolitic and coated grain sand bodies from the late Pleistocene highstand (120,000 YBP), the later Pleistocene lowstand (to 15,000 YBP), and the Holocene. Also shown is distribution of carbonate sediment on the Florida Shelf. GBB: Great Bahama Bank; LBB: Little Bahama Bank. (Data from Ball, 1967; Enos, 1977; Ginsburg and James, 1974; and authors' observations).

TYPES OF TIDAL-CURRENT–GENERATED OOLITIC SEDIMENT BODIES

In tide-dominated oolitic shoals, each tidal cycle agitates much of the shoal crests, flanks, and channel bottoms, permitting ooid growth in water depths ranging from intertidal to as much as 10 m (Dravis, 1979). Throughout mobile areas of active ooid generation, the majority of ooids have well-developed aragonitic coatings, and the outermost lamellae are relatively free of micritization (Dravis, 1979). These areas of active oolitic grainstone formation contain relatively few peloids or skeletal grains. In addition, the sand body exhibits preserved internal sedimentary structures dominated by bidirectional cross-stratification, reflecting the bidirectional tidal-current influence (Ball, 1967).

Four sediment body geometries are recognized in areas of tidal-current agitation. These are: (1) shelf-margin-parallel marine sand belts (Ball, 1967); (2) shelf-margin-parallel tidal bar belts (Ball, 1967); (3) flood tidal deltas (Halley et al., 1983); and (4) shoaled flats with ooid generation on flanks and in channels (Harris, 1979). Variations in geometry are caused by variations in tidal current strength, preexisting topographic setting, and evolving coexisting topographic setting.

(1) Marine Sand Belts. Oolitic marine sands belts parallel to the platform margin occur in areas of moderate sediment production and moderate tidal currents (Cat Cay oolitic belt, western Great Bahama Bank, Figure 3). Cat Cay oolitic sand belt is oriented parallel to the platform margin, and the shoal crests are, in places, emergent at low tide. The oolitic sand belt is 1–2 km in width and is separated from the platform margin by about 1 km of coralgal sand, but there is no reef development. Channels cut across the axis of the belt but shallow and digitate on the seaward and bankward flanks. The oolitic sand belt extends for more than 180 km along the western margin of Great Bahama Bank (Ball, 1967).

Sedimentary structures across this active sand belt consist of (a) sand waves and megaripples migrating across the belt (bankward or eastward dipping), (b) pronounced sand spillover lobes extending bankward from the poorly defined channels (strata dipping bankward), and (c) a seaward-flank mix of spillover lobes (dipping seaward or westward) and wave-generated sand waves (dipping bankward) (Ball, 1967). Smaller ripples cover the surface of most of the sand body and are active daily.

Bankward from the oolitic sand belt, the bottom deepens to 5–8 m and is largely stabilized by sea grass (Newell et al., 1960). Ooids moved bankward during storms are a significant constituent for the first several kilometers bankward of the shoal. Further platformward, the oolitic grainstone yields to a peloidal packstone to wackestone (Purdy, 1963; Enos, 1977; Wanless et al., 1981).

The Pleistocene oolitic facies of the Miami Limestone along the southeastern coast of Florida is thought to be a fossil equivalent (Hoffmeister et al., 1967).

(2) Oolitic Tidal Bar Belts. Oolitic tidal bar belts form in areas of strong tidal flow. This is most common at the ends of deep ocean reentrants (on Great Bahama Bank at the southern end of the Tongue of the Ocean and the northern end of Exuma Sound; Figure 3). In these settings, oolitic sands are forming from the margin platformward for 15–20 km. Strong tidal currents organize the sediment body into elongate sand ridges and channels extending roughly perpendicular to the platform margin. Individual ridges and channels may be 10–20 km in length and separated by channels 1 km in width and 8–10 m in depth. Although strong tidal flow moves onto and off of the platform through these channels, an important tidal flow component is across the ridges, producing cross-stratification within the tidal bars that dips parallel to the strike of the margin (perpendicular to that described for [1]; Ball, 1967).

A coralgal sediment with oolitic coatings may dominate the outer kilometer of the platform margin, but barrier reefal growth is not common. Some of the oolitic bars are straight, but many are sinuous with flood and ebb chutes. The surfaces of the channels and bars have areas of mobile oolitic sand, algal and seagrass-stabilized sand, and marine hardground formation, including growth of hardened, columnar, normal marine stromatolites (Dravis, 1983).

The active oolitic tidal bar belt at the north end of Exuma Sound extends about 15 km onto the platform. Further platformward is a stabilized continuation of this oolitic shoal called Yellow Bank (Taft et al., 1968). This appears to have formed as intense currents, generated during major hurricanes, swept sand platformward from the active oolitic tidal bar belt. Active cross-bank circulation has promoted extensive marine cementation on both the active tidal bar belt and Yellow Bank (Taft et al., 1968; Dravis, 1979).

The oolitic limestones comprising Andros Island and the lower Florida keys are Pleistocene examples of tidal bar belts (Hoffmeister et al., 1967; Evans, 1987).

(3) Oolitic Flood-Tidal Deltas. Oolitic flood-tidal deltas form adjacent to channels between Pleistocene limestone islands (i.e., west side of Exuma Sound; Figure 3; Halley et al., 1983). These form individual fan-shaped sediment wedges extending 0.5–1 km platformward of passes through the emergent topography. Large, marine cemented stromatolites are associated with some of these oolitic deltas (Dill et al., 1986).

(4) Ooid Sediment Production Restricted Through Shoaling. Broad oolitic sand complexes may have accumulated to sea level, modifying the pattern of oolitic sand production (Joulter's Cay area). Joulters Cay oolitic complex is situated on the eastern side of Great Bahama Bank, just north of Andros Island (Figure 3). Here an earlier belt of oolitic sand has built to sea level, and a series of oolitic sand islands have formed within the last one thousand years (Harris, 1979; Halley et al., 1983). Much of the area has shallowed to an algal- and seagrass-stabilized sand flat. Present ooid production occurs on the shoals to the east and north and on the island beaches. Ooids are moved bankward by storms (Imbrie and Buchanan, 1965), further building the islands and expanding the stabilized sand flat.

CAICOS PLATFORM

Setting

Caicos Platform is the southernmost platform of the Bahamian archipelago that has significant emergent islands (Figure 1). The platform is located between 71°30′–72°30′ W and 21–22°N and is situated 850 km southeast of Florida and 200 km north of Hispaniola. The platform is steep sided, 100 km in length north–south, and 70 km in width east–west.

Importantly, the platform is influenced by brisk prevailing easterly trade winds and associated wind-generated waves and currents and is situated within the Antilles Current. Waves and currents promote effective bottom agitation and set up effective westerly cross-bank circulation. Water depth over most of the open platform interior is less than 5 m, and the bottom is frequently agitated and rippled.

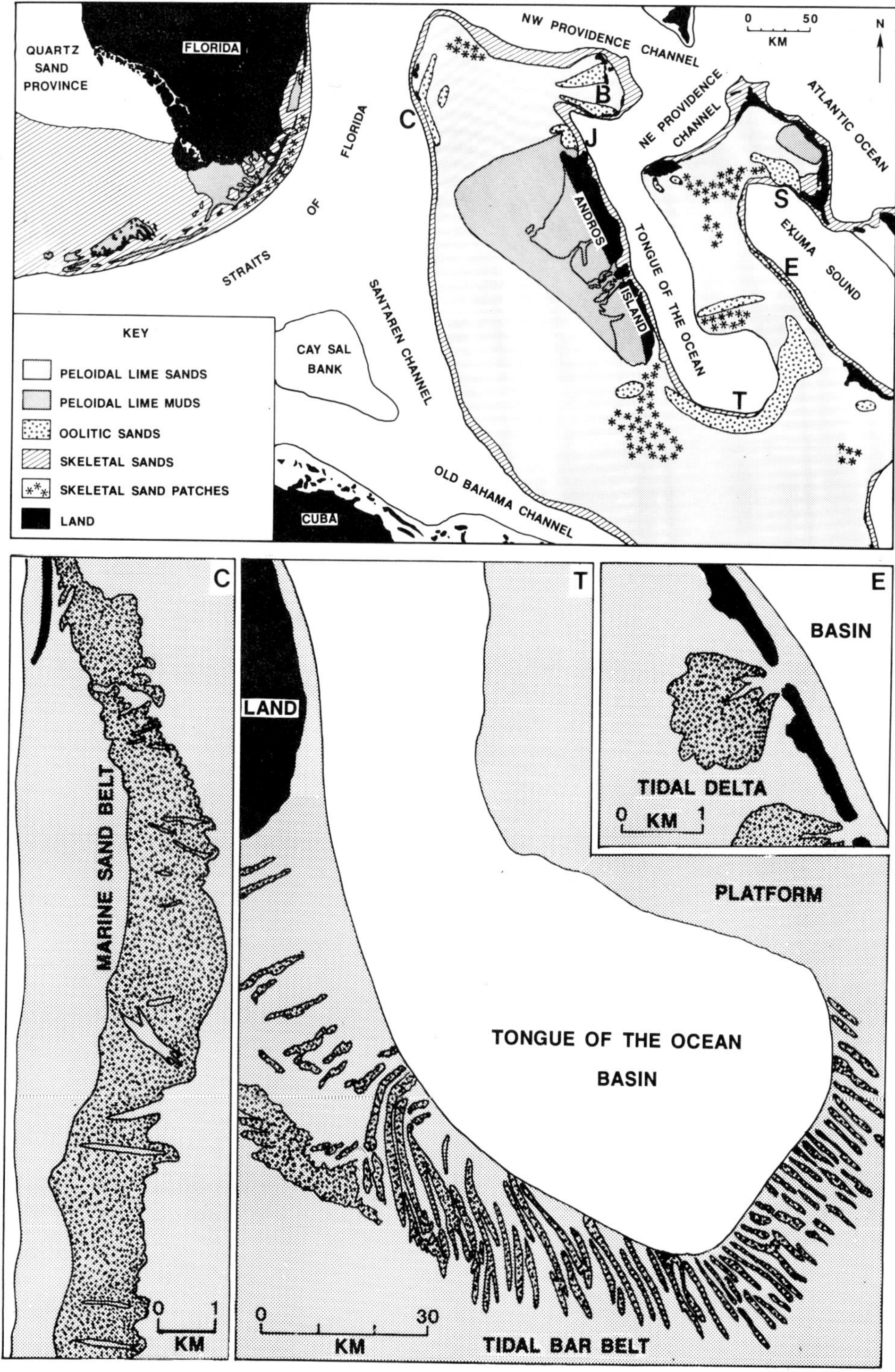

Figure 3. Upper: Map of the general distribution of surficial sediment types in south Florida and Great Bahama Bank (modified from Enos [1977] and Wanless et al. [1981]). Holocene ooid shoals are indicated by letters: B = Berry Islands oolitic shoals; C = Cat Cay oolitic marine sand belt; E = Oolitic tidal deltas at island gaps on margin of Exuma Sound; J = Joulters Cays oolitic band complex; S = Schooner Cays oolitic tidal bar belt; and T = oolitic tidal bar belt at south end of Tongue of the Ocean. Twenty-meter-depth contour defines platform margins (from Wanless et al., 1981). Lower: Oolitic sediment body geometries of (C) Cat Cay oolitic marine sand belt; (T) oolitic tidal bar belt at south end of Tongue of the Ocean; and (E) oolitic tidal deltas in Exumas. Note scale change. Sketched from aerial photographs and satellite images.

Physiography

The predominant physiographic feature on Caicos Platform is the string of high Pleistocene islands occurring within a few kilometers of the northwestern, northern, and northeastern platform margins (Figure 4). These islands, formed predominantly of oolitic dune ridges and beach ridges, create a nearly continuous barrier to tidal flow, wave energy and circulation along the northern margin and effectively isolate the northern skeletal sand rim from the nonskeletal sands of the platform interior (Figure 5). On the northeastern and northwestern margins, emergent islands form a discontinuous barrier and focus sites of on-bank and off-bank circulation and sediment transport through breaks in the island chain. An emergent Pleistocene reefal rim occurs along portions of the western margins of Providenciales (at Northwest Point on Figure 10) and West Caicos (Figure 9).

Five positive topographic/bathymetric features define the present physiography of Caicos Platform (Figure 4), the first three of which produce a rimmed platform margin to the northern, western, and much of the southern margins: (1) a reefal barrier rim on the northern, northwestern, and northeastern margins; (2) emergent Pleistocene and Holocene islands, of skeletal and oolitic grainstone, inland from the reefal barrier; (3) a broad subtidal storm levee along the open leeward (southwestern) portions of Caicos

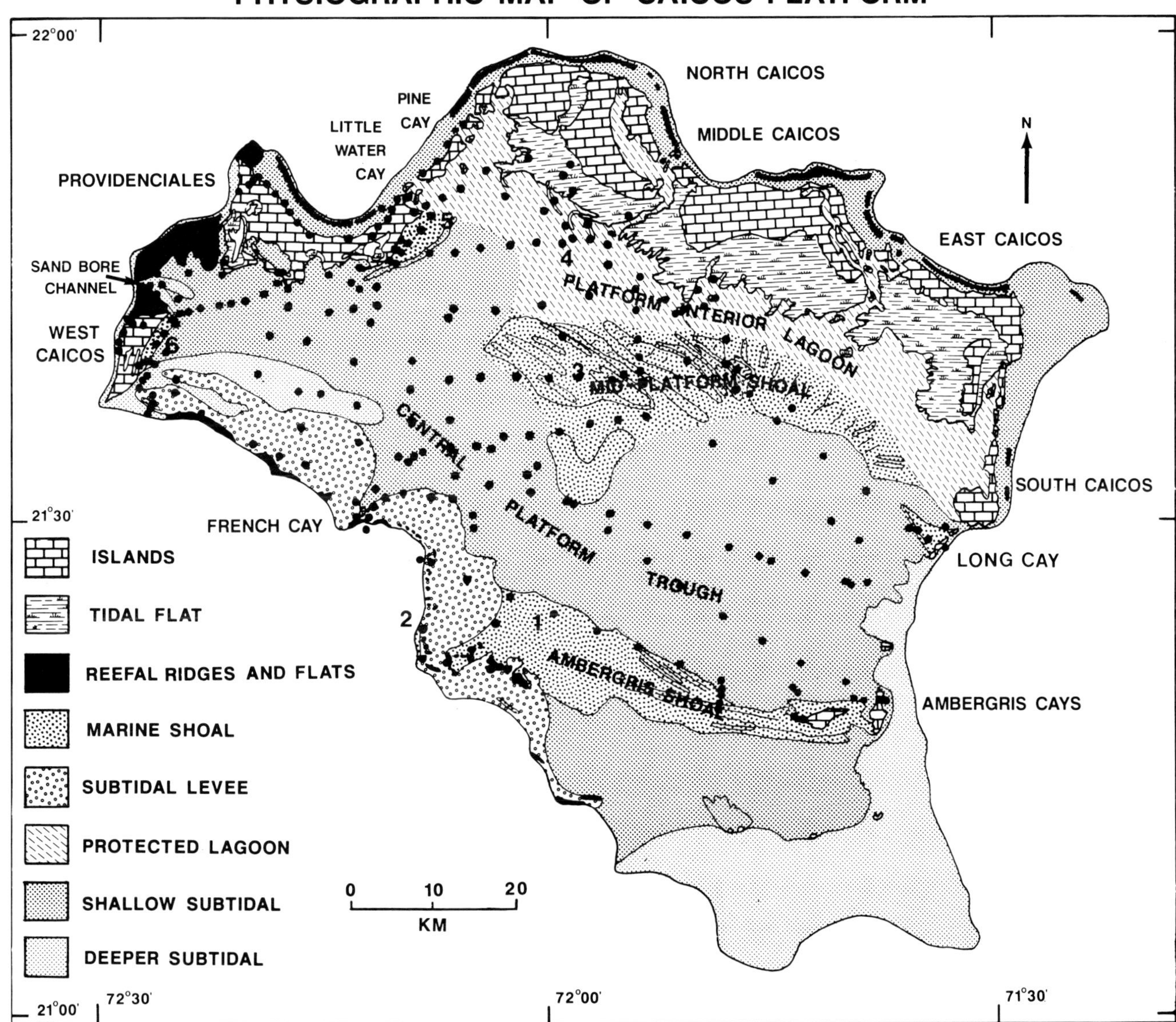

Figure 4. Map of Caicos Platform physiographic elements. Dots are location of surface sediment samples collected, thin-sectioned, and analyzed. 1 = Ambergris Shoal; 2 = West Spit; 3 = Mid-Platform Shoal; 4 = Platform-interior lagoon; 5 = Long Bay strand plain; 6 = West Caicos strand plain (from Wanless et al., 1989).

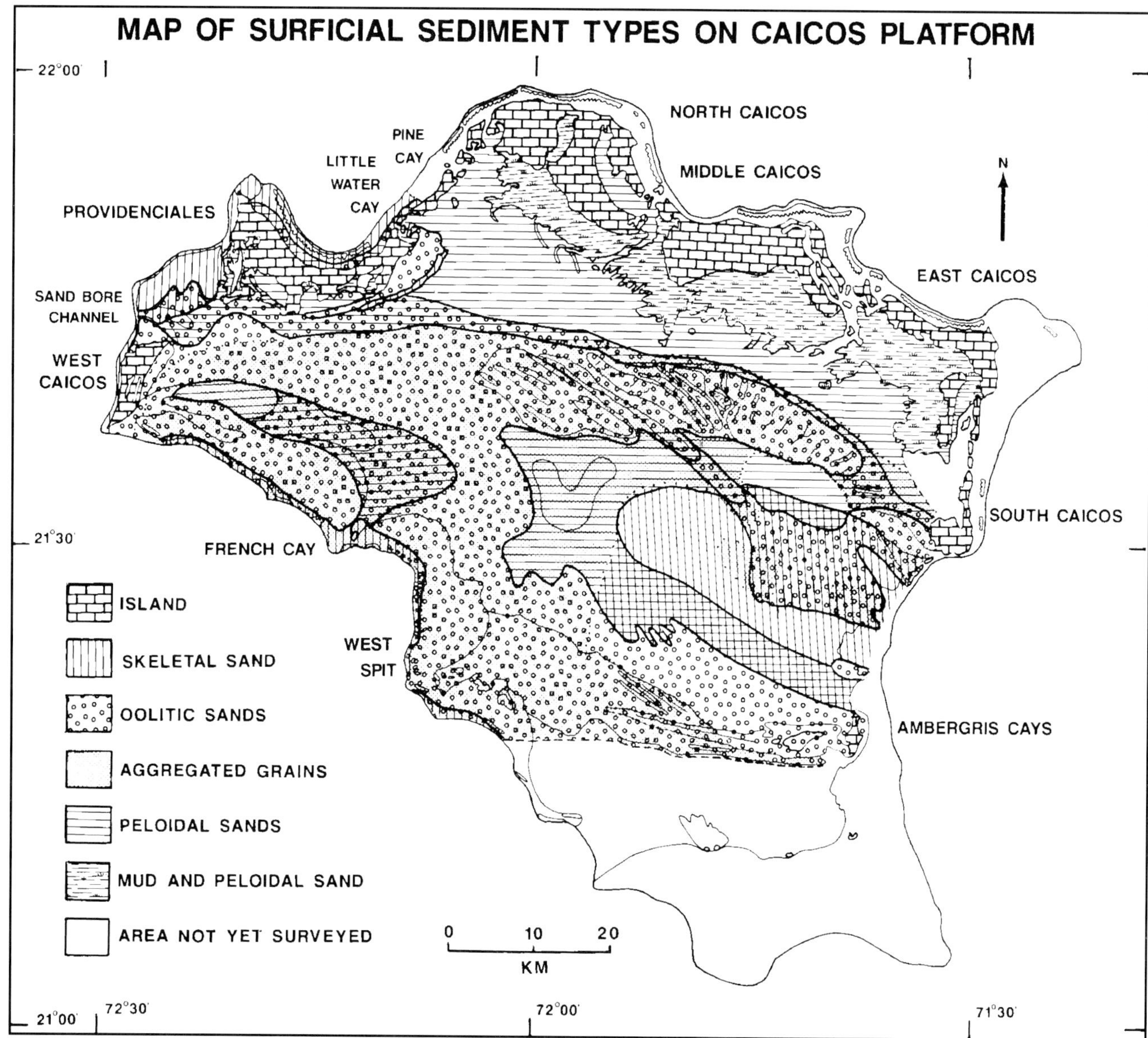

Figure 5. Map of Holocene surficial sediment types on Caicos Platform (from Wanless et al., 1989).

Platform; (4) an elongate Mid-Platform Shoal of Holocene grainstone extending east–west across the north central portion of Caicos Platform interior; and (5) an elongate Ambergris Shoal of Holocene grainstone extending westward across the platform from Ambergris Cay.

A broad Holocene evaporitic tidal flat forms a sediment wedge extending 3–10 km southward into the platform interior from the emergent islands along the northern margin (Figure 5).

These five positive topographic features define four sedimentary depressions on the platform (Figure 4): (1) a northern back-reef lagoon between the barrier reef and the emergent islands; (2) a very shallow protected platform-interior lagoon between the southern margin of the tidal flats and the Mid-Platform Shoal of oolitic and grapestone grainstone; (3) a central platform trough defined by the Mid-Platform Shoal and Providenciales on the north and the Ambergris Shoal of oolitic grainstone and the subtidal levee rim at the southern margin; and (4) the deeper, storm-swept southeastern portion of the platform, where topographic features have not provided sufficient protection for accumulation of significant Holocene sediment (Figure 4).

Sediment Types

Holocene

Skeletal Grainstone. Coralgal framestones, rudstones, and skeletal and oolitic grainstones dominate the northern and eastern margins of Caicos Platform (Figure 5). On the northern margins, reef-derived sands and rubble are swept across the back-reef lagoon toward the shore and contribute to lagoonal infilling and strand plain growth. Heavily biocorroded skeletal grainstones dominate the eastern portion of the central-platform trough. On the southern and western platform margins, skeletal sand is a minor component (10–20%) except near a few marginal reefs.

Peloidal Grainstones and Packstones. Peloidal grainstones to low-mud packstones dominate the most northern portion of the protected platform interior and the adjacent tidal flats. Peloidal grainstones grade southward into the grapestone and irregular ooid grainstones of the Mid-Platform Shoal and irregular ooid grainstones of the western platform interior (Figure 5). Peloids also dominate the deeper parts of the western platform interior. Most peloids are hardened. These environments are dominated by high-relief callianassid burrow mounds, sparse *Thalassia* and *Syringodium*, sparse calcareous algae, and patches of *Batophora* and *Acetabularia*. Carbonate mud is accumulating in significant amounts only in the most protected environments adjacent to the tidal flats.

Oolitic Grainstones. Oolitic grainstones are the dominant sediment type on Caicos Platform. Oolitic grainstones dominate the nearshore, beach, and dune sediments of east-facing (windward) coastlines in the platform interior, some south-facing beaches on the platform interior, Ambergris Shoal, the Mid-Platform Shoal, and much of the open subtidal sediment of the central and western platform interior (Figure 5). Oolitic sands are also an important component of those beaches facing the northern, eastern, and western platform margins that receive strong wave agitation but are not overwhelmed by reef-derived sands. Transport of ooids and grapestone grains from the platform interior provides an important contribution to the leeward storm levee sediments along the southern and western platform margins. Ooids vary from regular concentric to irregular in growth form depending on agitation setting (see next section).

Grapestone Grainstones. Grapestone grain aggregates are an abundant sediment component on the Mid-Platform Shoal, flanks of Ambergris Shoal, portions of the central-platform trough, and the subtidal levee rim on the southwestern and southern margins (Figure 5). Marine cementation occurs throughout areas of grapestone.

Pleistocene. Pleistocene islands are dominated by oolitic grainstones. These ooids are regular concentric and have very well-defined growth lamellae and essentially no micritization.

Emergent barrier reefs of skeletal framestone, boundstone, and grainstone occur at the western margin of West Caicos and Providenciales.

Pleistocene sediments beneath the platform interior have not yet been drilled.

Types of Ooids

There are two basic types of Holocene ooids on Caicos: regular concentric ooids and irregular ooids.

Regular concentric ooids grow toward spherical by the addition of oolitic lamellae of tangentially oriented aragonite needles onto a nucleus (Figure 6). Boring and micritization may or may not affect growth surfaces of the ooids, but they do not affect the growth form. Growth lamellae in regular concentric ooids thus vary from well-preserved (e.g., B2 in Figure 6) to poorly preserved (e.g., B1 in Figure 6) depending on the intensity of micritization.

Irregular ooids are those that tend to become more irregularly shaped as they grow larger (Figure 7). They tend to grow away from spherical. Processes of grain degradation and/or aggregation are always associated with irregular ooids and cause the irregular growth morphology. Grain corrosion and degradation include microborings, macroborings, and grain fracture (Figures 7E–G). Grain aggregation includes encrustation and cementation by skeletal agglutination, and organism binding (e.g., algal and sponge) (Figures 7A–F). This grain modification produces positive and negative changes in the surface topography of the ooid grains that affect the form and continuity of subsequent oolitic lamellae. Changes in grain form by aggregation and breakage produce new surfaces to which subsequent lamellae conform. Macroborings may produce sharp, narrow embayments in ooid grain surfaces. These embayments become sites of disruption for subsequent growth, and the lamellae may take on a lumpy or discontinuous lamellar growth form, with individual domes ending at grain surface disruptions. In some cases, additional lamellae may develop a continuous coating over the irregular surface (Figures 7A, B, F, G); in others, disruptions persist during subsequent growth (Figures 7C–E).

Ooid form appears to be related to frequency of bottom agitation. Three frequencies of bottom agitation are inferred: (1) continuous (bottom agitation occurring on a daily basis); (2) persistent (bottom agitation, but punctuated by quiescent periods ranging from days to weeks); (3) intermittent (bottom agitation interrupted by times and areas having weeks to months of bottom stability).

Regular concentric ooids occur where there is continuous to persistent bottom agitation and where grain aggregation and macro-scale grain degradation are minor. Areas of more continuous bottom agitation have ooids with the best preserved lamellae; areas of persistent agitation (not continuous) have more micritized ooids.

Irregular ooids occur in areas of persistent to intermittent agitation where there is some degree of grain aggregation and/or macro-degradation occurring. Grain aggregation and degradation occur in areas and/or times of bottom stability.

Environments of Holocene Ooid Formation and Occurrence

Regular concentric ooids occur (A) on wave-agitated shorefaces and (B) on wave-agitated shallow banks and the crests of shoals in the platform interior (A and B in Figure 8). Wave agitation may be from wind-generated waves or ocean swells. In addition to

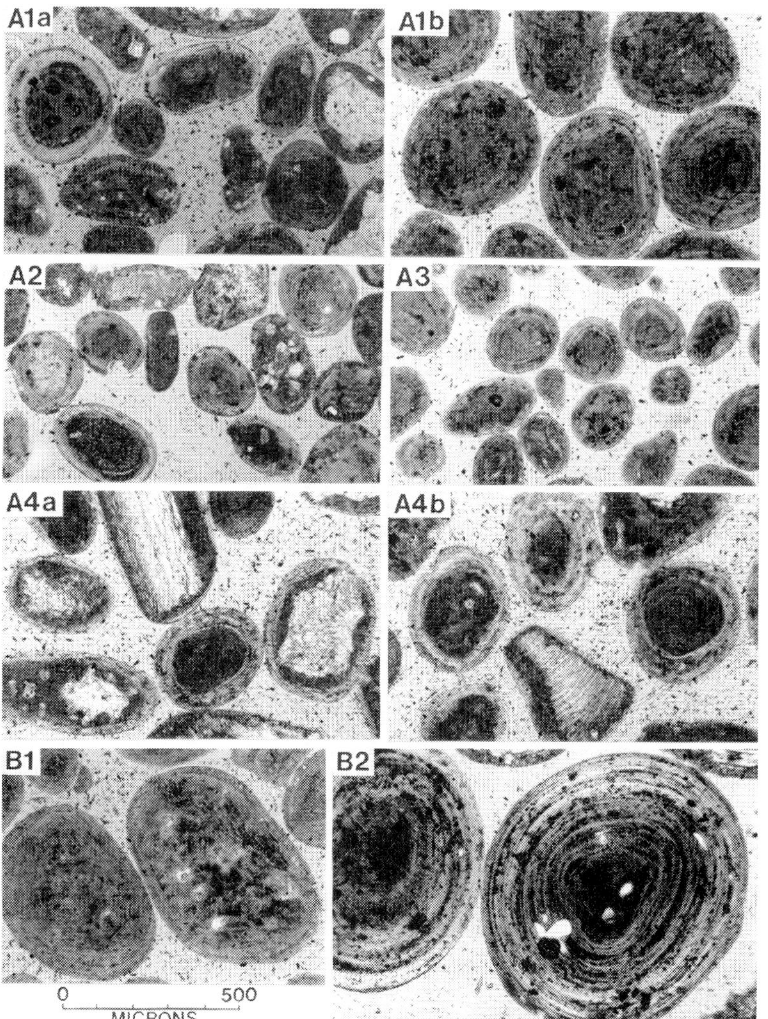

Figure 6. Thin-section photomicrographs of regular concentric ooids on Caicos Platform. Plane polarized light. Photo labels are keyed to text subheadings. See Figure 8 for locations. (A1a) Medium- to fine-grained ooids from the beach on the east coast of West Caicos. Oolitic lamellae are well to poorly preserved. These ooids are part of a 1-km-wide strand plain. (A1b) Medium-grained ooids from east-facing beach at Long Bay, Providenciales. Ooids have built a 100-m-wide strand plain. Oolitic lamellae are somewhat micritized. (A2) Medium- to fine-grained ooids from east-facing, platform-margin beach at Ambergris Cay. Oolitic lamellae are mostly well preserved. One to four Holocene beach ridges of oolite have formed on this east coast. (A3) Fine-grained ooids from protected south-facing beach in Taylor Bay, an embayment in the south shore of Providenciales. Lamellae are well to poorly preserved. One to six oolitic beach ridges have formed. (A4a) Medium-grained ooids from the platform-margin-facing north shore of northeast Providenciales adjacent to reef barrier. Ooids have built a Holocene strand plain as much as 1 km in width here. (A4b) Medium-grained ooids from northwest shore of West Caicos near reefal margin. Ooids have built a Holocene strand plain 0.5 km in width. (B1) Medium- to coarse-grained ooids from deeper subtidal elongate bar south of Providenciales. Oolitic lamellae are modified by micritization. (B2) Medium- to coarse-grained ooids with well-defined lamellae, from agitated crest of Ambergris Shoal.

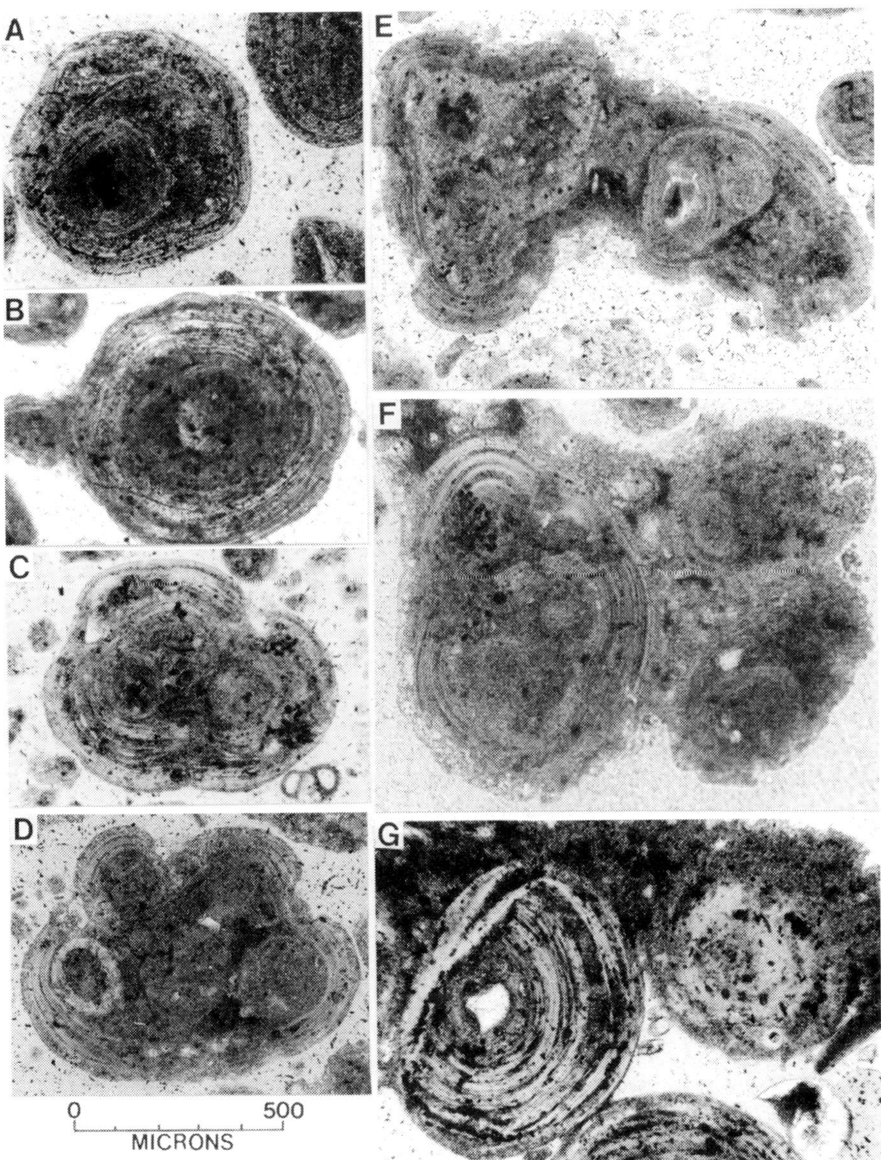

Figure 7. Thin-section photomicrographs of irregular ooids. Plane polarized light. (A) Irregular, lumpy growth from small patches of aggregation on grain surface (open western interior of platform, 5 m water depth). (B) Lumpy surface resulting from discontinuous growth of oolitic lamellae (Mid-Platform Shoal, less than 1 m depth). (C) Persisting irregular growth after aggregation associated with borings (open western interior of platform). (D) Discontinuous growth lamellae on grain aggregate. Growth lamellae are well defined on convex surfaces but poorly defined or absent on concave surfaces (Mid-Platform Shoal). (E) An irregular grain aggregate onto which an irregular oolitic coating has formed. As before, lamellae are well formed on convex surfaces and poorly defined to absent on convex surfaces (Mid-Platform Shoal). (F) Irregular ooid showing several stages of grain aggregation and growth of oolitic lamellae (Mid-Platform Shoal). (G) Ooids in small piece of marine-cemented intraclast from flank of Ambergris Shoal. Irregular ooid at right has previously undergone phase of grain destruction with subsequent oolitic lamellae on unconformable surface. Grain at right is presently undergoing partial destruction by microboring disruption of grain surface.

providing a setting for ooid generation, bottom agitation by waves and wave-generated currents influences the form of oolitic sediment bodies. Wave-generated currents may organize ooids into elongate sediment bodies (B and C in Figure 8). Storm waves and wind may either organize ooids into beach ridges and dunes adjacent to the shore or produce longshore currents that sweep shore-generated ooids away from the site of origin.

Irregular ooids form on persistently to intermittently agitated bottoms throughout much of the interior of Caicos Platform. Irregular ooids occur and appear to be forming on both (C) platform-interior shoal complexes and (D) the open platform interior (C and D in Figure 8).

The characteristics and occurrences of these four environments of ooid formation are given in the following sections.

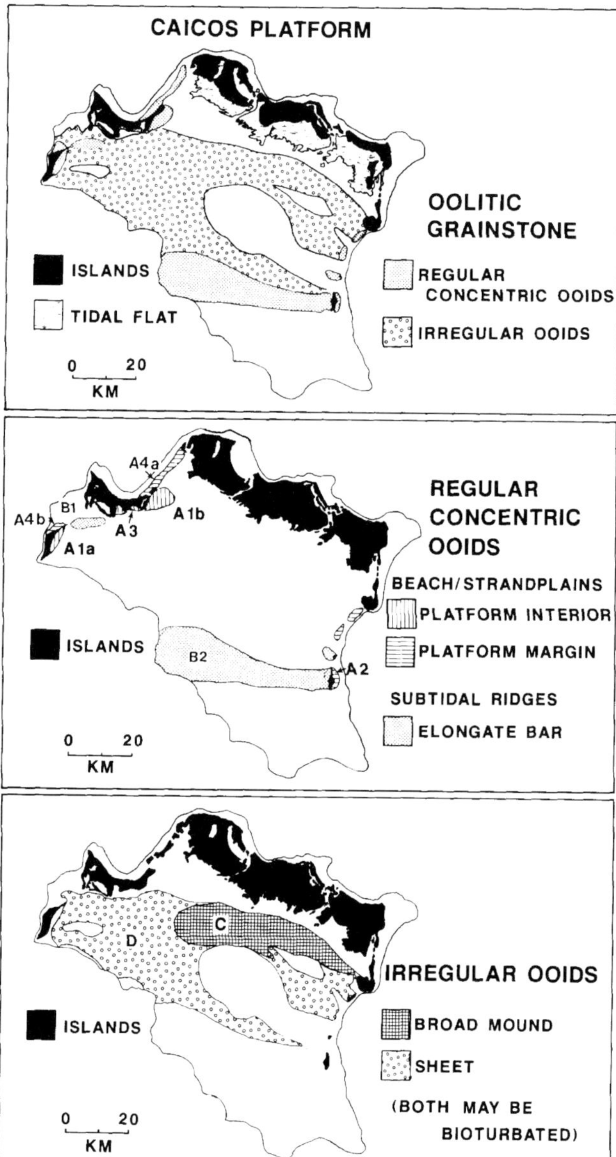

Figure 8. Maps of Caicos Platform showing distribution of ooid grain types. Upper panel shows distribution of regular concentric and irregular ooids. Middle panel shows locations of ooids generated in association with shorelines on the (A1) windward platform interior, (A2) windward platform margin, (A3) leeward-facing platform interior, and (A4) open-ocean-facing platform margin, and also on (B) the east–west-trending Ambergris Shoal. Lower panel shows locations of sediment bodies dominated by irregular ooids, including (C) the Mid-Platform Shoal and (D) much of the open platform interior.

(A) Prograding Strand Plains; Regular Concentric Ooids

Bottom agitation by waves, focused at the shoreface, provides a variety of settings for generation of regular concentric oolitic sands on coastlines facing both the platform interior and the platform margin. Storms have organized these shore-generated sands into extensive strand plains of low beach ridges and/or large dunes. Shoreface agitation by wind-generated waves is actively producing ooids on windward (east-facing) shorelines of (1) the platform interior and (2) the platform margin. In addition, (3) beaches on some south-facing embayments of the platform interior are generating fine-grained ooids and associated strand plains. (4) On the exposed northern coastlines, shoreface agitation by ocean swells is producing oolitic sands and strand plains in those areas where reef-derived skeletal sands are not being transported to the coast.

(A1) Windward Shores on the Platform Interior. Oolitic sands are forming at and producing extensive strand plains on the windward shorelines of West Caicos and Providenciales (Long Bay) (Figures 8, 9, 10). These shorelines face the platform interior. Ooids on these coastlines are regular concentric ooids and are mixed with peloids (Lloyd et al., 1987). Some of the ooids are highly micritized (Lloyd et al., 1987), and many of the peloids are simply completely micritized ooids.

(A1a) West Caicos. On the eastern coast of West Caicos, ooids are forming in the swash zone and shallow offshore (Figure 6A1a), an area subjected to persistent wind-generated wave agitation and without significant tidal influence. Lloyd et al. (1987) radiocarbon dated the beach ooids on the east shore of West Caicos and found them to date at 790 years before present (YBP), younger than those in the adjacent dunes and very young for a bulk sample. They concluded these ooids formed in the wave swash zone. Indeed, the beach zone is well agitated during much of the year. Lloyd et al. (1987) argue that the sands could not have come from further out on the platform because the adjacent offshore is stabilized by sea grass beds. Sea grass beds occur 50–100 m from the shoreline in 2–3 m of water along the southern part of this coast; Pleistocene limestone occurs in 1–2 m of water seaward of within 50 m of the shore along the northern part of the coast. Thus, there is neither a

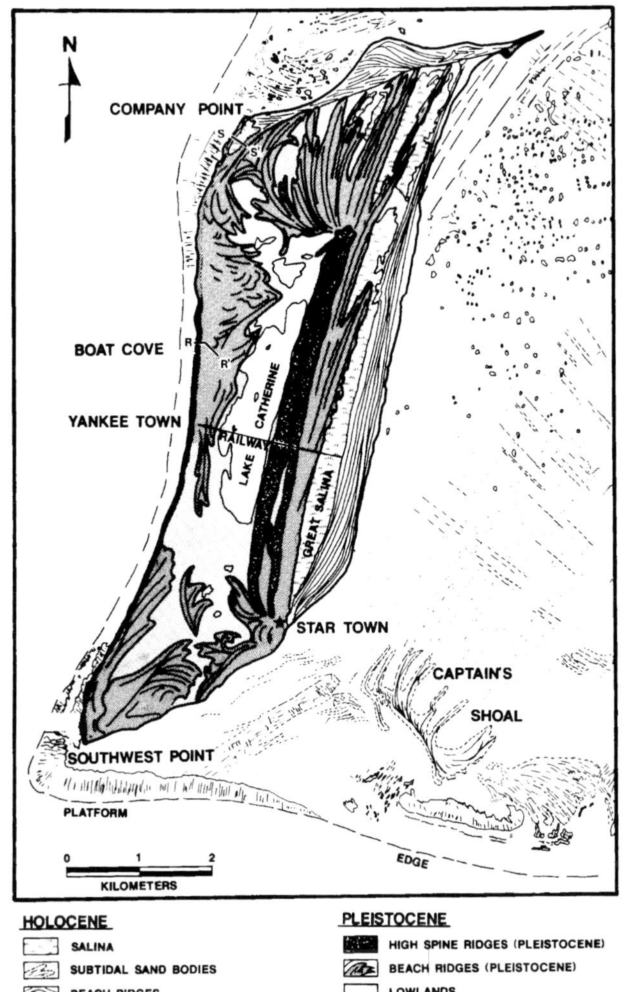

Figure 9. Map of West Caicos (Figure 4) showing main physiographic features and locations of drilling transects R–R′ and S–S′. The Holocene and Pleistocene strand plains are predominantly oolitic and have caused gradual enlargement of this leeward-margin island and blocking of cross-bank transport (from Wanless et al., 1989).

source for ooids further seaward nor an environment for bed load transport (Wanless, 1981).

The scale of this Holocene oolitic dune and beach ridge complex (up to 0.5 km in width, 7 km in length, with ridges 2–10 m in elevation and overlying a limestone surface 1 to more than 5 m below sea level) attests to the large volume of Holocene oolitic sand that has formed along this windward, platform-interior beach (Figure 9). The strand plain topography has created a broad impounded depression between Holocene sands and adjacent Pleistocene ridges. This depression is a salina (Great Salina; Figure 9) in which magnesium-charged brines are forming and gypsum and dolomite are precipitating (Leaver, 1985; Rossoff et al., 1989).

(A1b) Long Bay, Providenciales. Oolitic grainstones at Long Bay, adjacent to the southeast coast of Providenciales (Figures 5, 6A1b, 10), form at the windward beach (Lloyd et al., 1987) and on a broad, shallow shoal exposed to prevailing wave agitation but protected from significant tidal flow. Water depths are less than 2 m in this area, and shallow SW–NE-trending sand ridges extend across the subtidal shoal (Figure 10). Wind waves generated by prevailing easterlies are a persistent and effective cause of bottom agitation and sediment movement both at the beach and across the very shallow offshore platform. Lloyd et al. (1987) found that ooids from the present beach dated younger (690 YBP) than those from the agitated offshore (970 YBP) and concluded active ooid generation in the beach swash zone.

The agitated shallow marine bottom extends over 1 km seaward from the shoreline and is covered with active oolitic sand organized into westward-moving small sand waves and megaripples, with smaller oscillation ripples and ripple trains superimposed on them. This is likely to be preserved internally as unidirectional cross-stratification. Storms may generate upper flow regime planar stratification (Imbrie and Buchanan, 1965). Portions of the subtidal oolitic grainstone are covered and partially stabilized with a surficial algal scum mat. Hardground areas and subtidal, normal marine, grainstone stromatolites are scattered throughout this oolitic shoal (Wanless et al., 1989). Many of these hardgrounds initiated on a very coarse storm-rubble sheet conglomerate that is commonly less than 30 cm beneath the sediment-water interface.

A prograding strand plain of low beach ridges and higher dunes, 0.5 km or less in width, has formed adjacent to the beach. Some sand is lost to the northern platform interior by southeastward longshore drift (Figure 10).

(A2) Windward Shores of the Platform Margin. At least two of the islands on the eastern margin of Caicos Platform have oolitic sands forming in the exposed beach and nearshore zone and accumulating as beach ridge and dune deposits along the eastern side of the islands (Long Cay and Ambergris Cay; Figures 4, 5, 8). Ooids mostly have well-defined lamellae, have skeletal and peloidal nuclei, and are mixed with skeletal grains. Agitation varies from continuous at the shore to persistent in the adjacent marine environment.

The eastern margin of Ambergris Cay has several 8- to 12-m-high Holocene oolitic grainstone dunes fronted by a narrow zone of low beach ridges. The seaward shelf is narrow (less than 1 km) and has only scattered reefs, and the windward shore has little skeletal sand. Medium- to fine-grained ooids have regular concentric lamellae coating a variety of skeletal and non-skeletal nuclei (Figure 6A2).

Long Cay is a narrow island nearly at the eastern platform margin just south of South Caicos (Figures 4, 11). It appears to be entirely Holocene. It is basically one dune, 10–15 m in height, with steeply dipping avalanche strata on the platformward margin. It is cemented by meteoric and ocean spray waters and presently has a sheer, erosional seaward face. On a

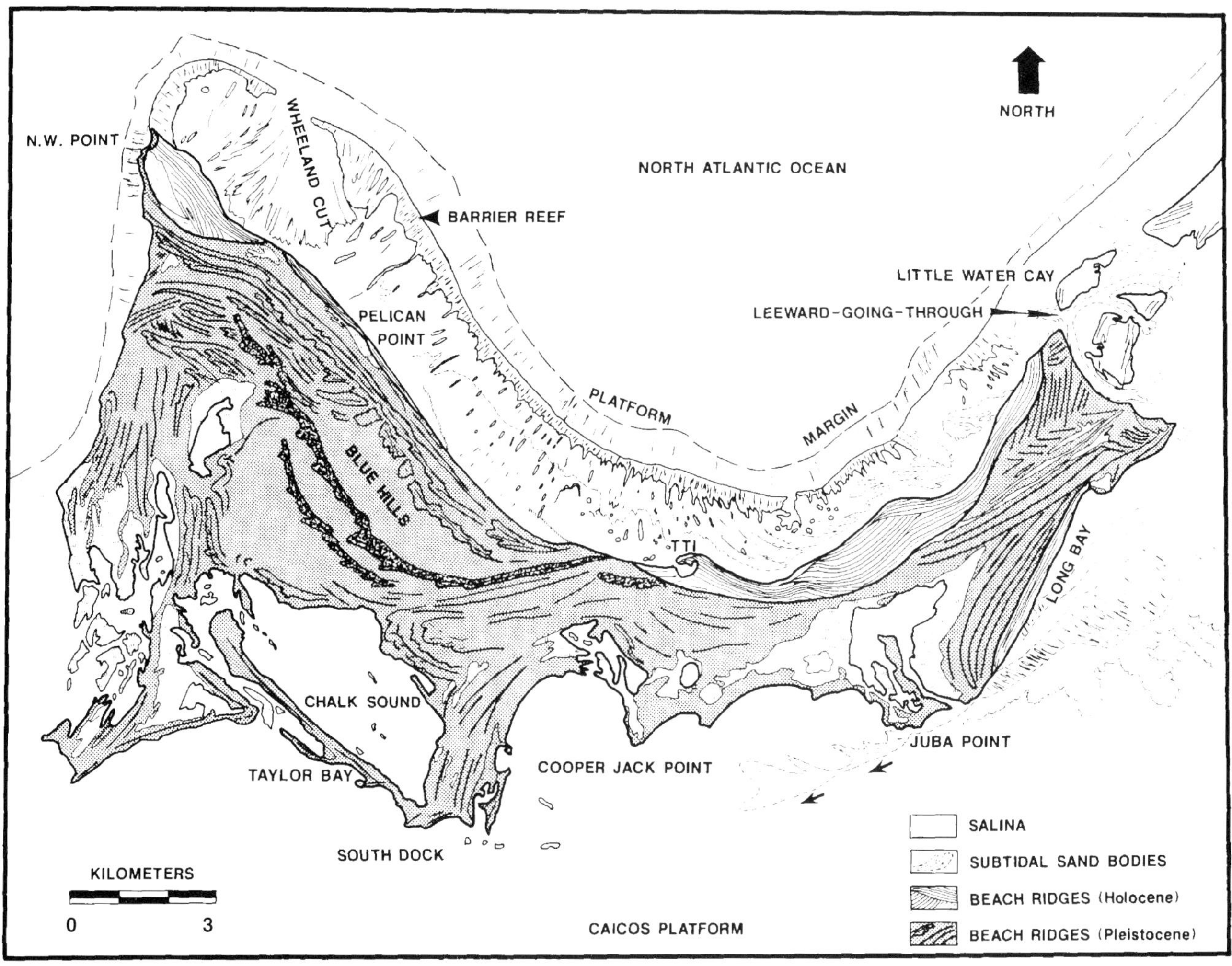

Figure 10. Map of Providenciales (Figure 4) showing physiographic features. Pleistocene and Holocene dunes and beach ridges are predominantly oolitic and record a history of ooid sedimentation on the north, south, and east sides of the island. Holocene ooids are forming on the windward, wave-agitated beach and offshore shoal at Long Bay, on the leeward, platform-interior-facing beaches between Taylor Bay and Juba Point, and on the open-ocean-facing beach on the eastern part of the north shore (see Figure 12) (from Wanless et al., 1989).

day with brisk easterly winds (about 15 knots) oolitic sand was being blown up the seaward face of the dune and across the top.

The sediment comprising the island is predominantly regular concentric ooids with well-defined lamellae; skeletal grains are a secondary component. As the island is at the windward margin, the only production area available for ooids is the seaward beach and nearshore zone. This is a setting where the shore zone is not overwhelmed by skeletal sand influx.

(A3) Leeward Shores of the Platform Interior. Small ooids occur along the beach of several south-facing, platform-interior beaches on Caicos Platform. These are on the south side of the island of Providenciales (e.g., Taylor Bay; Figure 10). The ooids are regular concentric but tend to be less than 200 μm in diameter (Figure 6A3). Nuclei are predominantly peloids and micritized grains. The ooids occur on the sloping shoreface and adjacent very shallow bay bottom but decrease in abundance seaward across the embayment. The beach ridges comprising the strand plain adjacent to the beaches are predominantly fine ooids with skeletal grains in the coarser fraction.

(A4) Open-Ocean-Facing Shores of the Platform Margin. The exposed northern margin of Providenciales and the northwestern margin of West Caicos have regular concentric ooids as a predominant component of shore sands along those portions of the shore where (a) oceanic swells generate continuous to persistent shore agitation and (b) skeletal sands from the seaward reefs are not being transported to the shoreline (Figures 8, 12). Oolitic grainstones are also an important component of the smaller islands to the east of Providenciales (e.g., Little Water Cay; Figure 10). Most striking is the 1- to 2-km-wide

Figure 11. Photograph of the windward, east-facing side of Long Cay at the eastern margin of Caicos Platform, looking south (see Figure 4). The narrow island is over 10 m in height and composed predominantly of Holocene oolitic sands with incipient marine, spray, and freshwater cements. The intensely agitated shoreface appears to be a site of continued ooid growth even though this island shore is erosional. Other eastern shorefaces (Ambergris Cay) have gently sloping accretionary beaches.

strand plain of low beach ridges of oolitic grainstone that has formed along the northern side of eastern Providenciales (Figure 10). These Holocene ridges are sufficiently cemented by meteoric diagenesis that they cannot be sampled by hand coring techniques.

Ooids on these shorelines agitated by ocean swells and adjacent to coralgal margins are regular concentric, medium-grained and have an abundance of skeletal nuclei (Figures 6A4a, 6A4b). Lamellae vary from well-preserved to significantly micritized.

(B) Elongate Subtidal Banks Capped by Regular Ooids

Subtidal shoals of regular concentric ooids form elongate banks that are *not* associated with the margin of Caicos Platform.

(B1) Platform-Interior Deeper Subtidal Bank. Ooids form a deeper subtidal bank in an platform-interior area south of Providenciales (B1 in Figures 6 and 8). Water depths are 2.5–4 m, and the bottom has abundant sand waves. These regular concentric ooids are coarser than many of the previously described beach ooids. Nuclei are peloidal to micritic, reflecting the platform-interior setting. Lamellae are somewhat micritized, indicating that the setting is not constantly agitated (Figure 6B1).

(B2) Ambergris Shoal. Elongate oolitic sand shoals extend westward across Caicos Platform downdrift from isolated emergent islands. Most dramatic is Ambergris Shoal, which extends westward from Ambergris Cay completely across Caicos Platform (Figures 4, 13). Other oolitic shoals extend westward partway across the platform (i.e., bankward of Long Cay; Figure 4). Although tidal movement may aid in agitation at the crest of the shoal, agitation by wind-generated waves and the east-to-west cross-platform current appear to be the dominating influence controlling shoal morphology. The orientation of the shoals is the result of east-to-west transport.

Regular concentric ooids with well-defined lamellae and little micritization characterize the constantly active crest of Ambergris Shoal (Figure 6B2). These ooids are medium- to coarse-grained and predominantly have peloidal nuclei. Irregular ooids and micritized regular concentric ooids dominate the persistently to intermittently agitated flanks of the shoal (Figure 7G). Surficial marine hardgrounds are present throughout this shoal, and rounded marine-cemented clasts are abundant on the flanks.

Little Ambergris Cay is a Holocene oolitic grainstone island that has emerged on top of a portion of Ambergris Shoal (Figure 4). The island is made of low beach ridges that impound small salinas.

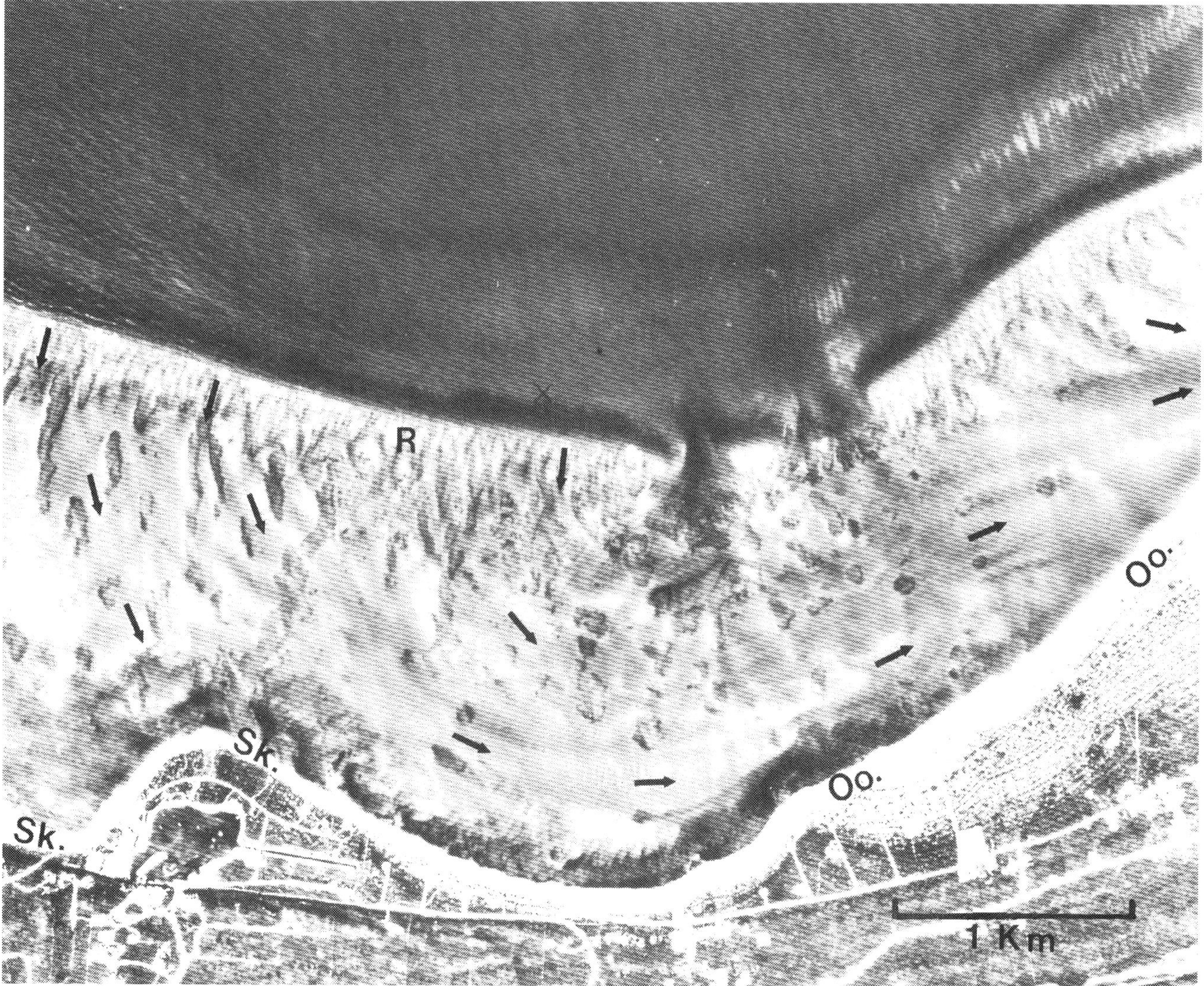

Figure 12. Vertical aerial photograph of the northern shore of eastern Providenciales with north at top (see Figure 10 for location). The eastern beaches (to the right) are dominated by oolitic sands (Oo.) because reefal sands do not transport to those shores but the persistent oceanic wave agitation does. The barrier reef (R) provides large volumes of skeletal sand into the back-reef lagoon, but this skeletal sand only reaches the shore where there is direct onshore transport (Sk.). During storm transport, strong shore-parallel current flow moves much of the skeletal sand parallel to the shore as indicated by bottom fabric of the back-reef lagoon (arrows).

Oolitic sand is moving off of the platform at the western, downcurrent margin of Ambergris Shoal. The focus of this off-platform transport to the west is defined by the position of islands on the eastern margin of the platform. At the western platform margin is a narrow, north–south-trending, oolitic sand spit (West Spit) paralleling the platform margin (Figure 5). Although the oolitic sand is moving off the platform here, onshore waves (from the west) maintain an oolitic sand ridge at this leeward margin.

(C) Irregular Ooids on Platform-Interior Shoal Complexes

The Mid-Platform Shoal is a very shallow grainstone shoal that extends east–west along the northern part of the platform interior but is separated from the tidal flats by a shallow peloidal grainstone lagoon, 5–10 km in width (Figures 4, 14). The Mid-Platform Shoal is cut by channels 0.5–2 m in depth at low water. The shoal areas are mostly less than 0.5 m in depth at low water, and the channels commonly have intertidal grainstone levees with washover lobes along their southwestern margins. One small island occurs on the crest of a shoal and is surrounded by a complex array of beachrock trends.

The surface of the Mid-Platform Shoal is covered with a very coarse grainstone, and the crests of some shoals have low mobile sand waves and patchy to continuous hardgrounds. Some hardgrounds are composed of submerged beachrock; others are areas of subtidal marine cementation or coral-algal and encrusting worm tube growth. Sponges, brown algae,

Figure 13. Oblique aerial photograph taken looking west along the Ambergris Shoal (see Figure 4 for location). Crest is about 100 m across. The crest is constantly agitated by wave and tidal agitation and contains regular concentric ooids. The flanks contain abundant marine-cemented oolitic grainstone intraclasts, and ooids become increasingly bored, micritized, and irregular. Dark patches are cloud shadows. (Photograph by A. Saller.)

Batophora, and *Acetabularia* are abundant on the flanks of the shoals. Sea grasses flourish on the deeper shoals and in channels. Burrow excavation processes are active throughout these shoals. Sea urchins, including the heart urchin, are abundant.

Repetitive occurrence of hardgrounds prohibits deep coring. Jet probing, however, did show as much as 4 m of Holocene sediment toward the northern flank.

The sediment is a mixture of irregular ooids (Figures 7B, D, E, F), grapestone, peloids, and minor amounts of skeletal grains. Microboring and micritization of grains and hardgrounds vary from minor to pervasive. Coarse skeletal sands are not abundant but may be concentrated by storms in subsurface burrow fills called tubular tempestites (Wanless et al., 1989) or at the base of surficial tempestite layers. Although ripple and small sand waves are present on the surface of portions of the Mid-Platform Shoal, excavating burrowers (e.g., *Callianassa*) and mixing burrowers (e.g., heart urchins) are abundant, and the sequence appears to be predominately bioturbated. Some hardground areas are colonized by coral patches.

Channels cut the western portion of the Mid-Platform Shoal and trend WNW–ESE and probably result from storm surges. North–south-trending ridges and swales are present along the central portion of the Mid-Platform Shoal. These appear to be widely spaced transverse sand waves, somewhat modified by channel scour, trending transverse to the shoal trend (Figure 14).

(D) Open Platform Interior

Regular concentric to irregular ooid grainstone dominates the sediment in the open interior of Caicos Platform except in the eastern portion (where there are abundant patch reefs and intense biocorrosion) and in the deepest areas (where there is insufficient agitation). The open platform interior is 3–5 m in depth, mostly covered by active to recently active ripples and low sand waves that indicate persistent to intermittent bottom agitation. In many areas the bottom is largely barren, and burrow-excavation mounds have little relief (Figure 15). Sea grasses or calcareous algae (the pioneer substrate recolonizer) form a sparse, patchy cover in some areas. Thin algal scum mats cover much of the surface, but the mat does not appear to be an effective stabilizer. There is an abundance of the open marine burrowers—heart urchins, the conch *Strombus gigas* (prior to recent overfishing), and blue crabs (which came to the surface during vibracoring). In this grainstone environment, *Callianassa* tubes tend to be mud-lined, and storm fillings are coarse grainstones.

Somewhat irregularly laminated ooids predominate across significant portions of the central interior of Caicos Platform (Figures 7A, C). Ooids range from nearly regular concentric to highly irregular depending on the amount of grain aggregation and biocorrosion. Oolitically coated grains are as common in 5 m of water as in the shallower zones. Peloids, grapestone, and skeletal grains are minor components. Grain micritization is common, and in the eastern part of the open platform interior, micritization tends to destroy much of the internal grain structure.

Pleistocene Oolitic Grainstones

Research to date on Pleistocene limestones of Caicos Platform is limited to exposures on the islands and shallow core borings on the western side of West Caicos (Wanless et al., 1989). Five important observations, however, emerge from this information.

First, the core borings show that oolites are an important part of the last two Pleistocene marine sequences at the western margin of Caicos Platform (Figure 16). Exposures on West Caicos, Providenciales, and North Caicos indicate that ooids were

Figure 14. Oblique aerial photograph taken looking west along Mid-Platform Shoal. View is about 3 km in width across middle of photo and is all on the shoal (<0.5 m at low water). The intertidal shoal crest extends as large patches up the middle of the photograph. Light-colored margins on north and east (right and front) are grainstone berms maintained by strong easterlies. Darker channels cutting obliquely across photograph are only 1.0–1.5 m in depth at low water. Marine-cemented areas and surfaces occur throughout the shoal and prohibit hand coring.

a dominant component of the last two Pleistocene marine sequences over a significant area of northern and western Caicos Platform and were the dominant component in building the Pleistocene beach ridges and dunes that comprise these islands.

Second, the Pleistocene ooid grains are regular concentric, have extremely well-defined lamellae, and have little or no boring or micritization (Figure 17). This suggests that the environments of ooid formation were constantly energetic.

Third, oolitic grainstones dominate the shore deposits facing the platform interior and the platform margin. The dominance along the platform margins suggests that reefs may not have been as effective sediment producers in some of the previous Pleistocene highstands. It is likely that rapidly rising sea level may have left the margin too deep to have had effective reefal growth at highstand. Hubbard (personal communication, 1987), studying the modern reefs in the Virgin Islands, has shown that there is a rapid decrease in rate of calcification below 12 m of water depth. The Sangamon 120,000 YBP highstand reached about 4 m above present sea level in the southern Bahamas (Wanless et al., 1989) to 7 m in the northern Bahamas and south Florida. There is evidence in many other Caribbean areas that Holocene reefal growth has produced much of the present barrier reef topography (Lighty et al., 1978; Macintyre, 1988) so that it is likely that the Pleistocene Sangamon topography of the northern and eastern margins of Caicos Platform were greater than 12 m in depth during the Sangamon highstand and were not a setting where reefal growth could have taken hold and caught up to sea level (Westphall, 1986; Wanless et al., 1989). If so, a barrier reef would not have formed to provide a flood of sediment to the shore and would not have effectively blocked oceanic wave agitation. The island shoreface and nearshore zones would have been constantly agitated and ideal sites for ooid formation. Similarly, the depth of the platform interior would have been much greater (10–15 m in depth with the Holocene sequence removed and sea level raised 4 m). Platform interior shorelines would have experienced higher energy and agitation than today.

Fourth, there is evidence that both water and wind were important in producing the beach ridge and dune

Figure 15. Underwater photograph of the oolitic grainstone bottom in 5 m of water in the central platform interior. Both physical and biogenic bottom features are present. Transverse wave ripples have about 20 cm wavelength. Also visible are biogenic tracks (across upper left part of photograph), burrow mounds (center), and burrow depressions (upper left and upper right). Area is too persistently agitated to permit significant colonization by sea grasses. Regular and irregular ooids occur (see Figure 6D). (Photograph by A. Saller.)

deposits. Laminae with keystone fenestrae can be traced up individual low-angle beach strata for a vertical distance of 2–3 m. They are interpreted to represent shore deposition during times of strong storm swash by moderate to large storm events. Avalanche lobes on the higher Pleistocene dunes on Providenciales are oriented westward regardless of the orientation of the dune. This indicates that dune growth is, in part, oriented by, and is a product of, the wind.

Fifth, where emergent islands blocked off-bank sediment transport across leeward margins, barrier reefs developed but were subsequently smothered by an influx of oolitic sand (Figures 16, 18). Three reef-to-ooid sequences have been documented on the western margin of West Caicos—two in the Pleistocene and one in the Holocene. Each reef-to-ooid sequence represents one marine highstand sequence (Wanless et al., 1989).

In some areas, the transition from reef to oolite is a depositional smothering event. In other cases, however, a physical marine erosional event appears to have separated the period of reefal growth from ooid deposition. This was presumably sand and rubble scour on the former reefal surface. In some areas, large head corals at the top of the reef sequence were planed off. This erosional surface was then both bored and encrusted with low-relief encrustations of the coral *Porites astreoides* and the mollusk *Chama macerophylla*. Coral-oolite conglomerates and shoaling-upward oolitic grainstones then buried the reefs (Wanless et al., 1989).

In each Pleistocene sequence, several lines of evidence indicate that the reef-to-oolite sequence formed during a single sea level highstand. The upper Pleistocene sequence provides the best exposure of this relation. First, ooids are present within the skeletal microstructure of some corals in the upper part of the reef sequence. Second, ooids are present in the grainstone matrix of the accumulating reefal sequence in the upper portion of the sequence. Third, there is no evidence of a subaerial exposure surface between the reef and ooids. Fourth, the conglomerate at the base of the oolitic sequence contains no calcrete clasts, and corals are unaltered.

For the Holocene, the reef-to-oolite sequence is undoubtedly contemporary (Figure 18).

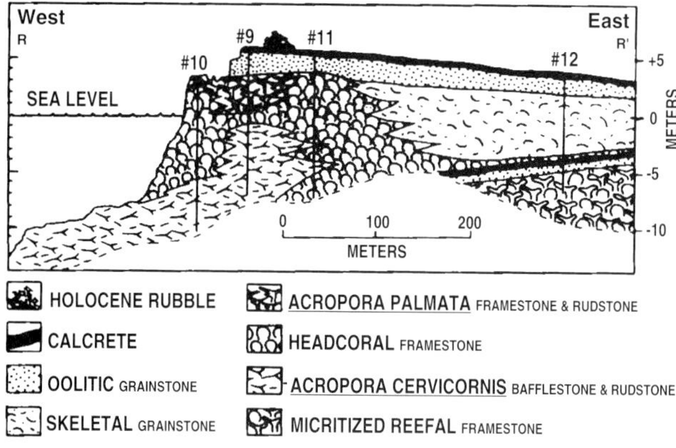

Figure 16. Lithostratigraphic cross section R–R′ (see Figure 9 for location) at the western margin of West Caicos based on cores of Pleistocene limestone recovered at four drill sites along transect and in exposed and underwater outcrops. Two shallowing-upward reef-to-ooid sequences are present (adapted from Waltz, 1988; from Wanless et al., 1989).

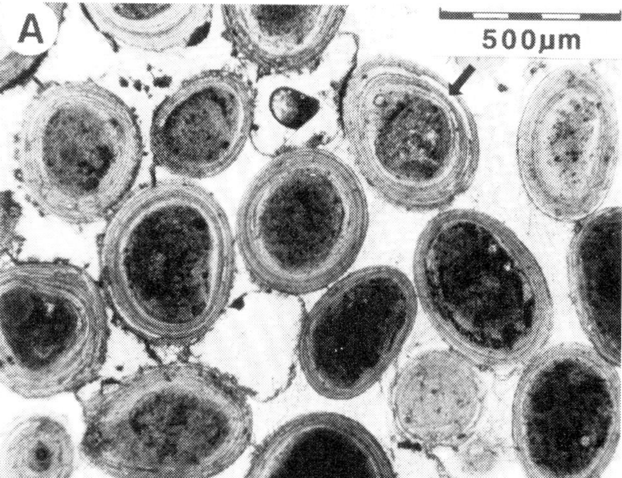

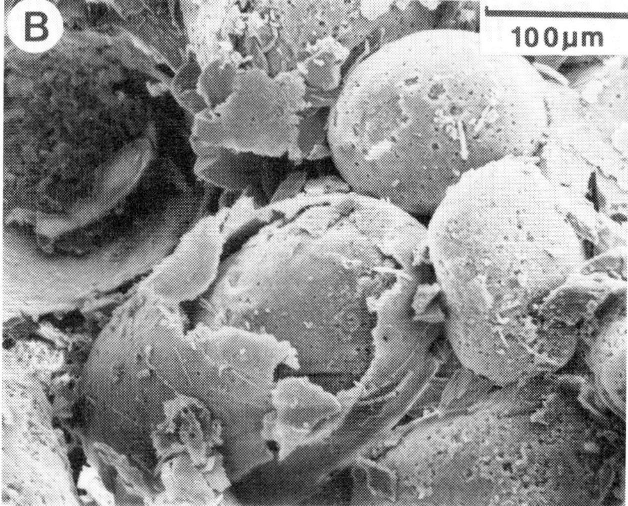

Figure 17. (A) Thin-section photomicrographs of late Pleistocene oolitic grainstone from West Caicos. Outer oolitic lamellae are very well defined, but have been subjected to lamellae-selective dissolution by meteoric waters (arrow). (B) Scanning electron photomicrograph of same Pleistocene oolitic grainstone showing selective dissolution of lamellae, microborings, and laths of vadose calcite cement (from Wanless et al., 1989).

Core borings on the northern part of West Caicos, where cross-bank transport was not blocked by emergent ridges during the later Pleistocene, contain a record of repetitive shallowing-upward oolitic grainstone sequences and no coral (Figure 19).

The shallowing-upward oolitic sequences have the following five facies from bottom to top, whether capping reefal deposits or entirely oolitic (Waltz, 1988).

1. Burrowed oolitic grainstone is moderately sorted and dominated by burrow structures or burrow-mottled fabric. Lined *Ophiomorpha* burrows (the ichnofossil of *Callianassa* in grainstones) dominate the upper, more sparsely burrowed portions of this lithology. Smaller subhorizontal to randomly oriented sand-filled burrow tubes are prevalent in the lower, more intensely burrowed zone, and can yield to intensely bioturbated, mottled fabrics. Burrowed oolitic grainstones represent a stabilized subtidal sand environment.
2. Planar and trough cross-stratified oolitic grainstone is composed of medium- to coarse-grained, moderately well-sorted oolite. Physical structures, together with scattered *Ophiomorpha* burrow structures and burrowing sea anemone (*Phylactis*) structures (Shinn, 1968) indicate a mobile, high-energy, shallow-subtidal depositional environment. Trough cross-bedded units are commonly exposed in plan view and record longshore sediment transport. Cross-stratification in larger planar cross-stratification generally has an onshore dip direction.
3. Oolitic/coral block conglomerate forms wedges and layers at the base of and within the oolitic sequences (Figure 20). Conglomerates are composed of subrounded to rounded oolitic blocks, some containing keystone fenestrae (beachrock-derived conglomerate), and/or coral blocks in an oolitic to oolitic-skeletal grainstone matrix. These conglomerates may be incorporated in an oolitic grainstone sequence, commonly occurring just below the beach deposits, or may directly cap an erosional surface on top of the reefal sequence and at the base of the oolitic sequence.

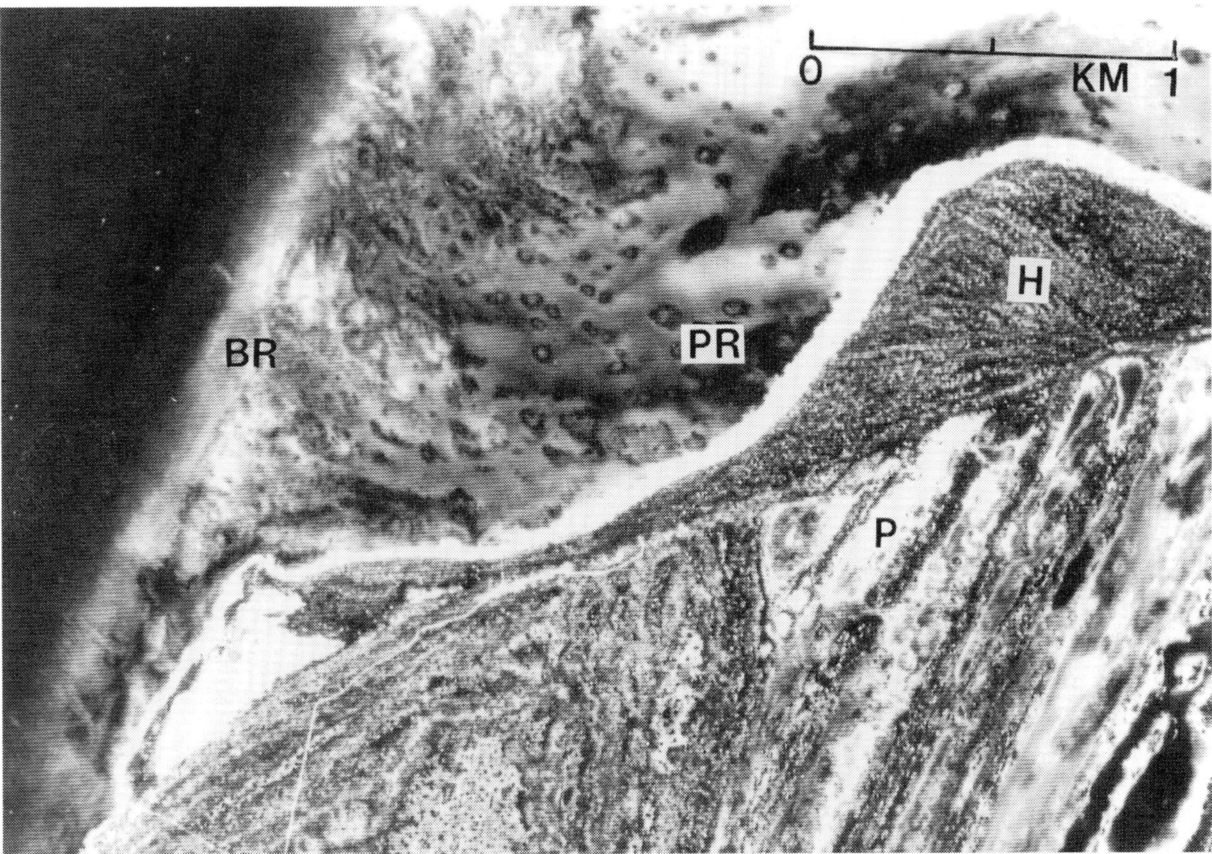

Figure 18. Aerial photograph of northwest West Caicos showing Holocene oolitic sand strand plain (H) prograding seaward across a living patch reef complex (PR) in the back-reef lagoon. With time this will prograde seaward and across the seaward barrier reef (BR). (P) is Pleistocene portion of island.

4. Layered oolitic grainstone is dominated by low-angle, seaward-dipping planar beds of well-sorted, medium-grained ooids. Keystone fenestrae, parting lineations, coarse basal skeletal lags at the base of the sequence, swash markings and lines, and rill markings indicate deposition in a beach environment.
5. Fine-grained oolitic grainstone is composed of fine-grained ooids that are well-sorted and layered to root-mottled. It is interpreted as a wind-deposited back beach and dune deposit.

The reef-to-ooid sequences on West Caicos appear to be very similar to those recently described by White and Curran (1987) for the leeward margin of Great Inagua Island, Bahamas.

Sediment Body Geometry and Internal Structure

Regular concentric ooids are forming (A) shore-parallel wedges of prograding strand plains of beach ridges and/or dunes (Figure 21), (A1) elongate bank and beach deposits derived by downcurrent transport from beach sources, (B) elongate subtidal shoals that roughly parallel the direction of the platform currents (and prevailing wind), and (B1) a focus of off-bank sediment transport where shoals extend to the leeward margin (Figures 4, 8). Irregular ooids are forming (C) broad subtidal banks cut by channels and (D) widespread subtidal sheet deposits (Figure 21).

(A) Shore-Parallel Strand Plains; Regular Concentric Ooids

Holocene ooids forming at the shoreface (and/or being transported to the shoreface) have constructed shore-parallel, seaward-thickening grainstone wedges. The sequences shallow upward and are capped by an emergent strand plain of beach ridges and dunes. Where actively prograding, strand plains are mostly low beach ridges. Where the shore is stable to erosional, higher dunes have formed in those areas where the shore faces to windward. Holocene oolitic strand plains vary from a single beach ridge to nearly 2 km in width and from 1–7 km in length (the extent of island coastlines). The prograding sediment wedge thickens seaward and may extend as much as 8 m below and 1–12 m above sea level at the shoreline. The strand plain trend may be straight or highly arcuate in embayments. Oolitic shore deposits of ooids face all directions and occur on shores facing the platform interior as well as margin. Highest dunes are on most exposed windward shores. Most extensive progradation is on those highly agitated shores where sand is not lost by longshore drift

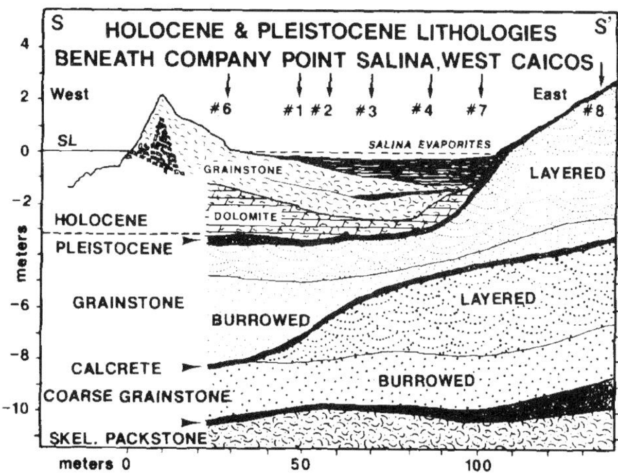

Figure 19. Lithostratigraphic cross section S–S′ (see Figure 9 for location) of Pleistocene burrowed to layered oolitic grainstone sequences recovered in drill cores along transect across Company Point Salina, northwest West Caicos. Each represents a shallowing sequence. Three exposure calcretes were penetrated. Holocene beach ridge and salina sequence is at top (from Wanless et al., 1989).

(northeast shore of Providenciales and east shore of West Caicos; Figures 4, 9, 10).

On east-, north-, and south-facing shores, the seaward extent of oolitic sands may coincide closely with the seaward limit of persistent agitation (or vegetative stabilization). On west-facing shores (e.g., northwest West Caicos), oolitic sand extends seaward through a reefal complex to the leeward platform margin.

More exposed shore sequences are largely stratified with deeper subtidal cross-bedding shallowing-upward to shore-parallel trough cross-stratification, beach and storm washover stratification, and wind-blown back-beach and dune stratification. On less exposed shorelines, sea grass mottling and animal bioturbation will partly to completely disturb subtidal stratification. Aeolian stratification may be mottled by vegetative rooting.

Oolitic sand produced in the platform interior or on windward shores may move by longshore drift to the north or south end of the island and then be swept westward as an elongate subtidal spit or as a westerly trending beach ridge of dune. This transport was important (a) in generating the Pleistocene beach ridges that extended West Caicos northward and (b) in feeding oolitic sand to the leeward margin where it smothered Pleistocene reefs. The volumetric role of transport vs. local production on the downcurrent shoals and beaches is not resolved.

(B) Wind-Aligned Shoals; Regular Concentric and Irregular Ooids

Elongate subtidal shoals occur downdrift from precursor islands. Shoals are 3–30 km in length and 0.5–3 km in width, with a narrow shoal crest. Shoal flanks are less agitated and should be preserved as a bioturbated grainstone with hardgrounds, coarse marine-cemented clast layers, and a mixture of regular and irregular ooids. The shoal crest is agitated and stratified and is dominated by regular ooids. Emergent islands on parts of the shoal may produce stratified beach deposits dipping to the flanks. Early diagenesis by meteoric and brine waters will modify the sequence beneath islands.

(C) Platform-Interior Shoal Complexes; Irregular Ooids

The Mid-Platform Shoal is a grainstone bank built and modified by episodic high-energy events but providing a setting for irregular ooid growth. The shoal is about 40 km in length, 6–12 km in width, forms a sediment body that is at least 3–5 m in thickness, and has built to the intertidal level. It is cut by a series of shallow channels that extend WNW–ESE and are 0.5–2 m deeper than the adjacent shoals. These channels appear to be the result of storm surges. There is also a series of north–south-trending sand waves, spaced 1–2 km apart, along the eastern half of the Mid-Platform Shoal. These sand waves are mostly stabilized and are thought to represent bottom organization and transport during times of strong cross-bank transport (very strong easterlies and hurricanes).

The complex shallow bathymetry on the Mid-Platform Shoal produces (a) areas of non-agitated bottom that may be further stabilized by algal mat, sea grass, and/or marine cementation and (b) widespread areas of very shallow bottom that may receive gentle persistent agitation (by wind-generated waves) but not stronger energy.

Although there are a variety of environments on the Mid-Platform Shoal, most of the sequences are bioturbated. Portions of the shoal appear to have widespread hardground sheets that are a product of marine cementation.

(D) Open Platform Interior; Irregular Ooids

The open platform interior has formed a bioturbated sheet deposit of regular concentric and irregular ooids that is 20 km × 40 km or larger and greater than 4 m in thickness. Its size is limited by the size of the platform. Although grapestone and grain aggregation are widespread throughout the platform interior, marine hardgrounds were not encountered in probing and coring in the western portion of the platform interior. Marine hardgrounds were common to the east, toward the Mid-Platform Shoal and toward the southern leeward-margin storm levee. These are areas of less persistent agitation and/or more intense biocorrosion.

Ball (1967), working in the northern Bahamas, recognized that "wind-driven currents" on the open platform interiors were, in some areas, sufficient to prevent deposition of mud and generate superficial ooids. He also hypothesized a setting with higher wind-driven current levels that would produce a sheet or blanket deposit of cross-stratified grainstone.

Figure 20. Photograph on West Caicos taken looking shoreward of Pleistocene reef boundstone spur (with hammer) smothered by Pleistocene oolitic grainstone containing layers of oolite-block and coral-head conglomerate. Marine burrowers are present in grainstone between two conglomerate layers. Top of photograph is fossil beach.

From Caicos, it is now clear that it is wind-generated bottom agitation that can produce this widespread environment of cross-stratification—and of ooid generation. It can also be concluded that, as the agitation increases, there will be an increase in the abundance of ooids, concentric oolitic form regularity, and preserved stratification.

APPLICATION TO ANCIENT LIMESTONE SEQUENCES

Although Caicos Platform is an isolated platform of moderate size, its subenvironments and climatic setting of brisk prevailing trade winds can provide important models for interpretation of ancient oolitic sequences. The principles of ooid sedimentation derived from Caicos Platform have direct application to ancient limestones in settings where persistent wave agitation is present, whatever the physiographic setting (small to large platform, carbonate shelf or ramp, epeiric sea or enclosed lake) (Figure 21).

Settings of persistent wave agitation occur in three situations: (1) in the Trade Wind belt (presently between 12° and 23° latitude) or other settings of persistent wind; (2) in areas subjected to bottom and/or shore agitation by oceanic swells (thus not necessarily in a setting of locally generated wind waves); and (3) on shorelines receiving renewed supersaturated waters. As ooids take time to form, these settings must not be overwhelmed by input of other sediment. In addition, as wave agitation is commonly focused at the shoreline, (4) it should be common for wave-generated ooids to accumulate as strand plains and dunes in which part of the sequence is of aeolian origin, in the subaerial environment, and subjected to meteoric diagenesis.

(1) Trade Wind Belt. There are many oolitic grainstones throughout the geologic record that contain an abundance of irregular ooids. Many of these form widespread deposits and are not focused at the platform margin (Cambrian, North China Platform, Feng et al., 1989). Others form shelf-wide deposits and are organized into parallel sand waves that are not oriented with respect to either the shoreline or the platform margin (Mississippian, Mission Canyon Formation; Hendricks, 1989).

(2) Ocean Swells. The recognition that ooids are forming on shores facing platform margins that are

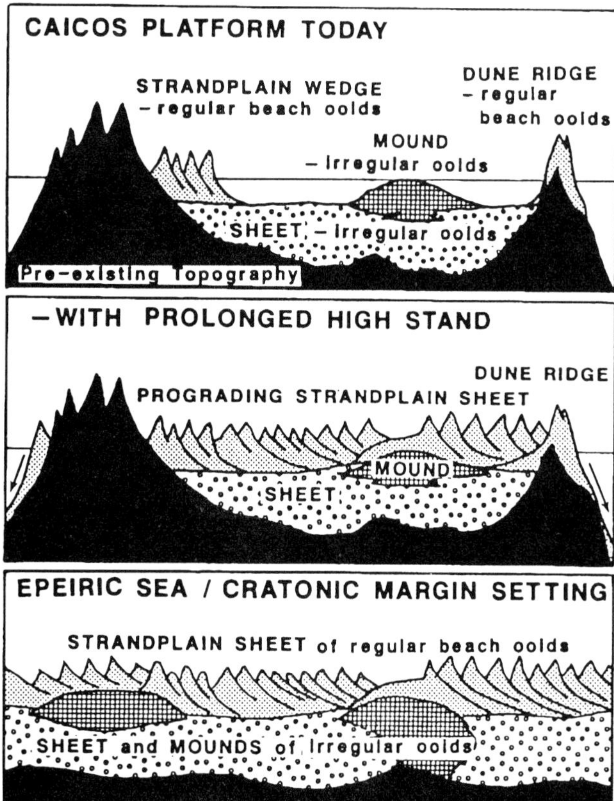

Figure 21. Upper: Schematic east–west cross section across Caicos Platform today showing preexisting topography (black) and type and form of oolitic grainstone deposits occurring across the platform interior and margin. Middle: With time, prograding strand plains can be expected to cover much of Caicos Platform as more and more ooids are generated by wave agitation at shorelines. This could form a sheet of cross-bedded grainstone capping the marine cycle of sedimentation. Arrows indicate off-platform shedding. Lower: Sediment bodies generated by wind-wave–generated ooids are not restricted to platform settings, but can also occur in shallow epeiric/inland seas and on broad cratonic margin seas, where they could form widespread sheet deposits dominating or capping marine cycles of sedimentation.

exposed to open ocean swells is very important. First, it provides an explanation for the oolitic and coated-grain grainstones that make up the high Pleistocene islands that occur on the ocean-facing margins along nearly the entire length of the Bahamian archipelago (Figure 1; Abaco Island on Little Bahama Bank; New Providence, Eleuthera, Cat Island, and Long Island on eastern Great Bahama Bank; Acklin Islands and Caicos Islands).

In the previous Pleistocene highstands, rapid sea level rise across the exposed seaward margin left the margins too deep for reefal growth to effectively initiate. Hubbard (personal communication, 1987) has shown that reef calcification rapidly diminishes below about 12 m. Rapid rises of sea level during the Pleistocene, which left the platform margin in 15–20 m or more of water, created ideal settings for widespread ooid sedimentation on exposed shorefaces. There was neither a major offshore source for skeletal sand nor a reefal block to incoming ocean wave energy. Thus, the Pleistocene highstand beaches were well agitated and a setting for effective ooid and coated grain growth. These same shorelines are now receiving an abundance of skeletal sand from nearshore reefs, and the Holocene dunes and strand plains are mostly skeletal.

(3) Beach Agitation. The ooids forming on the low-energy, south-facing beaches on Providenciales are forming mostly in the intertidal swash zone, but are producing enough ooids to build a small strand plain of fine-grained ooids. This shows that wave agitation can generate oolitic sediment bodies even in low-energy settings if other conditions are ideal (sufficiently supersaturated waters, water renewal, and low influx of other grain types). The shores of Great Salt Lake and other hypersaline lakes are common sites of ooid genesis (Eardley, 1938). These observations suggest the wave-generated ooids should be an expected part of early transgressive or late regressive (shallowing-upward) sedimentation in areas where there is adequate supersaturation and an absence of other sediment input. The Silurian and Devonian Wills Creek and Tonoloway limestones of western Maryland commonly contain ooids in the transgressive base to shallowing-upward sequences (authors' observations). These ooids could be locally generated on transgressive beaches rather than swept in from some distant tidal shoal.

Small islands scattered through a shallow marine environment could provide shore agitation sites for ooid generation even in areas of moderate to gentle winds.

(4) Oolitic Strand Plain Sequences. Recognizing that many wave-generated ooid sequences are forming emergent dunes and strand plains suggests oolitic strand plain and aeolian dune deposits should be common in many ancient sequences. Indeed, Hunter (1989 and this volume) reevaluates some of the Mississippian Ste. Genevieve Limestone in southern Indiana. He provides evidence that a good portion of those sequences are aeolian in origin—the product of prograding strand plains. Other deposits of widespread oolitic grainstone need similar reevaluation.

It is important to recognize strand plain environments in oolitic sequences because they (a) are indicators of approximate sea level, (b) necessitate a very different interpretation of the sites, processes, and rates of oolitic growth, (c) are deposits that are subjected to early freshwater leaching and cementation, and (d) in more arid climates, define sites of evaporite precipitation from surficial and reflux brines derived from salinas developed between the strand plains.

SUMMARY

The wave-agitation origin of many of the oolitic sands on Caicos Platform (Lloyd et al., 1987) is in striking contrast to the classic tidal-current–dominated oolitic sand environments of the northern Bahamas. In tide-dominated oolitic shoals, each tidal cycle agitates much of the shoal crests, flanks, and channel bottoms, permitting ooid growth over a broad area and in water depths ranging from intertidal to at least 10 m. Throughout mobile areas of active ooid generation, the majority of ooids have well-developed aragonitic coatings with the outermost lamellae relatively free of micritization. These active oolitic grainstones contain relatively few peloids or skeletal grains. In addition, the sand body exhibits preserved internal sedimentary structures dominated by bidirectional cross-stratification, reflecting the bidirectional tidal-current influence (Ball, 1967).

In contrast, the micritized mixture of oolitic and peloidal sands along the shores of Long Bay and West Caicos (Lloyd et al., 1987) is a reflection of more intermittent wave agitation. In subtidal shoals dominated by wave agitation, as in Long Bay, surface cross-stratification is represented by westward-moving asymmetrical bed forms, likely to be preserved internally as unidirectional cross-stratification.

Oolitic sand deposition on Caicos Platform represents an alternate model to the tidal-current–dominated oolitic sands of the northern Bahamas (Figures 2, 3) and provides a new set of guidelines applicable to the interpretation of ancient oolitic facies. The ooid factories of Caicos Platform, like those of the Persian Gulf, may be a more applicable model for ancient shallow intracratonic basins and areas along coastlines, where tidal currents may not have been a major hydrographic factor governing ooid growth.

Numerous ancient oolitic sand systems are dominated by unidirectional cross-stratification with a grainstone to low-mud packstone texture, and which contain a mixture of ooids, peloids and some skeletal grains, all exhibiting a relatively high degree of micritization. These sands likely formed in a depositional environment influenced by more intermittent wind-wave agitation and not persistent tidal currents, a situation operative on Caicos Platform today.

ACKNOWLEDGMENTS

The authors acknowledge the close association and assistance given by Jeff Dravis, Victor Rossinsky, Art Saller, and Michael Waltz. Research was funded by grants from ARCO, Exxon Production Research Co., UNOCAL, and National Science Foundation grant OCE-86-04449. The physiographic map of Caicos Platform (Figure 4) was constructed with the help of satellite imagery supplied by UNOCAL, with image processing by J.H. Penny and A.E. Prelat. Barry Boyce, Jamie Emerson, George Grabowski, Tim Lyons, Randall Parkinson, Kenneth Tyrrell, and Trevor Wanless assisted in field research. We appreciate critical reviews by Paul M. Harris, Brian Keith, Eugene Shinn, and Charles Zuppann.

REFERENCES CITED

Ball, M.M., 1967, Carbonate sand bodies of Florida and the Bahamas: Journal Sedimentary Petrology, v. 37, p. 556-591.

Browne, K.M., H.R. Wanless, P.K. Swart, V. Rossinsky, and K.M. Tyrrell, 1989, Dolomite cementation and sediment dissolution in association with ponded brines, Caicos, B.W.I.: Geological Society America Annual Meeting, Abstracts, St. Louis, p. A220.

Dill, R.F., Shinn, E.A., Jones, A.T., Kelly, K., and Steinen, R.P., 1986, Giant subtidal stromatolites forming in normal salinity waters: Nature, v. 324, p. 55-58.

Dravis, J.J., 1977, Holocene sedimentary depositional environments of Eleuthera Bank, Bahamas: M.S. Thesis, Univ. Miami, Coral Gables, FL, 386 p.

Dravis, J.J., 1979, Rapid and widespread generation of recent oolitic hardgrounds on a high energy Bahamian platform, Eleuthera, Bahamas: Journal Sedimentary Petrology, v. 49, p. 195-208.

Dravis, J.J., 1983, Hardened subtidal stromatolites, Bahamas: Science, v. 219, p. 385-386.

Eardley, A.J., 1938, Sediments of Great Salt Lake, Utah: American Association Petroleum Geologists Bulletin, v. 22, p. 1305-1411.

Enos, P., 1977, Quaternary sedimentation in south Florida, Part 1, Holocene sediment accumulations of south Florida shelf margin: Geol. Soc. America Memoir 147, p. 1-130.

Evans, C.C., 1987, The relationship between the topography and internal structure of an ooid shoal sand complex: the upper Pleistocene Miami Limestone, *in* Maurrasse, F.J.M.R., (ed.), Symposium on South Florida Geology, Miami Geological Society Memoir 3, p. 18-41.

Feng, Z., J. Chen, and W. Shenghe, 1989, Lithofacies Paleogeography of Early Paleozoic of North China Platform: Book Series on Lithofacies Paleogeography of China, Geological Publishing House, Beijing, 270 p.

Ginsburg, R.N., and N.P. James, 1974. Holocene carbonate sediments of continental shelves. *in* Burke, C.A., and C.L. Drake (eds.), The Geology of Continental Margins, pp. 137-155.

Halley, R.B., P.M. Harris, and A.C. Hine, 1983, Bank margin environment, *in* P.A. Scholle, D.G. Bebout and C.H. Moore, eds., Carbonate Depositional Environments: American Association Petroleum Geologists Memoir 33, p. 463-506.

Harris, P.M., 1979, Facies anatomy and diagenesis of a Bahamian ooid shoal: Sedimenta VII, Comparative Sedimentology Laboratory, Univ. Miami, 160 p.

Hendricks, M.L., 1989, Upper Mission Canyon coated grain producing facies in Williston Basin: American Association Petroleum Geologists Bulletin, v. 73, p. 1033-1034.

Hine, A.C., and A.C. Neumann, 1977, Shallow carbonate bank margin growth and structure, Little Bahama Bank, Bahamas: American Association Petroleum Geologists Bulletin, v. 61, p. 376-406.

Hoffmeister, J.E., K.W. Stockman, and H.G. Multer, 1967, Miami Limestone of Florida and its recent Bahamian counterpart: Geological Society America Bulletin, v. 78, p. 175-190.

Hunter, R.E., 1989, Eolianites in the Ste. Genevieve Limestone of Southern Indiana, *in* D.D. Carr and R.E. Hunter, eds., Geometry and Depositional Environments of Ste. Genevieve (Mississippian) Oolite Bodies in Southern Indiana: Department of Natural Resources Geological Survey Guidebook, p. 1-19.

Imbrie, J. and H. Buchanan, 1965, Sedimentary structures in modern carbonate sands of the Bahamas, *in* G.V. Middleton, ed., Primary Sedimentary Structures and Their Hydrographic Interpretation: Society Economic Paleontologists Mineralogists Special Publication No. 12, p. 149-172.

Leaver, J., 1985, Sedimentology, Mineralogy and Porewater Chemistry of Schizohaline Pond Sediment, Turks and Caicos Islands, British West Indies: M.S. Thesis, Duke Univ., Durham, N.C., 76 p.

Lighty, R.G., I.G. Macintyre, and R. Stuckenrath, 1978, Submerged early Holocene fringing reef southeast Florida shelf: Nature, v. 276. p. 59-60.

Lloyd, R.M., R.D. Perkins, and S.D. Kerr, 1987, Beach and shoreface ooid deposition on shallow interior banks, Turks and Caicos Islands, British West Indies: Journal Sedimentary Petrology, v. 57, p. 976-982.

Loreau, J.P., 1973, Nouvelles observations sur la genese et la signification des oolithes: Sci. Terre, v. 18, p. 213-244.

Macintyre, I.G., 1988, Modern coral reefs of the Western Atlantic: New geological perspective: American Association Petroleum Geologists Bulletin, v. 72, p. 1360-1369.

Milliman, J.D., 1966, The Marine Geology of Hogsty Reef, a Bahamian Atoll: Ph.D. Dissertation, Univ. Miami, Coral Gables, FL, 292 p.

Neumann, C.J., G.W. Cry, E.L. Caso, and B.R. Jarvinen, 1978, Tropical Cyclones of the North Atlantic Ocean, 1871-1977: National Climatic Center, Ashville, N.C., U.S. Govt. Printing Office Stock No. 003-17-00425-2, 170 p.

Newell, N.D., E.G. Purdy, and J. Imbrie, 1960, Bahamian oolitic sand: Journal of Geology, v. 68, p. 481-497.

Palmer, M.S., 1979, Holocene Facies Geometry of the Leeward Bank Margin, Tongue of the Ocean, Bahamas: M.S. Thesis, Univ. Miami, Coral Gables, FL, 168 p.

Parker, G.C. et al., 1955, Water resources of southeastern Florida with special reference to the geology and ground water of the Miami area: U.S. Geological Survey Water Supply Paper 1255, 963 p.

Purdy, E.G., 1963, Recent calcium carbonate facies of the Great Bahama Bank: 1. petrography and reaction groups; and 2. sedimentary facies: Journal Geology, v. 71, p. 334-355 and p. 472-497.

Rossoff, D.B., G.S. Dwyer, J. Leaver, and R.D. Perkins, 1989, Holocene salina sediments of West Caicos, British West Indies: Stratigraphy, mineralogy and pore-water geochemistry: American Association Petroleum Geologists Bulletin, v. 73, p. 408.

Shinn, E.A., 1968, Burrowing in recent lime sediments of Florida and the Bahamas: Journal Paleontology, v. 42, p. 879-894.

Smith, C.L., 1940, The Great Bahama Bank: I. General hydrographic and chemical factors; II. Calcium carbonate precipitation: Jour. Marine Research, v. 3, p. 147-189.

Taft, W.H., F. Arrington, A. Haimovitz, C. MacDonald, and C. Woolheater, 1968, Lithification of modern carbonate sediments at Yellow Bank, Bahamas: Bulletin Marine Science Gulf Caribbean, v. 18, p. 762-828.

Thomas, T.M., 1974, A detailed analysis of climatological and hydrological records of south Florida with reference to man's influence upon ecosystem evolution, *in* P.J. Gleason, ed., Environments of South Florida: Present and Past: Miami Geological Society Memoir 2, p. 82-122.

Traverse, A. and R.N. Ginsburg, 1966, Palynology of the surface sediments of Great Bahama Bank as related to water movement and sedimentation: Marine Geology, v. 4, p. 417-459.

U.S. Dept. of Commerce, 1990, Tide tables and high and low water predictions: National Ocean Survey, 288 p.

Waltz, M.D., 1988, The Evolution of Shallowing-Upwards Reef to Oolite Sequences at the Leeward Margin of Caicos Platform, BWI: M.S. Thesis, Univ. Miami, Coral Gables, FL, 95 p.

Wanless, H.R., 1981, Fining-upwards sedimentary sequences generated in seagrass beds: Journal Sedimentary Petrology, v. 51, p. 445-454.

Wanless, H.R., E.A. Burton, and J. Dravis, 1981, Hydrodynamics of carbonate fecal pellets: Journal Sedimentary Petrology, v. 51, p. 27-36.

Wanless, H.R., L.P. Tedesco, V. Rossinsky, and J.J. Dravis, 1989, Carbonate Environments and Sequences of Caicos Platform: American Geophysical Union, 28th International Geological Congress Field Trip Guidebook T374, 75 p.

Wanless, H.R., L.P. Tedesco, and K.M. Tyrrell, 1988, Production of subtidal tubular and surficial tempestites by Hurricane Kate, Caicos Platform, British West Indies, Journal Sedimentary Petrology, v. 58, p. 739-750.

Warzeski, E.R., 1976, Storm sedimentation in the Biscayne Bay region, *in* A. Thorhaug and A Volker, eds., Biscayne Bay: Past/Present/Future: Univ. Miami Sea Grant Special Publication No. 5, p. 33-38.

Westphall, M.J., 1986, The History and Anatomy of a Ringed-Reef Complex, Belize, Central America: M.S. Thesis, Univ. Miami, Coral Gables, Fl, 225 p.

White, B., and H.A. Curran, 1987, Coral reef to eolianite transition in the Pleistocene rocks of Great Inagua Island, Bahamas, *in* H.A. Curran, ed., Proceedings of the Third Symposium on the Geology of the Bahamas: Ft. Lauderdale, College Consortium of the Finger Lakes Bahamian Field Station, p. 165-179.

Chapter 16

A Quaternary Analog for Interpretation of Mississippian Oolites

Mark R. Boardman
Geology Department, Miami University
Oxford, Ohio, USA

Cindy Carney
Department of Geological Sciences, Wright State University
Dayton, Ohio, USA

Paul M. Bergstrand
Geology Department, Miami University
Oxford, Ohio, USA

ABSTRACT

A Quaternary example of an ooid shoal complex comprised of northern Andros Island (Pleistocene) and Joulters Cays (Holocene) provides insights into ancient ooid shoal deposition and preservation. Criteria that are commonly used in the interpretation of fossil ooid shoals are preserved in the Quaternary example. These include grain composition, lateral variability of shoal subenvironments and adjacent carbonate environments, topography, and to some extent, sedimentary structures. Two subenvironments in particular are easily recognized: (1) the mobile fringe is characterized by high ooid concentrations and well-preserved sedimentary lamination and is topographically elevated, and (2) the stabilized sand flat contains fewer ooids and occurs at a lower elevation. Lagoons, reefs, and offshore skeletal sands are located laterally adjacent to ooid shoals. Muddy tidal flats, however, are not. Evidence of extensive exposure is also well preserved in the Quaternary analog. Sea-level fluctuations as little as 5 m have a profound effect on the site of ooid shoal accumulation and the creation of exposure surfaces.

A comparison of the Quaternary analog to Mississippian oolitic rocks reveals similar grain compositions, sedimentary structures, and to some extent, adjacent carbonate environments. The mobile fringe and stabilized sand flats can also be recognized. In contrast, original topography, exposure surfaces, and eolian features are seldom reported for Mississippian oolitic rocks.

INTRODUCTION

Along with presentations of stratigraphic and petrographic information, some studies of Mississippian oolites attempt to relate facies and models of deposition to modern analogs described from ooid sand shoals such as those in the Bahamas. Our understanding of modern ooid sand shoals in the Bahamas includes descriptions of the lateral distribution of ooid sands and the sedimentary structures and facies associated with them (Newell and Rigby, 1957; Newell et al., 1960; Purdy, 1963; Ball, 1967; Gebelein, 1974). These early studies have been complemented by three-dimensional examination of facies and their temporal associations using sediment cores and C-14 dating. Today the principal modern analogs of ooid sand shoals are Joulters Cays (Newell et al., 1960; Ball, 1967; Harris, 1979; Halley and Harris, 1979; Halley et al., 1983; Strasser and Davaud, 1986; Boardman and Carney, 1991) and Lily Bank (Hine, 1977). To a lesser extent, the ooid sands of Schooner Cays, Cat Cay, and Yellow Bank are used (Ball, 1967; Dravis, 1979; Palmer, 1979; Halley et al., 1983), and additional examples are becoming available (e.g., Caicos shelf, Lloyd et al., 1987; Wanless and Tedesco, this volume).

These analogs provide comparisons for selective petrographic and outcrop features and sedimentary processes. Lateral, vertical, and temporal variability similar to the modern examples should be recognized in the ancient, and if not recognized, further study seems warranted. Models of ooid deposition based on Holocene examples alone, however, are limited because they can only provide information of ooid deposition during the rise and near stillstand of sea level for the past few thousand years. What the next few thousand years of stillstand will produce and what will be recorded following the fall of sea level are not known. Prediction of future change is highly speculative and depends on further documentation and a better understanding of past variations in ooid shoal complexes, something this volume hopes to supply.

THE QUATERNARY ANALOG

Studies of Pleistocene ooid sand bodies are an important link between Holocene examples of ooids sands (e.g., Joulters Cays and Lily Bank) and the ancient oolites beautifully exposed in Mississippian outcrops. The Quaternary analog (Holocene plus Pleistocene) provides information not only on recognition criteria of many of the subenvironments of modern ooid shoals, but also on the preservation of these criteria. In addition, a model of the superposition and lateral variability of facies through a longer time frame than the Holocene is possible, incorporating a sea-level fluctuation that is fairly well known (oxygen isotope stage 5e). Depositional models of Quaternary ooid sand bodies can then be used to direct further inquiry into the spatial and temporal distribution of facies (stratigraphy) of ancient oolites.

An excellent Quaternary analog for ooid shoals is the suite of carbonate environments associated with northern Andros Island (Pleistocene) coupled with the adjacent environments of Joulters Cays (Holocene; Figure 1). This composite analog is an extension of direct comparisons of Pleistocene oolites to Holocene ooid sand bodies (e.g., Hoffmeister et al., 1967) because it links two episodes of ooid accumulation and because it includes microscopic petrographic criteria. It examines the effects of short-term (thousands of years) and longer term (hundreds of thousands of years) fluctuations of sea level as recorded in the formation of exposure surfaces, the preservation of lateral and vertical facies geometries (mosaics) of ooid sand shoals, and associations of other nearby carbonate facies such as reefs, tidal flats, and lagoons.

This chapter describes this Quaternary analog and explores some possibilities for its use, in particular, the extent to which modern analogs are appropriate to Mississippian oolites. Northern Andros Island contains many of the features seen on Joulters Cays, but also contains several features not yet developed (but likely to develop in the future) on Joulters. The lateral juxtaposition of northern Andros with Joulters exemplifies the difficulties that can be anticipated in determining lateral and vertical continuity in Mississippian oolite deposits.

Information on Joulters Cays is largely taken from the work of Harris (1977, 1979) and is supplemented by our own field observations and laboratory analyses of sediments and rocks collected from Joulters Cays. The interpretation of northern Andros Island is based on megascopic and thin-section analysis of 86 samples of surficial rocks selected from a suite of 157 shallow (approximately 50-cm-deep) rock cores (Bergstrand, 1991). Mississippian examples are taken from the Ste. Genevieve of southern Indiana (Carr, 1973, 1989), the Ste. Genevieve and Warix Run limestones of northeastern Kentucky (Ettensohn, 1975; Dever, 1980; Ettensohn et al., 1984; Lierman, 1984; Carney, 1987) and the Greenbrier Limestone of West Virginia (Leonard, 1968; Carney, 1987; Yeilding and Dennison, 1986; Sullivan and Textoris, 1988; Carney and Smosna, 1989; Carney, this volume; Kelleher and Smosna, this volume; Smosna and Koehler, this volume).

QUATERNARY OOID SAND SHOALS

A comparison between Joulters Cays (Holocene) and Andros Island (Pleistocene) shows similarities and differences between these two ooid shoals and also illustrates the influence of a preexisting shoal on succeeding depositional environments. Included in the comparison are major grain types present and their lateral distribution (subenvironments of ooid shoals), the lateral and temporal association of the two Quaternary ooid shoals with adjacent carbonate

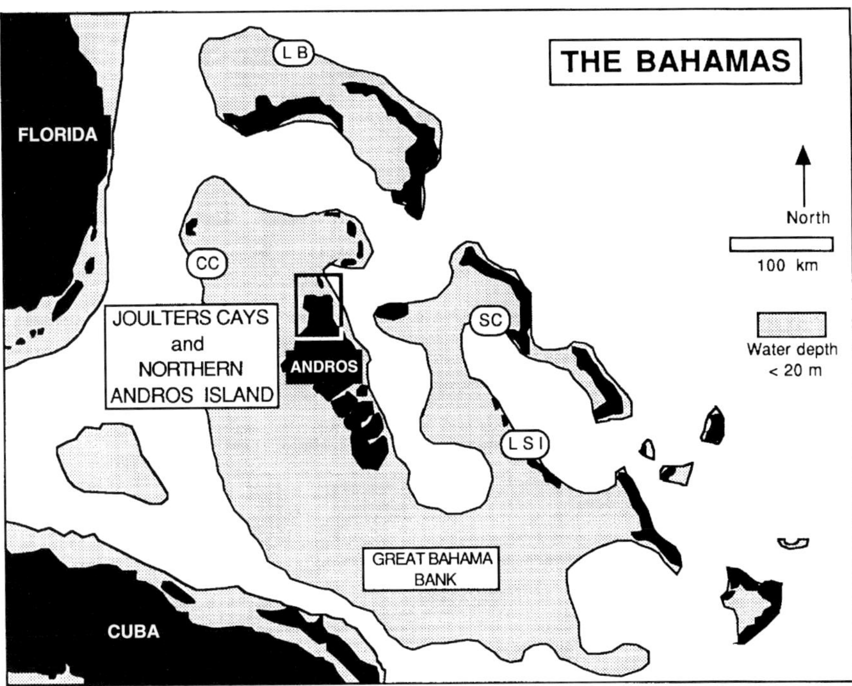

Figure 1. Location of Joulters Cays and northern Andros Island, Bahamas. Area in box refers to Figures 2 and 6. The location of several other ooid sand shoals that have been studied include Lily Bank (LB), Schooner Cays (SC), Cat Cays (CC), and Lee Stocking Island (LSI). Caicos Island is off the map to the southeast.

facies (reefs, tidal flats, and lagoons) and with each other, and the general size and location of the Quaternary ooid shoals.

Ooid Distribution

While ooids may receive attention because they are intriguing and make beautiful photographs, they are not always the dominant grain type in what are termed ooid shoals. This is true for "oolites" of many ages and prompts us to question how "oolitic" ooid shoals are formed.

It is difficult to define the limits of an ooid shoal, but one might follow the lead of Folk (1962) and use the term *oolitic* to include rocks and sediment with more than 25% ooids. Both northern Andros Island and Joulters Cays have ooid concentrations greater than 25% with similar lateral distributions (Figure 2). On Joulters, the area with greater than 25% ooids is approximately 660 km^2, and the section of north Andros included for study contains 880 km^2 of oolitic (>25% ooids) rock. In the central portions of both Joulters and Andros, the ooid concentration is greater than 50%. On Joulters, this area is approximately 370 km^2 (56% of the total area), while only 10% of the area on Andros has greater than 50% ooids. Environments with ooid concentrations greater than 75% are present along the eastern (windward) margin of Joulters (but not Andros) and comprise 80 km^2 (12% of the total area; Figure 2). Such high proportions of ooids on Joulters are not unusual for Holocene Bahamian ooid shoals (Newell et al., 1960) and typify the very narrow "mobile fringe" subenvironment (also called "marginal sand shoal") where the ooid concentration averages 83% (maximum = 98%; Harris, 1977, 1979). The mobile fringe includes areas where several important recognition criteria of ooid shoals have been described, including good sorting, high roundness, high sphericity, and cross-lamination (Harris, 1977, 1979).

The mobile fringe of Joulters Cays is relatively small, being only 0.5–2 km wide and 12 km long, and is surrounded by sands with a lower percentage of ooids (Figure 2), poorer sorting coefficients, and little in the way of preserved bedding. There is marked asymmetry to the distribution of ooids at Joulters Cays. On the ocean-facing (high-energy, windward) margin, ooid concentrations drop from 85% to 0% in less than 5 km (Harris, 1977, 1979). On the platform-facing (low-energy, leeward) margin, the change is more gradual, and ooid concentrations drop from 85% to 20% over a distance of approximately 40 km (Figure 2). Most of Joulters is a moderately oolitic sand (~50% ooids) with moderate sorting and little lamination.

The pattern of ooid distribution on Joulters is also found on northern Andros. Apparently, environmental conditions (waves, currents, storm frequency and intensity) favorable for ooid accumulation are/were similar enough between Joulters and Andros to pro-

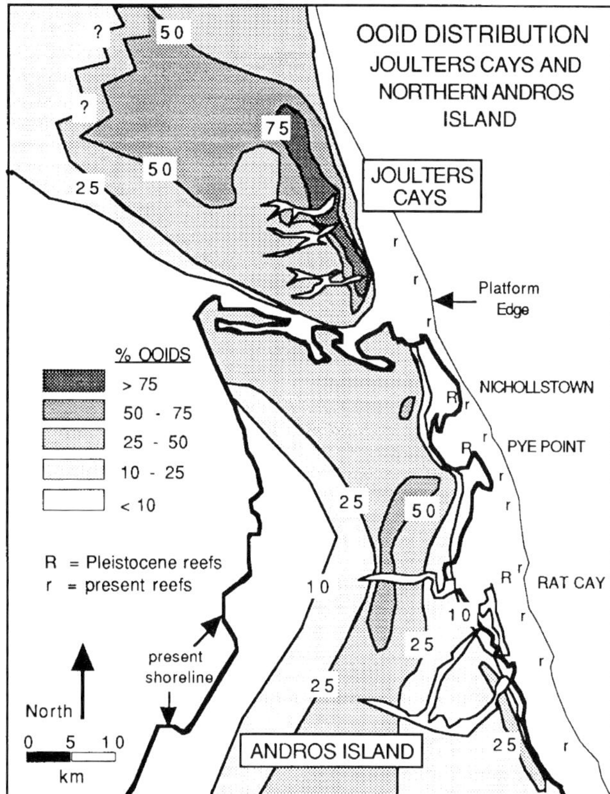

Figure 2. Map showing the lateral distribution of ooids on Joulters Cays and linear trends of ooid concentration on northern Andros Island.

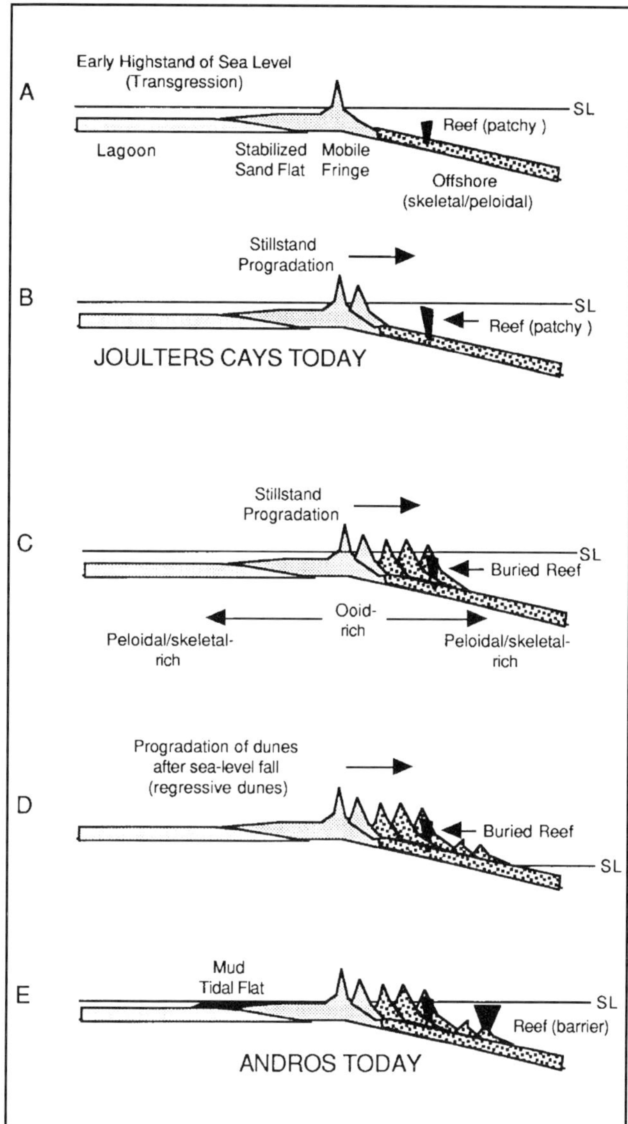

Figure 3. Cross section showing the development of an ooid sand shoal. (A) Ooids can accumulate rapidly accompanying initial rise of sea level and flooding of the platform (transgression). (B) Progradation of ooid dunes and stranded beach ridges (strand plains) can occur during a relative stillstand of sea level. Joulters Cays has experienced such progradation. (C) With time, seaward progradation can occur during a stillstand with the dominant grain type comprised of peloidal/skeletal-rich sands derived from adjacent offshore areas. (D) A fall of sea level can also produce dunes comprising peloidal skeletal-rich sands. Offshore reefs can be entombed by this progradation. (E) The Holocene rise of sea level has flooded the former (Pleistocene) stabilized sand flat. The former ooid ridges and peloidal/skeletal ridges have created an effective barrier that affords adequate protection on its western side to permit the accumulation of muddy tidal flats as well as blocking offbank transport of lagoonal water, thus creating conditions more suitable for growth of barrier reefs.

duce ooids and also to distribute ooids in a predictable fashion. However, important differences in the distribution of ooids do exist. The windward-leeward asymmetry of the gradient of percentage ooids is not as clear on Andros. A larger area of ooid-poor sediment is present on the high-energy margin of Andros. This difference may occur because Andros is a more "mature" ooid shoal. That is, in addition to deposition during a rise of sea level (transgression; Figure 3A), Andros contains oolitic sediments deposited accompanying a sea-level stillstand and fall (Figure 3B). During the stillstand and fall of sea level, the relatively ooid-poor sediment windward of the main ooid shoal may have been mobilized and plastered onto the earlier-formed Andros ooid shoal (Figures 3C, D). Additionally, Andros may have formed during a more protracted highstand of sea level which allowed greater accumulation of ooid-poor, windward margin sediment. The presence of elevated Pleistocene reefs encased in ooid-poor sediment on the windward margin of Andros supports this idea (Figure 3E).

Subenvironments—Lithology

On both Joulters Cays and Andros Island, the principal grain types are ooids, peloids and aggregates, and skeletal fragments (Figure 4). Grain type is not

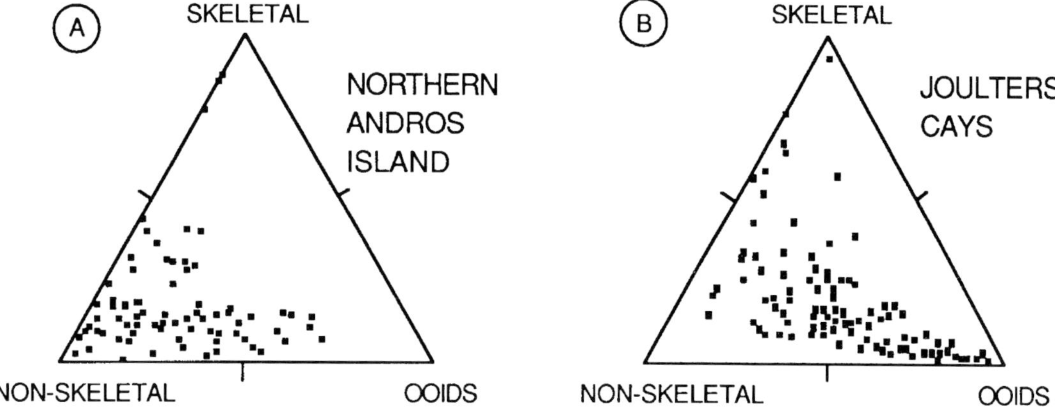

Figure 4. Ternary diagrams of grain composition—all facies. (A) Andros Island. (B) Joulters Cays.

particularly diagnostic or mutually exclusive for the several subenvironments recognized for Joulters by Harris (1977, 1979). High concentrations of ooids (>75%) are found in the mobile fringe as well as from the stabilized sand flat, tidal channels, platform margin shelves (open platform margin and windward margin shelf), and supratidal dunes (Figure 5). The data from Joulters and Andros suggests, however, that two subenvironments can be recognized based on ooid concentration, regional distribution (Figures 2, 5, 6), and to some extent, sedimentary structures. The first consists of highly oolitic sediments occurring on the more windward, high-energy margin. Using the classification of Wilson (1971), these sediments are termed "winnowed platform-edge sands." On Joulters, these sediments include the marginal sand shoal, supratidal dunes, portions of platform margin shelves, and some tidal channels (mobile fringe). The second environment contains a lower proportion of ooids and occurs on the more leeward portion of the shoal. Using Wilson's classification, one might lump these sediments with "open marine platform" sediments. But in order to separate lagoonal sediments (with <25% ooids) from this second oolitic environment of Joulters and Andros (with >25% ooids), we prefer to use "stabilized sand flat," which is the largest subenvironment recognized by Harris (1977, 1979).

The highest ooid concentration found on northern Andros was 80%, which is lower than that found on Joulters, where the average of the mobile fringe is 83% ooids and the maximum ooid concentration is 98% (Figures 2, 4; Harris, 1977). Why Andros has a lower concentration of ooids than Joulters is not clear. The sampling coverage on Andros should have uncovered higher concentrations of ooids (if they exist) because the highest density of samples comes from the more central and windward portions of the ooid body. Perhaps the Andros shoal was more peloid-rich, or some of the peloids in samples from northern Andros are micritized ooids. Also, the samples are surficial rocks and may represent an ooid-poor cap resulting from greater mixing of windward-derived sediment during a protracted highstand of sea level or a sea level fall (Figure 3). Other Pleistocene ooid shoals also contain relatively low concentrations of ooids in their "high-ooid" facies (e.g., ~45% for the Miami Oolite; Evans, 1983; maximum of 50% for Grand Bahama Island; Gerhardt, 1983). This difference should be explored further because, like Andros, Mississippian units often contain a much lower proportion of ooids than Joulters.

The percentage of mud (defined here as primary depositional matrix <62 μm in size) is generally low on both Andros Island (always <5%) and Joulters Cays (range = 0.2–8%). On Joulters Cays, the concentration of mud is greatest in the sediments of the stabilized sand flat. Because of the uniformly low concentration of mud and the difficulty of distinguishing depositional mud from diagenetic micritic cement in the Andros rocks, percentage of mud cannot be used as a criteria for distinguishing mobile fringe and stabilized sand flat environments.

Subenvironments—Topography

Shoal Asymmetry

An additional characteristic of Joulters Cays that might be useful for interpreting lateral variability of subenvironments is topography. A large portion of the high-energy sands of the ooid shoal is at a higher elevation (subaerial or intertidal) than the lower energy, stabilized sand flat and lagoon (Figures 6, 7). A profile along a west to east (leeward to windward) transect shows that the topography of Joulters and Andros is asymmetric with the higher elevations on the high-energy (windward) margin. The elevation differences are on the order of meters, and although it might be difficult to recognize in ancient rocks because traceable outcrop exposures are limited, this topographic asymmetry should be useful in determining energy direction and in aiding environmental interpretations of Mississippian oolites.

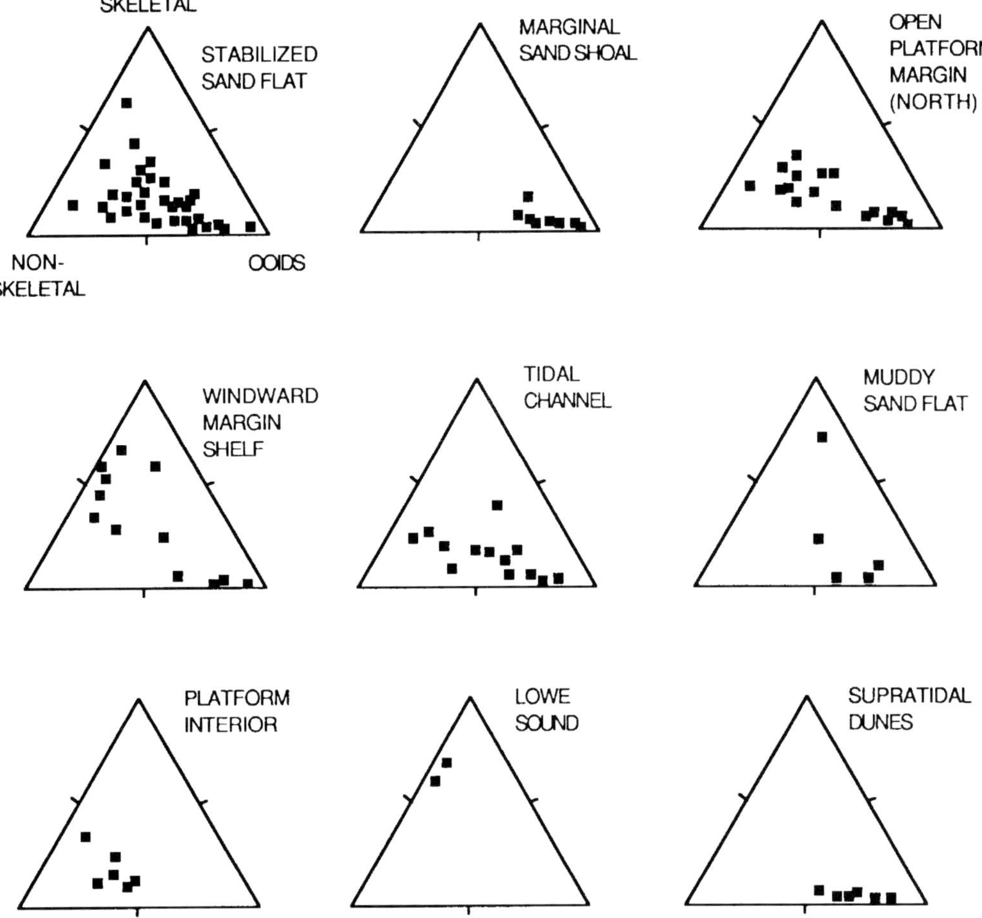

Figure 5. Ternary diagrams of subenvironments of Joulters (all data from Harris, 1977, except supratidal dunes from Bergstrand, 1991).

Channels

In addition to topographic asymmetry, the tidal channels (up to 5 m deep and up to 100 m wide) of Joulters and Andros (presently near sea level) are also located on the high-energy side. These channels are oriented perpendicular to the linear trend of higher elevation (Figure 6). On Andros the channels are currently intertidal or slightly above sea level. They are being filled with a poorly sorted skeletal sand or by muddy organic-rich sediments stabilized by mangroves. On Joulters Cays, most channels are still active and swept clean. One channel (between Middle Joulters and North Joulters) is presently choked with oolitic sand (and some mud) and is an intertidal sand flat. The margins of some channels are colonized by mangroves.

Eolian Dunes

Strand plains and eolian deposits are common subaerial features at Joulters as well as other Bahamian oolitic settings (Harris, 1979; Halley and Harris, 1979; Halley et al., 1983; Strasser and Davaud, 1986; Lloyd et al., 1987). These features are long, linear, and parallel with the shoreline. They contain excellent lamination. Elevations up to 4 m above present sea level are common.

The influence of eolian deposits on the original topography of a Pleistocene ooid shoal is faithfully preserved in the present-day topography of Andros Island. Arcuate, elongate features that parallel the shoreline are interpreted as island/strand plain accumulations. They are truncated by paleotidal channels oriented perpendicular to the trend of higher elevation (Figure 6). The elevation of the Andros ridges is generally 5–10 m above present sea level (maximum is 20 m) vs. 3–4 m above sea level for Joulters. The reason the Andros ridges are now 5–10 m above sea level is partly the result of formation during or following a higher stand of the sea which, according to evidence from elevated reefs and intertidal notches, was about 5 m higher than today (Neumann and

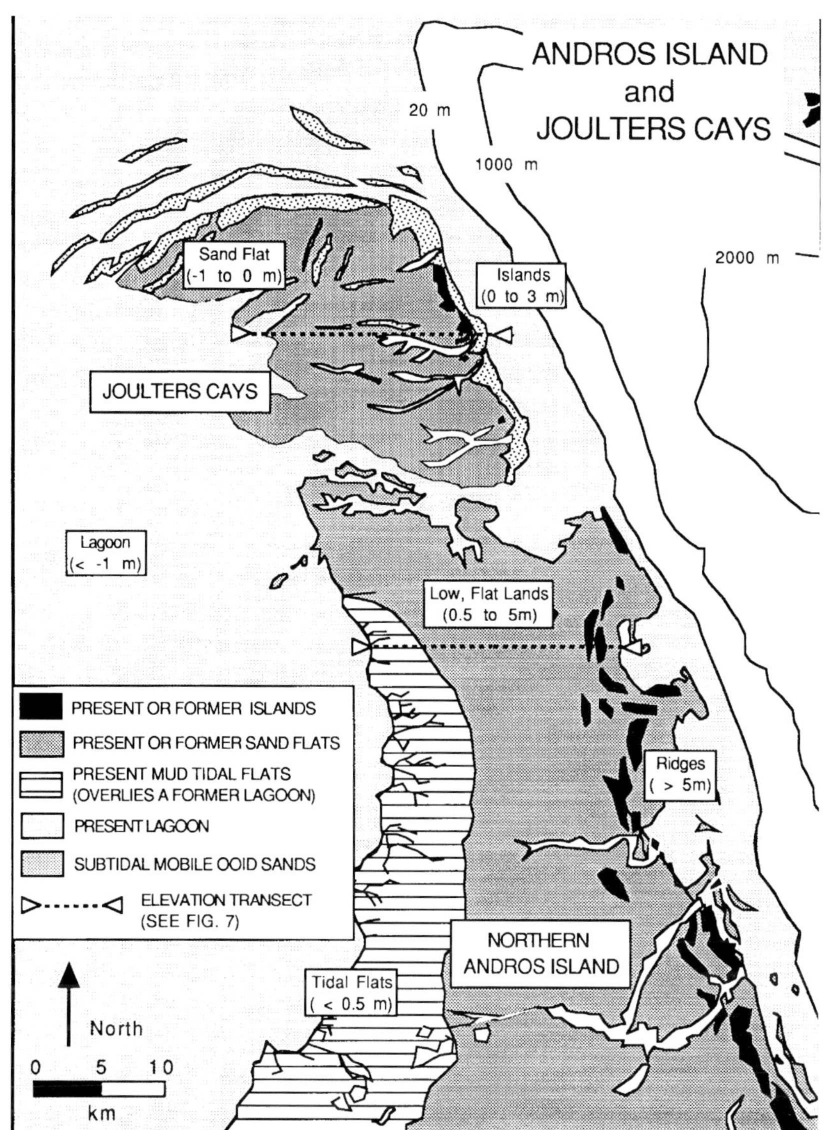

Figure 6. Topography of Joulters Cays and northern Andros Island showing linear ridges and broad flat areas.

Moore, 1975). The more protected stabilized sand flat occupies areas to the west (leeward, lower energy) of the ridges (Figures 6, 7).

Other Quaternary strand plains and eolian features of carbonate environments are found in Bimini (Strasser and Davaud, 1986), southern Florida (Evans, 1983), San Salvador (Carew and Mylroie, 1985; Boardman et al., 1987), Little San Salvador (Wilbur, 1987), Caicos (Lloyd et al., 1987), Grand Bahama Island (Gerhardt, 1983), New Providence Island (Garrett and Gould, 1984), Yucatan (Ward and Brady, 1979), Exuma (Wallis et al., 1989), Cat Cay (Lind, 1969), Bermuda (Mackenzie, 1964; Land, et al., 1967), and the Persian Gulf (Shinn, 1973). Eolian and strand plain deposits, though seldom recorded, should be common in ancient oolites as well.

Laterally Adjacent Carbonate Environments

Modern and ancient oolites exist in lateral (and vertical) continuity with other carbonate environments such as reefs, tidal flats, lagoons, and perhaps deep, off-shelf (periplatform) environments. The absence of these environments in ancient rocks should be noteworthy and warrant explanation and further research.

Reefs

On northern Andros Island, elevated Pleistocene reefs and skeletal-rich sediment are found seaward of the ooid-rich core of the island (Figure 2). The reef at Nichollstown (Andros Island) is about 2 m above present sea level, and two samples of the reef have been

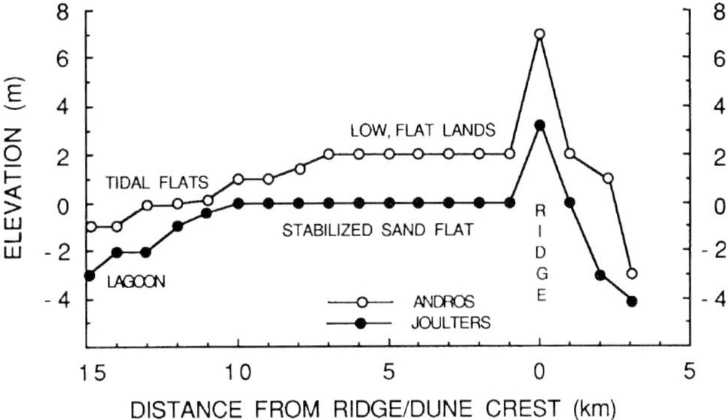

Figure 7. Topographic cross sections (west to east/leeward to windward) of Joulters and northern Andros Island are similar but offset by approximately 5 m. While the maximum elevation of Andros is approximately 20 m, most ridges are 10 m or less in elevation.

dated (U-Th, alpha counting) at 120,000 and 128,000 years B.P. (Neumann and Moore, 1975). Other elevated reefs with partially recrystallized corals are found along the eastern margin of northern Andros Island (e.g., Pye Point and Rat Cay). Surrounding these reefs and partially enclosing them are peloidal/skeletal-rich, ooid-poor grainstones. Pleistocene reefs are not found in the interior of northern Andros.

Reefs grow best on the seaward side of islands (especially windward islands) but are virtually absent from areas of active sand movement or from areas where the water is not continuously warm, normal-marine, and clear. On leeward margins of the Bahamas where offbank transport is unimpeded by barriers such as islands, early Holocene reefs have been smothered and buried by sediment movement (Hine and Neumann, 1977; Hine et al., 1981). Today, a nearly continuous barrier reef proliferates 0.5–2 km seaward of Andros Island. Apparently, the preexisting topography seaward of Andros Island provided a suitable substrate, and the island itself provides adequate protection from offbank transport of sediment and water harmful to reef growth. A more patchy distribution of reefs that have generally not yet reached the surface is located a similar distance seaward of the intertidal and supratidal portions of Joulters Cays. Perhaps the development of islands or ooid shoals has reduced offbank transport of sediment or offbank transport of water with characteristics inimical to reef growth (temperature or salinity extremes, turbidity; Gebelein, 1974).

The discontinuous distribution of elevated Pleistocene reefs seaward of the elevated Pleistocene ooid shoal of northern Andros Island is similar to the distribution of Holocene reefs seaward of Joulters. Both sets of reefs probably developed after the shoal complex grew to sea level to form a more effective barrier (e.g., islands) to offbank transport (Figure 3; Harris, 1977). In addition to the location and continuity of reefs, the seaward environments of Joulters ooid shoal and the seaward margin of the elevated Andros ooid shoal contain a high proportion of peloidal/skeletal material and a low proportion of ooids (~0%). This facies relationship of reefs, skeletal sands, and ooid shoals exists in the Quaternary analog presented here, and a similar facies relation should be expected in ancient oolites.

Tidal Flats

There is no mud tidal flat associated with either the Pleistocene ooid shoal of northern Andros Island or the Holocene ooid shoal of Joulters Cays. As part of the search for mud tidal flat lithologies on Andros, samples of all possible elevations including the rocks underlying the present-day (Holocene) mud tidal flat sediments on west Andros were taken. The upper surface of these rocks is 1–2 m below present sea level. The lithologies of rocks underlying the mud tidal flat sediments are peloidal grainstones (7% ooids, 15% skeletal, 78% peloidal), which are similar to the adjacent rocks on Andros interpreted as lagoon and stabilized sand flat and to sediments accumulating west of Joulters today. The Holocene tidal flat west of Andros is dominated by mud and exists because of the extraordinary protection afforded by Andros Island on the east and the great distance of Great Bahama Bank to the west.

Although Joulters Cay ooid shoal complex is not associated with a muddy tidal flat, the islands and intertidal shoals do create protected areas with quiet-water conditions, and Harris (1979) reports a subenvironment called a "muddy sand flat" that contains 3% mud (<62 μm) and 60% ooid sand. His stabilized sand flat environment contains an average of 4% mud. In neither of these subenvironments does the mud content approach that of typical Holocene tidal flats (Hardie, 1977; Hardie and Shinn, 1986).

The muddy tidal flat nearest to Joulters is 20 km away on the west side of northern Andros Island (Shinn et al., 1969; Hardie and Shinn, 1986). While mud tidal flats are apparently not associated with Bahamian ooid shoals (Hardie and Shinn, 1986; Ball, 1967), lime mud (as clasts and in layers and/or lenses) has recently been found associated with the ooid sand shoals near Lee Stocking Island (Kendall and Dill, 1987; Dill et al., 1989; Shinn et al., 1989; Boardman and Carney, 1991). This mud (of which >80% of the sediment is <62μm; Boardman and Carney, 1991) is not accumulated as tidal flat sediment, but is found as mud intraclasts and as thin layers or lenses (a few centimeters thick, undetermined lateral extent) deposited in tidal channels. The mud may have been formed and deposited during times when the channel was closed off by inlet migration and spit accretion (Boardman and Carney, 1991), or the mud may have been eroded from lagoons and transported (and accumulated) during storms (Dill et al., 1989; Shinn et al., 1989). Similar mud deposits have been found in every tidal channel of Joulters Cays.

Patches of mud have been found in the Miami Limestone (Pleistocene; Lasemi and Sandberg, 1984; Lasemi et al., 1990) but have not yet been found on Andros Island. Like Andros, there is no muddy tidal flat associated with the Pleistocene ooid sand shoal of the Miami Limestone (Hoffmeister et al., 1967). Muddy lithofacies (including mud intraclasts, layers, and lenses) within oolites may be characteristic of tidal channels, but muddy tidal flat deposits are not laterally adjacent to the ooid sand bodies of the Florida–Bahamas region. Muddy lithofacies of Mississippian oolites should be closely evaluated to determine the environment of deposition and temporal relation to ooid sand deposition.

Lagoons

The major lagoon associated with Andros and Joulters is located west and north (leeward) of the ooid shoal. This "lagoon" is the enormous (~100 km wide; 4 m deep) platform interior of Great Bahama Bank and contains abundant aggregates (grapestone) and peloids in addition to skeletal grains and mud (Cloud, 1962; Purdy, 1963).

As shown by the asymmetry of ooid distribution on Joulters (Figure 2), ooids are moved to the west (lagoonward) by storms and to the north by longshore transport. Even 20 km west (lagoonward) of the site of the mobile fringe (>75% ooids), ooids are still a volumetrically important grain constituent (~20%). Here, the stabilized sand flat (<2 m deep) grades into what is termed the "platform interior" by Harris (1977, 1979). The transition from ooid shoal to lagoon occurs over a long distance. Because ooids form in high-energy environments, shoreward transport and mixing with lagoonal sediments should be expected along with longshore transport of a highly oolitic sediment.

Based on low ooid content (<25%) and lack of lamination, a large portion of Andros Island is considered lagoonal (Figure 2). Lagoonal rocks of Andros are primarily located on the leeward, lower elevation side of the island (Figures 2, 6). The gradient of ooid concentration is similar to that of Joulters (Figure 2), and no sharp boundary between the ooid shoal and lagoon was found to the south or north of Andros. Large leeward lagoons with low concentrations of ooids should also be recognizable in ancient rocks.

Vertical Sequences

Ooid shoals grade laterally into lagoons found leeward of and in deeper water than the main ooid shoal and grade laterally into skeletal sands and reefs found seaward of and in deeper water than the main ooid shoal. Thus, vertical gradations of ooid shoals, lagoons, and skeletal sands and reefs should accompany growth and lateral shifting of the ooid shoal. However, mud tidal flats are not laterally adjacent to ooid shoals in our Quaternary analog. Thus, these two carbonate facies cannot be vertically juxtaposed without the presence of an exposure surface.

Location Of Ooid Shoals

The distribution of ooid sand bodies indicates that ooid shoals such as Joulters are restricted to high-energy zones (especially areas of high tidal flow) along open bank margins (Ball, 1967; Halley et al., 1983). In addition to chemical considerations associated with mixing of waters along the shallow bank margin, one requirement of ooid formation today may be that energy (currents and waves) is so high that biogenic sediment formation is inhibited, and ooids form and accumulate by default (Perkins, 1986). The concentration of energy can result from focusing by islands, by an abrupt change in slope, or by an embayment (Halley et al., 1983). The importance of high tidal flow is supported by the present-day distribution of ooids along the Andros Island–Joulters margin. Ooid sediments are not forming along the seaward margin of Andros. Reefs and narrow lagoons with skeletal sands form instead. However, just north of Andros, ooids are formed in a restricted area along the margin (Joulters Cays) and are swept onto the bank (during storms) and along the bank margin (by longshore transport) from this point source. During the late Pleistocene (120,000 to 128,000 years BP, when sea level was approximately 5 m higher than today), ooids accumulated along this margin as the Andros Island ooid shoal, so the locus of ooid accumulation has shifted northward from Andros to Joulters (Figure 8).

In ancient rocks an equivalent association may not be easily recognized. The lithologies and lateral trends of sedimentation on Andros and Joulters are very similar to each other, and if these units were to be viewed in an ancient sequence, a reasonable conclusion would be that the two ooid shoals are one continuous ooid shoal or at least one shoal undergoing continuous progradation. But the progradation is not continuous; rather it is a laterally shifting locus of deposition controlled by small changes of sea level. Andros developed during a sea-level highstand 5 m higher than the sea level producing Joulters (Figure 8). The growth and stabilization of the succeeding ooid shoal (Joulters) may have been controlled partly by a concentration of tidal flow at the north end of Andros Island (regenerative feedback of Ball, 1967).

An abrupt shifting of the locus of deposition is a significant modification of previous models of deposition in which slow, steady changes of sea level create continuous progradation or regression of adjacent facies (Irwin, 1965; Ahr, 1973; James, 1984; Read, 1985). Although the abrupt shift of the site of ooid accumulation from Andros Island to Joulters Cays occurred during the late Quaternary when sea-level fluctuations were very large, the significant shift of sea level was only a net abrupt drop of about 5 m. This order of sea-level fluctuation is not unreasonable for any time in the past. A similar model of deposition involving abrupt lateral shifts of sites of deposition was developed by Pratt and James (1986) for some ancient tidal flat sequences.

Exposure Surfaces

Ooid shoals are intertidal to shallow subtidal, and portions of them are supratidal. Even minor fluctuations of sea level should produce exposure surfaces

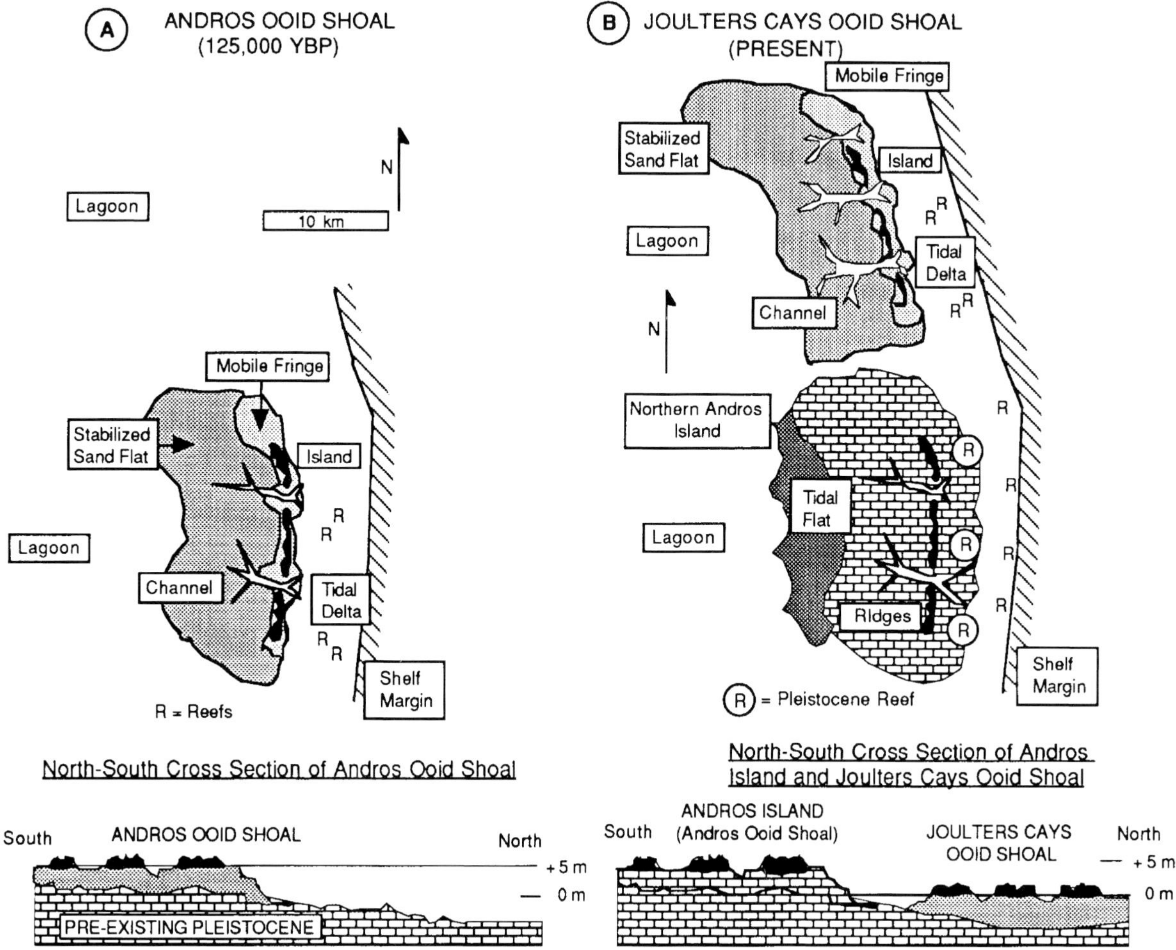

Figure 8. Model of deposition of Quaternary ooid shoals. A 5-m drop of sea level causes abrupt lateral shifting of the site of the ooid shoal. Tidal flats develop leeward of the elevated ooid shoal only.

that may include caliche, karst, and/or soil formation. Joulters Cays has experienced very little exposure modification because only a small portion is supratidal, and that portion is only about a thousand years old (Halley and Harris, 1979). On Andros, however, extensive evidence of exposure is found. Karst features (such as blueholes tens of meters across and tens of meters deep), paleosols, caliche crusts, and sediment-filled fissures are common (McCartney and Boardman, 1987; Daugherty et al., 1987; Pelle and Boardman, 1989). These exposure features developed in about 100,000 years. Exposure surfaces should also be commonly identified in ancient oolitic limestones.

COMPARISON WITH MISSISSIPPIAN EXAMPLES

Selected Mississippian oolites are next examined to determine the extent to which they can be compared to the Quaternary analog described in previous sections. As has already been shown with the Holocene and Pleistocene examples, some features are lost and others gained with added time. Problems of preservation and interpretation should be greater for Paleozoic limestones.

Ooid Distribution

The proportion of ooids in Mississippian oolitic limestones such as the Ste. Genevieve Limestone of Indiana and the Greenbrier Limestone of West Virginia appears to be similar to the shoals of northern Andros Island, Bahamas. At an outcrop near Orleans, Indiana, the Ste. Genevieve is comprised of about 75% ooids, including grains with only one or two coatings (superficial ooids; Carr, 1973, 1989). In the Greenbrier Limestone, only 4–8% of the rock section (which varies from 91 to nearly 213 m in total thickness) consists of oolitic limestones with greater than 50% ooids. However, as much as 30% of the section is comprised of limestones with greater than 25% ooids (Leonard, 1968; Carney, 1987; Carney, this volume). Like Andros, these Mississippian units do not contain ooid concentrations greater than 90% (maximum is 86%) such as those found on Joulters Cays. Joulters, then, is apparently much more oolitic than ancient counterparts. Of course, the area of Joulters with extremely high percentages of ooids is relatively small (see Figure 2), and similar areas of the ancient may either not be preserved or are not recognized. The Warix Run Limestone of northeastern Kentucky contains an average of only 4% ooids (maximum of 14%; Lierman, 1984; Carney, 1987) and therefore should not be considered an oolitic rock at all.

Spatial distributions of highly oolitic areas in the Mississippian are poorly known. What is known, however, suggests that the sizes of individual sand bodies may be comparable to the Quaternary model. The oolite body near Orleans, Indiana, for example, is reported to be approximately 3.2 km long and 6.4 km wide (Carr, 1989). Subsurface investigation of sand bodies in the Greenbrier of West Virginia indicates widths on the order of 1–2 km and lengths of 14–30 km (Kelleher, 1990; Kellerher and Smosna, this volume). By comparison, Joulters Cay shoal complex is approximately 15 km wide and 20 km long, and the area of winnowed platform-edge sands is about 0.5–3 km wide and 15 km long. On Andros, the area where ooids comprise >50% of the sediment is approximately 3 km wide and 15 km long.

Oolitic limestones (>25% ooids) of the Mississippian appear to be thicker than their modern analogs, with thicknesses ranging from 4 to 8 m in the Ste. Genevieve of Indiana to 18 m in the Greenbrier of West Virginia compared to 1–5 m for Joulters (Harris, 1977). Thicknesses of the Mississippian oolites may represent multiple oolitic bodies accumulated through several minor sea-level fluctuations.

Oolite Shoal Subenvironments

Studies of Mississippian oolites generally do not distinguish subenvironments of ooid shoals. Often, a single subenvironment equivalent to the stabilized sand flat seen in the Quaternary analog is described, but the laterally adjacent environments are not reported. Studies of the Ste. Genevieve of southern Indiana make no distinction among subenvironments of oolitic deposition (Carr, 1973, 1989). Studies of the Greenbrier distinguish an "oolitic" environment (no lamination, "lower" content of ooids) and an "ooid shoal" environment (laminations, "higher" ooid content). These may be the equivalents of the stabilized sand flat and mobile fringe. If so, the criteria needed to distinguish these two broad subenvironments appear to be preserved within Mississippian rocks. Recognition of many more of the subenvironments reported by Harris (1977, 1979) should be an important goal of petrographic and stratigraphic investigations of ancient rocks. Delineation of the lateral facies of ooid bodies is important to the interpretation of the history of deposition and diagenesis so critical to exploration.

As in the Quaternary analog, the dominant grain types present in Mississippian oolitic limestones are ooids, peloids, and skeletal fragments. The Ste. Genevieve of Indiana contains a high percentage of peloids and aggregates (22%) and a very low percentage of skeletal grains (3%; Carr, 1973, 1989). The Greenbrier Limestone contains a much lower percentage of peloids and aggregates (3–5%) and a higher percentage of skeletal grains (25%; Carney, 1987, this volume). The ooid shoals of the Ste. Genevieve of Indiana may be more similar to Andros Island, then, and the Greenbrier more similar to Joulters Cays in composition (Figure 4). The reasons for this are not clear. The percentage of mud is low in oolitic limestones of the Mississippian as in the Quaternary models.

Characteristics such as original topography are nearly impossible to evaluate, and strand plains and eolian features (island, dunes) are seldom described. Eolian features are very common in the modern, and their scarcity in the ancient may simply be a problem of recognition. Recently, Hunter (1989) reported eolian features in a few sections of the Ste. Genevieve of southern Indiana. As better diagnostic criteria become available and more critical and detailed examination of outcrops is attempted, these features may be identified more frequently in association with ancient oolitic deposits.

Laterally Adjacent Carbonate Environments

As discussed earlier, the important carbonate environments found adjacent to oolitic deposits in the Quaternary include reefs, tidal flats, lagoons, and deeper offshore environments. Reefs are not commonly recognized in association with the Mississippian examples discussed here. Recently, however, a small patch reef was described from an oolitic portion of the Greenbrier of southern West Virginia (Wulff, 1990). Tidal flats and lagoons, however, are commonly described (Leonard, 1968; Carr, 1973; Ettensohn, 1975; Dever, 1980; Lierman, 1984; Yeilding and Dennison, 1986; Carney, 1987; Sullivan and Textoris, 1988; Carney and Smosna, 1989; Carr, 1989). Often, interbedded mud-rich and ooid-rich limestones are interpreted to represent shifting of ooid shoals over lagoon and/or tidal flats during storms or sea level changes (Leonard, 1968; Carney and Smosna, 1989). Based on the Quaternary analog, however, only lagoons should be found in gradational contact with ooid shoals. Tidal flats should be associated with ancient oolites only when accompanied by an exposure surface. Storms move ooid shoals over relatively small portions of lagoons, but to distribute ooids as oolitic shoals (>25% ooids) over a large platform would require an accompanying sea-level rise (e.g., Hine, 1977; Hine and Neumann, 1977; Hine et al., 1981). The Quaternary model predicts that progradation of ooid-rich sediment would be greatest in a longshore direction rather than an onshore direction.

Location of Ooid Shoals

Unlike the Bahamian analog, Mississippian ooid shoals are not interpreted as forming on open bank margins. Oolitic shoals of the Ste. Genevieve of

Indiana, for example, are thought to have formed on a gently dipping ramp over topographic highs produced by buildup of skeletal material (Carr, 1973, 1989). In this case, ooid deposition in the Persian Gulf may be more analogous to Mississippian oolites. There, ooids accumulate close to shore, often as ebb-tidal deltas, along beaches, and as eolian dunes (Wagner and van der Togt, 1973; Purser and Evans, 1973; Loreau and Purser, 1973). The Greenbrier shoals are thought to have developed along high-energy shorelines around features like the West Virginia dome (an island or submarine high approximately 100 × >50 km; Carney, this volume; Smosna and Koehler, this volume) and as halos around irregularities in the underlying erosional surface (Youse, 1964; Leonard, 1968). Some analogies to Bahamian models may be appropriate, but the lack of a deeper-water facies is puzzling. The Warix Run of Kentucky is thought to have accumulated in topographic lows (Ettensohn, 1975; Dever, 1980; Lierman, 1984). This geomorphic setting may be more analogous to that of Lily Bank, where it is thought that ooids are accumulating because of higher energy between two topographic highs, (i.e., the ooids form in a reentrant) (Hine, 1977).

Exposure Surfaces

The Quaternary model illustrates that ooid shoals develop at, or near, sea level, and minor net changes of sea level (5 m) have profound effects on the site of accumulation and the creation of exposure surfaces. The exposure surface of Andros Island required only approximately 100,000 years to form (McCartney and Boardman, 1987). Exposure surfaces, therefore, should be commonly preserved in oolitic Mississippian limestones. "Intraformational" exposure surfaces have been recognized by Dever (1980) and Ettensohn et al. (1988) in the Ste. Genevieve and Warix Run limestones of northeastern Kentucky. Exposure surfaces were also found associated with eolian features in the Ste. Genevieve of southern Indiana (Hunter, 1989). Similar features have not as yet been recognized in the thick oolitic limestones of the Greenbrier of West Virginia.

CONCLUSIONS

Use of the composite Quaternary analog as presented here will allow a new and more extensive look at ancient oolites. A number of features characteristic of Holocene ooid shoals like Joulters are preserved into the Pleistocene and may be useful for evaluation of Mississippian limestones. These features include the proportion of ooids, lateral gradients of ooid concentration, lateral associations of adjacent carbonate environments, shoal topography, and strand plains/dunes. The various subenvironments of modern ooid shoals need to be more completely described and criteria for their identification in the ancient better developed.

Muddy tidal flats are not adjacent to Holocene ooid shoals, but if they are associated with Mississippian oolites, they should be closely examined as evidence of sea level change and/or island formation. Regional topography is an important control of Holocene ooid accumulation, and evidence for an adjacent deeper water environment should be carefully examined.

Sea level changes like those experienced by modern ooid shoals in the Bahamas (±5 m) may have influenced the compositions and geometry of ooid shoals in the Mississippian as well. Changes caused by minor fluctuations of sea level include creation of exposure surfaces, abrupt lateral shifting of the site of ooid shoal development, and possible formation of muddy tidal flats.

ACKNOWLEDGMENTS

We are grateful to Mitch Harris, Bob Halley, Charlie Zuppann, and Brian Keith for helpful comments that improved this manuscript. This research was supported by grants from Gulf Oil Company and Chevron Oil (to C. Carney) and Sigma Xi (to P. Bergstrand). Additional funding (for field work in the Bahamas) was provided by the annual summer workshop "Modern Carbonate Deposition" (directed by M. Boardman).

REFERENCES CITED

Ahr, W.M., 1973, Carbonate ramp: An alternative to the shelf model: Trans. Gulf Coast Ass. Geol. Soc., v. 23, p. 221-225.

Ball, M.M., 1967, Carbonate sand bodies of Florida and Bahamas: Journal of Sedimentary Petrology, v. 37, p. 556-591.

Bergstrand, P.M., 1991, The surficial geology and depositional environments of sediments which formed the Pleistocene age rocks of northern Andros Island, Bahamas: Unpubl. M.S. thesis, Miami University, Oxford, OH, 199 p.

Boardman M.R., J.L. Carew, and J.E. Mylroie, 1987, Holocene deposition of transgressive sand on San Salvador, Bahamas: GSA Abstracts with Programs, v. 19, p. 593.

Boardman, M.R., and C. Carney, 1991, Origin and accumulation of lime mud in ooid tidal channels, Bahamas: Journal of Sedimentary Petrology, v. 61, p. 661–680

Carew, J.L., and J.E. Mylroie, 1985, The Pleistocene and Holocene stratigraphy of San Salvador Island, Bahamas, with reference to marine and terrestrial lithofacies at French Bay, *in* Curran, H.A., ed., Pleistocene and Holocene carbonate environments on San Salvador Island, Bahamas: Guidebook for Geological Soc. Amer. Orlando annual meeting field trip: Ft. Lauderdale, FL, CCFL Bahamian Field Station, p. 11-61.

Carney, C., 1987, Petrology and diagenesis of the Upper Mississippian Greenbrier Limestone in the central Appalachian Basin and of the Lower Carboniferous Great Limestone in northern England: Unpubl. Ph.D. dissert., West Virginia University, Morgantown, WV, 398 p.

Carney, C., and R. Smosna, 1989, Carbonate deposition in a shallow marine gulf, the Mississippian Greenbrier Limestone of the central Appalachian Basin: Southeastern Geology, v. 30, p. 25-48.

Carr, D.D., 1973, Geometry and origin of oolite bodies in the Ste. Genevieve Limestone (Mississippian) in the Illinois Basin: Indiana Geological Survey Bulletin 48, 81 p.

Carr, D.D., 1989, Geometry and depositional environments of oolite bodies in the Ste. Genevieve Limestone (Mississippian) in southern Indiana: Indiana Department of Natural Resources Geological Survey Guidebook, p. 1-14.

Cloud, P.E., 1962, Environment of calcium carbonate deposition west of Andros Island, Bahamas: United States Geol Surv. Professional Paper 350, p. 1-138.

Daugherty, D.R., M.R. Boardman, and C.V. Metzler, 1987, Characteristics and origin of sedimentary dikes, Bahamas, *in* Curran, H.A., ed., Proceedings of the third symposium of the geology of the Bahamas: CCFL Bahamian Field Station, Ft. Lauderdale, FL, p. 45-56.

Dever, G.R. Jr., 1980, The Newman Limestone—An indicator of Mississippian tectonic activity in northeastern Kentucky, *in* Luther, M., ed., Proceedings of the Technical Sessions of the Kentucky Oil and Gas Association 36th and 37th Annual Meetings, 1972 and 1973: Kentucky Geol. Survey Spec. Publ. 5, 210 p.

Dill, R.F., C.G.St.C. Kendall, and E.A. Shinn, 1989, Giant subtidal stromatolites and related sedimentary features: Field Trip Guidebook T373, 28th International Geological Congress, American Geophysical Union, Washington, D.C., 33 p.

Dravis, J.J., 1979, Rapid and widespread generation of Recent oolitic hardgrounds on a high energy Bahamian platform, Eleuthera Bank: Journal of Sedimentary Petrology, v. 49, p. 195-208.

Ettensohn, F.R., 1975, Stratigraphic and paleoenvironmental aspects of Upper Mississippian rocks (Upper Newman Group), east-central Kentucky: Unpubl. Ph.D. dissert., University of Cincinnati, Cincinnati, OH, 320 p.

Ettensohn, F.R., C.L. Rice, G.R. Dever, Jr., and D.R. Chesnut, 1984, Slade and Paragon Formations—new stratigraphic nomenclature for Mississippian rocks along the Cumberland Escarpment in Kentucky: U. S. Geol. Survey Bull., 1605-B, 37 p.

Ettensohn, F.R., G.R. Dever, and J.S. Grow, 1988, A paleosol interpretation for profiles exhibiting subaerial exposure "crusts" from the Mississippian of the Appalachian Basin: GSA Special Paper 216, p. 49-79.

Evans, C.E., 1983, Aspects of the depositional and diagenetic history of the Miami Limestone: Control of primary sedimentary fabric over early cementation and porosity development: Unpubl. M.S. thesis, University of Miami, Coral Gables, FL, 233 p.

Folk, R.L., 1962, Spectral subdivision of limestone types, *in* Ham, W.E., ed., Classification of Carbonate Rocks—A Symposium: AAPG Memoir 1, 62-84.

Garrett, P. and S.J. Gould, 1984, Geology of New Providence Island, Bahamas: GSA Bull., v. 95, p. 209-220.

Gebelein, C.D., 1974, Guidebook for modern Bahamian platform environments: GSA field trip, 97 p.

Gerhardt, D.J.G., 1983, The anatomy and history of a Pleistocene strand: Unpubl. M.S. thesis, University of Miami, Coral Gables, FL, 170 p.

Halley R.B. and P.M. Harris, 1979, Fresh-water cementation of a 1000 year old oolite: Journal of Sedimentary Petrology, v. 49, p. 969-988.

Halley, R.B., P.M. Harris, and A.C.Hine, 1983, Bank margin environment, *in* Scholle, P.A., D.G. Bebout, and C.H. Moore, eds., Carbonate Depositional Environments: AAPG Memoir 33, p. 463-506.

Hardie, L.A., 1977, Sedimentation of the modern carbonate tidal flats of northwest Andros Island, Bahamas: Johns Hopkins Univ. Studies in Geology 22, 202 p.

Hardie, L.A., and E.S. Shinn, 1986, Carbonate depositional environments, modern and ancient; Part 3: Tidal flats: Colorado School of Mines Quarterly, v. 81, p. 1-74.

Harris, P.M., 1977, Sedimentology of the Joulters Cays ooid sand shoal, Great Bahama Bank: Unpubl. Ph.D. dissert., University of Miami, Coral Gables, FL, 452 p.

Harris, P.M., 1979, Facies anatomy and diagenesis of a Bahamian ooid shoal: Sedimenta VII, The Comparative Sedimentology Laboratory, University of Miami, Miami, Florida, 163 p.

Hine, A.C., 1977, Lily Bank, Bahamas: History of an active oolite sand shoal: Journal of Sedimentary Petrology, v. 47, p. 1554-1581.

Hine, A.C., and A.C. Neumann, 1977, Shallow carbonate bank margin growth and structure, Little Bahama Bank, Bahamas: AAPG Bulletin, v. 61, p. 376-406.

Hine, A.C., R.J. Wilbur, J.M. Bane, A.C. Neumann, and K.R. Lorenson, 1981, Offbank transport of carbonate sands along open, leeward margins, northern Bahamas: Marine Geology, v. 42, p. 327-348.

Hoffmeister, J.E., K.E. Stockman, and H.G. Multer, 1967, Miami Limestone of Florida and its Recent Bahamian counterpart: GSA Bull., v. 78, p. 175-190.

Hunter, R.E., 1989, Eolianites in the Ste. Genevieve Limestone of southern Indiana: Department of Natural Resources Geological Survey Guidebook, p. 1-19.

Irwin, M.L., 1965, General theory of epeiric clear water sedimentation: AAPG Bull., v. 49, p. 445-459.

James, N.P., 1984, Introduction to carbonate facies models, *in* Walker, R.G., ed., Facies Models: Geoscience Canada, Reprint series 1, p. 209-211.

Kelleher, G.T., 1990, Stratigraphy and diagenesis of the Mississippian Greenbrier Group in the Rhodell

Field Area of southern West Virginia, Unpubl. M.S. thesis, Morgantown, WV, 80 p.
Kendall, C.G.St.C., and R.F. Dill, 1987, Guidebook to the giant subtidal stromatolites and carbonate facies of Lee Stocking Island, Bahamas: Dept. of Geol., Univ. South Carolina, Columbia, SC, 151 p.
Land, L.S., F.T. Mackenzie, and S.J. Gould, 1967, The Pleistocene history of Bermuda: GSA Bulletin., v. 78, p.993-1006.
Lasemi, Z., and P.A. Sandberg, 1984, Transformation of aragonite-dominated lime muds to microcrystalline limestones: Geology, v. 12, p. 420-423.
Lasemi, Z., P.A. Sandberg, and M.R. Boardman, 1990, New microtextural criterion for differentiation of compaction and early cementation in fine-grained limestones: Geology, v. 18, p. 370-373.
Leonard, A.D., 1968, The petrology and stratigraphy of Upper Mississippian Greenbrier Limestones of eastern West Virginia: Unpubl. Ph.D. dissert., West Virginia University, Morgantown, WV, 219 p.
Lierman, R.T., 1984, Environment of deposition and diagenetic history of the Warix Run and Paoli-Beaver Bend Limestones (Upper Mississippian) of east central Kentucky: Unpubl. M.S. thesis, Miami University, Oxford, OH, 209 p.
Lind, A.O., 1969, Coastal landforms of Cat Island, Bahamas: A study of Holocene accretionary topography and sea-level change: Department of Geography, Research Paper 122, University of Chicago, Chicago, 155 p.
Lloyd, R.M., R.D. Perkins, and S.D. Kerr, 1987, Beach and shoreface ooid deposition on shallow interior banks, Turks and Caicos Islands, British West Indies: Journal of Sedimentary Petrology, v. 57, p. 976-982.
Loreau, J.-P., and B.H. Purser, 1973, Distribution and ultrastructure of Holocene ooids in the Persian Gulf, *in* The Persian Gulf (ed. by Purser, B.H.), Springer-Verlag, New York, NY, p. 279-328.
McCartney, R.F., and M.R. Boardman, 1987, Bahamian paleosols: Implications for stratigraphic correlation, *in* Curran, H.A., ed., Proceedings of the third symposium of the geology of the Bahamas: CCFL Bahamian Field Station, Ft. Lauderdale, FL, p. 99-108.
MacKenzie, F.T., 1964, Bermuda Pleistocene eolianites and paleowinds: Sedimentology, v. 3, p. 52-64.
Newell, N.D., and J.K. Rigby, 1957; Geological studies on the Great Bahama Bank, *in* LeBlank, R.J., and J.G. Breeding, eds., Regional aspects of carboante ceposition: a Symposium with discussion: Society of Paleontologists and Mineralogists Special Publication 5, Tulsa, OK, p. 15-72.
Newell, N.D., E.G. Purdy, and J. Imbrie, 1960, Bahamian oolitic sand: Journal of Geology, v. 68, p. 481-497.
Neumann, A.C., and W.S. Moore, 1975, Sea level events and Pleistocene coral ages in the northern Bahamas: Quaternary Research, v. 5, p. 215-224.
Palmer, M.S., 1979, Holocene facies geometry of the leeward bank margin, Tongue of the Ocean, Bahamas: Unpubl. M.S. thesis, University of Miami, Coral Gables, FL, 199 p.
Pelle, R.C., and M.R. Boardman, 1989, Stratigraphic distribution and associations of trace elements in vadose-altered multicomponent carbonate assemblages: GSA Abstracts with Programs, v. 21, p. 258.
Perkins, R.D., 1986, The default concept in carbonate sedimentation: SEPM annual mid-year meeting abstracts, v. 3, p. 88.
Pratt, B.R., and N.P. James, N.P., 1986, The St. George (Lower Ordovician) of western Newfoundland: Tidal flat island model for carbonate sedimentation in shallow epeiric seas: Sedimentology, v. 33, p. 313-344.
Purdy, E.G., 1963, Recent calcium carbonate facies of the Great Bahama Bank, part 2, sedimentary facies: Journal of Geology, v. 71, p. 472-497.
Purser, B.H., and G. Evans, 1973, Regional sedimentation along the Trucial Coast, SE Persian Gulf, *in* Purser, B.H., ed., The Persian Gulf: Springer-Verlag, New York, NY, p. 211 -232.
Read, J.F., 1985, Carbonate platform facies models: AAPG Bull., v. 69, p. 1-21.
Shinn, E.A., 1973, Sedimentary accretion along the leeward, SE coast of Quatar Peninsula, Persian Gulf, *in* Purser, B.H., ed., The Persian Gulf: Springer-Verlag, New York, p. 199- 209.
Shinn, E.A., R.M. Lloyd, and R.N. Ginsburg, 1969, Anatomy of a modern carbonate tidal flat, Andros Island, Bahamas: Journal of Sedimentay Petrology, v. 39, p. 1202-1228.
Shinn, E.A., R.P. Steinen, B.H. Lidz, and P.K. Swart, 1989, "Whitings" a sedimentologic dilemma: Journal of Sedimentary Petrology, v. 59, p. 147-161.
Strasser, A., and E. Davaud, 1986, Formation of Holocene limestone sequences by progradation, cementation, and erosion: Two examples from the Bahamas: Journal of Sedimentary Petrology, v. 56, p. 422-428.
Sullivan, E.M., and D.A. Textoris, 1988, Microfacies and paleoenvironments of the Mississippian Denmar Formation, eastern West Virginia: Southeastern Geology, v. 28, p. 133-152.
Wagner, C.W., and C. van der Togt, 1973, Holocene sediment types and their distribution in the southern Persian Gulf, *in* The Persian Gulf (ed. by Purser, B.H.), Springer-Verlag, New York, NY, p. 123-156.
Ward, W.C. and M.J. Brady, 1979, Strandline sedimentation of carbonate grainstones, Upper Pleistocene, Yucatan Peninsula, Mexico: AAPG, Bull., v. 63 , p. 362-369.
Wallis, T.N., H.L. Vacher, M.T. Stewart, P.J. Hearty, M.J. Wightman, and R.V. Cant, 1989, Holocene strandplain aquifer at Ocean Bight, Great Exuma Island, Bahamas: GSA Abstracts with Programs, v. 21, p. 242.
Wilbur, R.J., 1987, Geology of Little San Salvador and West Plana Cay: Preliminary findings with implications for Bahamian island stratigraphy, *in* Curran, H.A., ed., Proceedings of the third symposium of the geology of the Bahamas: CCFL Bahamian Field Station, Ft. Lauderdale, FL, p. 181-204.

Wilson, J.L., 1971, Carbonate facies in geologic history: Springer-Verlag, New York, 471 p.

Wulff, J.I., 1990, An upper Mississippian reef in southern West Virginia: GSA Abstracts with Programs, v. 22(5), p. 49.

Yeilding, C.A., and J.M. Dennison, 1986, Sedimentary response to Mississippian tectonic activity at the east end of the 38th Parallel fracture zone: Geology, v. 14, p. 621-624.

Youse, A.C., 1964, Gas producing zones of the Greenbrier Limestone—southern West Virginia and eastern Kentucky: AAPG Bull., v. 48, p. 465-486.

Chapter 17

Eolian Structures and Textures in Oolitic-Skeletal Calcarenites from the Quaternary of San Salvador Island, Bahamas: A New Perspective on Eolian Limestones

Mario V. Caputo
Department of Geology and Geography
Mississippi State University
Mississippi State, Mississippi, USA

ABSTRACT

Calcarenites on San Salvador Island, Bahamas, preserve a history of ooids and skeletal grains reworked by wind. French Bay and Cockburn Town members of the Pleistocene Grotto Beach Formation and comparable Holocene strata contain eolian sandflow, grainfall, and wind-ripple structures and textures similar to those of siliciclastic eolian deposits. In thin section, well-cemented, closely packed, fine-grained strata correspond with ledges in outcrop; poorly cemented, loosely packed, medium-grained strata correspond with recesses. Wind-ripple strata form topsets, brinksets, and bottomsets. White, ledgy, fine-grained, basal laminae are overlain by gray, recessed, medium-grained laminae to form one wind-ripple stratum. Sandflow cross-strata are tabular in longitudinal sections, lenticular in transverse sections, and wedge out into grainfall and wind-ripple bottomsets. They coarsen upward from white, ledge-forming, fine-grained laminations to gray, recess-forming, medium-grained laminations. Grainfall forms weakly stratified, fine-grained, ledge-forming units of variable thickness and geometry. They are associated with sandflow cross-strata and thicken downslope as bottomsets, especially in cross-bed sets thinner than 2 m (7 ft). Grainfall strata are difficult to recognize where structural relations between foreset and bottomset beds are hidden.

INTRODUCTION

History of Eolian Work

Some of the earliest studies on cross-stratified sandstones discriminated eolian features on the basis of scale and inclination of cross-beds and on grain texture (Knight, 1929; McKee, 1934a, b; Reiche, 1938). Subsequent reports on eolian quartzarenites refined a collection of physical and biogenic criteria, which, when combined with vertical and lateral facies associations, strongly supported an eolian origin. The work of Hunter (1977a, b, 1981) is particularly important because it detailed the structure of individual eolian bedding units, namely sandflow, grainfall, and ripple stratification, known as eolian fine structure (Hunter, 1985), in quartz-rich deposits. Eolian stratification in modern and ancient deposits can now be identified with a high degree of confidence.

Some Pleistocene and Holocene cross-bedded calcarenites, recognized throughout tropical and subtropical regions of the world, are associated with a host of subaerial features and represent marine sediment reworked by wind (Illing, 1954; Newell et al., 1960; Ball, 1967; Mackenzie, 1964a, b; Ward, 1970, 1973, 1975; McKee and Ward, 1983). An eolian origin of cross-stratified calcarenites has received little attention because of the unequivocal marine character of constituent, framework grains. Ooid and other high-energy facies in carbonate deposits commonly form in marine nearshore shelf environments (Illing, 1954; Newell et al., 1960; Irwin, 1965; Ball, 1967; Scholle et al., 1983); ooid facies are not uncommon in terrestrial aqueous environments (Dean and Fouch, 1983). However, unconsolidated carbonate grains can become exposed subaerially on tidal flats and beaches and become mechanically reworked by wind to form eolian calcarenites or eolianites.

The recognition of distinctive eolian stratification in limestones has been applied recently to Holocene (White and Curran, 1985, 1986, 1988) and Pleistocene (Caputo, 1989) calcarenites on San Salvador Island, Bahamas. Geologists have begun to reexamine older sedimentary sections for overlooked eolian limestone among dominantly marine beds and have uncovered eolian fine structure in grainstones of Pennsylvanian age in southeastern Utah (Loope, 1986; Loope and Haverland, 1988; Boubin and Loope, 1989), of Mississippian age in southern Indiana (Hunter, 1988 and this volume) and in southwestern Kansas (Handford, 1990), and of Pennsylvanian and Permian age in Arizona and Nevada (Rice and Loope, 1991).

Purpose

One purpose of this chapter is to show that sedimentary features of some Quaternary eolian calcarenites are analogous to those of modern and ancient eolian siliciclastic deposits.

Another purpose is to offer an additional perspective on eolian structure and texture, perhaps as a basis for comparison with other Phanerozoic eolian limestones.

An examination of the cross-bedded limestones on San Salvador Island suggests that eolian processes, stratification, grain texture, diagenesis, and weathering are successively interrelated. Wind interacts with sediment on dunes to generate specific subprocesses such as grainflow, grainfall, and wind-ripple migration. Resulting strata are distinguishable from one another by spatial arrangement, geometry, grain packing, and grain size, the latter two of which control the degree of cementation. The textural and cementation patterns are visible in thin section and are reflected in weathering patterns on outcrop surfaces. Consequently, the correspondence between weathering patterns and type of eolian stratification enhances the identification of architectural components within the carbonate eolian sand body. In a study on sedimentation of quartz sand in a wind tunnel by Schenk (1983), a similar correlation was drawn between depositional processes and structural and textural aspects of the resultant stratification.

Setting

San Salvador is one among a group of irregular carbonate islands that emerge from a platform system along the southeastern continental margin of North America (Curran, 1985) (Figure 1). The island is built of Pleistocene and Holocene oolitic, skeletal, and peloidal grainstones that crop out as low ridges separated by shallow swales. The limestones preserve a history of eustatic sea level fluctuations influenced by the advance and retreat of continental glaciers during the Pleistocene Epoch.

The Quaternary System of rocks on San Salvador has been organized into a lithostratigraphic and geochronologic framework within which there are three formations (Carew et al., 1984; Carew and Mylroie, 1985, 1987). In ascending order, they are the Owl's Hole Formation, Grotto Beach Formation, and the Rice Bay Formation (Figure 2).

The units of interest in this report are the French Bay Member and the Cockburn Town Member, the lower and middle members, respectively, of the Grotto Beach Formation (Figure 2). They contain bedding features that demonstrate the relationship among eolian dune processes, stratification, grain texture, cementation, and weathering patterns. The French Bay Member was examined in sea cliff exposures between Sandy Point and a small wooden pier called Government Dock, near western French Bay on the south side of the island (Figure 3). Other rocks,

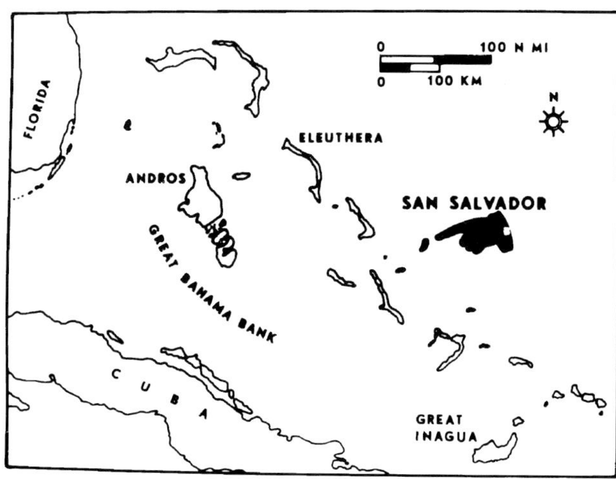

Figure 1. Location map of San Salvador Island and the other Bahamian Islands relative to Florida and Cuba (redrawn from Curran, 1985). Scale is in nautical miles (N MI) and kilometers (KM).

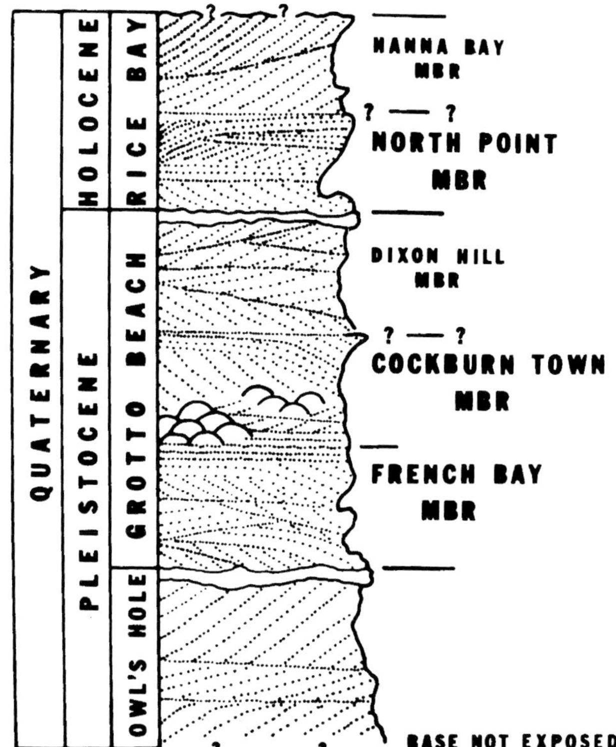

Figure 2. Stratigraphic nomenclature of Pleistocene and Holocene sedimentary rocks on San Salvador Island (redrawn from Carew and Mylroie, 1985). Explanation: cross-bedding (inclined stippling), flat laminations (parallel stippling), reef structures (hemispheres), cross-bed set boundaries (cross-cutting stippled lines), paleosols (unstippled intervals).

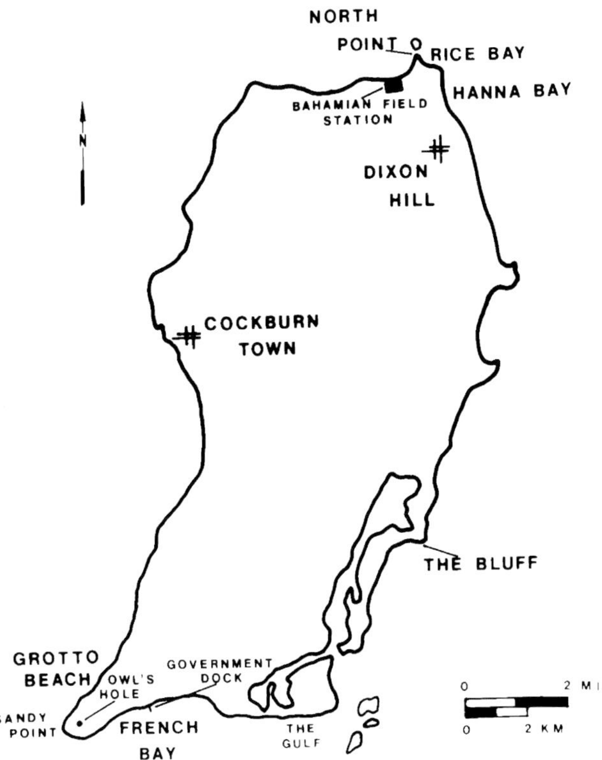

Figure 3. The island of San Salvador, Bahamas. Geographic names refer to study sites and type localities for Pleistocene and Holocene sedimentary rocks in Figure 2.

tentatively assigned to the Cockburn Town Member, were scrutinized for eolian bedding in sea cliffs between French Bay and the Gulf, located also on the southern margin of the island (Figure 3).

The occurrence of eolian fine structure in the French Bay Member and equivalent beds in the Cockburn Town Member on southern San Salvador was noted by Caputo (1989) and is further described in this chapter. However, much of the true Cockburn Town Member on the western side of the island is a bank/barrier reef and shoreline deposit (White et al., 1984; Curran and White, 1984, 1985). Beds in the Owl's Hole Formation and the Dixon Hill Member, the uppermost member of the Grotto Beach Formation (Figure 2), have been interpreted provisionally as eolian (Carew and Mylroie, 1985) until they are studied in greater detail. White and Curran (1985, 1986, 1988) have identified eolian and other subaerial features in the North Point and Hanna Bay members, the basal and upper members respectively, of the Rice Bay Formation (Figure 2). Three-dimensional exposures of the North Point Member are observable at North Point, which forms the northernmost projection of the island proper at a study site located near the Bahamian Field Station (Figure 3). Well-preserved and exposed eolian strata in the North Point Member are compared in this chapter with features described from the Grotto Beach Formation. They serve to reinforce the association among stratification, diagenesis, and weathering.

SEDIMENTARY PETROLOGY

Eolian Stratification in Outcrop

Sandflow, grainfall, and ripple strata are common in quartz-rich eolian dunes. They are generally distinguishable from one another by spatial relations, geometry, thickness, grain size, and grain packing. Low- and high-angle stratification with structural and textural traits resembling those in quartzose eolian deposits is present in the Pleistocene and Holocene calcarenites examined herein. The weathering of bedding units is influenced by grain texture and cementation, such that an outcrop expression of alternating positive-relief ledges and negative-relief recesses is the result. Weathering, then, may be useful in further discriminating eolian strata in Quaternary limestones. Comparable weathering patterns are recognized in exposures of other Quaternary eolian calcarenites (McKee and Ward, 1983).

The French Bay and Cockburn Town members of the Pleistocene Grotto Beach Formation comprise a complex array of wedge and tabular sets of wind-ripple laminations and cross-beds. Subhorizontal wind-ripple strata form amalgamated, low-angle topset and backset beds 0.3–1.5 m (1.0–4.5 ft) thick, and grade downcurrent into gently dipping brinkset beds composed of wind-ripple and possibly grainfall strata. Foreset beds dip between 25° and 32° and consist primarily of sandflow beds 2–10 cm (0.8–4.0 in.) thick and localized grainfall and wind-ripple strata up to 2 cm (0.8 in.) thick. Foreset beds ultimately grade downcurrent into poorly exposed bottomset beds that are either wind-ripple or grainfall strata, or a mixture of both (Figure 4).

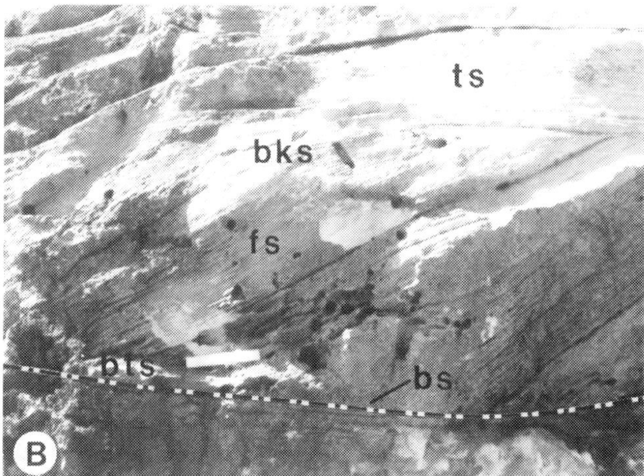

Figure 4. Longitudinal views of downcurrent succession of eolian bedding: subhorizontal topset wind-ripple (ts), brinkset wind-ripple (bks), foreset sandflow (fs), and bottomset wind-ripple or grainfall (bts) strata. (A) Large-scale set. Lower bounding surface not exposed. Scale (black arrow) is in centimeters and inches. (B) Medium-scale set. Dashed line is lower bounding surface (bs). White scale on bottomset is 15 cm (6 in.) long. Cockburn Town Member above sea cliffs east of French Bay.

Wind-Ripple Strata

In modern and ancient eolian quartz sand deposits, climbing translatent strata (Hunter, 1977a, b) or wind-ripple strata are recognized by their thin, even, cyclic occurrence (hence, pin-stripe laminations of Fryberger and Schenk, 1988), poorly developed foresets between bounding surfaces, and a general upward-coarsening grain-size trend between lower and upper bounding surfaces (Hunter, 1977a, b, 1981; Fryberger and Schenk, 1981; Kocurek and Dott, 1981; Schenck, 1983). Wind-ripple strata in rocks are generally so thin that foreset laminations and inverse grading are not clearly discernible. Consequently, the thin, even, cyclic nature of the bedding may be the only diagnostic criterion for recognizing wind-ripple strata (Hunter, 1977a; Fryberger and Schenk, 1981; Kocurek and Dott, 1981).

Pleistocene wind-ripple strata on San Salvador Island form composite or amalgamated beds. Unlike the low-angle stratification of interdune (between dunes), sand sheet (at the margins of dune fields), or dune apron (at the base of a dune slip face) areas (Kocurek, 1986), these low-angle wind-ripple strata were deposited as topset and, in certain exposures, backset beds on the stoss side near the crest of a former dune. They may comprise yet another type of low-angle eolian stratification in addition to those recognized by Kocurek (1986).

A wind-ripple stratum in eolian beds of the Grotto Beach Formation is 2–5 mm (0.1–0.2 in.) thick and was formed by the migration of a single wind ripple. Inverse textural grading is very well developed and is visible in oblique exposures that exaggerate the thickness of each ripple-produced stratum (Figure 5). A single wind-ripple stratum is composed of an apparent paired or twofold layering that forms a couplet. Each of the two layers or laminae of the couplet is distinguishable by color, grain size, grain packing, cementation, and weathering pattern.

The basal layer of each couplet is characterized by moderately to well-packed, well-cemented, fine-grained sand. It weathers to form a white, positive-relief ledge on the outcrop (Figure 6). An upper layer of a wind-ripple stratum consists of moderately to poorly packed, poorly cemented, medium-grained sand that weathers to form a gray, negative-relief recess on the outcrop. Collectively, the white and gray laminations form a succession of closely spaced, cyclic, parallel ledges and recesses, which are interrupted locally by low-angle truncation surfaces (Figure 6).

Other criteria suggest that the thin, even, parallel strata in the Grotto Beach Formation are the product of migrating wind ripples. Parting lineation is absent. This structure is a weathering feature related to the presence of plane beds or laminations, which are deposited by streaks or sheets of sand under high-velocity, upper flow-regime conditions. Ripple marks with a high wavelength-to-amplitude ratio (ripple index), a common feature of eolian ripples, are locally preserved and exposed on bedding plane surfaces (Figure 7).

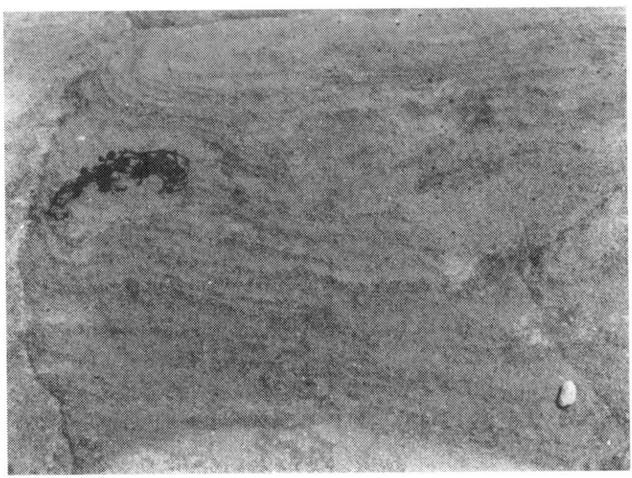

Figure 5. Oblique exposure of wind-ripple strata. Dark bands correspond with medium-grained sand in upper laminae and light bands with fine-grained sand in basal laminae for each wind-ripple stratum. Cockburn Town Member in sea cliffs east of French Bay. Bit of Sargassum weed, about 20 cm long (8 in.), for scale.

Figure 6. Subhorizontal wind-ripple strata (low-angle stratification) forming topsets. (A) Alternating white, well-cemented, ledge-forming basal laminae and gray, poorly-cemented, recess-forming upper laminae. (B) Amalgamated wind-ripple beds separated by low-angle truncation surfaces (black arrows). Cockburn Town Member in sea cliffs east of French Bay. Scale is in centimeters and inches.

Sandflow Cross-Strata

In general, medium- and large-scale eolian cross-stratification forms on steep lee slopes or slipfaces of dunes and consists of a series of sandflow strata, which are sometimes interbedded with grainfall and wind-ripple strata (Hunter, 1977a, 1981; Kocurek and Dott, 1981). Wind-transported traction and suspension loads settle onto the lee slope to accumulate as grainfall strata. Sandflow strata result from the failing and remobilizing of previously deposited grainfall and wind-ripple sediment on the lee slopes of dunes. Wind ripples may migrate temporarily across the lee slope under the influence of wind-eddy currents during a pause in dune migration. Later, wind-ripple laminae are buried by successive sandflow and grainfall deposits and become part of the foreset strata.

Sandflow strata are tongues of sediment produced during grainflow, a type of sediment gravity flow. When preexisting grainfall and ripple deposits on the upper lee slope of a dune build up to a critical slope angle and fail, they lose their original character and are transformed into an avalanche or grainflow of noncohesive sand moving down the lee slope. Depending on the slipface angle, grain cohesion, sediment volume, and grain momentum, flows can extend to the base of a dune, where they wedge out over bottomset strata and later become buried by additional bottomset strata.

In longitudinal view, parallel to the paleocurrent direction, the main body of most sandflow cross-strata is generally tabular; upper and lower boundaries converge downslope to form a wedge at the toe of the bed (Figure 8). However, there are variations in size, shape, and bedding relations of sandflow strata as a function of dune height (Hunter, 1977a, 1985). For dune amplitudes of 2 m (7 ft) or less, sandflow cross-strata wedge out and terminate downslope into bottomsets of either grainfall or wind-ripple strata and upslope into brinksets of similar strata. A significant feature of small dunes is that grainfall strata are not destroyed completely during a grainflow event. Therefore, sandflow tongues are separated and surrounded by grainfall deposits. Consequently, in transverse sections normal to the paleocurrent direction, sandflow cross-strata appear as lenses arranged in a rhomboidal pattern among grainfall cross-strata (Hunter, 1977a) (Figure 8).

The downslope length, along-strike width, and vertical thickness of sandflow cross-strata increase as the height of a dune increases (Hunter, 1977a, 1985; Hunter and Kocurek, 1986). For dune heights of 2 m (7 ft) or greater, sandflow beds generally extend the

Figure 7. Bedding-plane exposure of faintly preserved wind-ripple marks. Slightly darkened, parallel bands of medium-grained sand (perpendicular to 15 cm [6 in.] white scale) correspond with upper lee face of ripples. Cockburn Town Member in sea cliffs east of French Bay.

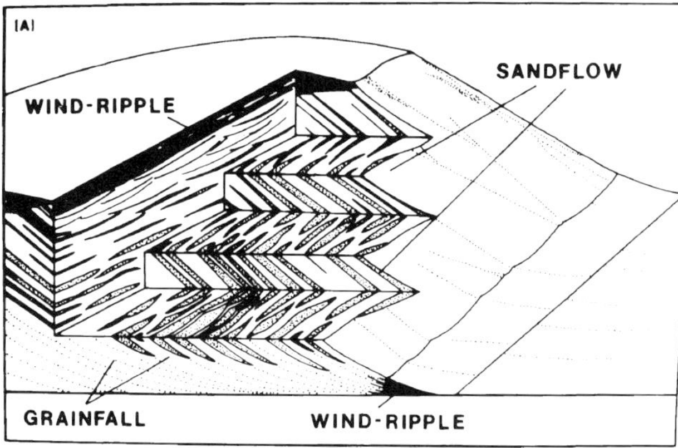

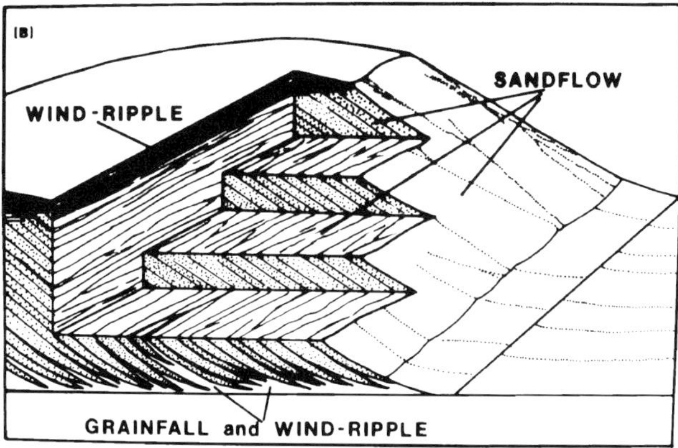

Figure 8. Cutaway diagrams of the interior of simple dunes in longitudinal view (parallel to cross-bed dip direction) and transverse view (horizontal cuts intersecting strike of bedding). (A) Interbedded sandflow and grainfall in dunes 2 m (7 ft) or less in height. (B) Contiguous sandflows wedging downslope into bottomset grainfall or wind-ripple strata in dunes 2 m (7 ft) or greater in height. (Adapted and redrawn from Hunter, 1977a, 1985.)

entire length of the lee slope. They are in contact with one another, are tabular or broadly lenticular in geometry, and are wider than the thickness of a cross-bed set that contains them. Any preexisting grainfall or wind-ripple deposits on the lee slope are consumed during grainflow events. Where these deposits are preserved as bottomsets, they may extend partly up the lower lee slope and form a mutual intertonguing with sandflow wedges (Figure 8).

The foregoing aspects of sandflow cross-strata and their association with grainfall and wind-ripple strata are relevant to the Quaternary limestones of San Salvador Island. In longitudinal outcrop sections, sandflow cross-strata are tabular units, which are set apart by thin grainfall units and marked by ledges and recesses in weathering profile. They are further recognized in the lower part of a cross-bed set by their downslope terminations, which wedge out into bottomsets of either grainfall or wind-ripple strata, or a complex of both (Figure 9). This relationship is especially true for cross-bed sets no thicker than 1–2 m (3–7 ft) in the Pleistocene and much of the Holocene section where bottomsets and the lower bounding surfaces are exposed. Sandflow lenses less than 15 cm (6 in.) wide enclosed in thin grainfall deposits are an additional feature of small- and medium-scale sets (Figure 10). The upslope terminus of a sandflow cross-stratum grades into brinkset beds of either grainfall or wind-ripple strata (see Holocene examples, Figure 16).

More than 80% of the Pleistocene section consists of large-scale cross-bed sets, which are as thick as 6 m (20 ft) and in which the lower bounding surfaces are not exposed. In these sets, straight, tabular cross-strata are typified by fine-grained, ledgy units interbedded irregularly with medium-grained recessed units. A similar bedding structure is also visible along transverse outcrop sections (Figure 11). Thus, the absence of a rhomboidal pattern of sandflow lenses suggests wide sandflow cross-strata and undeveloped grainfall deposits.

Textural trends in sandflow cross-strata are evident in outcrop and in thin section. In cross section, sandflow strata display an inverse grading or upward-coarsening in grain size from fine- to medium-grained sand between lower and upper boundaries. There is a less obvious downslope coarsening from fine- to medium-grained sand along the dip of the bed. The wedge-shape toe of the bed may be characterized by weak normal grading from medium- to fine-grained sand.

The above textural trends are accentuated as either well-cemented ledges or poorly cemented recesses. A typical sandflow cross-stratum in the Grotto Beach Formation is composed of a layer of well-cemented, fine-grained sand concentrated near the lower boundary; it is marked by a white, positive-relief ledge on

Figure 9. Detail of longitudinal view of cross-bed set in Figure 4B. Wedge-shaped sandflow toes (sf) pinch out into bottomset strata of probable grainfall strata (gf). Cockburn Town Member in sea cliffs east of French Bay. Scale is 15 cm (6 in.) long.

Figure 10. Transverse view of bed (between arrows) of small, recessed sandflow lenses outlined by ledgy grainfall laminae. Cockburn Town Member in sea cliffs east of French Bay. Scale is 15 cm (6 in.) long.

Figure 11. Partial longitudinal (A) and transverse (B) exposures of foreset bedding of probable grainflow origin. Fine-grained, well-cemented ledges alternate with medium-grained, poorly cemented recesses. Set is greater than 2 m (7 ft) thick in (A); scale is in centimeters and inches in (B). Undifferentiated Grotto Beach Formation in sea cliffs at the Bluff, east San Salvador Island (Figure 3).

weathered outcrops. The fine sand grades vertically into the medium sand of the upper part of the sandflow, which is poorly cemented and forms a gray, negative-relief recess (Figures 9, 10).

Grainfall Strata

The overall formation of eolian grainfall deposits is related to zones of flow detachment directly ahead of a dune. Here, the wind flow is too calm to produce traction transport of sediment in the form of ripples or any other bed-relief feature (Hunter, 1977a; Kocurek and Dott, 1981). As sediment arrives at the brink of a dune by ripple migration or suspension, it loses momentum as well as the boundary support of the dune and settles along the lee slope. The newly formed grainfall deposit is either buried by sandflow or other grainfall strata, or destroyed by grainflow.

In dunes 2 m (7 ft) or less in amplitude (Hunter, 1977a), grainfall strata form between grainflow events, separate and enclose sandflow cross-strata, and thicken downslope to form bottomset strata (Figure 8), whereas, in dunes greater than 2 m (7 ft) high, grainfall deposits on upper lee slopes are usually destroyed by grainflow and may be preserved only as bottomsets (Figure 8). The thickness of a grainfall deposit is variable and depends on the duration of the wind event (Schenk, 1983). Grainfall stratification is usually indistinct because of poor or variable grain segregation (Hunter, 1977a; Schenk, 1983). However, accelerating gusts of wind can produce inverse size-grading and decelerating wind can produce normal size-grading (Schenk, 1983; Fryberger and Schenk, 1988).

Grainfall intervals have been poorly developed and preserved in the Pleistocene French Bay and Cockburn Town members of the Grotto Beach Formation (Figures 9, 10). However, they are distinct among sandflow beds in the Holocene North Point Member of the Rice Bay Formation. They are identifiable in longitudinal exposures of small- and medium-scale cross-bed sets 1–2 m (3–7 ft) thick. A few centimeters above the lower bounding surface, recessed, medium-grained sandflow beds wedge out downslope into weakly stratified, ledgy, fine-grained bottomset units of probable grainfall origin (Figure 9). The fine-grained, ledge-forming units can be traced partly upslope to where they thin and wedge out among sandflow cross-strata. If the bottomset units are evenly spaced, cyclic laminations, they are wind-ripple strata; if they are weakly or indistinctly laminated, they are grainfall strata. In transverse outcrop sections, if the fine-grained units are indistinctly laminated and separate sandflow lenses arranged in a rhomboidal pattern, they are probably grainfall strata (Figure 10). These structural relationships are exemplified well in the Holocene North Point Member (Figures 13, 14, 15).

Distinguishing Sandflow and Grainfall Strata

The distinction between sandflow and grainfall strata in the Grotto Beach Formation is an unresolved problem where foreset-bottomset relationships in the lower half of thick cross-bed sets are obscured by weathering, burial, or submergence below sea level. In most exposures, cross-bedding is seen as a succession of alternating ledgy and recessed strata of varied thicknesses (Figure 11). Without a clear view of the structural relationships between foresets and bottomsets, such an array of foreset strata could have formed by grainflow, grainfall, or by combinations of both, punctuated by short episodes of wind-ripple migration. Recent studies have indicated that the grainfall mechanism can produce normal grading in decelerating gusts of wind and reverse grading in accelerating wind (Fryberger and Schenk, 1988). The probability that grainfall strata may be inversely graded like sandflow beds further complicates the problem of distinguishing sandflow from grainfall strata in limited exposures.

The origin of the cross-bedding described in the paragraph above can be interpreted in terms of textural and weathering patterns, and grainfall-sandflow association as a function of original dune height (Hunter 1977a, 1985) and, therefore, cross-bed set thickness. A grainflow origin for all the foreset strata is suggested by: (1) their occurrence in cross-bed sets 2 m (7 ft) or greater in thickness, (2) a grain-size trend that coarsens upward from a well-cemented, ledge-forming, fine-grained bed to a poorly cemented, recess-forming, medium-grained bed, (3) sharp contacts between the top of the recessed unit and the overlying ledgy unit, and (4) the absence of poorly or weakly stratified, fine-grained laminations of probable grainfall origin.

The apparent difference in thickness of each recess and ledge is a function of where the plane of the outcrop intersects the individual sandflow lenses in a cross-bed set. This geometric relationship holds true for a succession of either pure sandflow cross-strata or interbedded sandflow and grainfall strata, independent of the width of the sandflow tongues (Figure 8).

Comparison with Holocene Examples

Eolian fine structure in the North Point Member of the Holocene Rice Bay Formation is significantly better developed, preserved, and exposed relative to that in the Grotto Beach Formation. Backset or topset strata are well represented by amalgamated sets of wind-ripple strata separated by low-angle discordances. The fine-grained/medium-grained couplets or pinstripe laminations have been emphasized by weathering as cyclic alternations of white ledgy bands and gray recessed bands. The cyclic bands correspond respectively with a basal unit of fine sand and an upper unit of medium sand of a complete ripple stratum (Figure 12).

Abundant lenticular sandflow lenses are visible in transverse views. They collectively form recessed rhomboidal configurations and are encased in ledge-forming grainfall strata (Figure 13). The close association of both stratification types is suggestive of small- to medium-scale dunes.

In longitudinal views, foreset bedding consists of alternating ledge- and recess-forming units that correspond, respectively, with well-cemented grainfall strata and poorly cemented sandflow strata. Sandflow tongues commonly wedge out into bottomsets composed of wind-ripple or grainfall deposits and are not in contact with lower bounding surfaces (Figures 14, 15). Where preservation and exposure have been exceptional, the updip terminations of

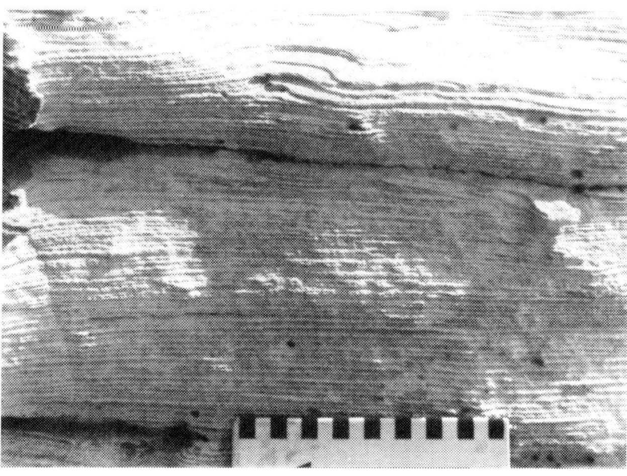

Figure 12. Wind-ripple strata. Cyclic alternations of white, ledgy, fine-grained basal laminae and gray, recessed, medium-grained upper laminae. North Point Member, Rice Bay Formation at North Point. Scale is in centimeters.

Figure 13. Oblique, downdip view (in direction of arrow on scale) of bedset composed of recessed sandflow lenses arranged in rhomboidal pattern enclosed in light-colored, positive-relief grainfall strata. Topset interval of wind-ripple strata (wr); bottomset complex of grainfall and wind-ripple strata (gf/wr). Dashed line is lower bounding surface. North Point Member, Rice Bay Formation at North Point. Scale is in centimeters and inches.

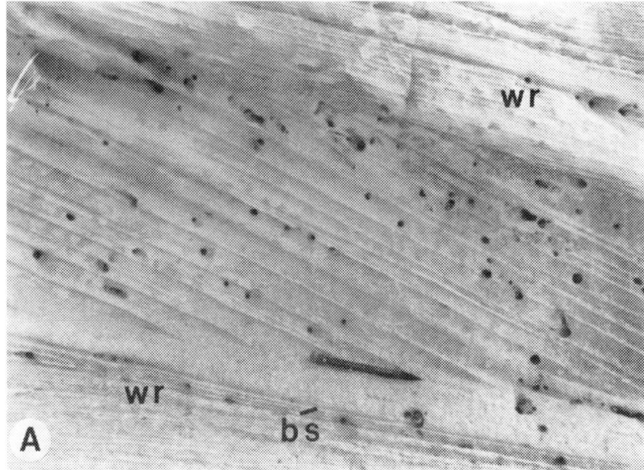

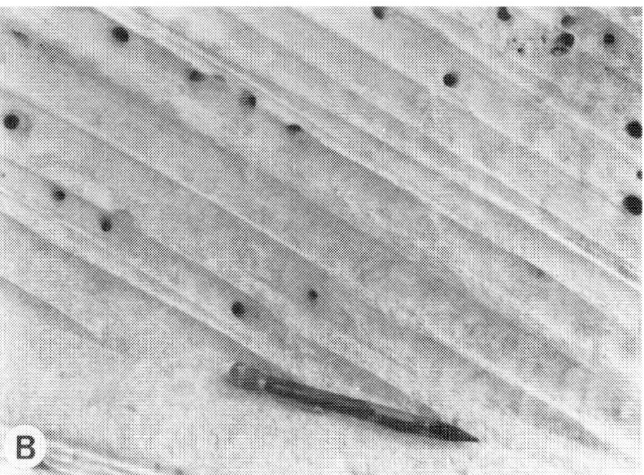

Figure 14. Longitudinal views of medium-scale cross-bed set. (A) Sandflow tongues wedge-out into bottomset of grainfall strata above lower bounding surface (bs). Wind-ripple strata (wr). (B) Detail of sandflow-grainfall strata in (A). North Point Member, Rice Bay Formation at North Point. Mechanical pencil, for scale, is 15 cm (6 in.) long.

sandflow beds wedge out into brinksets of distinctive grainfall and wind-ripple strata (Figure 16). In cross-bed sets greater than 2 m (7 ft) in thickness, grainfall deposits as foreset bedding are generally absent.

Eolian Stratification in Thin Section

Grain Composition and Texture

The framework grains of the Grotto Beach Formation are carbonate allochemical grains, which have been thoroughly micritized. They are composed of ooids, superficial ooids or coated grains, peloids, and skeletal fragments of mollusks, echinoderms, corals, calcareous algae, and foraminifera of unquestionable marine origin (Figure 17).

The grain size of eolian beds in the French Bay and Cockburn Town members is within the fine- to medium-sand size range, which is similar to that for eolian siliciclastic sediment. A finer grain size for sediment composed of aragonite and calcite would be expected to compensate for the greater density of aragonite (2.93 grams per cubic centimeter) and calcite (2.72 grams per cubic centimeter) relative to that of quartz (2.65 grams per cubic centimeter). The greater mineral density of aragonite and calcite is offset probably by intraparticle porosity inherent especially among skeletal grains. Consequently, the grain size of eolian quartzarenites and calcarenites is similar.

The degree of sorting for eolian beds in the Grotto Beach Formation depends upon the scale at which the rock is examined. Individual laminations or very thin and thin beds are well-sorted. However, bedsets are less well-sorted because of grain size differences between each lamination or bed. The distribution of grain sizes among strata has been controlled by specific mechanisms of eolian sedimentation.

Details of the grain size, packing, and cementation patterns associated with the eolian strata were examined in oriented thin sections. Twenty-two oriented samples were collected and examined from the Pleistocene French Bay and Cockburn Town members of the Grotto Beach Formation; six of them are of wind-ripple strata, 14 are of sandflow beds, and two are of mixed sandflow and grainfall strata. Both transverse and longitudinal orientations of foreset bedding were examined. Seven samples from the Holocene North Point Member of the Rice Bay Formation were collected and examined in a similar manner for comparison with microscopic features in the Grotto Beach Formation. Structure and fabric in

Figure 15. Longitudinal view of foreset bedding composed of recessed sandflow tongues wedging out into ledgy, bottomset wind-ripple strata (wr). Further updip, thin ledges are fine-grained basal units of successive sandflow beds. North Point Member, Rice Bay Formation at North Point.

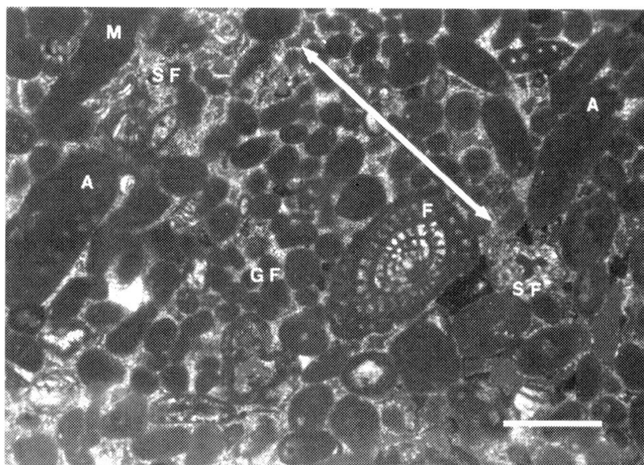

Figure 17. Photomicrograph of marine peloids, coated grains, and skeletal fragments in sandflow (SF) and grainfall (GF) units; mollusk (M), foraminifer (F), green alga (A). Doubly pointed line is approximate thickness of grainfall stratum. Crossed nicols. Cockburn Town Member in sea cliffs east of French Bay. Bar is 0.5 mm long.

Figure 16. Longitudinal view of beds consisting of topset wind-ripple (twr), brinkset wind-ripple (bwr), upper foreset grainfall (gf), and foreset sandflow (sf) strata. Note updip, wedge-shaped terminations of slightly recessed sandflow beds. North Point Member, Rice Bay Formation at North Point. Scale is in centimeters and inches.

thin section were examined in three samples of wind-ripple strata, three samples of sandflow beds, and one sample of grainfall laminations. Observations from both Pleistocene and Holocene examples are combined in the following discussion.

The ideas presented here are based on the notion that grain size and packing differences affect the fluid behavior in intergranular pores and the resulting cementation pattern. Laminations and beds consisting of closely packed, fine-grained particles favor the retention of water in the small pore spaces because of enhanced fluid surface tension and capillarity (Ward, 1975). A relationship among grain size, fabric, water retention, and effective cementation has been observed in marine (Halley and Harris, 1979; Harris et al., 1985; Strasser and Davaud, 1986; Bathurst, 1987; Evans and Ginsburg, 1987) and eolian (Ward, 1975; McKee and Ward, 1983; White and Curran, 1988; Caputo, 1989) limestones and in unconsolidated quartz-rich eolian sand (Fryberger and Schenk, 1988).

Wind-Ripple Strata

A wind-ripple stratum in the Grotto Beach Formation consists of a lower fine-grained lamina that alternates cyclically with an upper, medium-grained lamina. Each of the two constitutes a discrete, laterally persistent band and is distinguished by differences in grain size, grain packing, pore size, and cementation.

Bagnold (1941) observed that the coarsest grains in wind ripples are transferred to the crest area by creep motion caused by ballistic impacts of saltating grains. At the crest position, the coarse grains are sheltered from further saltation impacts and become buried by fine grains with the passage of a ripple trough as a train of ripples moves downcurrent. In effect, the grain segregation produces an upward-coarsening grain-size trend. This texture is clearly visible in thin section (Figure 18). The lower lamina is 1–5 grain diameters thick and is characterized by fine-grained sand with an average diameter of 0.15 mm. Point and long grain contacts (Pettijohn et al., 1987) reflect a moderate to high degree of packing. Calcite cement forms local isopachous rims and blocky, equant crystals, which become enlarged and interlock to completely fill the pore (Figure 19). A high degree of

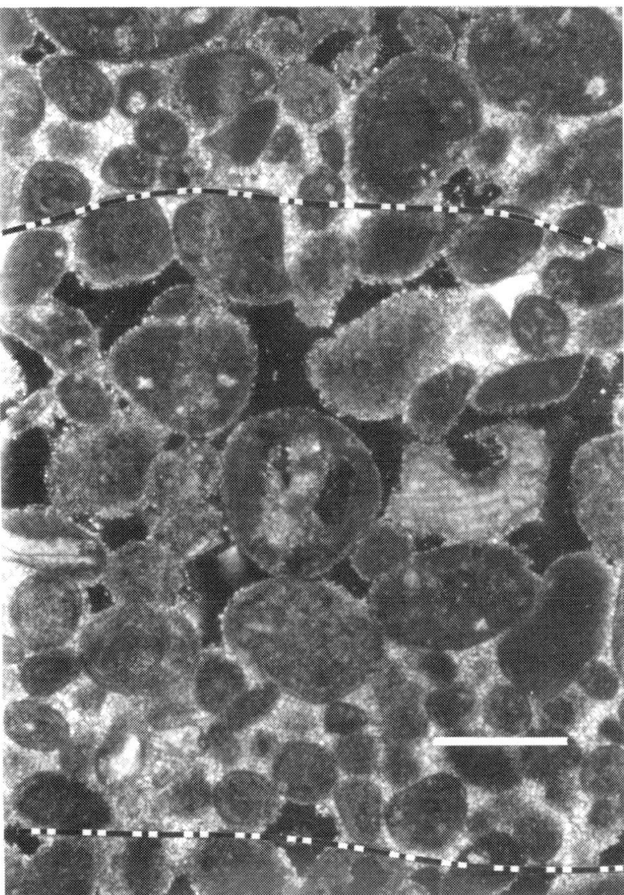

Figure 18. Photomicrograph of micritized coated and skeletal grains in one complete wind-ripple cycle (enclosed between dashed lines). Twofold layering is evident in well-cemented, fine-grained basal unit and poorly cemented, medium-grained upper unit. Lamina above upper line is fine-grained basal unit of next overlying wind-ripple stratum. Dark areas are pores; bright areas are cement. Crossed nicols. French Bay Member in sea cliffs west of French Bay. Bar is 0.5 mm long.

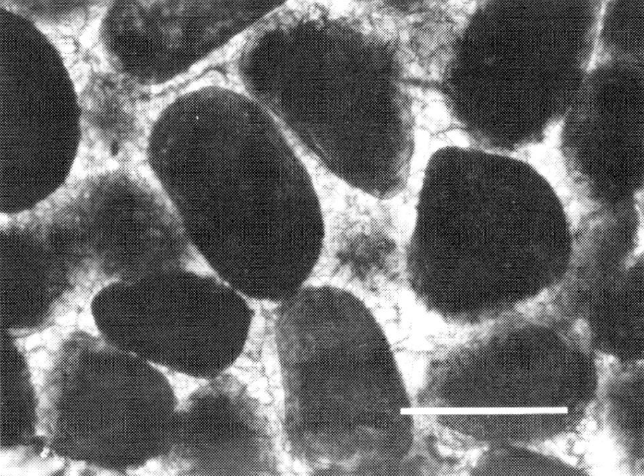

Figure 19. Photomicrograph of coated grains and cement in basal lamina of wind-ripple stratum. Pores contain interlocking isopachous-rim and pore-filling calcite cement. Crossed nicols. French Bay Member in sea cliffs west of French Bay. Bar is 0.2 mm long.

cementation is related to the fine grain size and close packing and, therefore, small pores. Thus, the lower lamina of a wind-ripple stratum weathers to form a positive-relief ledge on the outcrop (Figure 6).

The upper lamina is at least 3–4 grain diameters thick and contains medium-grained sand with an average diameter of 0.28 mm. Floating and point contacts exhibit a less compacted framework around large pores. Poor cementation is indicated by isopachous and meniscus cement and the lack of complete pore-filling cement textures. Consequently, the upper lamina of a wind-ripple deposit weathers to form a negative-relief recess in outcrop (Figure 6).

Sandflow Cross-Strata

Oversized thin sections were used to capture one or two cycles of thin sandflow bedding in the Grotto Beach Formation. In thin section, sandflow beds are characterized by a fine-grained lower part and a medium-grained upper part, each of which corresponds with a ledge and a recess, respectively, on weathered outcrops. Successive sandflow cross-strata are set apart along sharp boundaries created by two adjacent laminae of abruptly different grain sizes (Figure 20).

The lower part of a sandflow bed consists of a concentration of fine-grained sand on the lower boundary. The grains average 0.16 mm in diameter and are moderately to closely packed as indicated by the point and long grain contacts. Entire intergranular pores are filled with calcite cement, which initially precipitated as an isopachous rim around grains and later continued to grow and enlarge toward the center of the pore. Hence, basal sandflow beds weather to form positive-relief, ledge-forming units in outcrop because of the well-cemented nature of the grains (Figure 15).

Carbonate particles in the upper parts of sandflow beds are medium-grained sand with an average diameter of 0.33 mm. They exhibit point and floating contacts because they are moderately to poorly packed. The grains are held together by equant to bladed isopachous and equant meniscus cement crystals that partly fill the intergranular pores (Figure 21). Consequently, the remainder of the sandflow above the well-cemented base weathers to form a negative-relief recess in outcrop (Figure 15).

Measurement of grain diameters along traverses normal to bedding indicates that the fine-grained sand gradually coarsens up to medium-grained sand in the upper part of a sandflow bed. In contrast, a wind-ripple stratum is composed of cyclically alter-

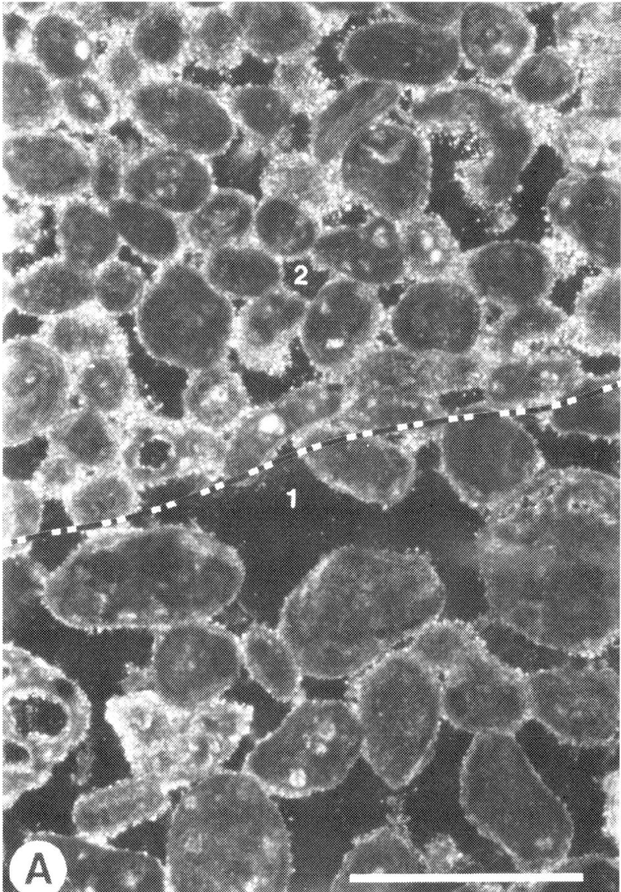

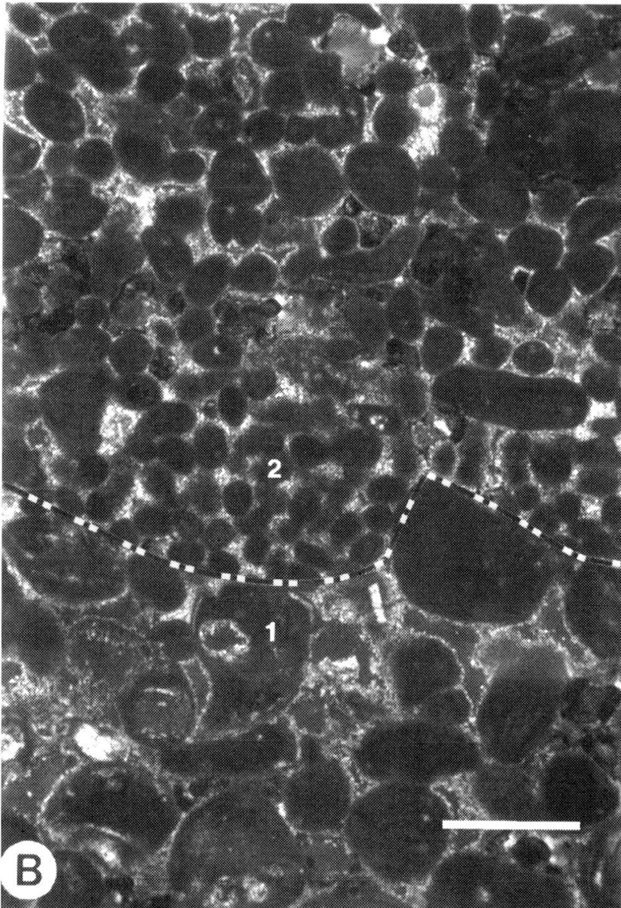

Figure 20. Photomicrograph of contact (dashed line) between two vertically adjacent sandflow beds, 1 and 2. Contact marked by abrupt grain-size change between fine sand in basal part of bed 2 and medium sand in upper part of bed 1. Note poorly cemented, open framework of bed 1 in (A) and gradual upward-coarsening of bed 2 in (B). Crossed nicols. Cockburn Town Member in sea cliffs east of French Bay. Bar is 0.5 mm long in both (A) and (B).

nating laminae, the boundaries of which are delineated by sharp contrasts in grain size, grain packing, pore size, and cementation (Figure 18). Furthermore, grainfall laminations show poor to no grain-size grading and are delimited abruptly above and below by medium-grained sand of thick, adjacent sandflow beds (Figure 22).

The gradual upward, and downslope (described earlier), increase in grain size of sandflow beds in the Grotto Beach Formation suggests that grain segregation was the result of grainflow mechanics. Experimental studies have shown that during grainflow motion, a residue of fine grains, associated with shearing, accumulates at or near the base of the flow. The dispersive pressure of grain-to-grain collisions during flowage causes large grains to rise and approach the top (Bagnold, 1954; Inman et al., 1966; Middleton, 1970; Yaalon and Laronne, 1971; Fryberger and Schenk, 1981; Hunter, 1985; Hunter and Kocurek, 1986). The fine grains either remain behind on the plane of shearing or are incorporated into the base of the sandflow stratum when the flow ceases (Schenk, 1983). The coarse grains that were buoyed to the top of the flow by grain collisions have, because of their large size and momentum, outrun the fine grains at the head of the flow. Later they become buried by the trailing fine grains. The result is a coarsening downslope (Bagnold, 1941; Hunter, 1977a; Schenk, 1983) and coarsening upward in sandflow toes (Hunter, 1985).

Grainfall Strata

Grainfall deposits in Quaternary limestones on San Salvador Island are positively identified by their poor stratification, occurrence in bottomsets, and upslope interfingering with recognizable sandflow cross-strata in small- and medium-scale cross-bed sets. The following are some generalized features of grainfall deposits in thin section.

Grainfall laminations sampled from middle levels in a cross-bed set are present as bands or laminations 1–2 mm thick in thin section (Figure 22). They consist of well-cemented, fine-grained sand set apart by thicker, poorly cemented, medium-grained sandflow

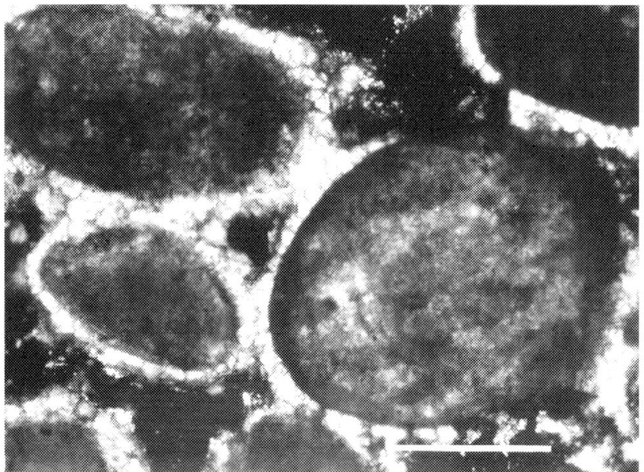

Figure 21. Photomicrograph of equant, blocky crystals of isopachous and meniscus calcite cement in partly filled pores between micritized coated grains. Medium-grained upper part of sandflow bed. Crossed nicols. French Bay Member in sea cliffs west of French Bay. Bar is 0.2 mm long.

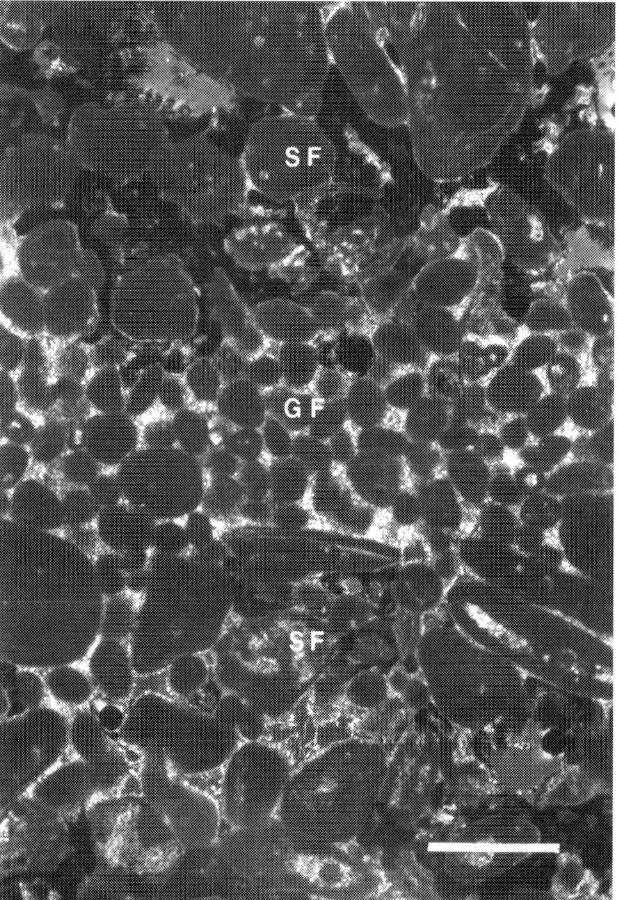

Figure 22. Photomicrograph of well-cemented, fine-grained grainfall lamination (GF) abruptly interlayered between medium-grained, poorly cemented sandflow units (SF). Note absence of vertical size grading within grainfall unit and between grainfall and adjacent sandflow units. Crossed nicols. Cockburn Town Member in sea cliffs east of French Bay. Bar is 0.5 mm long.

beds. Bottomsets of grainfall strata were not sampled. Point and long grain contacts suggest a moderate to high degree of packing. Isopachous and complete pore-filling cement and positive-relief ledges on weathered outcrops are features similar to those of fine-grained, basal laminae in sandflow and wind-ripple strata. The average grain diameter is 0.10 mm, a value that is not significantly different from the average grain diameter of 0.15 mm for a basal unit in a wind-ripple stratum and the average grain diameter of 0.16 mm for a basal unit in a sandflow bed.

Lower and upper bedding contacts are marked abruptly by an upward shift in grain size from medium to fine sand and fine to medium sand, respectively, between adjacent sandflow beds. Vertical grain-size grading is poor or absent as suggested by measurements of grain diameters along traverses normal to bedding. However, until further sampling can validate this feature, it is not considered diagnostic.

Grainfall strata in thin section closely resemble the basal laminae of wind-ripple and sandflow strata in grain size, grain packing, pore size, and cementation. In outcrop, they display the same ledge-form weathering pattern. Therefore, they are difficult to distinguish in thin section alone without a clear view of bedding characteristics and structural relationships in the lower part of a cross-bed set.

Cement Textures

Most of the Quaternary eolian limestones on San Salvador Island have been cemented with early, low-magnesium calcite precipitated from meteoric water (Hutto and Carew, 1984; White and White, 1990). The cements are usually characterized by meniscus, pendant, needle-fiber, blocky, and drusy textures typical of a vadose diagenetic environment. However, portions of Pleistocene and Holocene eolianites on San Salvador, which have been submerged after the late Holocene rise in sea level, have been recemented by submarine aragonite cement (White and White, 1990).

In unconsolidated sedimentary beds, fluid surface tension and capillarity are enhanced by small pore space related to close grain packing (Bagnold, 1938; Ward, 1975). These conditions create a micro-environment at the scale of a thin bed or lamination that resembles that of a phreatic diagenetic environment where the pore is filled entirely with liquid. Calcite cements formed in freshwater phreatic environments are generally characterized by a foundation of isopachous, bladed calcite crystals that surround grains, become coarser toward the center of the pores, fill the pores, and form an interlocking array of crystals (Longman, 1980). Meniscus calcite cements are usually precipitated in a freshwater vadose diagenetic environment where drops of water having meniscus or hour-glass shapes adhere between grains (Longman, 1980)

In the French Bay and Cockburn Town members of the Grotto Beach Formation, the three eolian strata types, sandflow, grainfall, and wind-ripple, are either partly or wholly cemented as a function of grain size and grain packing. The two component layers of an individual wind-ripple or sandflow stratum show different degrees of cementation. Only grainfall strata appear uniformly cemented. The occurrence of isopachous and complete pore-filling cements amid dominant meniscus and incomplete pore-filling cement in the eolian beds of the Grotto Beach Formation suggests localized phreatic diagenetic conditions in an overall vadose diagenetic environment (Figures 19–22). The effect is a diagenetic micro-environment on the scale of a bed or lamination, in which localized conditions related to eolian processes and grain texture created an apparent alternation between vadose and phreatic characteristics.

The association of meniscus vadose cement with isopachous and complete pore-filling cement in the same bedding interval suggests a particular sequence of diagenetic events in the Grotto Beach Formation. Initially, there were probably several episodes when the intergranular pores within all the strata were filled temporarily and completely with water from which an isopachous cement was precipitated. Later the water was drained by seepage pressure as the level of water saturation, or water table, fell either through evaporation or decreased rainfall, a mechanism described in terms of gravity water by Manning (1987). In grainfall strata, and in the basal layer of sandflow and wind-ripple strata in the Grotto Beach Formation, the closely packed, fine-grained sand efficiently retained the pore water through fluid surface tension and capillarity and consequently became cemented until the pores were completely filled. The ability to retain pore water within the medium-grained upper units of wind ripple and sandflow strata was diminished by loose packing and large pore space and, therefore, decreased fluid surface tension and capillarity. Water adhered between grains in meniscus form. Consequently, cements only partially filled intergranular pores.

SUMMARY

With the exception of the biogenic, carbonate framework grains, the structural and textural characteristics of eolian calcarenites in the Pleistocene Grotto Beach Formation and in the Holocene Rice Bay Formation are equivalent to those in modern and ancient siliciclastic eolian deposits. Marine sediment composed of ooids, coated grains, peloids, and skeletal fragments was generated when most of the San Salvador platform was submerged during a Pleistocene interglacial event (Carew et al., 1984; Carew and Mylroie, 1985, 1987). Nearshore currents distributed the sediment along margins of beaches, tidal flats, estuaries, and lagoons where it was remobilized by wind. Sinuous dune ridges were built by easterly wind.

Grainflow, grainfall, and ripple migration are subprocesses on modern eolian dunes. Resultant sandflow, grainfall, and wind-ripple strata in modern and ancient deposits can be discriminated on the basis of spatial arrangement, geometry, grain size, grain packing, and grain segregation and grading. Eolian structures in the Grotto Beach Formation can be further distinguished by cementation and weathering trends (Table 1). Fine-grained laminations of basal wind-ripple and sandflow strata, and entire grainfall strata were selectively well-cemented because of fine grain size, close packing, small pores, and high retention of cement-generating pore water. These units, then, form ledges on weathered outcrops. In contrast, the upper laminae of wind-ripple and sandflow strata were poorly cemented because of medium grain size, loose packing, large pores, and low retention of cementing pore water. These units are not as well-cemented and resistant as the fine-grained units, so they form recesses on weathered outcrops.

Texture, cementation, and weathering patterns are shown to enhance the recognition of ripple and sandflow strata as structural elements of some Pleistocene and Holocene eolian calcarenites. Grain texture and cementation could be generally useful in detecting these eolian structures in thin sections from core samples. However, analysis of texture and diagenesis is not effective in the identification of grainfall strata in thin section without the aid of clear bedding relationships in outcrop.

The description of features in this chapter is an initial attempt to assemble structural and textural information on some Pleistocene Bahamian limestones. It is hoped that the report contributes to the study of other Quaternary eolian limestones on San Salvador Island, throughout the Bahamas, and in other regions where carbonate sediment has been retransported by wind to form dunes. The correlation with analogous features known from modern and ancient siliciclastic eolian deposits may be useful in the study of older Phanerozoic eolian limestones.

Further work is required to understand the timing of cementation and the architecture and morphology of carbonate dunes. Additional analysis of permeability and capillarity in samples from specific stratification types will verify the behavior of fluid during diagenesis for individual strata and, perhaps, predict the behavior of fluid in eolian-carbonate sand bodies as oil, gas, and water reservoirs.

ACKNOWLEDGMENTS

Financial or material support for this project was provided generously by John Mylroie (Department of Geology and Geography, Mississippi State University), Donald Gerace (Bahamian Field Station, San Salvador Island), Karen Cook (Prentice Hall), Del Gann (Department of Geology, Millsaps College), Jim Harrington (The Thin Section Lab), and Gordon

Table 1. Summary of structural and textural characteristics of eolian stratification in the French Bay and Cockburn Town members of the Pleistocene Grotto Beach Formation, San Salvador Island.

	Eolian Stratification		
	Wind-Ripple	Sandflow	Grainfall
Process:	ripple migration	avalanching	grain-settling
Occurrence:	topset, brinkset, foreset, bottomset	foreset	foreset, bottomset
Geometry:	tabular, wedge	lenticular, wedge	variable
Outcrop expression:	upper recess basal ledge	upper recess basal ledge	ledge
Sand size:	fine- to medium-grained	fine- to medium-grained	fine-grained
Grading:	upward coarsening	upward coarsening; downslope coarsening; upward-fining near toes	weak to none
Grain contacts (packing):	floating, point, and long	floating, point, and long	point and long
Cement texture:	equant: bladed isopachous and meniscus	equant: bladed isopachous and meniscus	equant: bladed isopachous

Carroll (Advanced Scientific, Inc.). The following reviewers offered their time and suggestions to help improve the contents of the paper: Ralph Hunter (USGS), John Mylroie (Mississippi State University), Doug Jordan (ARCO Research), Charles Zuppann (Indiana Geological Survey), and Brian Keith (Indiana Geological Survey).

REFERENCES CITED

Bagnold, R. A., 1938, Grain structure of sand dunes and its relation to their water content: Nature, v. 142, p. 403-404.

Bagnold, R. A., 1941, The physics of blown sand and desert dunes: London, Methuen, 265 p.

Bagnold, R. A., 1954, Experiments on a gravity-free dispersion of large solid spheres in a Newtonian fluid under shear: Proceedings Royal Society London, 225 A, p. 49-63.

Ball, M. M., 1967, Carbonate sand bodies of Florida and the Bahamas: Journal of Sedimentary Petrology, v. 37, no. 2, p. 556-591.

Bathurst, R. G. C., 1987, Diagenetically enhanced bedding in argillaceous platform limestones: stratified cementation and selective compaction: Sedimentology, v. 34, no. 5, p. 749-778.

Boubin, M. A., and D. B. Loope, 1989, Petrology of eolian carbonates, upper Hermosa Formation (Pennsylvanian), southeastern Utah (abs.): American Association of Petroleum Geologists Bulletin, v. 73, no. 3, p. 336.

Caputo, M. V., 1989, Selective cementation of eolian stratification in Pleistocene calcarenites, San Salvador Island, Bahamas, *in* J. E. Mylroie, ed., Proceedings of the fourth symposium on the geology of the Bahamas: Ft. Lauderdale, College Center of the Finger Lakes, Bahamian Field Station, p. 61-72.

Carew, J. L., J. E. Mylroie, J. F. Wehmiller, and R. A. Lively, 1984, Estimates of late Pleistocene sea level high stands from San Salvador, Bahamas, *in* J. W. Teeter, ed., Proceedings of the second symposium on the geology of the Bahamas: Ft. Lauderdale, College Center of the Finger Lakes, Bahamian Field Station, p. 153-176.

Carew, J. L., and J. E. Mylroie, 1985, The Pleistocene and Holocene stratigraphy of San Salvador Island, Bahamas, with reference to marine and terrestrial lithofacies at French Bay, *in* H. A. Curran, ed., Pleistocene and Holocene carbonate environments on San Salvador Island, Bahamas, field trip guidebook for Geological Society of America: Ft. Lauderdale, College Center of the Finger Lakes, Bahamian Field Station, p. 11-62.

Carew, J. L., and J. E. Mylroie, 1987, A refined geochronology for San Salvador Island, Bahamas, *in* H. A. Curran, ed., Proceedings of the third symposium on the geology of the Bahamas: Fort Lauderdale, College Center of the Finger Lakes, Bahamian Field Station, p. 35-44.

Curran, H. A., 1985, Introduction to the geology of the Bahamas and San Salvador Island with an overflight guide, *in* H. A. Curran, ed., Pleistocene and Holocene carbonate environments on San Salvador Island, Bahamas, fieldtrip guidebook for Geological Society of America: Ft. Lauderdale, College Center of the Finger Lakes, Bahamian Field Station, p. 1-10.

Curran, H. A., and B. White, 1984, Field guide to the Cockburn Town fossil coral reef, San Salvador, Bahamas, *in* J.W. Teeter, ed., Proceedings of the second symposium on geology of the Bahamas: Ft. Lauderdale, College Center of the Finger Lakes, Bahamian Field Station, p. 71-96.

Curran, H. A., and B. White, 1985, The Cockburn Town fossil coral reef, *in* H. A. Curran, ed., Pleistocene and Holocene carbonate environments on San Salvador Island, Bahamas, fieldtrip guidebook for Geological Society of America: Ft. Lauderdale, College Center of the Finger Lakes, Bahamian Field Station, p. 95-120.

Dean, W. E., and T. D. Fouch, 1983, Lacustrine, *in* P. A. Scholle, D. G. Bebout, and C. W. Moore, eds., Carbonate Depositional Environments: Tulsa, American Association of Petroleum Geologists, Memoir 33, P. 97-130.

Evans, C. C., and R. N. Ginsburg, 1987, Fabric-selective diagenesis in the Late Pleistocene Miami Limestone: Journal of Sedimentary Petrology, v. 57, no. 2, p. 311-318.

Fryberger, S. G., and C. J. Schenk, 1981, Wind sedimentation tunnel experiments on the origins of aeolian strata: Sedimentology, v. 28, no. 6, p. 805-822.

Fryberger, S. G., and C. J. Schenk, 1988, Pinstripe lamination: a distinctive feature of modern and ancient eolian sediments: Sedimentary Geology, v. 55, p. 1-15.

Halley, R. B., and P. M. Harris, 1979, Fresh-water cementation of a 1,000-year-old oolite: Journal of Sedimentary Petrology, v. 49, no. 3, p. 969-988.

Handford, C. R., 1990, Mississippian carbonate eolianites in southwestern Kansas (abs.): American Association of Petroleum Geologists Bulletin, v. 74, no. 5, p. 669.

Harris, P. M., C. G. St. C. Kendall, and I. Lerche, 1985, Carbonate cementation—a brief review, *in* N. Schneidermann and P. M. Harris, eds., Carbonate Cements: Tulsa, SEPM Spec. Pub. 36, p. 79-95.

Hunter, R. E., 1977a, Basic types of stratification in small eolian dunes: Sedimentology, v. 24, p. 361-387.

Hunter, R. E., 1977b, Terminology of cross-stratified sedimentary layers and climbing-ripple structures: Journal of Sedimentary Petrology, v. 47, p. 697-706.

Hunter, R. E., 1981, Stratification styles in eolian sandstones: some Pennsylvanian to Jurassic examples from the western interior U.S.A., *in* F. G. Ethridge, and R. M. Flores, eds., Recent and ancient nonmarine depositional environments: models for exploration: Tulsa, SEPM Spec. Pub. 31, p. 315-330.

Hunter, R. E., 1985, Subaqueous sand-flow cross-strata: Journal of Sedimentary Petrology, v. 55, no. 6, p. 886-894.

Hunter, R. E., 1988, Eolianites in the Ste. Genevieve limestone (Mississippian) of southern Indiana (abs.), *in* Abstracts: SEPM Annual Midyear Meeting, Abstracts, v. 5, p. 26-27.

Hunter, R. E., and G. Kocurek, 1986, An experimental study of subaqueous slipface deposition: Journal of Sedimentary Petrology, v. 56, no. 3, p. 387-394.

Hutto, T., and J. L. Carew, 1984, Petrologyofeolian calcarenites, San Salvador Island Bahamas, *in* J. W. Teeter, ed., Proceedings of second symposium on the geology of the Bahamas: Ft. Lauderdale, College Center of the Finger Lakes, Bahamian Field Station, p. 197-207.

Illing, L. V., 1954, Bahamian calcareous sands: American Association of Petroleum Geologists Bulletin, v. 38, p. 1-95.

Inman, D. L., G. C. Ewing, and J. B. Corliss, 1966, Coastal sand dunes of Guerrero Negro, Baja California, Mexico: Geological Society of America Bulletin, v. 77, p. 787-802.

Irwin, M. L., 1965, General theory of epeiric clear water sedimentation: American Association of Petroleum Geologists Bulletin, v. 49, p. 445-459.

Knight, S. H., 1929, The Fountain and Casper Formations of the Laramie Basin—a study on genesis of sediments: University of Wyoming Special Publications in Geology, v. 1, no. 1, p. 1-82.

Kocurek, G., 1986, Origins of low-angle stratification in aeolian deposits, *in* W. G. Nickling, ed., Aeolian Geomorphology, Proceedings of the 17th annual Binghamton geomorphology symposium: Boston, Allen and Unwin, p. 177-193.

Kocurek, G., and R. H. Dott, Jr., 1981, Distinctions and uses of stratification types in the interpretation of eolian sand: Journal of Sedimentary Petrology, v. 51, no. 2, p. 579-595.

Longman, M. W., 1980, Carbonate diagenetic textures from near surface diagenetic environments: American Association of Petroleum Geologists Bulletin, v. 64, no. 4, p. 461-487.

Loope, D. B., 1986, Pennsylvanian eolian limestones, Paradox Basin, U.S.A., (abs.), *in* Abstracts, Sediments down under: International Association of Sedimentologists, 12th International Sedimentological Congress, p. 189-190.

Loope, D. B., and Z. E. Haverland, 1988, Giant desiccation fissures filled with calcareous eolian sand, Hermosa Formation (Pennsylvanian), southeastern Utah, *in* G. Kocurek, ed., Late Paleozoic and Mesozoic eolian deposits of the Western Interior of the United States: Sedimentary Geology, v. 56, p. 403-413.

Mackenzie, F. T., 1964a, Geometry of Bermuda calcareous dune cross-bedding: Science, v. 144, p. 1449-1450.

Mackenzie, F. T., 1964b, Bermuda Pleistocene eolianite sand paleowinds: Sedimentology, v. 3, no. 1, p. 52-64.

Manning, J. C., 1987, Applied principles of hydrology: Columbus, Ohio, Merrill Publishing Company, 278 p.

McKee, E. D., 1934a, An investigation of the light-colored cross-bedded sandstones of Canyon de Chelley, Arizona: American Journal of Science, 5th ser., v. 28, no. 165, p. 219-233.

McKee, E. D., 1934b, The Coconino Sandstone—its history and origin: Washington, Carnegie Institute, Pub. 440, p. 77-115.

McKee, E. D., and W. C. Ward, 1983, Eolian, *in* P. A. Scholle, D. G. Bebout, and C. H. Moore, eds., Carbonate Depositional Environments: Tulsa, American Association of Petroleum Geologists, Memoir 33, p. 131-170.

Middleton, G. V., 1970, Experimental studies related to the problems of flysch sedimentation, *in* J. Lajoie, ed., Flysch sedimentology in North America: Geological Society of Canada, Special Pub. 7, p. 253-272.

Newell, N. D., E. G. Purdy, and J. Imbrie, 1960, Bahamian oolitic sand: Journal of Geology, v. 68, p. 481-497.

Pettijohn, F. J., P. E. Potter, and R. Siever, 1987, Sand and Sandstone: New York, Springer-Verlag, 549 p.

Reiche, P., 1938, An analysis of cross-lamination of the Coconino Sandstone, Journal of Geology, v. 46, p. 905-932.

Rice, J. A., and D. B. Loope, 1991, Wind-reworked carbonates, Permo-Pennsylvanian of Arizona and Nevada: Geological Society of America Bulletin, v. 103, no. 2, p. 254-267.

Schenk, C. J., 1983, Textural and structural characteristics of some experimentally formed eolian strata, *in* M. E. Brookfield, and T. S. Ahlbrandt, eds., Eolian sediments and processes: Amsterdam, Elsevier Scientific Publishers, p. 41-49.

Scholle, P. A., D. G. Bebout, and C. H. Moore, eds., 1983, Carbonate depositional environments: Tulsa, American Association of Petroleum Geologists Memoir 33, 708 p.

Strasser, A., and E. Davaud, 1986, Formation of Holocene limestone sequences by progradation, cementation, and erosion: two examples from the Bahamas: Journal of Sedimentary Petrology, v. 56, no. 3, p. 422-428.

Ward, W. C., 1970, Diagenesis of Quaternary eolianites of N.E. Quintana Rio, Mexico: Houston, Rice University, unpub. Ph.D. dissert., 206 p.

Ward, W. C., 1973, Influence of climate on the early diagenesis of carbonate eolianites: Geology, v. 1, p. 171-174.

Ward, W. C., 1975, Petrology and diagenesis of carbonate eolianites of northwestern Yucatan Peninsula, Mexico, *in* K. F. Wantland and W. C. Pusey, eds., Belize shelf carbonate sediments, clastic sediments and ecology: American Association of Petroleum Geologists, Studies in Geology, no. 2, p. 500-571.

White, B., and H. A. Curran, 1985, The Holocene carbonate eolianites of North Point and the modern marine environments between North Point and Cut Cay, *in* H. A. Curran, ed., Pleistocene and Holocene carbonate environments on San Salvador Island, Bahamas, fieldtrip guidebook for Geological Society of America: Ft. Lauderdale, College Center of the Finger Lakes, Bahamian Field Station, p. 73-93.

White, B., and H. A. Curran, 1986, Holocene carbonate eolianites from San Salvador, Bahamas, *in* Abstracts, Sediments down-under: International Association of Sedimentologists, 12th International Sedimentological Congress, p. 330.

White, B., and H. A. Curran, 1988, Mesoscale physical sedimentary structures and trace fossils in Holocene carbonate eolianites from San Salvador Island, Bahamas: Sedimentary Geology, v. 55, p. 163-184.

White, W., K. A. Kurkjy, and H. A. Curran, 1984, A shallowing-upward sequence in a Pleistocene coral reef and associated facies, San Salvador, Bahamas, *in* J. W. Teeter, ed., Proceedings of the second symposium on the geology of the Bahamas: Ft. Lauderdale, College Center of the Finger Lakes, Bahamian Field Station, p. 53-70.

White, K. S., and B. White, 1990, The effects of Holocene sea level rise on the diagenesis of Quaternary carbonate eolianites, San Salvador Island, Bahamas (abs.), *in* Abstracts and Programs, fifth symposium on geology of Bahamas: Bahamian Field Station, p. 19.

Yaalon, D, H., and J. Laronne, 1971, Internal structures in eolianites and paleowinds, Mediterranean coast, Israel: Journal of Sedimentary Petrology, v. 41, p. 1059-1064.

INDEX